Principles of Gas Path Analysis

Principles of Gas Path Analysis offers a self-contained reference of the concept and theory of gas path analysis (GPA), as both a diagnostic and prognostic methodology for gas turbine engines. It provides a chronological account of the methodology as it evolved over the past 50 years.

Before expanding into specific GPA concepts, the book begins by covering the basics that are generic to diagnostics and prognostics as well as a discussion of engine health management (EHM) and how GPA can contribute to this strategy. The text further introduces essential parameter corrections important for understanding the foundational principles of GPA. Additionally, advanced topics such as information fusion and ambiguity resolution are explored to highlight potential future advancements in the field. Examples are provided using simulated data generated from a fictional high bypass turbofan engine model. The book contains a comprehensive set of appendices with detailed treatment of the mathematical derivations and statistics behind GPA as well as specialized constructs for relevant methods such as neural networks and fuzzy logic.

The book is intended for professional engineers engaged in the gas turbine industry and EHM, including aircraft operators and maintainers. It will also benefit researchers studying gas turbine engine diagnostics and prognostics.

Principles of Gas Path Analysis

Allan J. Volponi
and Liang Tang

CRC Press
Taylor & Francis Group
Boca Raton London New York

CRC Press is an imprint of the
Taylor & Francis Group, an informa business

First edition published 2026
by CRC Press
2385 NW Executive Center Drive, Suite 320, Boca Raton FL 33431

and by CRC Press
4 Park Square, Milton Park, Abingdon, Oxon, OX14 4RN

CRC Press is an imprint of Taylor & Francis Group, LLC

Library of Congress Cataloging-in-Publication Data
Names: Volponi, Allan J. author | Tang, Liang (Engineer) author
Title: Principles of gas path analysis / Allan J. Volponi and Liang Tang.
Description: First edition. | Boca Raton, FL : CRC Press, 2026. | Includes
bibliographical references and index.
Identifiers: LCCN 2025026326 (print) | LCCN 2025026327 (ebook) | ISBN
9781041092674 hbk | ISBN 9781041092681 pbk | ISBN 9781003649274 ebk
Subjects: LCSH: Gas-turbines--Performance--Mathematical models
Classification: LCC TJ778 .V65 2026 (print) | LCC TJ778 (ebook)
LC record available at https://lccn.loc.gov/2025026326
LC ebook record available at https://lccn.loc.gov/2025026327

ISBN: 978-1-041-09267-4 (hbk)
ISBN: 978-1-041-09268-1 (pbk)
ISBN: 978-1-003-64927-4 (ebk)

DOI: 10.1201/9781003649274

Typeset in Times
by Deanta Global Publishing Services, Chennai, India

Dedicated to the late Louis A. Urban (1928–2019), whose seminal work in gas turbine engine diagnostics remains a cornerstone of the discipline and a lasting source of inspiration for the global engineering research community.

Contents

ADDENDUMS

About the Authors

Allan J. Volponi is a retired Senior Fellow from the Controls & Diagnostic Systems organization at Pratt & Whitney. He received his BS and MS degrees from Pratt Institute (1971–1972) and his PhD from the Adelphi University (1977), all in mathematics. He spent 40 years with United Technologies, initially with the Hamilton Sundstrand Division as a Senior Principal Engineer in the engine controls group before transferring to P&W in 2000. His interests are in propulsion health management, in which he has been active in the development of engine performance diagnostic systems. Dr. Volponi has been the recipient of numerous awards, including the 1992 Manly Memorial Medal by the SAE, the 2006 Aircraft Engine Technology Award by the ASME, the Silver Specialist and Sir Roy Fedden Awards from the Royal Aeronautical Society, and the 2013 Scholar Award from the ASME/IGTI. Dr. Volponi has been an active member of the International Gas Turbine Institute (IGTI) and a past Chairman of its Controls, Diagnostics and Instrumentation Committee and is an ASME Fellow. He holds 17 patents, has two pending. He is the author of numerous technical papers on gas turbine diagnostics and prognostics health management.

Liang Tang is an expert in jet engine health management, controls, and the application of artificial intelligence (AI) and machine learning (ML) in aerospace. Since 2012, he has served in several technical leadership positions with two major engine OEMs, focusing on engine health management, diagnostics and prognostics, and the application of AI/ML to inspection and process improvements. Dr. Tang is recognized for extending gas path analysis from snapshot-based approaches to the full-flight domain and for applying ML techniques to resolve diagnostic ambiguity. Since 2005, he has also led multiple government-funded research programs as Principal Investigator for agencies including NASA and the DoD, spanning health management, controls, navigation, and autonomy. Dr. Tang earned his Ph.D. from Shanghai Jiao Tong University in 1999 and was a postdoctoral research fellow at the Georgia Institute of Technology from 2003 to 2004. He has published more than 60 papers and holds multiple patents related to diagnostics, prognostics and health management for aerospace systems. He remains actively engaged in technical committees and standards organizations such as the SAE and ASME, where he has held several leadership positions, including chairing technical committees over the past decade.

Preface

Assessing the general health of an operational gas turbine engine is an activity that is nearly as old as the machine itself. It has had a long and diverse history, which spans many decades and encompasses many technical disciplines. If we had to categorize the diagnostic methodologies that have materialized over this course of time, three general classifications would emerge: *instrumentation*, *mechanical systems,* and *performance* diagnostics. The instrumentation category addresses the faults or failures in sensors and measurement systems that are installed for control or health monitoring purposes. Mechanical system diagnostics deals with mechanical actuation, vibration and sub-system mechanical wear, and debris accumulation, while performance diagnostics concerns itself with faults and failures that have a direct and observable impact on the machine's thermodynamic performance. Clearly, any viable health management system must contain all classes to ensure appropriate diagnostic coverage utilizing a plurality of onboard and off-board solutions. Nevertheless, *this text will focus on performance diagnostics* and, in particular, a methodology known as *gas path analysis.* This methodology has had a long history of development (approximately five decades) and has a proven track record as an effective approach to engine diagnostics and prognostics. This quality, combined with its (relatively) simplistic implementation requirements, has contributed to its longevity as a core element in engine health management (EHM) systems. As a result, it has been the focus of considerable research within the academic, industrial, and government arenas. Although, there are many hundreds upon hundreds of papers that have been written, and continue to be written, on this general topic and its applications, there presently exists no single source that provides a reference to the many aspects that underlie the methodology, from its historical beginning to current evolving capability. It is the goal of this text to provide such a reference. This is timely, as over the past five decades, a clear picture of the mathematical foundation, and associated machinery, can be assembled to provide a self-contained account of the theory that has been ardently developed and meticulously matured by many researchers. To be clear, this text is not intended, nor does it provide, a blueprint for the design of an effective diagnostic system but is, rather, a chronological account and mathematical foundation for what already appears in the general literature utilizing a single consistent set of symbols and nomenclature to enhance readability.

Allan J. Volponi, *West Simsbury, CT, USA*
Liang Tang, *Glastonbury, CT, USA*

Introduction

Any discourse on gas path analysis (GPA) would not be complete without acknowledging the personage responsible for its initial development as a diagnostic method and whose work and early publications, has inspired its continued exploration and application worldwide. Louis Andrew Urban was educated at Columbia University in Mechanical Engineering (BS in 1950, MS in 1954) and began his professional career at the Westinghouse Electric Corporation at their Combustion Turbine Systems Division in south Philadelphia in 1951. He later transferred to the Hamilton Standard division of the then United Aircraft Corporation (later United Technologies Corporation, and now Raytheon Technologies Corporation) in 1954, where he worked on the design of numerous gas turbine hydro-mechanical fuel controls. In 1969, he published a unique monograph [1], which has, subsequently, been distributed worldwide, on analytic techniques to *provide a rapid and simple means of estimating steady-state or transient variations in engine and control performance parameters for any of the various types of gas turbine engines and control modes under most conceivable sets of input conditions.* Many consider this monograph to be the genesis of the analytic mindset that would later evolve and form the mathematical foundation of GPA. Mr. Urban's pioneering work on this topic [2] appearing in 1972 in the AIAA *Journal of Aircraft*, launched a nearly five-decade interest in the research and practice of this approach ranging across aero-propulsion and stationary gas turbines alike. His continued efforts and contributions, up until his retirement in 1987, have provided an inspirational legacy for those who followed.

The practice of GPA began in earnest with the emergence of the commercial wide-body aircraft in the 1970s. The Boeing 747, McDonnell Douglas DC-10, and Lockheed L-1011 wide-body aircraft introduced propulsion systems that were modular and allowed the possibility to partially overhaul the engine through repair of individual modules (e.g., a compressor unit or turbine unit) to restore performance, thereby negating the need to perform a full engine overhaul, which was common practice for the smaller, low-bypass engines that preceded them. These modular engines (Pratt & Whitney JT9D, GE CF650, and the Rolls-Royce RB211) introduced another innovation: the advent of the digital engine control in commercial aviation. Initially, these controls took the form of *supervisory* electronic units that provided control trims in support of the tried-and-true main hydro-mechanical control units (HMUs) that had become the industry standard. These rapidly evolved to full authority digital engine controls (FADECs) in the '80s, replacing the HMUs, and along with them, introduced the need for additional gas path information, i.e., sensors. These took the form of temperatures and pressures at various stages along the engine's gas path and augmented the classic *core measurements* of spool speeds (N_1, N_2), fuel flow (w_f), and exhaust gas temperature (EGT). While the core measurement suite provided a picture of overall engine performance health, it was largely insufficient to determine individual engine module health. The newly added gas path (control) sensors, however, delivered greater visibility on individual modules and opened the door for potential assistance in guiding partial engine overhaul. What was needed was a reliable and automated means to analyze the data collected.

One must keep in mind that up until this time, data was typically collected manually by the pilot during steady-state cruise conditions by reading analog cockpit gauges and writing down their values on so-called *flight coupons* that were collected after each flight and transcribed into computer systems by powerplant engineers for subsequent trending and analysis. It should be fairly evident that this manual technique of capturing flight data was prone to possible error in determining repeatable steady-state conditions, reading gauges, and transcribing data to a ground-based computer system. The addition of new sensors just compounded this issue. Fortunately, another innovation was being introduced at this time: automated engine monitoring via electronic boxes dedicated to collecting and storing salient engine parameters during takeoff and stable cruise operation. These systems, airborne integrated data system (AIDS), later renamed airborne integrated monitoring system (AIMS) when the earlier acronym took on a different meaning, consisted of a series of electronic hardware that would monitor a plurality of flight parameters (e.g., altitude, Mach [M_n], total air temperature [TAT and/or engine inlet temperature T_2], ambient pressure [P_{amb}, and or total engine inlet pressure P_2], aircraft angle of attack, pitch, roll, and yaw rates) as well as available engine specific parameters (e.g., N_1, N_2, EGT, w_f, oil system temperature, and pressure and vibration amplitudes). At first, only *core* engine control parameters were available in the gas path, but later, other gas path parameters were included as FADEC control functions became more complex. At first, these additional parameters were added through arrangement between the airline and the engine manufacturer (OEM) on a case-by-case basis, requiring special sensor instrumentation purchase, installation, and multiplexing hardware to route the signals to the AIMS hardware. Much of this was driven and specified by the individual airlines (in concurrence with the OEM) for different aircraft/engine model combinations.

The net result was that large masses of data became available with no immediate method of utilization. The initial response to this dilemma was to utilize the computational capability of the AIMS to reduce the data and generate a variety of onboard reports to capture critical parameters for ground trending and annunciate any exceedances that might have occurred. Although useful, these measures did not exploit the full potential of the data, especially as it might aid in determining engine *module* health in support of individual module refurbishment. It was this mindset that became the catalyst for creating and adopting what we would call, in hindsight, the first generation of GPA programs. Unlike modern diagnostic, prognostic and health management (DPHM) strategies, there were no system-level requirements stated in terms of now familiar metrics such as diagnosis accuracy, error rates, false alarms, and the like, to drive sensor selection and their attendant accuracies. What existed was a more pedestrian thought process of what can we do to utilize the *already existing* bill of material (BOM) sensor information, used for control, to enhance our engine health visibility. What surfaced was GPA in its most elemental form. Given the high cost of maintenance for these large, high-bypass engines, any early success in detecting and isolating a fault provided the needed motivation to further develop the processes and approach to attain greater accuracy and fewer false alarms. Those efforts have not diminished for the four decades to follow, and GPA continues to be a vibrant area of research for academics, industrial practitioners, and government agencies alike.

While the primary focus of this text is the theory and application of GPA as both a diagnostic and prognostic methodology, we intend this work to be as self-contained as possible. Thus, before launching into specific GPA concepts, Chapters 1–3 are devoted to some of the *basics* that are generic to diagnostics and prognostics as well as a discussion of engine health management (EHM) and how GPA can contribute to this strategy. These chapters are recommended reading for all persons, GPA knowledgeable or otherwise. For the reader who is unfamiliar with GPA, it is advised that these three chapters be read before venturing into Chapter 4 and beyond. There are no specialized prerequisites necessary to understand the material presented outside of a rudimentary knowledge of gas turbine operation, thermodynamics, calculus, and linear algebra. Needed concepts are defined and/or derived throughout the text. When more advanced topics such as neural networks, or more generally, machine learning and artificial intelligence techniques are applied, fundamental concepts and references are provided to facilitate the understanding.

The structure of the book is as follows. Chapter 1 provides a broad discussion of EHM and how GPA contributes to its effectiveness. Chapter 2 reviews some of the relevant general concepts appearing in diagnostic and prognostic systems and provides many of the fundamental definitions that will be used throughout the text. This will include a dialog on the requisite basics pertaining to anomaly and fault detection as well as fault isolation and identification. Chapter 3 provides the basics of gas turbine parameter corrections that will be used throughout the text and are fundamental in describing the performance monitoring and engine modeling utilized in the classical gas path performance diagnostic methods described in subsequent chapters. Chapter 4 begins our discussion of GPA with a general overview of the methodology presented in a chronological fashion, with Chapter 5 continuing on the development of the GPA process for the analysis of time series data as well as other enhancements to the methodology. The extension to higher sampling rate data (full flight) begins in Chapter 6, with a discussion on modeling methods specific to the needs of this type of data stream. This will be followed by an in-depth discussion, in Chapter 7, of the extensions made to legacy GPA that are necessitated in order to perform the basic GPA functions on the high bandwidth data and potential real-time environment. Chapter 8 focuses on the topic of information fusion and the role it played in the development and evolution of the GPA. Chapter 9 continues with exploring intelligent reasoning and information fusion to enhance GPA analysis through ambiguity resolution. Chapter 10 finishes our GPA presentation with a discussion of prognostics and future trends.

In our attempt to make this text as self-contained as possible and to not clutter the main text with too much mathematical detail, a ***sizeable Appendix*** has been included, in which Chapters 2–9 have addendums containing derivations, ancillary information, and support notes. As such, the Appendix and its Addendums are *central* to understanding GPA and *not*, as Webster defines it as simply a "section of extra information added to the end of a book." Reference sections are included at the end of each chapter for the interested reader.

The chapters present a rough chronological development of GPA, wherein each chapter builds upon the results of the preceding chapters. Finally, while the algorithms discussed in this book can, for the most part, be applied to both aero-engines and stationary gas turbines alike, it should be noted that the general flavor of the discourse will emphasize aero-engines, especially in discussions involving applications and examples. It is also worth noting that relative to examples, we have *not* been able to include real engine data due to proprietary and ownership restrictions. Alternatively, we have used a high fidelity non-linear, closed-loop simulation of a notional high-bypass twin spool aero-engine developed by NASA called the commercial modular aero-propulsion system simulation (C-MAPSS) for the purpose of generating representative data for deteriorated engine modules, faults, and model numerics needed for demonstrating GPA principles.[1]

One last comment we will make is in regard to the notation used in this text. There are many references to papers written on the general subject of GPA by many authors over many decades. Notation varies widely amongst these expositions. Unfortunately, after nearly five decades, a standard notation has not emerged. As a result, we will be adopting the notation used by the authors in their publications, which has changed little over the years. Nevertheless, the reader is urged to refer to the "Nomenclature" section of this book for definition of all the symbols employed throughout the ensuing chapters and appendices. It should be noted that some alternative notation, such as N1 for N_1, N2 for N_2, Wf for w_f, etc., is used in some figures and tables for readability. Please refer to the "Nomenclature" section for alternate symbology. Finally, the symbol (⇒) appears throughout the text in many of the derivations. This symbol means *which implies* or *implying*. As it is understood that the relationships used by engineers are (often by their nature) approximations, we will use the symbol approximately equal (≈) interchangeably with mathematically equal (=) throughout our presentation as a matter of convenience.

REFERENCES

1. Urban, L.A., 1969, *Gas Turbine Engine Parameter Interrelationships*, HSD UTC, Windsor Locks, CT. (1st ed. 1967, 2nd ed. 1969)
2. Urban, L.A., 1972, *Gas Path Analysis Applied to Turbine Engine Conditioning Monitoring*, AIAA/SAE Paper 72-1082.

NOTE

1. See "Addendum to Introduction, A0: Commercial Modular Aero-Propulsion System Simulation"

Nomenclature

Acronyms

AANN	Auto associative neural networks
ACARS	Airborne communications addressing and reporting system
ACMF	Aircraft condition monitoring function
AC	Advisory Circular
ADEM	Advanced diagnostics and engine management
AI	Artificial intelligence
AIAA	American Institute of Aeronautics and Astronautics
AIDS	Airborne integrated data system
AIMS	Airborne integrated monitoring system
ANN	Artificial neural network
ARINC	Aeronautical Radio, Inc.
ARMA	Autoregressive moving average model
BBN	Bayesian belief networks
BOM	Bill of material
CAMP	Continuous airworthiness maintenance program
CBR	Case-based reasoning
CFR	Code of Federal Regulations
C-MAPSS	Commercial aero-propulsion system simulation
CPT	Conditional probability table
CRA	Cumulative relative accuracy
DPHM	Diagnostics, prognostics, and health management
D&C	Delays and cancellations
EAP	Electrified aircraft propulsion
EASA	European Aviation Safety Agency
ECM	Engine condition monitoring
EHM	Engine health monitoring or engine health management
EKF	Extended Kalman filter
EoL	End of life
eSTORM	Enhanced STORM
ETOPS	Extended twin-engine operations
FAA	Federal Aviation Authority
FADEC	Full authority engine control
FAN	FAN module, also referred to as Fan module
FAR	Federal Acquisition Regulations
FDA	Fault detection accuracy
FDI	Fault detection and isolation
FOQA	Flight operations quality assurance
FMP	Fleet management program
GA	Genetic algorithm

GMM	Gaussian mixture model
GPA	Gas path analysis
GUI	Graphical user interface
HECTM	Honeywell engine condition trend monitoring
HMU	Hydro-mechanical unit
HPC	High-pressure compressor module
HPT	High-pressure turbine module
IC	Influence coefficients
IFSD	In-flight shutdown
ISA	International standard atmosphere
LPC	Low-pressure compressor module
LPT	Low-pressure turbine module
LSQ	Least square
MCS	Monte Carlo simulation
MEA	More electric aircraft
MFA	Multiple fault analysis
ML	Machine learning
MLP	Multi-layer perceptron
MOM	Mean of max defuzzification
MPA	Module performance analysis
MRO	Maintenance, repair, and overhaul
NACA	National Advisory Committee for Aeronautics
NLP	Natural language processing
NPSS	Numerical propulsion system simulation
OEI	One engine inoperative
OEM	Original engine (or equipment) manufacturer
PF	Particle filter
PHM	Prognostics health management
PNN	Probabilistic neural network
QAR	Quick access recorder
QED	Quad erat demonstratum (*Latin*: *what was to be shown*)
QSS	Quasi-steady state
ROC	Receiver operator curve
RUL	Remaining useful life
SAT	Static air temperature
SFA	Single fault analysis
STORM	Self-tuning onboard real-time model
SVD	Singular value decomposition
TCC	Turbine case cooling
TRA	Throttle resolver angle (power lever)
UBL	Usage-based lifing
UER	Unanticipated engine removal
UKF	Unscented Kalman filter
WebECTM	Web-based engine condition trend monitoring

WLSQ	Weighted least square
WOW	Weight on wheels

Symbols

$A>0$	Matrix A is positive definite
ALT	Altitude
α_x	Correction θ exponent for parameter X
$A^{(n)}{}_k$	n point exponential average at time k
AG	Ambiguity group
AGM	Ambiguity group matrix
A_4	High turbine nozzle area
A_{45}	Low turbine nozzle area
β_x	Correction δ exponent for parameter X
cp	Specific heat at constant pressure
$cp_{a,d}$	Specific heat of dry air at constant pressure (same as cp)
$cp_{a,w}$	Specific heat of wet air at constant pressure
cp_w	Specific heat of water vapor at constant pressure
cv	Specific heat at constant volume
$\mathcal{C}$	Discrete confusion matrix
C	Isolation confusion matrix
C^*	Augmented isolation confusion matrix
CF	Coupling factor (1% efficiency + CF% flow)
CIR	Correct isolation rate
$deg\ K$	Degrees Kelvin
$deg\ R$	Degrees Rankine
$\det(A)$	Determinant of matrix A
D	Combined engine and sensor diagnostic matrix ($D^T = [D_e^T \vdots D_s^T]$)
D_e	Engine diagnostic (inverse influence coefficient) matrix
D_s	Sensor diagnostic (inverse influence coefficient) matrix
δ	Pressure ratio relative to standard day
Δ	Delta, change in parameter (ΔP = change in parameter P)
$E(x)$	Expected value (mean) of random variable x (also denoted μ_x)
EGT	Exhaust gas temperature
EPR	Engine pressure ratio (P_5/P_2)
FN	Number of false negatives (missed detections)
FP	Number of false positives (false alarms)
FPM	Fault propagation matrix
FPR	False positive rate (false alarm rate)
η	Generic engine health parameter
γ	Ratio of specific heats $\gamma = cp/cv$
g	Grams, unit of measurement
h	Enthalpy

H	Combined engine and sensor fault matrix ($H = [H_e \vdots H_s]$)
H_e	Engine fault (influence coefficient) matrix
H_s	Sensor fault (influence coefficient) matrix
$IC(\mathrm{P}\vert\eta)_{\mathrm{Q=const}}$	Influence coefficient of $fault = \eta$ on $measurement = P$ holding $measurement\ Q = const$
m	Meters, unit of measurement
M_n	Mach No.
MIR	Mis-isolation rate $(1 - CIR)$
$\mathrm{N}(\mu,P)$	Normal distribution with mean μ and covariance P
μ_x	Humidity correction for parameter X
N_1	Low spool rotor speed (alternative notation N1)
N_2	High spool rotor speed (alternative notation N2)
Ω_k	Collection of QSS data points in k^{th} flight
P_2	inlet (total) pressure (alternative notation P2)
P_{25}	HPC inlet (total) pressure (alternative notation P25)
P_3	HPC exit (total) pressure (alternative notation P3)
P_{48}	LPT inlet pressure (alternative notation P48)
P_5	exhaust (total) pressure (alternative notation P5)
P_{amb}	Ambient pressure (alternative notation Pamb)
P_b	Burner pressure (alternative notation Pb)
P_{ISA}	ISA standard day pressure (14.696 psia, 101325.353 Pa)
P_T	Total pressure
P_S	Static pressure
P_{xx}	Covariance matrix for random vector x
P_x	Shorthand notation for covariance matrix P_{xx}
P_{xy}	Covariance matrix for random vectors x and y
Pa	Pascals unit of pressure
$psia$	Pounds per square inch absolute unit of pressure
PSP	Power setting parameter
R	Gas constant
$R_{a,d}$	Gas constant of dry air (same as R)
$R_{a,w}$	Gas constant of wet air
R_w	Gas constant of water vapor
S	Entropy
SFC	(Thrust) specific fuel consumption
SH	Specific humidity
$Sss(Y)_x$	Steady state slope of Y with respect to X
$\sigma 2_x$	Variance of random variable x [also denoted $Var(x)$]
T_2	Inlet (total) temperature (alternative notation T2)
T_{25}	HPC inlet pressure (alternative notation T25)
T_3	HPC exit pressure (same as T_{30}) (alternative notation T3)
T_{48}	LPT inlet temperature (alternative notation T48)
T_{50}	Exhaust temperature (same as EGT) (alternative notation T50)
T_{amb}	Ambient temperature

T_{ISA}	ISA standard day temperature (518.67 degR, 288.15 degK)
T_m	Metal temperature
SAT	Static air temperature
TAT	Total air temperature
TIT	Turbine inlet temperature
TN	Number of true negatives (correct non-detection)
TP	Number of true positives (correct detection)
TPR	True positive rate (true detection rate)
Trace(A)	Trace of matrix *A*. (sum of diagonal elements)
θ	Temperature ratio relative to standard day
$Var(x)$	Variance of random variable x [also denoted σ_x^2]
v	Velocity
w_a	Air (mass) flow (alternative notation Wa)
w_f	Fuel (mass) flow (alternative notation Wf)
war	Water to air ratio
X	Generic gas path parameter
X_c	Generic *corrected* gas path parameter
X_{Base}	Generic baseline reference for gas path parameter X. (same as X_{NOM})
X_{NOM}	Generic nominal reference for gas path parameter X. (same as X_{Base})
$X_c^{(PSP)}$	Generic power setting parameter
x_e	Vector of engine fault Δs
x_s	Vector of sensor fault Δs
Z	Vector of measurement Δs

Math Symbols

$\Rightarrow$	Implies
$\forall$	For all
$\exists$	There exists
$\ni$	Such that
$\in$	Is an element of
$\therefore$	Therefore
$\equiv$	Equivalent to
$\triangleq$	Defined as
$\propto$	Proportional to
∞	Infinity
nCk	Number of combinations of n items taken k at a time.
$n!$	n factorial = $n(n-1)(n-2)...1$

Subscripts

a	Air
a,d	Dry air
a,w	Wet air

Base	Baseline
c	Corrected (parameter)
e	Engine
f	Fuel
NOM	Nominal
S	Sensor
w	Water vapor

Superscripts

T	Matrix transpose
-1	Matrix inverse

1 Engine Health Management

1.1 OVERVIEW

In its broadest context, an engine health management (EHM) system deals with the monitoring, detection, isolation, predictive trending, and accommodation of engine degradation, faults, and failures. Its methods utilize a wide range of engine system disciplines, including materials, structures, and controls. Minimizing total cost of (engine) ownership and increasing asset availability and readiness, while ensuring safe operation, are key drivers for the implementation of an EHM system. Relative weighting of these factors varies in importance depending on the end-user's operating and business environment, as well as application-specific needs. Hence, there exists no specific approach for an EHM system but, rather, an array of capabilities that can be assembled to address the specific needs of an end-user. The range of capabilities employed in these systems include not only those that are engine specific but also draw from mathematical, statistical, and artificial intelligence methods that are more generic in nature and have potential cross-cutting application to other propulsion, marine, and power generation gas turbines as well as rocket and hypersonic propulsion systems. In this chapter, we will briefly explore the factors and benefits that drive the need for EHM and how it has been implemented in the aero-engine industry. The final section will discuss how gas path analysis (GPA) can fit into these EHM strategies and why it has become a core element activity in supporting EHM goals.

1.2 EHM LANDSCAPE

There are many acronyms that pervade the technical engineering disciplines. This is most assuredly true for the engine health discipline. It is almost inescapable to have a meaningful yet succinct technical discussion without using one in almost every other sentence. In an effort to remain concise, we cannot avoid them, but we have strived to minimize their use so that the reader can proceed with some semblance of clarity and minimized distraction. Nevertheless, a list of well-used acronyms pertinent to our discussions is included in the "Nomenclature" section of this book. We begin our discourse with one of the oldest acronyms in this discipline: EHM. A testament to its age is that it has changed its meaning over the years, albeit in a subtle manner. Five decades ago, this acronym surfaced with the meaning engine health *monitoring*. Two decades ago, the same acronym took on a different connotation, i.e., engine health *management*. Monitoring is, for the most part, a passive activity, whereas

DOI: 10.1201/9781003649274-1

management suggests a more proactive mindset. For our purposes, we will use the term EHM interchangeably.

Gas turbine engine health monitoring is as old as the engine itself. From its beginnings as simple monitoring practices performed by a line mechanic, advances in engine complexity, regulatory requirements, and economic pressures have advanced monitoring to impact flight operation, safety, and maintenance activities to an ever-increasing degree. As we alluded to in the introduction, EHM began in earnest with the advent of commercial high bypass turbofan engines in the 1970s. These modular-design machines provided an economic impetus for the development of analysis methods to track the performance health down to the module level in an attempt to reduce maintenance work scope and overall costs. With the advent of the full authority digital engine control (FADEC) in the early 1980s, flyable instrumentation intended for control functions became available for diagnostic purposes. With continued advances in low-cost computing, high-speed communication, and more sophisticated sensors, EHM systems are now found in a variety of applications. Because the need for EHM systems varies between applications, a one-size-fits-all solution is not likely to exist, and designs must be tailored to each application. EHM can be visualized as a portfolio of building blocks that can be used to create customized architectures that best meet individual user needs. These architectures may include both engine-hosted and ground-based elements as complementary features of an overall integrated system. Engine-hosted elements generate data from onboard sensors and perform basic fault isolation and failure prediction, supporting on-wing maintenance, while the ground-based elements support long-term degradation trending, providing planning information that can be used by fleet managers.

As we shall see, GPA has become a core element in the EHM strategies employed in the commercial and military sectors worldwide. *Please note* that throughout the book, we will be referring to gas path measurements by engine station number as depicted in the schematic rendering of a generic high bypass, twin spool turbofan engine shown in Figure 1.1.[1]

In the aero-engine world, EHM systems first evolved as specialized ground-based diagnostic systems supporting test cells. These systems were limited in scope, largely comprised of independent programs for relatively low bandwidth, time-averaged, steady-state gas path and vibration data and were typically performed during outbound post-overhaul acceptance test runs and occasionally for inbound test-as-received engine runs. From this test cell background, GPA was developed and matured. The literature abounds with descriptions of this type of diagnostic technique based on research from large and small businesses, as well as academia and government. We will attempt to provide a broad description of the methodologies that have evolved (chronologically) in subsequent chapters; however, the interested reader may refer to [1–14] in the general literature for additional historical insight. No matter how a particular EHM program is configured, data acquisition is a key concern. In a ground-based system, data availability (as to type and quantity, i.e., bandwidth) imposes certain restrictions that are not necessarily present for engine-hosted systems. Conversely, tracking engine health on a *fleet* basis is only possible with ground-based systems. Having an appropriate data infrastructure in place for

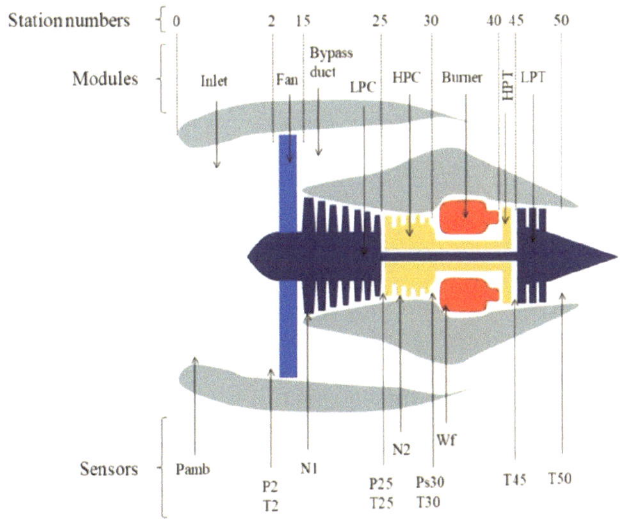

FIGURE 1.1 Typical gas path parameter locations

capturing in-flight engine data with seamless transfer to ground-hosted systems is a pivotal element for EHM to be successful. This must be accompanied by appropriate links to maintenance information, supply chain information, life cycle information as well as a suitable interface for the end user to access and utilize the information.

The primary purpose of the engine-hosted EHM system is to enhance engine safety against damage by providing early real-time detection of conditions that might progress to critical system failures, while also aiding in the reduction of maintenance support costs and aborted missions by minimizing the time from fault isolation of a failing component to installing a line replaceable unit. The purpose of the onboard system is to acquire data, monitor and assess current engine condition, and, if possible, predict future engine state. Subtle changes in combinations of some measured parameters can indicate early symptoms of a progression toward a failure condition. These changes in a parameter's characteristics are often difficult to detect by simply observing limit exceedances, as the parameter values typically remain within the normal operating range. Trend analysis of parameter values, taken over time at specific operating conditions, enables detection of shifts, abnormal rate of change, and abnormal scatter in the data. While emerging technologies are driving more and more comprehensive onboard capabilities, these latter functions are most easily accomplished off-board. As a result, EHM systems often employ a mixture of onboard and off-board analysis functions. An idealistic goal for EHM would be complete health assessment of the monitored engine onboard, with fleet tracking and management accomplished off-board in a ground system. Relative to GPA, we will be discussing some emerging technologies bringing us closer to realizing this goal in Chapters 8–10.

The adoption of an EHM stratagem has been driven primarily by two factors: cost and regulatory compliance. We will briefly discuss these in the next two sections.

1.3 COST BENEFIT AND FLEET MANAGEMENT PROGRAMS

It should come as no great surprise that engine life cycle cost has been a driving force behind the design and execution of EHM programs since their inception. Cost is a complex topic in and of itself, but for simplification, let us divide this into two general (albeit, not mutually exclusive) categories: *operational* cost and *maintenance* cost. To have appreciation of the magnitudes involved, in the commercial aviation sector alone, it is estimated that $69B was spent in 2018 on fleet maintenance, repair, and overhaul (MRO) activities for the worldwide fleet of 27,535 operational aircraft. This is projected to rise to $103B by 2028 [15]. Engines claim 47% of direct maintenance costs. These seemingly large numbers would be something just short of astronomical if extended to include all military aero-engine fleets worldwide as well as all of the land and marine propulsion applications and power generation and oil pumping stations in operation. Improvements in maintenance logistics, reductions in unscheduled events (and their consequences), and improvements in operational efficiency can have an enormous impact on reducing these costs. Diagnostics and prognostics have a significant role to play in contributing to these cost reductions.

Throughout this book, we will use the term *diagnostics* somewhat liberally to encompass *both* detection and isolation (i.e., identification) of engine-related faults and failures as well as the monitoring of engine system/sub-system degradation and deterioration prior to actual failure. From a diagnostic perspective, reductions in operational cost, for the aero-engine, can be achieved by avoiding unscheduled events – such as unanticipated engine removals (UERs); *extensive* line maintenance for failed sub-systems causing secondary damage; and in-flight shutdowns (IFSDs), which can drive an engine removal or, more catastrophically, result in aircraft damage or loss of life – and as well as detecting and identifying partial failures or malfunctions of sub-systems, which contribute to increased fuel consumption, such as engine bleed leaks/failures, active clearance control and variable geometry actuation abnormalities, to name a few. Any of these can contribute to an increased number of delays and cancellations (D&Cs) (in the aero-sector), which result in customer dissatisfaction and its associated costs. From a maintenance and fleet management perspective, diagnostics can impact maintenance logistics, repair scheduling, spare parts inventory, overhaul and repair work scope, etc.; all of which contribute to the overall life cycle cost of an individual engine as well as the entire fleet.

A related term, *prognostics*, deals with predicting the future (health) state of the engine and its accessories and has come into vogue in the past couple of decades. Prognostics strive to predict the future condition of a component or system of components. For the purposes of gas turbine engine prognostics, this definition is often further described in terms of hard failures of components or condition/degradation of performance-related problems. Failure prognostics is focused on the prediction of damage state or failure rate of a component or system of components in an engine. Failure prognostics is usually affected by the diagnosis of specific engine faults, depending on the level of impact the component experiences from the fault condition. Prognostic models are required to project to the future condition. Prognostics can also be associated with the slower degradation (wear-related) processes that an engine is exposed to throughout its life. It is usually associated with the diagnosis of fault(s) conditions and the capability of predicting when the symptoms of the identified fault(s) will reach an undesirable state in which system operation will be adversely affected. Prognostic models are required to project the future "path" of these identified fault(s) on total system performance or reliability.

An example of prognostics is the calculation of the remaining useful life of a life-limited part, component, or sub-system of the engine. Prognostics may or may not be preceded by the detection of a failure precursor. An example of a prognostic technique that is preceded by failure detection includes the detection of bearing spalling (a typical failure precursor) and then the estimation of remaining useful life of the bearing. Another example is effective cycle counting of life-limited parts such as disks, spacers, and shafts. While no failure precursor is evident in this case, a forecast of the remaining time before the part reaches the end of its useful life is generated and can be used to schedule its removal before it fails. Finally, it should be noted that it is not uncommon to see the term prognostics used in cases of "early fault diagnosis" in the industry. The argument is that, if early fault symptoms are detected and addressed, a real fault or failure could be mitigated. Therefore, in that sense, it is predictive. Strictly speaking, these are not prognostics, but we will leave it for the readers to decide.

From all of these observations and assertions, a philosophy on how we manage engine health slowly emerges. Simplistically, if we think about possible engine maintenance schema, there are three general philosophical approaches to consider as depicted below in Figure 1.2. *Operate to failure* is not a particularly attractive approach to take for many reasons. Aside from the safety aspects associated with engine failure, this would be the most expensive (to the operator) maintenance approach possible. Addressing maintenance on a *fixed service schedule,* wherein maintenance, part replacements, and overhaul are accomplished in accordance with a fixed time schedule dependent on engine operation hours, mitigates the safety concerns and expenses associated with failures. From the Original Equipment Manufacturer (OEM) perspective, this was a lucrative proposition for spare parts sales, as parts were replaced as dictated by schedule, thereby creating a steady and predictable income stream. For the operator, it was an expensive undertaking to replace parts that were still performing adequately. Nonetheless, it was the maintenance philosophy in play for commercial carriers and the military for many decades until EHM, supported by evolving diagnostic and prognostic technologies, opened the door to assessing the health condition of the engine, its components, and subsystems. Performing maintenance, *on condition*, was an attractive alternative for the operator in terms of cost. The cost to the aero-engine operator is really *cost per flight hour.* The last graph in Figure 1.2 displays two curves, the upper curve being fixed cost (for the engine or part purchase, which amortizes with time) and lower curve representing the effects of degradation, which generally increases cost per flight hour over time (e.g., increased fuel consumption). The sum of the two curves is the total

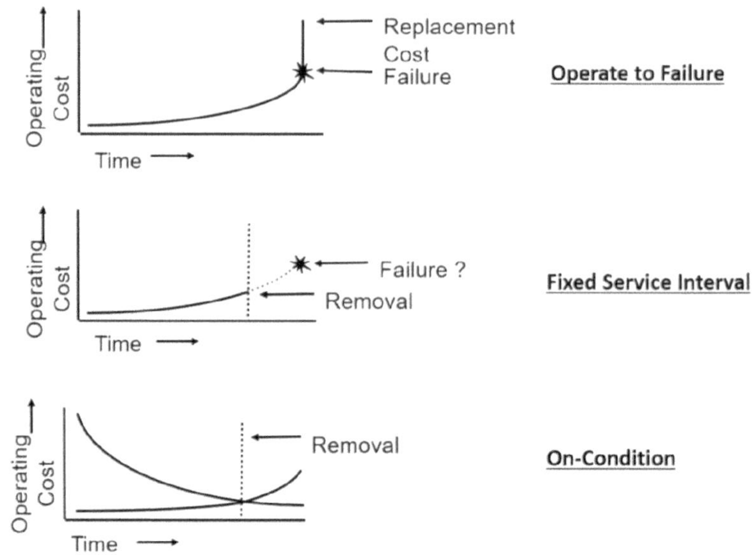

FIGURE 1.2 Possible maintenance philosophies

impact on cost per flight hour and has its minimum (in time) where the two intersect. This becomes the optimal time to perform the maintenance. The switch from fixed interval to on condition did not happen all at once. As you can imagine, it was not favored by the OEM, as it negatively impacted their spare parts sales. The transition spanned decades with the introduction and continued evolution of so-called fleet management programs (FMPs).

FMPs provided a means for the OEM to replace lost spare parts income with a *service*-related income stream. From a pedestrian viewpoint, an FMP can be aptly described by the popular phrase, *power by the hour*,[2] originally coined by Bristol Siddeley Engines Ltd, a British aero-engine manufacturer later purchased by Rolls-Royce. All major engine OEMs offer some form of an FMP to their airline customers, wherein, for a fixed dollar-per-flying-hour payment, a complete engine and accessory maintenance and overhaul service is provided, thus allowing the operator to forecast their operational costs with great accuracy. This also relieves them of the need to stockpile spare engines, parts, and accessories. The OEM effectively provides a maintenance program over an extended period of time, and the operators are assured of an accurate cost projection and avoid the costs associated with breakdowns. Knowing the current health of each engine in the fleet contributes to the OEM's ability to configure an FMP to support their maintenance business model and derive profit and satisfy their customers' needs. This is a win-win situation for all parties.

FMPs can be quite varied and complicated with many factors to consider. We will suffice ourselves with our simple overview, which was intended to show a connection to EHM, diagnostics, and prognostics at a very high level.[3] Hopefully, we have provided the reader with a glimpse as to *why* diagnostics and prognostics have the potential to reduce operational and maintenance cost. The question of *who* engages in these diagnostic activities is also worthy of mention. Commercial airlines and powerplant operators, military users, OEMs and independent maintenance and overhaul facilities all engage in a variety of diagnostic activities, ranging from engine monitoring, health tracking, fault diagnosis, and fleet health management. Some of these functions are accomplished onboard, some of them off-board, and some require a mixture of both. The role that diagnostics and prognostics play for the engine OEM has changed over time from one of *supporting* maintenance troubleshooting to a more *proactive* role. All engine OEMs today offer their own EHM services. For example, Rolls-Royce (RR) offers EHM and TotalCare. GE Aviation offers a suite of systems, including the Predix, ADEPT, ECM, and PHM. Pratt & Whitney offers ENGINEWISE™ and Advanced Diagnostics and Engine Management (ADEM™). Pratt & Whitney Canada offers WebECTM™ and HECTM® [19]. [4]

For a detailed presentation on cost benefit potential and determination, see SAE Standard Recommended Practices [20]. The reader may also wish to refer to SAE Standard Recommended Practices [21], [22] for guidance on the design and development of an EHM system.

1.4 REGULATORY FACTORS

In addition to the cost drivers, the stakeholders are also obligated to comply with regulations set by the aviation authorities. A civil aviation authority is a government authority in each country that maintains an aircraft registry and oversees the approval and regulation of civil aviation. The Federal Aviation Administration (FAA) and the European Aviation Safety Agency (EASA) are such authorities in the US and Europe, respectively. In the US, civil aviation regulations are part of Title 14 of the Code of Federal Regulations (CFR) that prescribe the rules governing all aviation activities in the country. The Federal Aviation Regulations are organized into sections, called parts, and are commonly referred to by the term "14 CFR Part X" instead of "FARs."[5] For example, "14 CFR Part 33" contains rules for aircraft engines airworthiness standards.

Pertaining to engine health monitoring and preventive maintenance programs, 14 CFR Part 21 requires that manufacturers, such as engine OEMs, provide complete instructions for continued airworthiness when the aircraft is delivered or when the first standard airworthiness certificate is issued. To implement these instructions, 14 CFR Part 121 (for large transports) assigns responsibility for performing maintenance, inspections, and alterations to the operators and prescribe regulations for contract maintenance if performed by third-party providers. Once certification requirements have been met and a standard airworthiness certification has been issued, the aircraft enters service. Operators and manufacturers are responsible for maintaining the product performance and reliability according to the requirements for continued airworthiness. The FAA is responsible for monitoring continued compliance [23]. In order to comply with these regulations and to guard against liability, most (if not all) operators of commercial aircraft, such as commercial airlines, develop and implement some sort of maintenance and preventive maintenance programs, which often include various levels of scheduled maintenance. The role of EHM systems in compliance is to improve these programs to assure continued airworthiness. This is achieved by enabling condition-based maintenance through active monitoring of engine health using techniques such as GPA. Examples are given in a later chapter.

In the following paragraphs, we will list a few FAA regulations pertaining to engine health monitoring, particularly for extended operations (ETOPS), for basic awareness.

1.4.1 14 CFR Part A33.3(C) ETOPS Requirements

Appendix A to Part 33 specifies requirements for the preparation of instructions for continued airworthiness. The ETOPS requirements for engine condition monitoring is stated as follows:

> The engine condition monitoring procedures must be able to determine prior to flight, whether an engine is capable of providing, within approved engine operating limits, maximum continuous power or thrust, bleed air, and power extraction required for a relevant engine inoperative diversion. For an engine to be installed on a two-engine airplane approved for ETOPS, the engine condition monitoring procedures must be validated before ETOPS eligibility is granted.[6]

1.4.2 14 CFR Part K25.1.5 Engine Condition Monitoring

CFR Appendix K to Part 25 specifies airworthiness requirements for the approval of an airplane–engine combination for ETOPS. Procedures for engine-condition monitoring must be specified and validated in accordance with Part 33, Appendix A, Paragraph A33.3(c).

1.4.3 14 CFR Part 121.374 Continuous Airworthiness Maintenance Program (CAMP) for Two-Engine ETOPS

As one of the "must include" elements of the ETOPS Continuous Airworthiness Maintenance Program (CAMP), engine condition monitoring must be provided to meet the following requirements.

The certificate holder must have an engine condition monitoring program to detect deterioration at an early stage and to allow for corrective action before safe operation is affected.

This program must describe the parameters to be monitored, the method of data collection, the method of analyzing data, and the process for taking corrective action.

The program must ensure that engine-limit margins are maintained so that a prolonged engine-inoperative diversion may be conducted at approved power levels and in all expected environmental conditions without exceeding approved engine limits. This includes approved limits for items such as rotor speeds and exhaust gas temperatures.

Another form of FAA publication of interest to this topic is the Advisory Circular (AC). ACs are published FAA directions for compliance with airworthiness regulations, operational standards, training standards, and any other rules within the 14 CFR Aeronautics and Space Title. They define acceptable means – but not the only means – of accomplishing or showing compliance with airworthiness regulations [24]. For instance, AC33.201-1, titled "Extended Operations (ETOPS) Eligibility For Turbine Engines," provides guidance and acceptable methods for demonstrating compliance with the ETOPS eligibility requirements in accordance with Title 14 of the CFRs (14 CFR) Part 33.

1.4.4 AC 43-218 Operational Authorization of Integrated Aircraft Health Management System

This AC, released in July 2022, provides guidance for developing an operator's integrated aircraft health management (IAHM) program (which includes EHM as an integral part of it). This AC describes an acceptable means – but not the only means – to comply with the applicable sections of Title 14 of the CFRs (14 CFR). A "certified for credit" concept is defined in this AC as the "means to give design approval to an IAHM system that adds to, replaces, or intervenes in industry-accepted maintenance practices currently performed by parts 65, 121, 135, and 145 certificated entities" [25]. Pertaining to EHM, the SAE ARP5987A, published in April 2024, provides a recommended process for utilizing EHM systems for airworthiness credit [26].

For further information on the interplay between maintenance philosophies in EHM and regulations and standards, the interested reader can consult [27]. Additional certification considerations for EHM were discussed in [28].

1.5 ENGINE MONITORING

As previously mentioned, engine condition monitoring and associated diagnostic practices are as old as the gas turbine itself. Monitoring exhaust gas temperature (*EGT*), spool rotational speed, and overall engine vibration levels[7] marked the beginning of monitoring methods to ensure safe operation by highlighting abnormal levels or exceedances. This rudimentary set of instrumentation was sufficient to *detect* deteriorated and undesirable operation; however, it did little to aid in the *identification* of the root cause of the abnormality. As the gas turbine became more intricate and the associated cost of maintenance and repair increased, the onboard instrumentation suite expanded to support its needs. As with most systems, reliability and cost were the main drivers. Engine control functions also drove the need for additional sensors, primarily in the engine's gas path. At first, this parametric information was only available in the cockpit for the flight crew to monitor the engine's operation. During stable cruise operation (for aero-engines), with the engine in a relative steady state condition, the flight crew would record parameter levels from cockpit gauge observation (typically spool speeds [N_1, N_2], *EGT*, fuel flow (w_f), vibration level, and oil system temperature and pressure) and submit a *flight coupon* to the ground crew upon landing for subsequent trending by the airline's powerplant engineers.

In the '70s, electronic engine condition monitoring hardware was introduced on wide-body aircraft, which would free the flight crew from recording flight coupons by automatically monitoring and capturing salient engine parameters during takeoff and stable cruise operation. These systems, airborne integrated data system (AIDS), later renamed to airborne integrated monitoring system (AIMS) when the earlier acronym took on a different meaning, consisted of a series of electronic hardware that would monitor a plurality of flight parameters (e.g., altitude, Mach, total air temperature [*TAT* and/or engine inlet temperature T_2], ambient pressure [P_{amb}] or total engine inlet pressure [P_2], and aircraft angle-of-attack, pitch, roll, and yaw rates) as well as available engine specific parameters (e.g., N_1, N_2, *EGT*, w_f, oil system temperature, and pressure and vibration amplitudes). At first, only engine control parameters were available in the gas path, but, later, additional gas path parameters were included. Initially, these supplementary parameters were added through arrangement between the airline and the OEM on a case-by-case basis, requiring special sensor instrumentation purchase, installation, and cabling hardware to route the signals to the AIMS hardware. This was typically specified by the individual airlines, in concurrence with the OEM, for a variety of aircraft/engine model combinations. Features in the AIMS differed from airline to airline and the software content and configuration for the AIMS was dictated and controlled by the individual airlines. As such, there was no *standard* AIMS configuration. There was, however, some functional similarity between systems. The AIMS would monitor the aircraft and engine parameters available to it and record parameter exceedances from stored thresholds during the flight.[8] Throughout cruise operation, the system would search for *stable* conditions,[9] perform a time average across all parameters within this *stable frame*, and store the data[10] for subsequent analyses on the ground. Likewise, during takeoff, the monitoring system would capture various parameters and record critical values, for example, max EGT, and record it for subsequent ground analyses. At the conclusion of the flight, these recordings were retrieved from the quick access recorder (QAR) and loaded into the airline's ground station computers for trending, *EGT* margin tracking, and other analysis to support fleet-wide health tracking supporting maintenance (line and depot) decisions and logistics. Figure 1.3 depicts a sample AIMS configuration.

As the digital age took hold in the '80s, with advances in aviation electronics and computers, greater sophistication in the AIMS hardware was possible. Full Authority Digital Engine Control (FADECs) replaced the hydro-mechanical engine controls with digital data buses Aeronautical Radio, Inc. (ARINC), enabling a wider range of parameters available to the monitoring system. With faster central processing unit (CPUs) and greater memory capacity available, the monitoring systems could perform a wider range of analysis in real-time during flight. The airborne communications addressing and reporting system (ACARS) slowly replaced the need for QARs, wherein, the recorded data could be radio transmitted to the ground-station

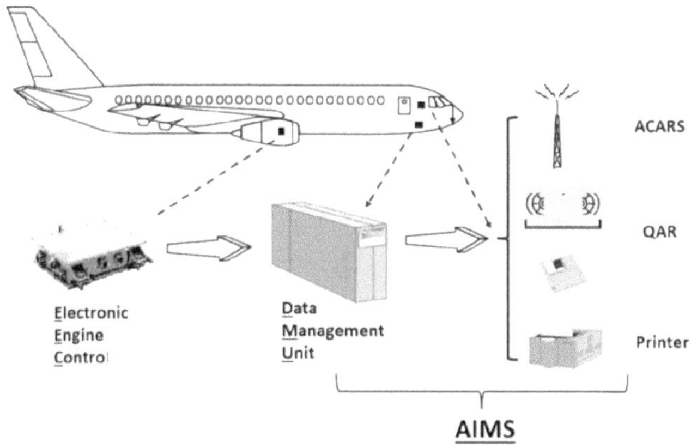

FIGURE 1.3 Sample AIMS configuration

computers *during* flight,[11] transcribed from its cryptic format to something more manageable, and quickly analyzed to provide greater informational direction to line maintenance crews in advance of the aircraft landing.[12] The added computational capacity of the onboard system enabled the generation of aircraft condition monitoring function (ACMF) flight reports whose content could provide indications of impending failures within the aircraft and engine systems. This permitted operators to initiate preventative maintenance actions to minimize the risk of impacts on dispatch reliability and overall safety.

Since its inception four decades ago, the ACMF has expanded to include more extensive reporting capability and is standard practice in current aero-engine monitoring. Typical reports in use today might include [29]:

- Engine takeoff report
- Engine climb report
- Engine cruise report
- Engine gas path advisory report
- Engine mechanical advisory report
- Engine start report
- Engine divergence report

The ACMF reports provide a wealth of information regarding the monitored engine's general health state as well as indications of emerging faults and failure conditions. These reports have expanded in scope and complexity as onboard computer capability has evolved over time, but this still requires the man-in-the-loop to digest and analyze the information, perform additional off-board analysis, and combine other sources of information to decide upon the appropriate action in maintaining the flight integrity of the engine. The reports generated in the '70s and '80s were subsets of those listed above and relied more heavily on the ground powerplant analyst to piece together the health profile of the engine. To aid in this task, a new analysis methodology was introduced in the early '70s known as *gas path analysis* (GPA). This method took advantage of *existing* instrumentation common to all gas turbines and provided a means of tracking overall performance degradation of the engine as well as degradation in each of the main engine modules and some accessories. The data required to support this analysis method was within the capability of the early AIMS to capture and store (and later transmit) and was generic in its approach to be applied to any configuration of gas turbine, i.e., single spool to triple spool, turbojet, and turbofan. Over the subsequent five decades, it has been the topic of much research, which has extended its capabilities from its early formulation and remains one of the core techniques supporting health management to this day.

In addition to ACMF reports (i.e., snapshot data), high cut rate streaming data (aka, full flight data) from revenue service flights are becoming available to the engine OEMs thanks to the recent advances in avionics. The avionics devices that perform this task have been marketed with various names, but they all typically offer high-speed wireless data transmission via cellular network or WiFi to a data system hosted either by the airline or the OEM. The advent of full flight data brings tremendous opportunities for advanced analytics, as well as challenges for storing and managing a huge amount of data. We will discuss potential GPA methods for full flight data in Chapter 8.

1.6 HOW GPA SUPPORTS EHM

EHM concerns itself with making intelligent, informed, and appropriate decisions about maintenance and logistics actions based on diagnostics and prognostics information, available resources, and operational demand. To be successful, this requires myriad diagnostic and prognostic technologies to address the health tracking of such a complex piece of machinery as the gas turbine. Because of its relative longevity, GPA is one of the more mature methodologies in the diagnostic/prognostic arsenal. As we shall see in subsequent chapters, it is both a diagnostic tool for detecting and isolating a wide range of engine component and sub-system faults as well as a prognostic tool for tracking long-term performance deterioration. In addition, it requires no special instrumentation and operates with data provided from the bill of material (BOM) sensors, already installed for control purposes, and is flexible, in the sense that it can be adapted to different sensor (i.e., measurement) configurations with little to no software modification. Finally, it can process data acquired from ground testing or data acquired during flight from the AIMS. Historically, GPA has been a ground based, off-board analysis system. However, relative to the previous statement, recent developments have paved the way for allowing some of its major diagnostic functions to be wholly contained and processed onboard for flight acquired data in the near future. This topic will be discussed in Chapters 8 and 10.

Some of the ways GPA contributes to EHM's goal in reducing maintenance costs are outlined in *Table 1.1.*

We will defer discussion of these elements to subsequent chapters after we have formally introduced GPA and its methodology. Instead, we end this section with some *hypothetical scenarios* that exemplify GPA's impact on EHM. The following

TABLE 1.1
GPA Potential Impact on Maintenance Cost Reduction

Area	GPA Impact
Shop visits	Reducing work scope on modules Extending time on wing (TOW) Avoiding secondary damage Time reduction to determine work scope
Schedule Interruption	Reduce D&Cs IFSD risk mitigation Reduce UERs
Line Maintenance	Extend maintenance intervals Reduce troubleshooting time

theoretical examples span multiple stages in the engine life cycle, including development test cell, revenue service, and shop visit. In the first example, the typical data flow and some basic elements of the GPA algorithms are briefly described to give the readers a system overview. They are omitted in the other examples for brevity.

1.6.1 Scenario 1: In-flight Detection and Isolation of Critical Major Module Fault

Stage: Revenue service.
Impact: IFSD risk mitigation; avoiding secondary damage.
Highlight: High pressure turbine (HPT) blade fracture detected during a long-haul flight.

When the airplane took off, a snapshot data (ACMF takeoff report) was captured and transmitted automatically through the ACARS network to the engine OEM. The ACMF report was then decoded by a ground station computer program and processed into the EHM system. The incoming data triggered a series of automated analytics, including GPA. Gas path parameters were compared with a steady state performance model to calculate the "delta parameters" that quantified the performance deviation from a nominal healthy engine. Next, the delta parameters were trended with legacy data for anomaly detection. Takeoff *EGT* margin was also calculated and trended. For this particular takeoff data point, no anomaly was detected. The data and calculated parameters were saved in the EHM database for future trending. No alerts were annunciated.

Later into the flight, a cruise report was captured and transmitted via ACARS to the EHM system, where it went through a similar process. Unlike the takeoff data, the cruise data exhibited abrupt shifts on several gas path (delta) parameters, accompanied by increased vibration on the high-speed spool, all detected by a suite of anomaly detection (alerting) algorithms. The collective delta parameter changes were subsequently analyzed by a fault isolation algorithm, which, in this case, pointed to a potential HPT fault. The alert and diagnostic information were highlighted in the system and presented to EHM personnel, who verified the diagnosis and notified the airline. The flight crew were contacted and informed of the potential fault. Procedures were followed to check engine parameters, and the flight continued smoothly to the destination, while the suspect engine was closely monitored. Ground maintenance staff were ready at the gate to conduct the recommended borescope inspection upon landing and found fractured HPT blades. The engine was subsequently removed and sent for overhaul.

Before concluding this example, we want to highlight a few things. First, because the takeoff analysis was normal, the fault must have occurred after takeoff. If the cruise report were also captured before the occurrence of the fault, it would go undetected. Moreover, because there was no cockpit warning triggered during the flight (all engine parameters were in range), the aircraft would have an increased risk of potential rejected takeoff, IFSD, or air turn back in the subsequent flights. This risk could be better mitigated if full flight data/analytics were available. Second, timely removal of the faulty engine helped to avoid secondary damages to the high and low-pressure turbines, which are expensive to repair. Finally, timely and accurate diagnosis, as in this case, helps the airline to make judicious decisions that minimize service interruption and optimize the logistics for operation and maintenance.

1.6.2 Scenario 2: Performance Assessment and Work Scope Recommendation for Shop Visit

Stage: Shop visit.

Impact: Time reduction to determine work scope; reducing work scope on modules.

Highlight: Major module performance assessed prior to shop induction.

Before the engine in Scenario 1 arrived at the shop, engine performance was evaluated using revenue service data to determine major module performance using the GPA method. This eliminated the need, and the cost, for performing a test-as-received. The analysis suggested that a heavy maintenance on the HPT and low pressure turbine (LPT) alone could recover takeoff *EGT* margin to an acceptable level; therefore, only a light maintenance was recommended for the high pressure compressor (HPC). The recommended work scope was accepted and performed after a tear-down inspection was made. The overhauled engine matched predicted performance recovery within a reasonable uncertainty bound.

1.6.3 Scenario 3: Diagnosis of Sub-system Faults to Support Line Maintenance

Stage: Revenue service.

Impact: Reduce troubleshooting time, extending time on wing.

Highlight: Detected and isolated maintenance-induced stability bleed actuator fault.

An engine in service was alerted for abrupt takeoff *EGT* margin drop, sudden increase of normalized *EGT* at cruise condition, and shifts on several other gas path parameters. The result of the fault isolation algorithm pointed to a stability bleed fault. To confirm this diagnosis, the EHM personnel queried the system for supporting evidence. While no additional parameters pertaining to the stability bleed system were available in the system, a maintenance record showed that the engine just underwent an off-wing maintenance prior to the performance shifts. The stability bleed system fault was likely maintenance induced. Because increased engine temperatures had a detrimental impact on turbine life, a notification was sent to the operator describing the diagnosis and suggested inspections. The inspection later confirmed an improperly rigged stability bleed actuator. Line maintenance was performed to address the issue and the performance trends recovered afterward. This example also shows that, although no direct measurements were taken from the stability bleed system, its faulty behavior could still be detected and isolated accurately based on its impact on gas path performance.

1.6.4 Scenario 4: Diagnosis of Sensor Faults to Support Line Maintenance

Stage: Revenue service.
Impact: Reduce troubleshooting time; reduce D&Cs.
Highlight: Detected and isolated rapid deterioration of burner pressure sensing system.

An engine in service was alerted for rapid deterioration in normalized burner pressure (P_b) as compared to other engines in the fleet. While no abrupt shifts were observed, the slope of the parameter was significantly "steeper" than its peers, triggering a slope-based alert. The fault isolation algorithm pointed to burner pressure sensing system, as no shifts or trends were observed on other gas path parameters. A notification was sent to the operator to inspect the burner pressure sensing system. The inspection confirmed leakage in the burner pressure sensor line. Because burner pressure measurement is instrumental to engine operability and control (pertaining to control limits and surge detection), this successful detection helped reduce the risk of a D&C, as a P_b sensor fault, if triggered by onboard fault detection and accommodation logic, is non-dispatchable for this engine type. The accurate diagnosis helped to reduce troubleshooting time, which, in turn, helps to minimize service interruption.

1.6.5 Scenario 5: Takeoff EGT Margin Prediction

Stage: Revenue service.

Impact: Reduce UERs, extending time on wing.
Highlight: Manage engine removal and placement based on estimated takeoff margin.

Takeoff *EGT* margin calculation is a basic EHM function that is implemented for most commercial jet engines. Prognostics algorithms can be applied to takeoff *EGT* margin data to predict the remaining cycles to reach zero margin. Takeoff *EGT* margin, together with the prognostics, is used by the airlines to plan engine removals to reduce the risk of rejected takeoffs while keeping the engines on wing as long as safe operation permits. Meanwhile, a low-margin engine can be placed on an aircraft that flies routes in cold areas or be thrust-rating downgraded (to a lower nameplate) so that the engine can stay on wing longer.

1.6.6 Scenario 6: Maximum Continuous EGT Margin for ETOPS Operation

Stage: Revenue service
Impact: Reduce UERs, extending time on wing.
Highlight: Manage engine removal and placement based on estimated maximum continuous margin for ETOPS routes.

In addition to monitoring takeoff *EGT* margin, maximum continuous *EGT* margin must be monitored for engines engaged in ETOPS flights as required by FAA regulations previously mentioned. This is to ensure that the engines maintain positive margin for ETOPS at maximum continuous rating in case of one engine inoperative (OEI) condition. The calculation of maximum continuous margin is similar to that of takeoff margin, and prognostics can be applied to predict remaining cycles to zero margin.

1.6.7 Scenario 7: Water Wash Optimization

Stage: Revenue service.
Impact: Extend maintenance intervals, extending time on wing.
Highlight: Optimizing water wash schedule based on estimated compressor performance.

The air that the engine breathes is filled with contaminants, such as dirt, oil vapor, soot, exhaust fumes, and industrial pollutants, especially in areas of severe air pollution. These contaminants enter the compressor with the air supply and build up on the compressor blades under the high pressure and temperature conditions over time. Water wash is a common practice to restore compressor efficiency, which recovers *EGT* margin and improves fuel burn. The water wash schedule can be optimized based on an analysis of compressor efficiency deterioration using revenue service data, and the recovery effects can be analyzed to determine best practice.

1.6.8 Scenario 8: GPA Applications for Engine Test Cell

Stage: Test cell (development or shop).
Impact: Time reduction to determine work scope, reduce troubleshooting time, etc.
Highlight: Engine test cells use GPA for various supporting functions.

As mentioned earlier, the earliest form of GPA applications was found in engine test cells as specialized ground-based diagnostic tools for engine performance analysis, thrust estimation, etc. At the development stage, the engine on the stand is often equipped with additional, non-BOM sensors, which provides improved visibility into module performance. In addition to the traditional use of GPA to evaluate major module performance, test cell GPA programs have been used in various innovative ways to identify test cell process and tooling issues, direct water wash, and isolate instrumentation issues.

NOTES

1. Reprinted with permission from the National Aeronautics and Space Administration.
2. For a contemporary account of this strategy, see [16].
3. For more information on FMPs, see [17], [18].
4. See "Nomenclature" for acronym definitions.
5. To avoid confusion with the Federal Acquisition Regulations (FAR).
6. Taken from FAA regulation Appendix A to Part 33.
7. Only amplitudes.
8. Data manipulation and calculations were accomplished in the data management unit (DMU) of the AIMS.
9. Initially, the logic was simple, (i.e., altitude, Mach, spool speeds, etc., not changing more than pre-specified limits) and became more complex as the gas turbine and its control grew in complexity over time.
10. Initially, because memory was limited, data storage format was cryptic and sparse, thereby requiring a subsequent transcription process on the ground system to transform it to a more human-friendly presentation.
11. The advent of ACARS relieved the need for onboard storage of data; however, expensive (per byte) transmission costs gave reason enough to continue with transmitting data in a sparse and cryptic format that still required transcription upon receipt by the ground-based systems.
12. In recent years, the use of commercial cellular network or WiFi for post-flight reporting is often seen as a low-cost alternative; however, it comes with a reporting latency that may affect fault dispositioning turnaround time.

REFERENCES

1. Urban, L.A., 1972, Gas Path Analysis Applied to Turbine Engine Conditioning Monitoring, AIAA/SAE Paper 72-1082.
2. Urban, L.A., 1974, Parameter Selection for Multiple Fault Diagnostics of Gas Turbine Engines, *AGARD Conference Proceedings* No. 165.

3. Urban, L. A., and Volponi, A. J., 1992, Mathematical Methods of Relative Engine Performance Diagnostics, SAE 1992 Transactions, *Journal of Aerospace*, Vol. 101. Technical Paper 922048.
4. Volponi, A. J., 1983, Gas Path Analysis - An Approach to Engine Diagnostics, Time Dependent Failure Mechanisms and Assessment Methodologies, *Proceedings of the 35th Symposium, Mechanical Failures Prevention Group*, Gaithersburg, MD, Cambridge University Press.
5. Andrenucci, M., and Lazzeretti, R., 1974, Problems in Fault Diagnostics and Prognostics for Engine Condition Monitoring, *AGARD Conference Proceedings* No. 165.
6. Stamatis, A., Mathioudakis, K., Berios, G., and Papailiou, K., 1989, *Jet Engine Fault Detection with Differential Gas Path Analysis at Discrete Operating Points*, ISABE 89-7133.
7. Stamatis, A., Mathioudakis, K., Berios, G., and Papailiou, K., 1991, Jet Engine Fault Detection with Discrete Operating Points Gas Path Analysis, *Journal of Propulsion*, vol. 7, no. 6, 1043–1048.
8. Ripoli, J. C., 1974, Implementation of Engine Condition Monitoring, *AGARD Conference Proceedings* No. 165.
9. Smetana, F., 1974, Turbojet Engine Gas Path Analysis—A Review, AGARD-CP-165
10. Doel, D. L., 1992, TEMPER - A Gas Path Analysis Tool for Commercial Jet Engines, ASME Paper 92-GT-315.
11. Doel, D. L., 1993, An Assessment of Weighted-Least-Squares Based Gas Path Analysis, ASME Paper 93-GT-119.
12. Merrington, G. L., 1988, Fault Diagnosis of Gas Turbine Engines from Transient Data, ASME Paper 88-GT-209.
13. Merrington, G. L., 1993, Fault Diagnosis in Gas Turbines Using a Model Based Technique, ASME Paper 93-GT-13.
14. Provost, M. J., 1988, COMPASS: A Generalized Ground Based Engine Monitoring System, AGARD-CP-448.
15. IATA, 2019, Airline Maintenance Cost Executive Commentary, IATA Maintenance Cost Technical Group, December.
16. Smith, D. J., 2013, Power-by-the-hour: the role of technology in reshaping business strategy at Rolls-Royce, *Technology Analysis & Strategic Management*, vol. 25, no. 8, 987–1007.
17. Kerr, C.I.V., and Ivey, P.C., 2001, A strategic review of the large civil aero engine market and the paradigm shift to a service, *The Aeronautical Journal,* vol. 105, no. 1051, 287–93.
18. Jardine, A.K., Lin, D., and Banjevic, D., 2006, A Review on Machinery Diagnostics and Prognostics Implementing Condition-Based Maintenance, *Mechanical Systems and Signal Processing*, vol. 20, no. 7, 1483–1510.
19. Walthall, R. and Rajamani, R., 2018, *The Role of PHM at Commercial Airlines: Fundamentals, Machine Learning, and the Internet of Things.* ISBN 9781119515326, ch. 18, pp 503–534.
20. SAE, 2014, Determination of Cost Benefits from Implementing an Integrated Vehicle Health Management System, SAE Standard, ARP6275.
21. SAE, 2016, Aircraft Gas Turbine Engine Health Management System Development and Integration Guide, SAE Standard, ARP5120.
22. SAE, 2013, Aircraft Gas Turbine Engine Health Management System Guide, SAE Standard ARP1587.
23. Thomas, E. M., Richard, E., Aubrey, M., Dennis, A. M., and Renee, M., 2000, Analysis of Regulatory Guidance for Health Monitoring. NASA/CR-2000-210643

24. Walen, D. B., 2010, Civil Aircraft System Safety and Electromagnetic Compatibility, PowerPoint slides. Presented to the National Research Council Electronic Vehicle Controls and Unintended Acceleration Study, Washington, DC.
25. AC 43-218, Operational Authorization of Integrated Aircraft Health Management System, 2022, FAA. https://www.faa.gov/documentLibrary/media/Advisory_Circular/AC_43-218_FAA_Web.pdf
26. ARP5987A: A Process for Utilizing Aerospace Propulsion Health Management Systems for Airworthiness Credit, 2024, the SAE International.
27. Goebel, K., and Rajamani, R., 2021, Policy, Regulations and Standards in Prognostics and Health Management, *International Journal of Prognostics and Health Management*, https://doi.org/10.36001/IJPHM.2021.v12i1.2908 Merriam International Journal of Prognostics and Health Management, ISSN 2153-2648, 2021.
28. Rajamani, R., and Waters, N., 2011, Certification of Engine Health Management Systems: Guidelines for Selecting Software Assurance Levels, SAE Technical Paper 2011-01-2704, 2011, https://doi.org/10.4271/2011-01-2704.
29. Verhufen, M., and Schwenke, M., 1994, Condition Monitoring System (ACMS) for Airbus A330/A340 - New Concept and Applications, *The 19th Congress of the International Council of Aeronautical Sciences,* ISBN: 1563470845.
30. Johnson, S. B., et al., 2011, *System Health Management with Aerospace Applications,* John Wiley & Sons, ISBN: 978-0-470-74133-7.

2 Diagnostics and Prognostics

2.1 OVERVIEW

Diagnostics concerns itself with the current time, whereas prognostics attempts to predict a future state. Within the realm of EHM, diagnostics pertains to the techniques and methods employed to determine the present health of the engine and the degree, if any, to which its numerous sub-systems are abnormal or failing. Prognostics, on the other hand, grapples with the problem of forecasting the level of deterioration or potential failure of the engine's sub-systems and components at a future time. As we shall see in due course, GPA can provide capability in *both* realms. In this chapter, we will focus on some basic concepts regarding *performance* diagnostics and prognostics that will serve as a prerequisite to our foray into GPA in the remaining chapters of the book. As such, the topics discussed will be fairly generic and in overview, as a preparation for the GPA chapters to come.

2.2 DIAGNOSTICS

Engine diagnostics depends on the ability to perceive *change*. This implies that you must have a *reference* that represents the nominal or normal state and an *observation* from the engine under consideration to compare to the reference. This is a quantitative calculation. A familiar example from medicine might be monitoring a person's temperature. We know the typical reference is 98.6 degrees (deg) Fahrenheit (F) (37 deg Celsius [C]), and if we observe a temperature of 101 deg F (38.33 deg C), we have a change of +2.4 deg F (1.33 deg C). Whether this is considered significant depends on having a defined *threshold* to apply to the calculated change (Δ) in order to note whether an exceedance has occurred. If it has, we have *detected* a problem, and a diagnostic process will ensue, no doubt involving more observations, to lead the physician to a final *identification* of the malady, i.e., a diagnosis. Engine diagnostics follows the same general paradigm. It begins with observations (data) and ends with a health assessment. The general generic process is depicted in Figure 2.1.

Referring to Figure 2.1, *data* represents our observations made via the various sensors mounted on the engine, airframe, test stand, etc., as well as visual inspection information. *Categorization and normalization* refers to determining the type of data observed and applying corrections to account for environmental and engine operating conditions in order to allow a compatible set of values to compare to the reference. The *analytical engines* are where the magic takes place. They consist of a host of algorithms and models that are applied to calculate the changes (Δs) observed

DOI: 10.1201/9781003649274-2

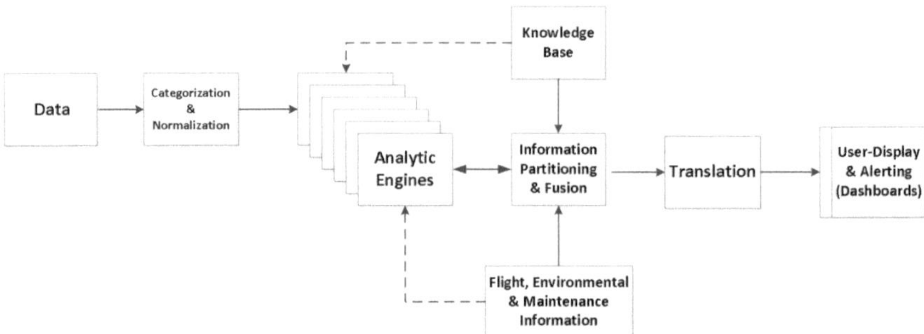

FIGURE 2.1 Generic engine health diagnostic process

and to calculate health condition indicators (other parameters) that quantify, in some manner, the health of the engine system, its major components, and sub-systems. *Information partitioning and fusion* is a process (one can think of this as an intelligent reasoner) that consolidates all of the calculated information to arrive at a final solution (or set of solutions) that then undergoes a *translation* process to ultimately display the health information in a form that is meaningful and understandable to the particular end user (the engine analyst or maintainer, in this case). The translation could be as simple as a change of engineering units that the end user is comfortable with, or it could be more complicated, such as translating a diagnosed drop in compressor efficiency (%) to its effect on cruise specific fuel consumption (*SFC*), or the loss of *EGT* margin at takeoff on a hot day. In summary, it begins with *data* and ends with meaningful and actionable *health information.* As a general *process*, it is no different from what the physician performs given his direct observations, patient responses, lab work, magnetic resonance imaging (MRI), computed axial tomography (CAT-scans), electrocardiogram (EKG), etc. The main difference lies in a knowledge base, transformations, and analytics that are guided by physics rather than biology. Another important differentiation is that the process for engine diagnostics is more amenable to automation than what we find in the medical example.

Diagnostics has two main stages to consider: *detection* of an abnormality (or fault), followed by the *isolation* (i.e., identification) of the fault. Of the two, detection is the easiest to perform. All that is required, in theory, is knowledge of what is considered nominal and observations to compare to the nominal. As in the medical example, it is a "relatively" simple matter to detect that the patient has a fever; however, it is quite a different matter to determine what is causing the fever, i.e., identification of the underlying condition. Nevertheless, detection does have its own challenges as well. The accuracy of the sensor providing the observation is one consideration. How repeatable is the observation, i.e., what is the precision error? What confidence level can we assign to declaring a fault detection? What is the false alarm rate associated with the detections mechanism employed? These are a few of the many issues to be considered when configuring a fault detection process. Isolation is a more difficult task and shares many of these concerns as well as some that are exclusive to the identification process. One such issue is ambiguity. Do the measured parameters at hand uniquely isolate to a single fault or a *set of possible faults* that *look alike* given the noise levels present in the data? These are important factors to be aware of and address (to the extent possible) in developing and implementing a performance diagnostic system (such as GPA), as a false detection or erroneous isolation can potentially launch an inappropriate remedial activity having consequential economic and safety ramifications.

We will begin a discussion of these topics in this chapter but will reiterate and expand upon them in the chapters to follow. We begin with what is probably the simplest of these topics: how do we quantify *change*?

2.3 MEASUREMENT DELTAS

Diagnostic and prognostic analyses are driven by observable and, hence, measurable changes in the parameters that we monitor. The quantification of these *changes* carries implicit information regarding the health of the engine; the role of any diagnostic analysis is to transform this implicit information to actionable health advisories and remedial actions. As such, these parameter *changes*, which we will often refer to as *measurement deltas*, are the basic currency from which to administer an effective and successful EHM program. Because of the central role they will play throughout this book, we will take a moment to discuss the form that *measurement deltas* will take, along with some properties and observations. This discussion will be of a general nature, without reference to any specific analysis methodology.

The word *change* (or *delta*) infers that there are two quantities at play, one being an *observation* and the other a *reference* level to compare to, from which we can quantify the difference. The reference can take different forms depending on the particular analysis being performed, but for our purposes, let us assume it represents the *nominal* or *healthy state*. If X represents the (parameter) observation, and X_{NOM} the comparable nominal or reference value,[1] the simplest delta (which we will denote by ΔX) can be computed as the arithmetic difference

$$\Delta X = X - X_{NOM} \tag{2.1}$$

Aside from being simple, it preserves the engineering units of the parameter X. For some, this property is desirable and familiar; for others, it is a nuisance to remember what units are being used. It can also disguise the seriousness of the delta because of the parameter units in place. For example, if $X = N$, rotor speed, which can reach operating values in excess of 10,000 rpm on some machines, a ΔN of 50 rpm is probably *not* as consequential as a smaller magnitude ΔP of 5 psia for $X = P$, an inter-stage pressure whose operating values may only be in the 100s of psia. Another potential issue arises with parameters of differing Δ magnitudes when taken *collectively* as a *vector* of Δs and subsequent vector-matrix manipulations are made as part of an analysis. That issue involves the propagation of error in complex matrix calculations. These last two concerns directly impact the algorithms employed in GPA; therefore, apart from a few exceptions, we will not be using Equation (2.1) to calculate ΔXs. We can remedy the situation by merely dividing Equation (2.1) by X_{NOM} to generate a *% of point* Δ as given in Equation (2.2).

$$\Delta X = 100\left(\frac{X - X_{NOM}}{X_{NOM}}\right) \tag{2.2}$$

Another minor advantage of Equation (2.2) is that it is of the form

$$\frac{\Delta X}{100} = \frac{dX}{X} \tag{2.3}$$

This allows us to derive hypothetical expressions for *change*, given a known parameter inter-relationship. For example, a fundamental relationship between the gas constant R, specific heats, cp and cv, and their ratio $\gamma = cp/cv$ is given by

$$\begin{aligned}\frac{R}{cp} &= \frac{cp - cv}{cp}\\ &= 1 - \frac{cv}{cp}\\ &= 1 - \frac{1}{\gamma}\\ &= \frac{\gamma - 1}{\gamma}\end{aligned}$$

Applying a useful manipulation of taking logs and differentiating (as described in Appendix A2.1), we obtain

$$\begin{aligned}\frac{dR}{R} - \frac{dcp}{cp} &= \frac{d\gamma}{\gamma - 1} - \frac{d\gamma}{\gamma}\\ &= \left(\frac{\gamma}{\gamma - 1}\right)\frac{d\gamma}{\gamma} - \frac{d\gamma}{\gamma}\\ &= \left(\frac{\gamma}{\gamma - 1} - \frac{\gamma - 1}{\gamma - 1}\right)\frac{d\gamma}{\gamma}\\ &= \left(\frac{1}{\gamma - 1}\right)\frac{d\gamma}{\gamma}\end{aligned}$$

Rearranging terms slightly, we have

$$\frac{d\gamma}{\gamma} = (\gamma - 1)\left(\frac{dR}{R} - \frac{dcp}{cp}\right) \tag{2.4}$$

Equation (2.4) offers us a theoretical expression of how % of point changes in R and/or cp induces % of point change in γ. In the subsequent chapters describing GPA, we will often use gas path parameter inter-relationships, and this manipulation to establish an expected framework of parameter Δ propagation that can be applied to observational parameter Δs calculated using Equation (2.2).

Before leaving this general discussion on measurement Δs, we would like to mention another manipulation that is made (countless times), which, on face value, seems to contradict our basic approach as described above. The manipulation involves the addition (or subtraction) of deltas when a time series of data is available for the purpose of initializing the diagnostic process or, perhaps, to allow estimation of a *temporal* change in performance. We will give a couple of representative examples for clarification, but first note that the (additive) combination of two % of point deltas is ***not necessarily*** itself a % of point delta. This can readily be seen by considering the difference between two delta points, $\{\Delta X_0 , \Delta X_1\}$, i.e.,

$$\Delta X_0 = 100\left(\frac{X_0 - X_{NOM}}{X_{NOM}}\right) \quad , \quad \Delta X_1 = 100\left(\frac{X_1 - X_{NOM}}{X_{NOM}}\right) \quad , \quad \Delta\Delta_1 = \Delta X_1 - \Delta X_0 \tag{2.5}$$

This yields

$$\begin{aligned}\Delta\Delta_1 &= 100\left(\frac{X_1 - X_{NOM}}{X_{NOM}}\right) - 100\left(\frac{X_0 - X_{NOM}}{X_{NOM}}\right) \\ &= 100\left(\frac{X_1 - X_0}{X_{NOM}}\right) \\ &= \underbrace{\underbrace{100\left(\frac{X_1 - X_0}{X_0}\right)}_{\%\,pt\ relative\ to\ X_0}\left(\frac{X_0}{X_{NOM}}\right)}_{not\%\,pt}\end{aligned} \tag{2.6}$$

The last expression in Equation (2.6) establishes the assertion. Of course, how much different the $\Delta\Delta_1$ will be from a true % of point will depend on how great the difference is between X_0 and X_{NOM}.

We will now mention two examples of delta manipulations that will be made routinely throughout our discussions on GPA, one of which violates the strict usage of % of point deltas as defined by Equation (2.2). The first concerns itself with *initialization* of the reference level and the second with *averaging* of deltas. For the case involving *initialization,* we will show that the difference from being a true % of point is very small, and for *averaging*, we will show that % of point is maintained.

2.3.1 Initialization

Whether we collect and process data acquired from ground run or from inflight, there are always *installation effects* that we will want to assess and use to modify the reference to which we compare in our analysis. These effects can be different for test cell operation versus flight data but, in general, contain items, such as test cell geometry, test nozzles, instrumentation biases, installed engine performance differences from nominal, nacelle effects, and so forth. Why do we care about such things? One reason is that, in some applications, it is preferable to know how an engine changes (over time) relative to itself, i.e., its installed state rather than a pure theoretical nominal state or perhaps how a particular engine changes relative to an average overhauled fleet of engines. For purposes like these, it is not necessary to itemize and assess all of the aforementioned *installation effects* individually (there are always exceptions) but, rather, to treat them collectively in a statistical fashion. For the purposes of the discussion at hand, let us assume that for each monitored parameter X, we have determined its installed level X_0. We can rewrite Equation (2.2) as

$$X_0 = X_{NOM}\left(1+\frac{\Delta X_0}{100}\right) \tag{2.7}$$

where ΔX_0 represents the % of point delta of the installed state relative to the nominal. If we let ΔX_k represent the % of point delta of datum X_k (acquired at a later discrete time k) relative to the installed state X_0, we have

$$X_k = X_0\left(1+\frac{\Delta X_k}{100}\right) \tag{2.8}$$

Combining equations (2.7) and (2.8), we have

$$\begin{aligned} X_k &= X_{NOM}\left(1+\frac{\Delta X_0}{100}\right)\left(1+\frac{\Delta X_k}{100}\right) \\ &= X_{NOM}\left(1+\frac{\Delta X_0}{100}+\frac{\Delta X_k}{100}+\underbrace{\frac{\Delta X_0\,\Delta X_k}{10000}}_{very\ small}\right) \\ &\approx X_{NOM}\left(1+\frac{\Delta X_0}{100}+\frac{\Delta X_k}{100}\right) \\ &\approx X_{NOM}\left(1+\frac{(\Delta X_0+\Delta X_k)}{100}\right) \end{aligned} \tag{2.9}$$

Equation (2.9) indicates that summing the two deltas $(\Delta X_0 + \Delta X_k)$ is *approximately* the % of point delta of X_k relative to X_{NOM}. Figure 2.2 illustrates this graphically.

We now consider the second delta manipulation, namely averaging.

2.3.2 Averaging

Averaging is typically employed to establish a temporal reference level when a sudden fault occurs and there is an attendant *large* shift in the observable parameter X. This would allow a *temporal* change (or $\Delta\Delta$) to be quantified for subsequent diagnostic analysis. The averaging process is typically a rolling average (or comparable

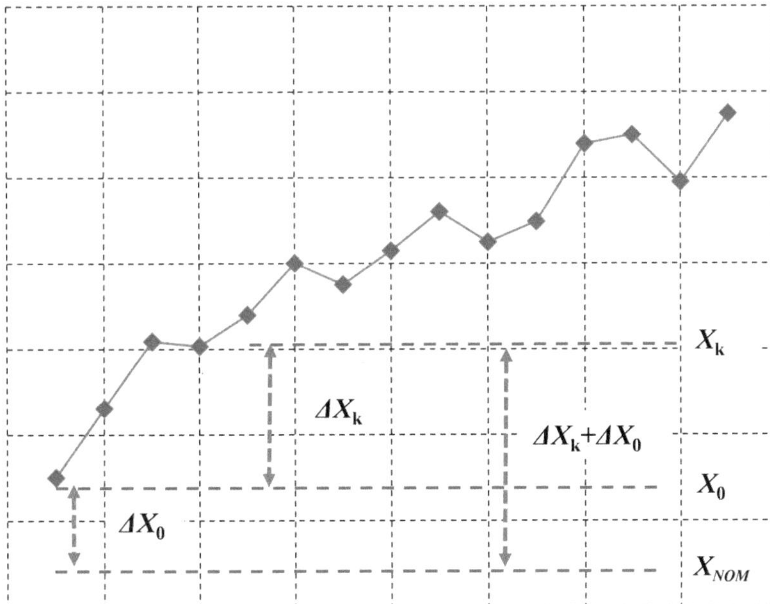

FIGURE 2.2 Hypothetical usage of summing % of point deltas

exponential average[2]) of a small number ($n \leq 10$) of points just prior to the detection of a large parameter delta shift (see Figure 2.3). Those n points would vary from each other by a small amount (in the absence of a sudden fault or failure), as these points represent the slow process of deterioration and, therefore, the difference is approximately[3] due to the non-repeatability (σ_X^2) of the sensor that is capturing the X parameter being monitored. For definiteness, let $\{X_1, X_2, \ldots, X_n\}$ be a (time) rolling set of n consecutive values wherein we have available the corresponding % of point delta set $\{\Delta X_1, \Delta X_2, \ldots, \Delta X_n\}$ where $Var(\Delta X_k) = \sigma_X^2$. Then

$$
\begin{aligned}
\Delta X_k &= 100\left(\frac{X_k - X_{NOM}}{X_{NOM}}\right) \quad \Rightarrow \\
\Delta \bar{X} &= \frac{1}{n}\sum_{k=1}^{n} \Delta X_k \\
&= \left(\frac{100}{X_{NOM}}\right)\frac{1}{n}\sum_{k=1}^{n}(X_k - X_{NOM}) \\
&= \left(\frac{100}{X_{NOM}}\right)(\bar{X} - X_{NOM}) \\
&= 100\left(\frac{\bar{X} - X_{NOM}}{X_{NOM}}\right)
\end{aligned}
\tag{2.10}
$$

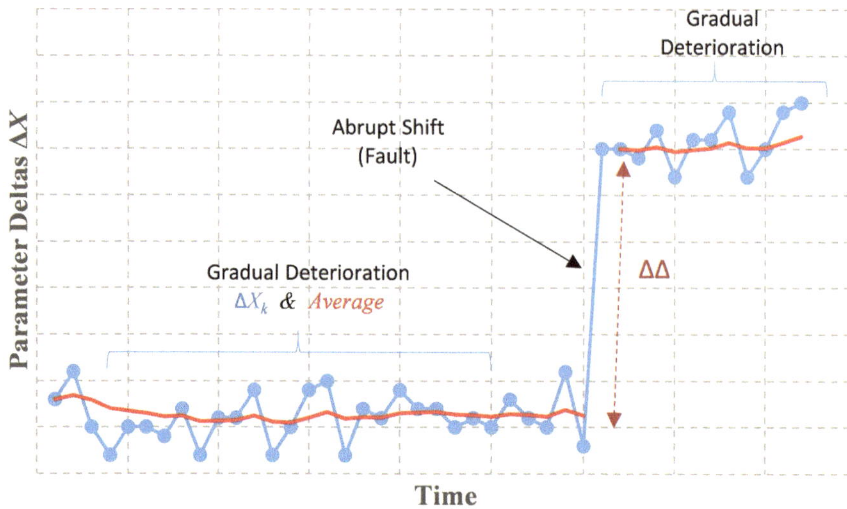

FIGURE 2.3 Averaging % of point deltas for hypothetical fault detection

Equation (2.10) indicates that rolling average of the deltas is the % of point delta of the mean level of the $\{X_k\}_{k=1}^{n}$ from the nominal and, furthermore, that its variance $Var(\Delta\bar{X})$ is a fraction of the ΔX precision uncertainty, i.e.,

$$Var(\Delta\bar{X}) = Var\left(\frac{1}{n}\sum_{k=1}^{n}\Delta X_k\right) = \frac{1}{n^2}\sum_{k=1}^{n}Var(\Delta X_k) = \frac{1}{n^2}\left(n\sigma_{\Delta X}^2\right) = \frac{\sigma_{\Delta X}^2}{n} \tag{2.11}$$

See Appendix Addendum 2.2 for a similar argument for *exponential averaging*.

2.4 MEASUREMENT UNCERTAINTY

As the reader is surely aware, when dealing with data, you must cope with the uncertainties that accompany it. Measurement uncertainty (i.e., measurement noise) is ubiquitous, and in diagnostics, where you use the data to arrive at a health assessment, the results will be corrupted if remedial measures are not employed. It is important to keep in mind that the principal quantity that is driving our diagnostic process is the *measurement delta,* and, as such, it is *measurement delta uncertainty* that is of key concern. There are many sources that can introduce error, and it is possible for small errors, in each contributing source, to accumulate to the point where the total error introduced exceeds the deviations induced by an underlying fault. In such a case, the fault becomes invisible and cannot be detected. In this section, we will consider some of these error sources and how to mitigate their effects where possible. This discussion will also be continued and expanded in later chapters after GPA has formally been introduced.

For the purpose of our discussion, we will use the terms *uncertainty* and *error* interchangeably and define an error in a given parameter value to be simply the

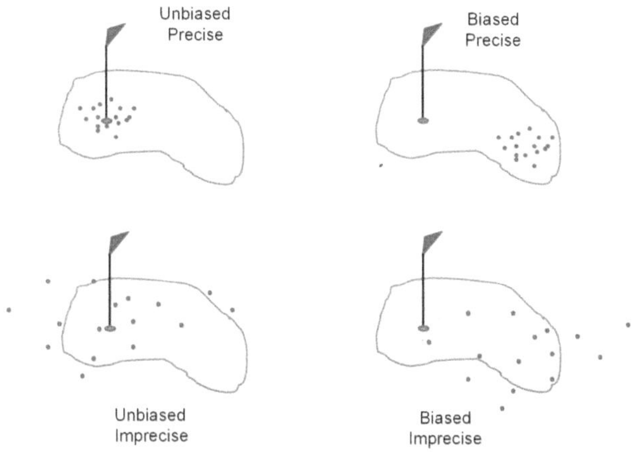

FIGURE 2.4 Types of uncertainty

amount by which that quantity differs from its true value. Generally, we will differentiate between two types of errors:

1) Random errors (non-repeatabilities or precision errors)
2) Fixed errors (biases or systematic errors)

Random errors constitute the variation observed between repeated measurements of a quantity and can be statistically represented by a sample standard deviation or variance. Fixed errors, on the other hand, are errors that appear in repeated measurements and represent the difference between the average measurement and the true value. Figure 2.4 provides a pictorial view of these errors (as they might be applied to the game of golf). The preference would be (and the golfer would agree) to have the situation depicted in the upper left of the figure, i.e., precise (or low non-repeatability) and no bias error.

Unfortunately, in real life, that is rarely the situation, and for engine diagnostics, it is more probable that we have one of the two scenarios on the right side of the figure, and among those, most likely the last (biased and imprecise). The literature abounds with many papers and books on the general problem of uncertainty and its effects, e.g., [1–5], and many that are more gas turbine centric, e.g., [6–11]. The gold standard reference for gas turbine measurement uncertainty is the *classic* work by Bob Abernethy [12], which we highly recommend to the interested reader.

2.4.1 Bias Error

Because we are dealing with measurement deltas, the sensor measurement X and the reference value $X_{NOM,}$ both contribute to the delta's uncertainty. For the former, the sensor, and its associated cabling, signal conditioning, data acquisition system processing, etc., contribute (potentially) both bias and precision errors. The latter is usually a model that is typically physics-based and, hence, repeatable and would, therefore, contribute possible bias error but not random error. This statement is not entirely true, in a practical sense, given that there are *sensed parameters* (with precision errors) driving the model response, which might then give the *appearance* of having a degree of non-repeatability or noise. In any event, the measurement delta, ΔX, generally inherits some level of both bias and non-repeatability that the diagnostic practitioner must contend with. For the most part, we cannot know exactly what the bias error is for a given ΔX. In fact, we do not have any way to know the *true value* for either X or X_{NOM}, which would be a prerequisite to evaluating ΔX bias. A few of the possible contributing factors might be:

- Unknown sensor drift or bias
- Inadequate signal processing
- Data acquisition issues
- Temperature and pressure profile effects
- Inadequate data reduction
- Test cell geometry effects
- Nacelle installation effects
- Model fidelity issues
- Unknown engine deviations

This is just a short list of broad categories, each of which having numerous underlying corruptive elements. In addition to this, we have the random error to cope with. If we had to address all the corruptive possibilities, it would be too overwhelming to proceed with developing a diagnostic methodology. Fortunately, we can pursue a more pragmatic, but efficacious, approach to this problem. For the *bias error* (in ΔX), we will (approximately) remove it through an *initialization* process, *collectively*, without assessing (or knowing, for that matter) any of the individual bias contributors. This has been alluded to in our discussion in the previous section concerning Figure 2.2, in which we would statistically determine a $\Delta X_{0,}$ which is essentially an average difference between the installed engine being monitored and the nominal reference model at time $t = 0$. By doing this, we remove (on the average) all biases, as well as absorbing into that estimate any performance deviations that might exist at $t = 0$. This is an acceptable concession, as at time $t = 0$, the engine is typically either new or newly overhauled. In effect, this practice makes X_0 the new reference level from which our performance tracking and health assessments will ensue. Once ΔX_0 is determined, Equation (2.8) can now be used to calculate measurement deltas ΔX_k at time $t = k > 0$. As noted earlier, the quantity $\Delta X_0 + \Delta X_k$ can now be used to represent the % of point measurement delta at time k relative to X_{NOM}. We will return to the topic of initialization in subsequent chapters on GPA.

2.4.2 Precision Error

Regrettably, random errors are not as easy to reconcile as bias. A large non-repeatability in our measurement delta inputs will translate to increased difficulty in detecting faults, increased ambiguity in isolating faults, and lower confidence levels in diagnostic alerting and health assessments. Unlike the bias situation, we cannot remove the error (not even approximately), but there are some occasions in which its effect can be mitigated. For example, *averaging* reduces random errors, and there are some instances in which it can be effectively deployed. We discussed one such case in the last section relative to Figure 2.3, concerning detecting a sudden measurement delta shift and quantifying the temporal $\Delta\Delta$ to assess the underlying fault. We mentioned two possibilities frequently employed – *rolling averages* and *exponential averages* – to stabilize the temporal frame of reference.[4] Equation (2.11) for the rolling average and Equations (A2.2.7) and (A2.2.10) for the exponential average indicate the effect on parameter variation. For a (commonly used) 10-point rolling average and its "equivalent" 10-point exponential average, variation in both cases is reduced by an order of magnitude. Other mitigation schemes involve deploying noise statistics directly in the diagnostic estimation algorithm (like GPA) to dampen the effects of the random noise in the data. We will be discussing these in detail in the subsequent chapters on GPA; however, we can give a generic example to complete this present overview. We can *weigh* our measured parameters by the inverse of their respective variations in the estimation algorithm. For example, if η represents an engine health parameter to be estimated from n measurement delta parameters $\{Z_1, Z_2, \cdots, Z_n\}$ having noise standard deviations $\{\sigma_1, \sigma_2, \cdots, \sigma_n\}$, respectively, and an estimation algorithm f, wherein

$$\hat{\eta} = f\left(Z_1, Z_2, \cdots, Z_n\right) \tag{2.12}$$

We can replace this with a weighted solution

$$\hat{\eta}' = f\left(\frac{Z_1}{\sigma_1}, \frac{Z_2}{\sigma_2}, \cdots, \frac{Z_n}{\sigma_n}\right) \tag{2.13}$$

that weighs precise parameters more heavily than those that are imprecise, which should provide at least a marginal gain in estimate accuracy. A familiar example is Mr. Gauss's least squares (LSQ) vs weighted least squares (WLSQ) solutions.

Although increasing sensor accuracy by using only the most accurate (and costly) instruments on the planet is beneficial, it does *not remove* the noise observed on ΔX. The reason is that model inaccuracy contributes significantly to $Var(\Delta X)$. Steady-state models are fairly decent and impose few problems; however, flight acquired data requires models with transient capability. In general, these models exhibit diminished accuracy (depending on the type, frequency, and duration of the transients), so the practitioner must be prepared to contend with *bad data*. In short, the algorithms employed to process the data (Figure 2.1) must be configured with this assumption in mind (as does GPA) and still be able to produce reliable health information.

2.5 FAULT DETECTION AND ISOLATION

One of the primary tasks of any EHM system is to recognize when a noteworthy change has occurred in the data stream available to the system. To accomplish this, data is trended over time. The intent of this exercise is to enable *fault detection* in the case that a fault should develop in the engine, its components, or sub-systems over time. Parameter deltas (ΔX) are trended instead of actual parameter values (X), as the latter is more susceptible to fluctuations due to normal changes in engine operation experienced in flight with no fault present. As we shall see, any process developed for fault detection is subject to a delicate balance between avoiding false alarms and missed detections. In general, there are three primary detection modes that are deployed as depicted in Figure 2.5. The first illustrates a sudden and *persistent* step change in the measurement delta beyond what its statistical noise level would dictate. *Persistency* is key to this test. When something fails or develops a fault, wholly or partially, it does not fix itself. Data outliers or anomalies, on the other hand, are typically transitory in nature. The second test involves *rate of change* in the parameter delta over time. It is not unusual for the deltas to trend up or down slowly over time in reflection of natural long-term deterioration in the engine components and systems. A rapid rate of change, however, is something to alert and investigate. The third mode is a simple threshold exceedance test. The thresholds for these latter two tests are chosen from engineering experience on a per application basis. The breach of these levels would be suspicious and drive an inspection for further information and guidance.

2.5.1 Trend Detection

The first of these modes, trend change, is by far the most popular and the one that appears most frequently in the literature. It lends itself to automation and is the most suitable for driving an *automated fault isolation* process as well. It has been used in almost all the GPA programs that have appeared in literature and is the fault detection mode we will concentrate on throughout our discussions on GPA. The specific mechanics and implementation of a trend detection methodology, however, varies widely among practitioners. *The devil is in the detail*, as the saying goes, and there is usually a good deal of fine tuning and the addition of ad hoc techniques to ensure a robust and reliable implementation. All the OEMs have their own specialized logic

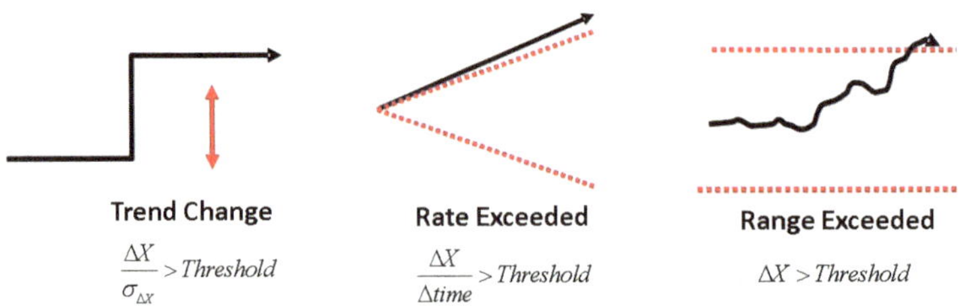

FIGURE 2.5 Primary detection modes

for trend detection and is considered proprietary information. That being said, there is still a good deal of high-level generic principles (sans proprietary detail) that apply and can be discussed.

The first and foremost is determining whether a *substantial* shift has occurred. This typically takes the form of comparing the current measurement delta to a *threshold* to determine whether there is an exceedance. The threshold is generally determined statistically from a (rolling) window of the prior n data points and takes the form as a multiple of the n-sample standard deviation.[5] Therefore, a data point shift from its exponential average level is considered *substantial* if it falls outside the range of the data non-repeatability exhibited in the rolling n-sample window. This simple test only flags the data point as being an outlier relative to its neighboring data points. In order to declare a trend shift, there is a fundamental requirement to show *persistency*. This is what separates data outliers from a fault-induced trend shift. This has a direct impact on the effects of data *outliers* and their potential contribution to false alarm rates. False alarms can spawn unnecessary repairs and take airplanes out of service. Too many false alarms may undermine the operator's confidence in the diagnostic accuracy and cause genuine faults to be ignored.[6] When hardware fails, hard or soft, the trend shift is persistent; whereas, data outliers are, by definition, short-lived. Outliers are a fact of life and are one manifestation of *bad data* whose effects we must mitigate. There are many causes for outliers, ranging from inadequate data acquisition, modeling errors, irregular flight maneuvers, insufficient data normalization, unmodeled effects from sub-systems and accessories, etc. The list is endless. Nevertheless, a good persistency check will catch a majority of them.

Data filters can be effective devices to aid in the detection of a sudden trend shift. Several sections back, we discussed *averaging* as a simple means to reduce variation and stabilize the reference level to allow an accurate temporal $\Delta\Delta$ to be quantified when a trend shift occurs, as depicted in our artist rendering of such an event in Figure 2.3. In fact, averaging can also be used as a main ingredient in the determination that a trend shift has occurred, i.e., *trend detection*. What is needed are two averages, each with a different reaction time. We will use exponential averaging in lieu of rolling averages for computational efficiency and select a 3-point and a 10-point exponential average having weighting factors $\alpha = 0.5$ and $\alpha = 0.82$, respectively.[7] Figure 2.6 offers an artist rendering of an idealized trend shift and the effect on the two averages. The 3-point average reacts quickly with the 10-point average being somewhat slower. Naturally, both will eventually converge to the new shifted level, as depicted.

Because the two averages have different time constants, the absolute difference between them following a trend shift, will *diverge quickly* and then *slowly decrease* back to zero, as portrayed in Figure 2.6 as the green dashed curve. We will refer to this difference as *divergence*[8], i.e.,

$$Divergence_k = \left| A_k^{(3)} - A_k^{(10)} \right| \tag{2.14}$$

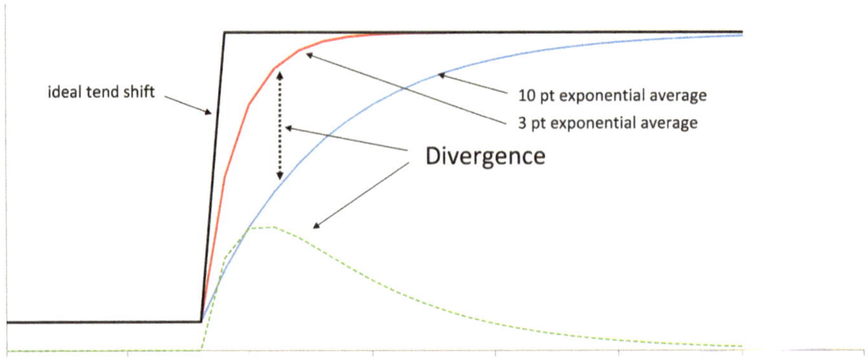

FIGURE 2.6 Ideal trend shift and divergence

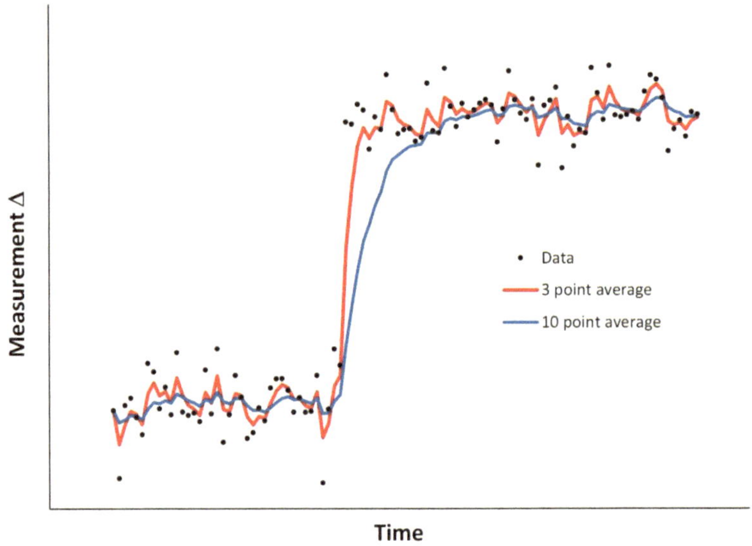

FIGURE 2.7A Noisy data with trend shift

where $A_k^{(3)}$and $A_k^{(10)}$are the 3- and 10-point exponential averages at time k, respectively. This (idealized) property of *divergence* relative to a sudden and persistent shift suggests a possible test for trend detection. The addition of measurement delta noise adds more challenge to the task, as seen in Figure 2.7a and 2.7b.

The *divergence* is notably less smooth than the hypothetical ideal trend case due to the measurement delta scatter; however, it still exhibits the salient quality (for *persistency*) that the rise is rapid followed by a slower decline back to its pre-shift level. What if this had been a *data outlier* condition instead of a true trend shift? Figures 2.8a and b depict such a situation. In this scenario, the *divergence* exhibits

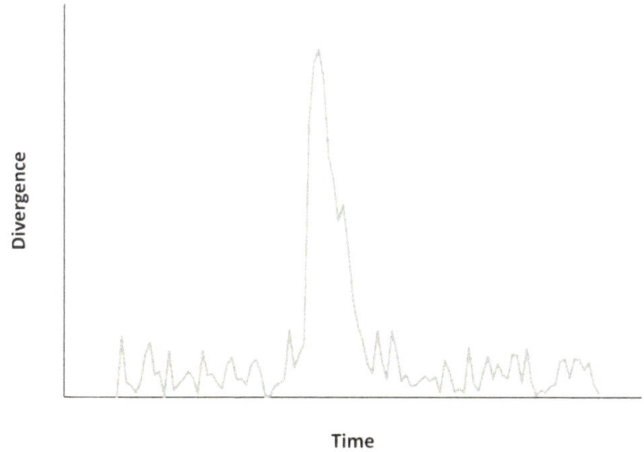

FIGURE 2.7B Divergence for noisy data with trend shift

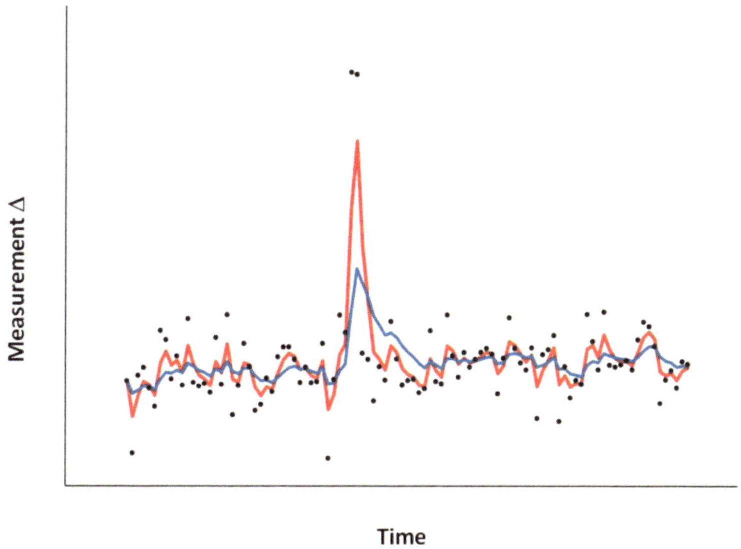

FIGURE 2.8A Noisy data with outliers

a different profile, i.e., the rapid rise is matched with a rapid decline unlike in the trend shift case.

So there appears to be hope for using *divergence* as a trend shift detection indicator. What is needed, of course, are all those ad hoc logic details we alluded to earlier to ensure robust and accurate interpretation of the *divergence* profile. We are not going to pursue this further here, as our intent was only to be introductory and acquaint the reader with the concept of *persistency*. The choice of using exponential

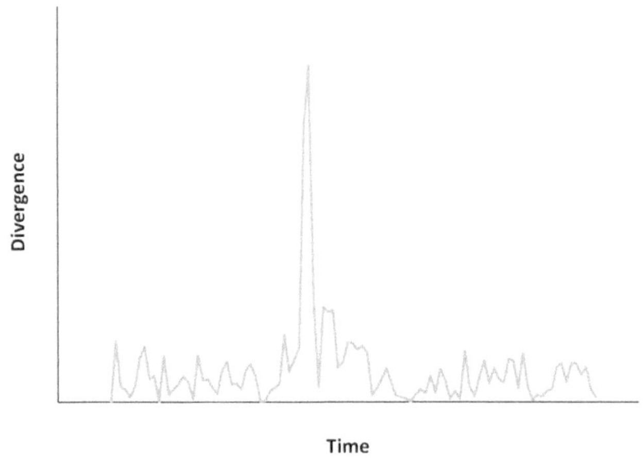

FIGURE 2.8B Divergence for noisy data with outliers

averages to detect trend shifts is certainly not universal. Median filters (of which there are many varieties) are another form used well-suited for this purpose and have favorable detection properties. Unfortunately, they are more computationally expensive to employ, which limits their practicality for automated systems processing large amounts of data as well as for implementation in onboard systems that must run in real-time. To learn about median filters and other alternatives for trend detection, we recommend an excellent book by Ganguli [20].[9]

Before leaving this topic, we would like to briefly discuss the notion of *accuracy* in detection. How do we quantify the accuracy attributes of a fault detection system or even compare two or more competing algorithms to determine which is better? To begin this discussion, we need to introduce two concepts:

1) Confusion matrix
2) Receiver operator characteristic (ROC) curve

2.5.2 Detection Confusion Matrix

We can think of our engine as having two states, *fault* and *no fault*, and our fault detection system as being a binary classifier that has two states, *fault detected* and *no fault detected*. Accordingly, there are four possibilities, as illustrated in the binary confusion matrix in Figure 2.9.

In our confusion matrix depiction, the areas that we would ideally like to have for our detector algorithm are labelled *TP* and *TN*; if a fault exists, we want to detect it, and if there is no fault present, we would expect no detection to be annunciated. Thus, *TP* is the number of *true positives* (true fault detection), while *TN* is the number of *true negatives* (no fault detected when no fault is present). The remaining classifications are the possible detection errors, wherein, *FN* is the number of *false negatives* (missed detections) and *FP* is the number of *false positives* (false

		Detected State	
		Detected	Not Detected
True State	Fault	**TP** (True Detection)	**FN** (Missed Detection)
	No Fault	**FP** (False Alarm)	**TN** (No Fault Detected)

FIGURE 2.9 Detection confusion matrix

alarms). Each of the latter categorizations carry with it a negative impact of varying severity. For example, false alarms can cause service interruptions, unnecessary inspections, and potential unneeded part replacement, all of which have a detrimental economic impact. Missed detections can be far more insidious, causing possible IFSDs, increased D&Cs, UERs, engine failure, and possible loss of aircraft and life.

From these four quantities, a number of useful metrics can be computed. For example, the true positive rate (TPR) is the proportion of fault conditions correctly detected, whereas the false positive rate (FPR) is the false alarm rate. Mathematically they are defined as

$$TPR = \frac{TP}{TP + FN} \tag{2.15}$$

and

$$FPR = \frac{FP}{FP + TN} \tag{2.16}$$

respectively. The total number of faulted cases is $TP + FN$, whereas the total number of no-fault cases is $FP + TN$. Overall, the total number of cases represented is $TP + FN + FP + TN$. Therefore, an *inclusive* metric representing the proportion of all conditions (both faulty and non-faulty) that are *correctly classified*, called the fault detection accuracy (FDA), can be defined as

$$FDA = \frac{TP + TN}{TP + FN + FP + TN} \tag{2.17}$$

The three metrics define by Equations (2.15–2.17) provide a basis to assess the performance and suitability of a fault detection methodology under consideration.

Broadly speaking, they provide measures of accuracy that can be used to compare competing detection algorithm approaches as well as whether the system under proposal meets the underlying design accuracy requirements. Naturally, these are not the only metrics available for these purposes, and we will discuss a few others later in this chapter relative to prognostics. The interested reader can consult references [21–24] for further information on diagnostic metrics and [24–27] for prognostic metrics. It is worth mentioning that while accuracy is important, there are other properties to consider when determining an algorithm's efficacy or comparing two or more potential designs. Addendum 2.3 renders a short discussion on this topic.

2.5.3 ROC Curve

A visual rendering of the metrics *TPR* and *FPR* is provided by the ROC, which plots true detections (*TPR*) versus false alarms (*FPR*). In its idealized form, the ROC for a given *fixed detection algorithm*, is the locus of points for *TPR* versus *FPR* as the *threshold* for detection is continuously varied. Figure 2.10 depicts two graphs. The first is the hypothetical *healthy state* and *faulted state* distributions wherein the abscissa of the plot is the measurement feature being tested by the detection algorithm. The vertical *green line* represents the threshold that (in theory) is being varied continuously from $(-\infty, +\infty)$, which produces a set of (*FPR*, *TPR*) pairs that, when plotted, form the ROC curve depicted in the second graph.

For a *fixed* detection algorithm, the shape of the ROC is dictated by the level of intersection between the two distributions. As the faulted state distribution moves toward the healthy state distribution, the ROC migrates to the right, as depicted in Figure 2.11. When the two distributions coincide, the ROC takes on its limiting form as a diagonal line. With the two distributions completely overlapped, it is essentially impossible to distinguish between healthy and faulty states with a simple threshold based on this feature. The ROC becomes a diagonal line, as shown in Figure 2.12. In this case, the diagnostic has no better odds of detecting a fault than a random flip of a coin.

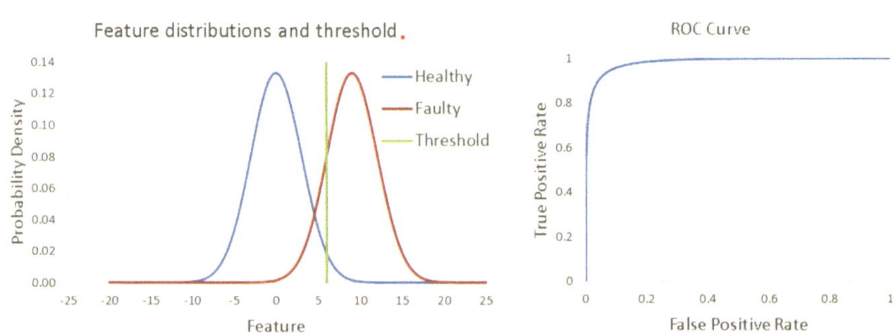

FIGURE 2.10 Hypothetical ROC (fault distribution Scenario 1)

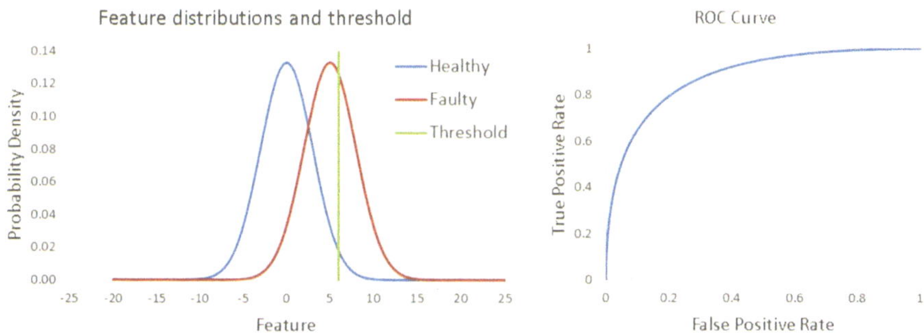

FIGURE 2.11 Hypothetical ROC (fault distribution Scenario 2)

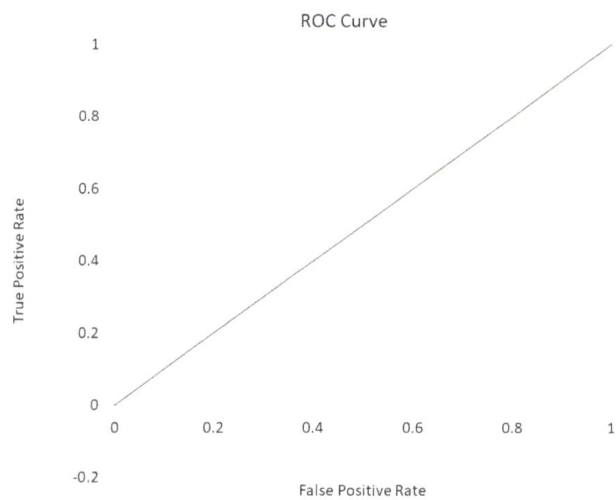

FIGURE 2.12 Diagonal ROC (fault distribution overlaps healthy state distribution)

2.5.3.1 Example

In practice, of course, we do not know the healthy state nor the faulted state distributions. Even if we did possess this knowledge, we certainly would not have the means to *move* the faulted distribution to allow a more favorable ROC response, i.e., better true detection and false alarm rates to meet our needs. We do, however, control the algorithm we use and the threshold being applied. With the distributions *fixed*, different detection algorithms can produce different ROC responses. To illustrate this effect, we have generated a very simple example, using the C-MAPSS model, of an HPT performance fault and a single measurement feature, ΔEGT in degrees C.[10] Figure 2.13 depicts the data generated for our purpose.

In keeping with our theme of simplicity for this example, we will apply just two algorithms: the first being a simple threshold check, and the second being the same threshold check along with a *persistency* criterion. Eleven thresholds (0, 1, 2, …, 10)

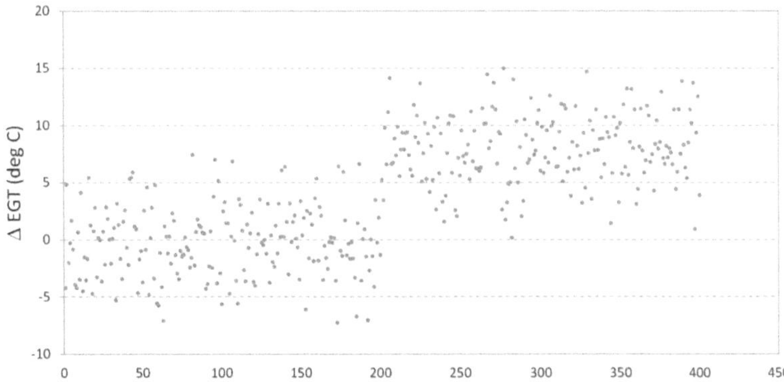

FIGURE 2.13 Simulation generated HPT performance fault

were applied to the data, *without* persistency (Algorithm A) and again *with* persistency (Algorithm B), and confusion matrices for each were calculated. Additionally, metrics *TPR*, *FPR* and *FDA*, defined in Equations (2.15)–(2.17), were generated and tabulated in Table 2.1a and 2.1b.

Plotting *TPR* and *FPR* for both algorithms, we obtain the ROC displayed in Figure 2.14.

It is clear from the ROC that adding a persistency check has produced a more favorable response. This is further substantiated by plotting overall accuracy, *FDA*, as shown in Figure 2.15, wherein it appears that Algorithm B achieves a higher accuracy (using a threshold = 2) over Algorithm A (using a threshold = 4). As mentioned previously, while results like this are of interest to an algorithm designer, they alone do not address the entire picture, and there are many other properties to consider. Addendum 2.3 briefly discusses some of these. Ultimately, the final metric is to what

TABLE 2.1A
ROC Metrics for Algorithm A – No Persistency

Threshold	FP	TN	TP	FN	FPR	TPR	FDA
0	95	105	200	0	0.475	1	0.7625
1	66	134	198	2	0.33	0.99	0.83
2	43	157	195	5	0.215	0.975	0.88
3	30	170	190	10	0.15	0.95	0.9
4	18	182	179	21	0.09	0.895	0.9025
5	14	186	171	29	0.07	0.855	0.8925
6	7	193	150	50	0.035	0.75	0.8575
7	1	199	124	76	0.005	0.62	0.8075
8	0	200	96	104	0	0.48	0.74
9	0	200	76	124	0	0.38	0.69
10	0	200	53	147	0	0.265	0.6325

TABLE 2.1B
ROC Metrics for Algorithm B – Persistency

Threshold	FP	TN	TP	FN	FPR	TPR	FDA
0	43	157	199	1	0.215	0.995	0.89
1	22	178	195	5	0.11	0.975	0.9325
2	7	193	189	11	0.035	0.945	0.955
3	5	195	180	20	0.025	0.9	0.9375
4	2	198	163	37	0.01	0.815	0.9025
5	2	198	151	49	0.01	0.755	0.8725
6	0	200	113	87	0	0.565	0.7825
7	0	200	70	130	0	0.35	0.675
8	0	200	36	164	0	0.18	0.59
9	0	200	22	178	0	0.11	0.555
10	0	200	12	188	0	0.06	0.53

extent the algorithms contribute to the total diagnostic system's ability to address the concerns outlined in Table 1.1.

2.5.4 Fault Isolation

While having the ability to determine (accurately) that a fault has occurred, it is clearly not as valuable as knowing *what* fault has occurred. Little remedial action can be taken unless we can identify (i.e., isolate) the fault in question. In the example above, ΔEGT was used as a measurement feature to drive the detection process that could perceive the *effect* of the *HPT* fault that was present. Unfortunately, a *single* measurement feature is generally insufficient to allow identification of the underlying fault. If it had been an *HPC* fault, or an engine bleed system fault, or a turbine case cooling (*TCC*) fault, ΔEGT would shift (positively) and the detection would be the same; without further information, we can only conclude a long list of possible causes. Thus, we can safely assume that if we have more than a single fault to contend with, which is certainly the case with gas turbines, we will need a plurality of measurement features at our disposal. Having a multi-dimensional feature vector provides the basis for fault isolation, which can be formulated either as a *classification* or an *estimation* problem. We will talk about both methods briefly.

Classification is an example of pattern recognition. When formulated as a classification problem, the fault isolation process aims to determine to which predefined category (i.e., fault mode) the new observation belongs, on the basis of a training dataset containing instances whose category membership is known. An algorithm that implements classification is known as a *classifier*. Figure 2.16 shows a simplified example in which the training data for various levels (1%–3%) of Fan, LPC, HPC, HPT, and LPT faults are generated from the simulation model at a fixed operating condition without consideration of noise. To visualize the data in 3D, delta N_2, delta

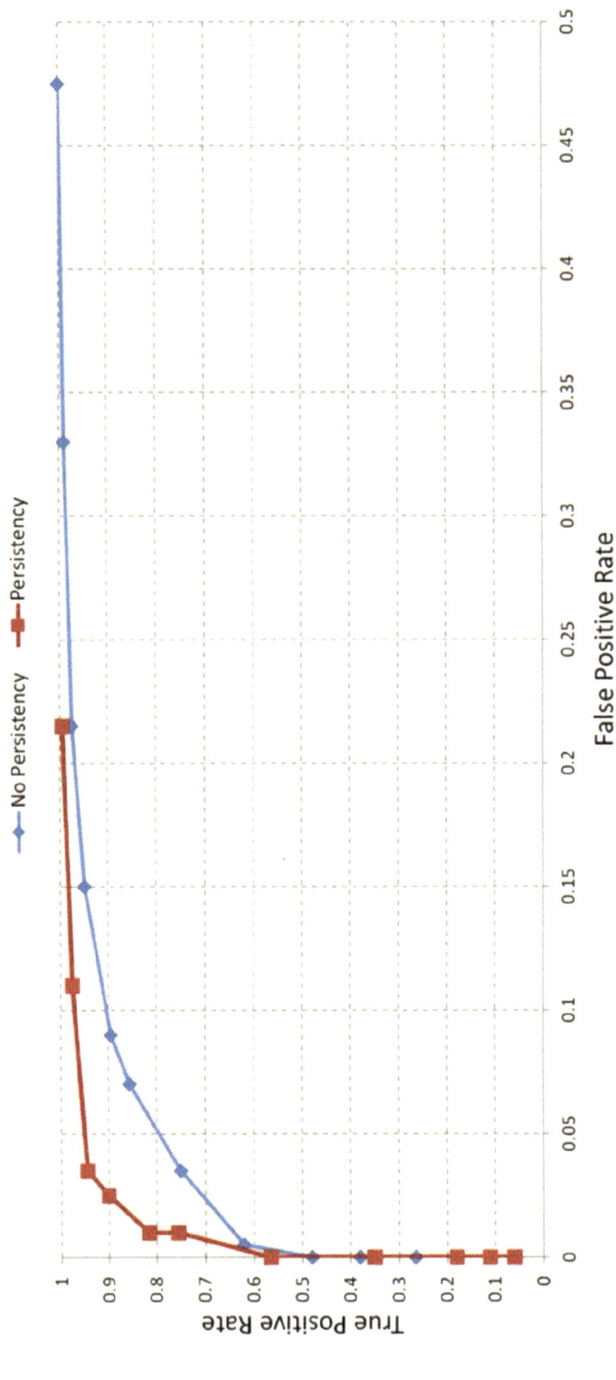

FIGURE 2.14 ROC for Algorithms A & B

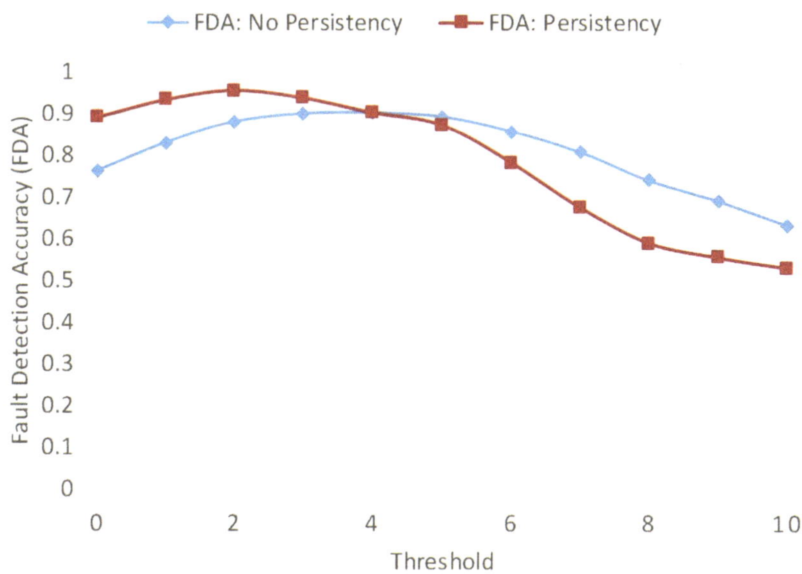

FIGURE 2.15 FDA for Algorithms A & B

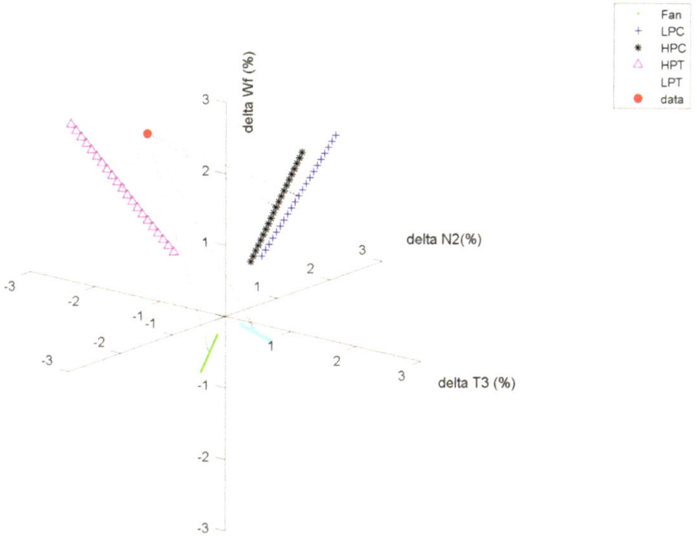

FIGURE 2.16 Fault categories and classification

T_3 and delta w_f data are plotted and only five fault categories (i.e., Fan, LPC, HPC, HPT, and LPT faults) are considered. For a given new observation (as shown by the red circle), the task of fault isolation is to determine to which category it belongs by means of similarity or distance, as shown in the figure.

A large toolkit of classification algorithms has been developed over the past decades based on various statistical and machine learning approaches. The most commonly used include linear classifiers, quadratic classifiers, support vector machines, k-nearest neighbor, decision trees, neural networks, learning vector quantization, etc. [28], [31], [32]. Applications of these classification algorithms to gas turbine fault isolation are also abundant. We do not intend to dive further into the details of classification algorithms in this chapter. Interested readers can find plenty of publications on this topic [33–36]. Regardless of the form of the classifier, a training dataset is needed and has to be collected either from fielded systems or simulation programs. Because faulty data from the field is often limited in quantity and incomplete in terms of fault mode coverage, many would resort to high fidelity models (such as numerical propulsion system simulation [NPSS] models or a program such as the C-MAPSS) to generate simulated faulty data. In this case, one would argue that instead of constructing a data-driven model based on simulated data, why not adopt a physics-based method directly? This brings us to the second approach.

When formulated as an *estimation* problem, the *estimator* not only points to the most likely fault, but it also gives an estimate of fault magnitude. Moreover, the same estimation framework can be applied to both (single) fault isolation and (multiple fault) module performance analysis.[11] As a simple example similar to the one given above, let us also assume a *single performance fault feature* associated with FAN, LPC, HPC, HPT, and LPT, respectively, denoted by $\Delta\eta$. We let x represent the vector of these five faults and z the vector of the $m > 1$ measurement features (Δs) , $\{z_i\}_{i=1}^m$, i.e.,

$$x = \begin{bmatrix} \Delta\eta_{FAN} \\ \Delta\eta_{LPC} \\ \Delta\eta_{HPC} \\ \Delta\eta_{HPT} \\ \Delta\eta_{LPT} \end{bmatrix} \quad and \quad z = \begin{bmatrix} z_1 \\ z_2 \\ z_3 \\ \vdots \\ z_m \end{bmatrix} \tag{2.18}$$

Because z is a vector of measurement features for x, we are assuming that there is some functional relationship $z = f(x)$, and if a faulty condition is detected by one or more of the z_i, we would seek to isolate the fault by determining $\hat{x} = f^{-1}(z)$. Because we are dealing with a gas turbine engine, f would be a non-linear function; to make things more manageable, we will assume the relationship to be *approximately linear*[12] and that we are at a *fixed steady state operating condition*. Thus, we have

$$z = H\,x \quad where\ H\ is\ a\ (m \times 5)\ matrix\ of\ maximal\ rank \tag{2.19}$$

There are three possibilities:

1) $m = 5$, in which case, the *unique* solution is $\hat{x} = H^{-1} z$

2) $m > 5$, in which case, there is *no unique* solution. In fact, there are infinitely many solutions. One such, was provided by Gauss, i.e., the LSQ solution given by $\hat{x}_{LS} = \left(H^T H\right)^{-1} H^T z$ (see Addendum A4.2, A4.2.18).
If we allow for measurement uncertainty, we can rewrite Equation (2.19) as

$$z = H\,x + \varphi \qquad where\ \varphi = (m \times 1)\ measurement\ noise\ vector \tag{2.20}$$

Without loss of generality, we can also assume that φ is normal distributed, with mean 0 and covariance R, i.e., $\varphi \in N(0, R)$ (see Addendum 4.1).
Doing so, we can generalize our solution to a *weighted least squares* solution $\hat{x}_{WLS}$ given by $\hat{x}_{WLS} = \left(H^T R^{-1} H\right)^{-1} H^T R^{-1} z$ (see Addendum A4.2, A4.2.18).

3) $m < 5$, like the preceding case, there is *no unique* solution and, likewise, has an infinite number of possible solutions. This is precisely where we find ourselves in classic GPA, and this will be discussed in great detail in Chapter 4 and subsequent chapters, so we will defer discussion of this case until that time.

Our discourse, thus far, is sufficient to allow us to introduce a construction often referred to as the *isolation confusion matrix*, which will play an important role in our subsequent exploration of GPA.

2.5.5 Isolation Confusion Matrix

The *isolation confusion matrix* (denoted C) is a rough analog for *fault isolation* as the *detection confusion matrix* is for *fault detection*. Whereas the latter is a (2 × 2) matrix, because detection is a binary classifier, i.e., the system is faulty or it is not faulty, the former is an (n × n) matrix, as for any one of the n faults present, the isolator can return a correct identification or an incorrect identification in (n - 1) possible ways, for a total of n classifications. The generic form of the matrix is depicted in Figure 2.17, where N_{ij} is the number of times the *Fault j* is isolated when the (actual) *Fault i* is present.

The sum of the diagonal, $\sum_i N_{ii}$, is the total number of correct identifications made in a total of $N = \sum_{i,j} N_{ij}$ classifications. Thus, the perfect isolator admits a diagonal confusion matrix.

Like the *detection confusion matrix*, the *isolation confusion matrix* has several metrics associated with it, the two most utilized being:

1) Correct isolation rate (*CIR*) defined as the rate of correct identifications, i.e.,

$$CIR = \frac{\sum_i N_{ii}}{\sum_{i,j} N_{ij}} = \frac{Trace(C)}{N} \tag{2.21}$$

		Isolated Fault			
		Fault 1	Fault 2	...	Fault n
True Fault	Fault 1	N_{11}	N_{12}	...	N_{1n}
	Fault 2	N_{21}	N_{22}	...	N_{2n}
	⋮	⋮	⋮	...	⋮
	Fault n	N_{n1}	N_{n2}	...	N_{nn}

FIGURE 2.17 Isolation confusion matrix

2) Mis-isolation rate (*MIR*) defined as the rate of incorrect identifications, i.e.,

$$MIR = 1 - CIR \tag{2.22}$$

Isolation is intimately bound to detection as a secondary process in the sense that, if nothing faulty is detected, there is nothing to isolate. Some would argue that the concept of isolation could be extended by stipulating that if nothing is detected, then the default isolation is automatically the healthy state. With that in mind, there is another (slightly different) form of C that includes the *healthy state* as *Fault* $n + 1$. This is the form we will use in subsequent chapters; we will refer to it as the *augmented isolation confusion matrix,* and denote it by C^*. It is depicted in Figure 2.18.

We mention in passing that the intimate connection between detection and isolation will introduce inter-relationships between these two processes' metrics.[13] A few simple examples would be

$$\begin{aligned} &Trace(C^*) \leq TP + TN \\ &\sum_{j=1}^{n} N_{n+1\,j} = FP \\ &N_{n+1\,n+1} = TN \\ &\sum_{i=1}^{n} N_{i\,n+1} = FN \\ &etc \end{aligned} \tag{2.23}$$

Before leaving this topic, we give a simple example based on the discussion at the beginning of this section in which we have $n = 5$ faults (given in Equation (2.18)), and we will choose $m = 8$ measurement features z as follows

		Isolated Fault				
		Fault 1	Fault 2	...	Fault n	Fault n+1
True Fault	Fault 1	N_{11}	N_{12}	...	N_{1n}	$N_{1\ n+1}$
	Fault 2	N_{21}	N_{22}	...	N_{2n}	$N_{2\ n+1}$
	⋮	⋮	⋮	...	⋮	⋮
	Fault n	N_{n1}	N_{n2}	...	N_{nn}	$N_{n\ n+1}$
	Fault n+1	$N_{n+1\ 1}$	$N_{n+1\ 2}$	...	$N_{n+1\ n}$	$N_{n+1\ n+1}$

FIGURE 2.18 Augmented isolation confusion matrix (fault n + 1 = healthy state)

$$x = \begin{bmatrix} \Delta\eta_{FAN} \\ \Delta\eta_{LPC} \\ \Delta\eta_{HPC} \\ \Delta\eta_{HPT} \\ \Delta\eta_{LPT} \end{bmatrix} \quad and \quad z = \begin{bmatrix} \Delta N_1 \\ \Delta N_2 \\ \Delta T_{25} \\ \Delta T_5 \\ \Delta P_3 \\ \Delta T_3 \\ \Delta P_{25} \\ \Delta w_f \end{bmatrix} \tag{2.24}$$

We also assume a linear measurement system given in Equation (2.20) with Matrix H given by

$$z = Hx + \varphi = \begin{bmatrix} -0.270 & -0.137 & -0.023 & -0.032 & 0.406 \\ 0.095 & -0.307 & -0.024 & 0.452 & -0.197 \\ -0.213 & 0.093 & -0.208 & -0.285 & 0.371 \\ 0.250 & -0.744 & -0.713 & -0.977 & -0.501 \\ -0.086 & 0.064 & 0.032 & 1.024 & 0.140 \\ 0.036 & -0.313 & -0.455 & 0.426 & -0.048 \\ -0.841 & 1.367 & -0.914 & -1.242 & 1.378 \\ 0.269 & -0.829 & -0.654 & -0.910 & 0.085 \end{bmatrix} x + \varphi \tag{2.25}$$

To generate the fault isolation matrix, we used the simulation to generate (at a fixed operating condition) noisy measurement data vectors (of eight measurements) for 100 random different level performance faults for each of the five major modules for

		Isolated Fault				
		Fan	LPC	HPC	HPT	LPT
True Fault	Fan	91	4	1	0	4
	LPC	0	89	9	0	2
	HPC	0	11	89	0	0
	HPT	0	1	0	98	1
	LPT	0	10	4	1	85

FIGURE 2.19 Sample isolation confusion matrix

a total of 500 trials. Baseline levels were used to calculate % measurement delta vectors z and analyzed by a simple *single fault* identification classifier. Healthy engine data was not generated for this simple example; thus, a standard (non-augmented) isolation confusion matrix was generated as depicted in Figure 2.19.

For this example, we see that $CIR = 90.4\%$ *and* $MIR = 9.6\%$. Most of the discrepancies (or confusion) seem to be in the LPC isolation classifications, which can be seen from the second column of the confusion matrix. This can occur if either the isolation methodology used is somehow biased toward LPC faults (not likely) or (more likely) that the LPC fault measurement patterns for the eight-measurement suite being used have close overlap with other fault measurement patterns such as HPC and LPT. This latter explanation is referred to as *ambiguity*, which we will discuss briefly in overview in the next section.

2.5.6 Ambiguity

Ambiguity arises in the process of identifying system faults that manifest themselves in identical or near-identical indications or observations, i.e., measurement features. This is possible given the ubiquity of measurement delta uncertainty. For instance, in our preceding example (Equation (2.25)), if

$$z^{(1)} = H \begin{bmatrix} 0 \\ 0 \\ \Delta\eta_{HPC} \\ 0 \\ 0 \end{bmatrix} + \varphi_1 \tag{2.26}$$

is a (noisy) measurement delta vector representing a single fault in the HPC of magnitude $\Delta\eta_{HPC}$, such that there exists a magnitude $\Delta\eta_{LPC} = \alpha$ where

$$\left\| z^{(1)} - H\begin{bmatrix} 0 \\ \alpha \\ 0 \\ 0 \\ 0 \end{bmatrix} \right\| < \left\| z^{(1)} - H\begin{bmatrix} 0 \\ 0 \\ \beta \\ 0 \\ 0 \end{bmatrix} \right\| \quad \forall\beta \tag{2.27}$$

the fault would be incorrectly identified as an LPC fault of magnitude α. Thus, for a given sample of data, the *columns* of the resultant *isolation confusion matrix* (or the *augmented isolation confusion matrix*) suggest a grouping of faults that can be ambiguous. For the preceding example that produced the confusion matrix in Figure 2.19, we note, for example, that when a Fan fault is introduced, it was correctly identified 92 out of the 100 trials, and furthermore, there were no misidentifications to a Fan fault when any of the four remaining faults were introduced. What this means is that *from an identification perspective, a Fan fault classification has no ambiguity.* In stark contrast, consider the second column in which all of the rows have non-zero entries. When an LPC fault was truly present, we identified it correctly in 89 (out of 100) trials. Unfortunately, when an HPC fault was present, we misidentified it 11 times out of 100 trials. Again, from the *analysis perspective,* when the system predicts an LPC fault, we really only have an 89% confidence that it is, in truth, an LPC fault, a 4% chance it is Fan, 11% chance it is HPC, and so forth. To be totally confident all of the time, C (or C^*) must be a diagonal matrix, in which case, all of the ambiguity groups (AGs) would be *singletons.* Thus, for Figure 2.19, the AGs are as follows

$$\begin{aligned} AG_{Fan} &= \{Fan\} \\ AG_{LPC} &= \{Fan,\ LPC, HPC, HPT, LPT\} \\ AG_{HPC} &= \{Fan,\ LPC, HPC, LPT\} \\ AG_{HPT} &= \{HPT, LPT\} \\ AG_{LPT} &= \{Fan,\ LPC, HPT, LPT\} \end{aligned} \tag{2.28}$$

With these arguments in mind, it is fair to say that an ambiguity group is, in effect, the anticipated consequence of a diagnostic fault isolation procedure. The essence of the ambiguity group is that it comprises a collection of failure modes for which the diagnostic procedure in play can detect and isolate the fault to that collection and no further, i.e., it cannot further isolate to any sub-set of that collection. Thus, the notion of ambiguity is intrinsically connected to the diagnostic methodology being employed and/or the measurement suit being used; change the diagnostic approach and/or the measurement suite, and you potentially change the ambiguity group structure.

In truth, there are many nuances to consider, but we intended this to be an overview introduction and hopefully our colloquial discourse on this matter will suffice for now. This discussion on diagnostic ambiguity is revisited in much greater detail in Chapter 9, where a methodology that can be employed to resolve at least some of the ambiguity we find in our GPA diagnostic system is introduced.

2.6 PROGNOSTICS

Prognostics and diagnostics are often mentioned together and occasionally used (indeed, misused) interchangeably in the industry. Academia tends to have a narrower, or more restrictive, definition for prognostics, which requires *remaining useful life* (RUL) as an output as well as the presence of a monitoring/diagnostic system that provides relevant parameters for prognosis. Moreover, prognostics is also distinguished from the traditional reliability and survivability methods. The industry, however, has associated it with any method capable of making a statement about a future state that would affect system functionality with at least some understanding of the timeframe in which it will happen. Some even consider "proactive diagnostics" (such as an impending failure flag) as prognostics, as it provides an opportunity to address the situation before it disrupts operations [37]. In essence, we will say diagnostics is about current state, while prognostics is focused on future state. In this book, we adopt the following definition for prognostics from SAE Aerospace Information Report (AIR7999):

> The forecast of future functional status and condition of a system, piece of equipment, component or subassembly based on current and accumulated inputs. The forecast can be expressed as remaining useful life, or time to reach a specific degradation level, or a time window to reach a point at which the risk of component failure is unacceptable. [24]

Regardless of how it is defined, the timeliness and accuracy of prognostics are critical to its intended business value. Figure 2.20 illustrates the timeliness considerations and the associated uncertainties that affect the value of prognostics for any application. The term *failure*, in this context, is characterized by the inability of a system, component, or sub-assembly to fulfill its design purpose, i.e., reaching the end of life (EoL). A *fault*, or incipient fault, is a deteriorated health condition, which may be indicative of an impending failure [24]. The prediction horizon is the amount of time made available (due to prognostics) to prepare maintenance, repair, and overhaul (MRO) logistics before the failure happens. One characteristic of a valuable prognostic algorithm is that the prediction horizon is greater than the minimum time window needed to take MRO actions without disrupting operations. Prognostics can arguably start anywhere in the timeline, but is only useful when the prediction is actionable, i.e., consistent prediction, reasonable RUL range, and allowing time to address the impending failure. Apparently, predicting that some part of an engine is going to fail sometime in the next 50 years with a high confidence has absolutely no value, as no action can be taken from this statement. Some measures of uncertainty, such as the distributions of EoL and current health state estimate, are illustrated in the figure. It should also be noted that, for many prognostics applications, the eventual failure time is often unknown due to part removal or maintenance activity (the purpose of prognostics is to prevent critical failure, after all). Therefore, the prediction horizon can only be estimated with historical or run-to-failure experimental data.

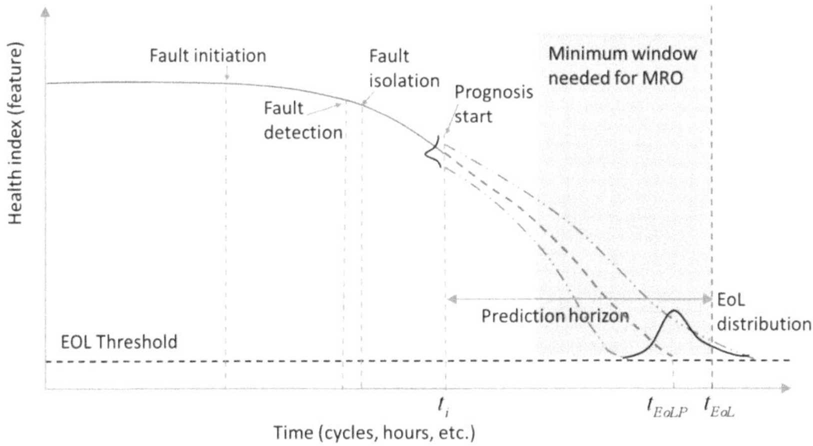

FIGURE 2.20 Timeline and uncertainties in prognostics

Figure 2.20 also illustrates a few other time indices. Fault initiation time is often unknown due to the lack of observable symptoms. Fault detection time and isolation time could either be identical or different depending on the nature of the fault and the capability of the monitoring system. Fault isolation is typically required for prognostics, as different fault modes propagate at different time scales/rates. However, there are reported studies using machine learning, artificial intelligence, or statistical methods that do not require fault isolation for prognostics [29], [30]. The EoL on the time axis marks the actual time when the health index reaches a threshold defined for end of useful life. At time t_i, the actual RUL (r_{actual}) is the time between t_i and t_{EoL}: $r_{actual}(t_i) = t_{EoL} - t_i$. Note that t_{EoL} is unknown at time t_i. The predicted trajectories from the prognostics are shown as the dotted lines (with confidence bounds) along with the predicted t_{EoLP}. Therefore, the estimated RUL is $r_{predicted}(t_i) = t_{EoLP} - t_i$. The performance metrics of prognostic algorithms are mostly defined around the deviation of $r_{predicted}(t_i)$ from $r_{actual}(t_i)$ as time moves on, as shown in Figure 2.21 [24].

Note that in Figure 2.21, the actual RUL vs time (solid line) becomes a straight line with a slope of negative one when RUL and time are in the same unit. Based on these two curves, some simple performance metrics, such as error and average bias, can be defined as follows (dropping the t in time index t_i for simplicity):

$$\textit{Error: } \Delta(i) = r_{actual}(i) - r_{predicted}(i) \tag{2.29}$$

$$\textit{Average Bias: } B_l = \frac{1}{\rho}\sum_{i=1}^{\rho} \Delta(i) \tag{2.30}$$

where ρ is the number of time indices in which predictions are made between t_P and t_{EoLP} for the lth unit under monitoring (in a fleet of units). In most applications, end users are primarily interested in how accurate an algorithm's predictions are as

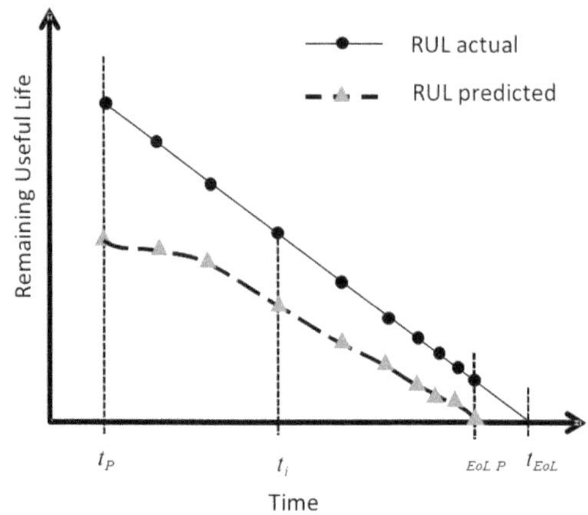

FIGURE 2.21 RUL vs time plot

a unit approaches EoL and the average accuracy measure among multiple units. In light of this, a modified definition that reflects average prediction bias across multiple (L) units at a fixed time horizon H prior to EoL is defined as:

$$\textit{Average Bias: } B_H = \frac{1}{L}\sum_{l=1}^{L}\Delta^l(EoL_l - H) \tag{2.31}$$

where L is the number of units in the fleet, EoL_l indicates the actual EoL for the lth unit, and H denotes the length of the fixed time horizon when prediction is made. With this definition, average bias is a function of the horizon specified by H, which may be the amount of time needed to make maintenance decisions.

The average bias metric, as it is formulated in Equation (2.31), does not account for the fact that positive and negative prediction errors cancel each other when averaged and, therefore, may not reflect high variance in the prediction [24]. Mean absolute error (MAE) and mean absolute percent error (MAPE) metrics are introduced to address this issue:

$$MAE_H = \frac{1}{L}\sum_{l=1}^{L}|\Delta^l(EoL_l - H)| \tag{2.32}$$

$$MAPE_H = \frac{1}{L}\sum_{l=1}^{L}\frac{\Delta^l(EoL_l - H)}{r^l_{actual}(EoL_l - H)} * 100 \tag{2.33}$$

For MAPE, because $\Delta^l(EoL_l - H)$ and $r^l_{actual}(EoL_l - H)$ are the RUL prediction bias and true RUL at time horizon specified by H, respectively, this metric essentially quantifies the percent deviations from the true RULs as a function of H.

A fundamental assumption made for these types of metrics is that RUL data (i.e., EoL time) is available which is only true in development phase (with historical or run to failure test data, or in a post-failure event analysis). These performance metrics cannot be calculated and utilized to guide the decision-making process in run time.

One would argue that if an algorithm/system has been verified and validated (using these performance metrics) with sufficient test data, it would produce trustworthy prognosis in the field. This argument is reasonable as long as the test setting is close enough to actual field situation and the data is sufficient enough to cover the failure modes occurring in the field.

Another popular way to quantify prognostics performance is to turn RUL prediction into a Boolean (or discrete) output based on a predicted time window. For example, if a prognostic algorithm is designed to "alert" if the system is about to fail 7 to 14 days ahead (allowing 7 days for maintenance logistics), then it essentially becomes a proactive diagnostic algorithm that can leverage the performance metrics designed for diagnostics (e.g., missed detection and false alarm).

As much as one can leverage from diagnostics methods to support prognostics work, notwithstanding, the practitioner still has to face the challenge of increased *uncertainty* associated with predicting the future state (which involves making all sorts of assumptions). Many papers have been published in the past two decades to address the quantification and management of prognostics uncertainty [26], [38], [39]. We will not explore these details in this book. In Table 2.2, we briefly introduce two additional uncertainty-related performance metrics, namely, relative accuracy (RA) and cumulative relative accuracy (CRA), which are utilized in Chapter 10. Interested readers can find a comprehensive set of diagnostics and prognostics performance metrics and reference materials in [24].

Although not as mature as diagnostics techniques, prognostics has found numerous applications in EHM, including performance analysis and prediction [40],

TABLE 2.2

Relative Accuracy & Cumulative Relative Accuracy

Metric	Definition	Purpose
Relative Accuracy	$RA_\lambda^l = 1 - \frac{\left|r_{actual}^l(\lambda) - r_{predicted}^l(\lambda)\right|}{r_{actual}^l(\lambda)}$ $\lambda = (t_\lambda - t_P)/(t_{EoL} - t_P)$ where: λ =normalized time window indicator; $r_{actual}^l(\lambda) =$ actual RUL at t_λ ; $r_{predicted}^l(\lambda) =$ predicted RUL at t_λ .	Quantifies the relative accuracy at a given instant in time (specified by normalized time indicator λ)
Cumulative Relative Accuracy	$CRA_\lambda^l = \frac{1}{\rho}\sum_{i=1}^{\rho} w(i) RA_\lambda^l(i)$ where: ρ= number of time instances in which RA is assessed; $RA_\lambda^l(i)$ = relative accuracy for the ith time instance; $w(i)$ = weight factor, $\sum_{i=1}^{\rho} w(i) = 1$.	A normalized weighted sum of relative accuracy (RA) levels at multiple time instances approaching EoL.

usage-based lifing (UBL) [41], [42], and mechanical systems (e.g., crack growth prediction) [43], [44], among others [45]. Because this is a book on GPA, we focus on performance analysis related examples in Chapter 10. In this context, the term *failure* does not necessarily refer to the inability of an engine to fulfill its design purpose (i.e., to provide required thrust or torque). The "failure" threshold can be defined to predict remaining time until a maintenance or overhaul has to be performed.

NOTES

1. The reference could be temporal change across a time series of deltas, i.e., $\Delta\Delta$.
2. See Appendix Addendum 2.2.
3. Because these are deltas, any error in determining X_{NOM}, i.e., modeling error, will contribute to the scatter.
4. Other filters have been explored as well. See [0]–[20].
5. For computational efficiency, *exponential* standard deviation can be used as an approximation. See Addendum 2.3.
6. The boy who cried wolf syndrome.
7. See Addendum A2.2.
8. See Volponi, A.J., 2005, *Data Fusion for Enhanced Aircraft Engine Prognostics and Health Management*, NASA Report CR-2005-214055, Glenn Research Center, December 2005.
9. Median filters of different varieties are explored, with examples, in six of the 12 chapters of the book.
10. We have deviated from using %Δ in this simple exercise solely for mental clarity.
11. Details to follow in Chapter 4.
12. In general, we would detect relatively small abnormalities in x ($\leq 5\%$) making a linear approximation assumption viable.
13. The definitions of *CIR* and *MIR* remain unchanged for C^*, except for notation, Equation (2.21), *Trace*(C) becomes *Trace*(C^*).

REFERENCES

1. Barnes, J.W., 1988, *Statistical Analysis for Engineers*, Prentice Hall.
2. Ronen, Y., 1988, *Uncertainty Analysis,* CRC Press Inc.
3. Fisher, G.H.,1973, *The Nature of Uncertainty,* Rept. P-5133, Rand Corp.
4. Brownlee, C.A., 1960, *Statistical Theory and Methodology in Science and Engineering*, John Wiley & Sons.
5. Adams, G.R., 1980, *Improvement of Parameter Estimation through Uncertainty Analysis*, ASME paper 80-GT-9.
6. Mathioudakis, K., Tsalavoutas, A., 2002, Uncertainty reduction in gas turbine performance diagnostics by accounting for humidity effects, *Journal of Engineering for Gas Turbines and Power* 124:801–808. doi: 10.1115/1.1470485.
7. Kosman, W., 2017, The influence of measurement inaccuracies on the assessment of the state of gas turbines in diagnostic systems, *Journal of Power Technologies* 97(2):142–148.
8. Chen, M., et al., 2015, An approach for optimal measurements selection on gas turbine engine fault diagnosis, *Journal of Engineering for Gas Turbines and Power* 137:071203. doi: 10.1115/1.4029171.

9. Ntantis, E., 2013, The impact of measurement noise in GPA diagnostic analysis of a gas turbine engine, *Journal of Aeronautics & Aerospace Engineering* 2(2):401–408. doi: 10.4172/2168-9792.1000109
10. Aretakis, N., Mathioudakis, K., Stamatis, A., 2004, Identification of sensor faults on turbofan engines using pattern recognition techniques, *Control Engineering Practice* 12(7):827–836.
11. Pinelli, M., Spina, P., 2002, Gas turbine field performance determination: sources of uncertainties, *Journal of Engineering for Gas Turbines and Power* 124(1):155–160.
12. Abernethy, R.B., 1973, *Handbook: Uncertainty in Gas Turbine Measurements*, AEDC-TR-73-5.
13. Ganguli, R., 2005, Adaptive myriad filter for improved gas turbine condition monitoring using transient data, *Journal of Engineering for Gas Turbines and Power* 127(2):329–339.
14. Ganguli, R., 2002, Data rectification and detection of trend shifts in jet engine gas path measurements using median filters and fuzzy logic, *ASME Transactions: Journal of Engineering for Gas Turbine and Power* 124(4):809–816.
15. Ganguli, R., Dan, B., 2004, Trend shift detection in jet engine gas path measurement using cascaded recursive median filter with gradient and Laplacian edge detector, *Journal of Engineering for Gas Turbine and Power* 126(1):55–61.
16. Ganguli, R., 2002, Noise and outlier removal from jet engine health signals using weighted FIR median hybrid filters, *Mechanical Systems and Signal Processing* 16(6):967–978.
17. Ganguli, R., 2003, Jet engine gas-path measurement filtering using center weighted idempotent median filters, *Journal of Propulsion and Power* 19(5):930–937.
18. Nieminen, A., et al, 1988, Algorithms for real time trend detection, *ICASSP-88., International Conference on Acoustics, Speech, and Signal Processing*, New York, NY, pp. 1530–1533, vol. 3, doi: 10.1109/ICASSP.1988.196895.
19. DePold, H., and Gass, F., 1999, The application of expert systems and neural networks to gas turbine prognostics and diagnostics, *Journal of Engineering for Gas Turbines and Power* 121:607–612.
20. Ganguli, R., 2012, *Gas Turbine Diagnostics: Signal Processing and Fault Isolation*, CRC Press, ISBN 9781466502727.
21. Davison, C.R. and Bird, J.W., June, 2008, *Review of Metrics and Assignment of Confidence Intervals for Health Management of Gas Turbine Engines*, ASME GT200850849, ASME Turbo Expo 2008, Berlin, Germany.
22. DePold, H., Siegel, J. and Hull, J., June, 2004, *Metrics for Evaluating the Accuracy of Diagnostic Fault Detection Systems*, ASME GT2004-54144, IGTI Turbo Expo 2004, Vienna, Austria.
23. Orsagh, R.F., et al., 2003, Development of performance and effectiveness metrics for gas turbine diagnostic techniques, *Aerospace, 2002 IEEE Conference Proceedings*, Big Sky, MT, USA vol. 6, pp. 2825–2834.
24. SAE, 2020, *Diagnostic and Prognostic Metrics for Aerospace Propulsion Health Management Systems*, SAE Standard AIR7999.
25. Kacprzynski, G.J., Hess, A.J., Begin, M., March 6–13, 2004, Metrics and development tools for prognostic algorithms, *IEEE Aerospace Conference*, Big Sky, MT.
26. Saxena, A., et al., October 6–9, 2008, Metrics for Evaluating Performance of Prognostic Techniques, *IEEE International Conference on Prognostics and Health Management*, Denver, CO.
27. Saxena, A., et al., March 7–14, 2009, Evaluating algorithm performance metrics tailored for prognostics, *IEEE Aerospace Conference*, Big Sky, MT.

28. Matloff N., 2017, *Statistical Regression and Classification: From Linear Models to Machine Learning*, CRC Press Inc.
29. Carvalho, T.P., et al., 2019, A systematic literature review of machine learning methods applied to predictive maintenance, *Computers & Industrial Engineering* 137, pp.106024-1–106024-10.
30. Schwabacher, M., Goebel, K., 2007, A survey of artificial intelligence for prognostics, *AAAI Fall Symposium: Artificial Intelligence for Prognostics*. pp. 108–115.
31. Alpaydin, E., 2010, *Introduction to Machine Learning*, MIT Press.
32. Gnanadesikan, R., 1977, *Methods for Statistical Data Analysis of Multivariate Observations*, Wiley.
33. Ogaji, S.O.T., Li, Y.G., Sampath, S., Singh, R., June 16–19, 2003, Gas path fault diagnosis of a turbofan engine from transient data using artificial neural networks, *Proceedings of the ASME Turbo Expo 2003, Volume 1: Turbo Expo 2003*. Atlanta, Georgia, pp. 405–414.
34. Sadough Vanini, Z.N., Meskin, N., Khorasani, K., 2014, Multiple-model sensor and components fault diagnosis in gas turbine engines using autoassociative neural networks, ASME. *The Journal of Engineering for Gas Turbines and Power* 136(9):091603.
35. Shen, Y., Khorasani, K., 2020, Hybrid multi-mode machine learning-based fault diagnosis strategies with application to aircraft gas turbine engines, *Neural Networks* 130:126–142.
36. Schwabacher, M., 2005, Machine learning for rocket propulsion health monitoring, *SAE Journal of Aerospace* 114:1192–1197.
37. SAE, 2020, Prognostics for aerospace propulsion systems, *SAE Aerospace Information Report*, AIR5871A.
38. Tang, L., Kacprzynski, G.J., Goebel, K., Vachtsevanos, G., March 7–14, 2009, Methodologies for uncertainty management in prognostics, *IEEE Aerospace Conference*, Big Sky, Montana.
39. Tang, L., Orchard, M.E., Goebel, K., Vachtsevanos, G., March 5–12, 2011, Novel metrics and methodologies for the verification and validation of prognostic algorithms, *IEEE Aerospace Conference*, Big Sky, Montana.
40. Ramasso, E., Saxena, A., 2014, Performance benchmarking and analysis of prognostic methods for CMAPSS datasets, *International Journal of Prognostics and Health Management* 5(2):1–15.
41. Vittal, S., Hajela, P., Joshi, A., 30 August 2004–01 September, 2004, Review of Approaches to Gas Turbine Life Management, *10th AIAA/ISSMO Multidisciplinary Analysis and Optimization Conference*, Albany, New York.
42. Matchett, R. A. 19–20 February 2001, Certification of engine usage monitoring systems, *Proceeding of the second DSTO International Conference on Health and Usage Monitoring*, Melbourne, Australia.
43. Patrick R., Orchard, M., Zhang, B., Koelemay, M., Kacprzynski, G., Ferri, A. Vachtsevanos, G. September 2007, An integrated approach to helicopter planetary gear fault diagnosis and failure prognosis, *42nd Annual Systems Readiness Technology Conference*, AUTOTESTCON 2007, Baltimore, MD.
44. von Flotow A., Tappert P., Mercadal M., Hardman B., Drumm M. 2004, Low cycle fatigue rotor burst prognostics using blade tip sensors crack detection on aero-engine hardware, *Proceedings of a Symposium of the Materials Science and Technology 2004 Conference*, New Orleans, LA, USA, pp. 209–216.
45. Roemer, M. J., Byington, C. S., Kacprzynski, G. J., Vachtsevanos, G., May 8–11, 2006, An overview of selected prognostic technologies with application to engine health management, *Proceedings of the ASME Turbo Expo 2006: Power for Land, Sea, and Air*, Barcelona, Spain.

3 Parameter Corrections

3.1 OVERVIEW

The various parameters appearing along an engine's gas path, such as flows, pressures, temperatures, speeds, etc., not only vary with power condition but also with the ambient conditions at the engine's inlet. Because a change in inlet temperature and/or pressure will contribute to an attendant change in a gas path parameter's value, it would be difficult to characterize the aero-thermodynamic relationships between gas turbine engine parameters, (even at a constant engine operating point) unless the ambient conditions are somehow accounted for. This is usually accomplished through the use of corrected engine parameters. In keeping with our goal of maintaining, as much as possible, a self-contained exposition of GPA, the purpose of this chapter is to furnish a summary of the commonly used corrections for the "major" gas path parameters that are used in performance analysis, diagnostics, and, in particular, within GPA. Corrected parameters will also play a central role in our modeling discussions, in particular, for transient real-time models, which will be discussed in Chapter 7. Although not typically used for the adjustment of data collected at altitude, humidity corrections are included for completeness, as they are important for the analysis of test stand data, especially in regions with high humidity levels. For the most part, corrections are presented without justification. We will, however, provide some abbreviated derivations where needed to support understanding as well as a short discourse on data-based methods. For those interested readers, a more extensive discussion on this general topic, complete with derivations, can be found in [1] and [11].

3.2 GENERAL FORM

It is well known that, at a given operating point, gas turbine gas path parameters, such as inter-stage pressures, temperatures, flows, etc., will vary with ambient conditions. As such, it is common practice to adjust these parameters for ambient temperature, pressure, and, to a lesser degree, humidity before performance calculations are preformed, especially in cases where performance is to be trended over a long period of time when it is likely that these conditions will vary. The adjustment methodology is commonly referred to as *standard day corrections*. Like many engineering practices, it is not a (mathematically) precise method but, rather, an *approximation* that is sufficient for the purpose under which it is used, which is primarily to remove the variation introduced solely by the ambient environment in which the gas turbine operates. We will mention, in passing, that the historical perspective that originally gave rise to these types of corrections involves a methodology that

DOI: 10.1201/9781003649274-3

is sometimes referred to as *dimensional analysis* and has its roots in the early part of the twentieth century with the work of an American physicist working at the US National Bureau of Standards named Edgar Buckingham. Among his contributions is a relationship that has come to be known as *Buckingham's π theorem* [2]. This celebrated result afforded a means to reduce the complexity of any calculation that needed to be performed by reducing the number of independent variables in physical inter-relationships, which was an important consideration before the advent of the digital computer. The adjusted parameters created in this reduction were the genesis for what we today call (standard day) *corrected parameters*. Nevertheless, with all due deference to history and his celebrated contribution, we will take a different approach and offer a simple *mathematical* definition for a *corrected parameter* that is consistent with our colloquial understanding of standard day corrections previously mentioned.

If we let X represent an arbitrary gas path parameter, the equivalent *corrected* parameter representation will be denoted by X_c throughout this discussion. In general, a change in the engine inlet conditions T_2 and P_2 will be accompanied by an attendant change in any downstream gas path parameter X. A *corrected* parameter X_c would be constant, regardless of the change in inlet condition, and represents the value the parameter X would have at some *fixed reference inlet condition*. This reference condition can be whatever you like; however, it is common practice to select the international standard atmosphere (ISA) *standard day* conditions ($T_2 = 518.67$ deg R, $P_2 = 14.696$ psia, or $T_2 = 288.15$ deg K, $P_2 = 101325.353$ Pa) for this purpose. Thus, we can assume without loss of generality that a *corrected* parameter quantity X_c is one that satisfies the following *implicit* relationship

$$X = f(X_c, T_2, P_2) \tag{3.1}$$

The interpretation of Equation (3.1) requires some explanation. Its intent is that a gas path parameter X will vary from its *corrected* form X_c **only** on account of changes in ambient condition as given by the inlet temperature and pressure. Implicit in this statement is that something is being held constant. For example, if power condition were to change (e.g., through a change in PLA, then that would certainly affect X as well. Indeed, so would X_c. So, in some sense, we are assuming from this implicit definition that the *state* of the engine has been held fixed. But what does this mean? Whatever it is, it needs to be independent of the engine control system, otherwise corrected parameters would *not* be independent of the engine model, configuration, and method of control. So we can rule out things such as EPR or thrust, etc. What is certain is that we need for the thermodynamic and aerodynamic *condition* of the engine (and its gas path parameters) to be fixed. Because the gas turbine is an air breathing machine, and air is the elemental fluid, whose properties are being measured by the gas path parameters that we wish to adjust (i.e., *correct*), *we will assume* that *local* Mach numbers, along the gas path have remained fixed. Because air flow, temperatures, pressures, and parameters derived from them are solely a function of geometry (fixed) and Mach number, this would appear to be a reasonable condition to ensure a fixed engine *state*. We will adopt this assumption throughout this discussion and make other simplifying approximations (e.g., averaging) where needed. Because the *corrections should only be considered approximations* themselves, not precise physical quantities, we can afford ourselves this convenience.

Equation (3.1) will be our working definition for what a *corrected* quantity represents. Through formal differentiation (and some sleight of hand), it follows that

$$dX = \left.\frac{\partial X}{\partial T_2}\right|_{\substack{P_2=const\\X_c=const}} dT_2 + \left.\frac{\partial X}{\partial P_2}\right|_{\substack{T_2=const\\X_c=const}} dP_2 + \left.\frac{\partial X}{\partial X_c}\right|_{\substack{P_2=const\\T_2=const}} dX_c \tag{3.2}$$

This implies that

$$\frac{dX}{X} = \left.\frac{\partial X / X}{\partial T_2 / T_2}\right|_{\substack{P_2=const\\X_c=const}} \frac{dT_2}{T_2} + \left.\frac{\partial X / X}{\partial P_2 / P_2}\right|_{\substack{T_2=const\\X_c=const}} \frac{dP_2}{P_2} + \left.\frac{\partial X / X}{\partial X_c / X_c}\right|_{\substack{P_2=const\\T_2=const}} \frac{dX_c}{X_c} \tag{3.3}$$

If we make the further *assumption* that the first two partials in Equation (3.3) *are constant*,[1] (the third is clearly unity by definition) we have

$$\begin{aligned}
&\frac{dX}{X} \approx \alpha\,\frac{dT_2}{T_2} + \beta\,\frac{dP_2}{P_2} + \frac{dX_c}{X_c} \quad \Rightarrow \quad \frac{dX_c}{X_c} \approx \frac{dX}{X} - \alpha\,\frac{dT_2}{T_2} - \beta\,\frac{dP_2}{P_2} \quad \Rightarrow \\
&\frac{dX_c}{X_c} \approx \frac{dX}{X} - \alpha\,\frac{d\theta}{\theta} - \beta\,\frac{d\delta}{\delta} \quad \Rightarrow \\
&\ln(X_c) = \ln(X) - \ln(\theta)^{\alpha} - \ln(\delta)^{\beta}
\end{aligned} \tag{3.4}$$

Note that the constant of integration in the above equation is zero.[2]

This suggests that the general form for a corrected parameter is

$$X_c = \frac{X}{\theta^{\alpha}\,\delta^{\beta}} \quad where \quad \theta = \frac{T_2}{T_{ISA}}, \quad \delta = \frac{P_2}{P_{ISA}} \tag{3.5}$$

where

$$T_{ISA} = 518.67 \deg R,\ P_{ISA} = 14.696\ psia \quad or \quad T_{ISA} = 288.15 \deg K,\ P_{ISA} = 101325.353\ Pa$$

depending on the units of measurement being used. Because both θ and δ are unitless, the units of measurement can be mixed, e.g., temperature in deg K and pressure in *psia* as needed to match those in the actual data stream, as long as it remains consistent in all calculations.

This provides a simple form from which to calculate an *approximate* standard day parameter correction. It is an approximate value for the parameter X that would be seen, under the same operating state but on a standard day atmospheric condition, i.e., 59 deg F (15 deg C) and 14.696 psia pressure. Of course, the *reference condition* can be something other than sea level static ISA conditions, e.g., a hot tropical day, for example, if we desired to approximate, say *EGT*, at a worst-case ambient temperature.

The *reference condition* can also refer to an engine station (location) other than the inlet (Station 2) for various applications. For example, we might be interested in correcting quantities local to a specific engine module (e.g., the HPC) to the engine station representing the inlet of the module under consideration (e.g., Station 25 for the HPC). There are various notations that are common in the industry to denote corrected parameters that refer to particular stations to which they are being corrected; however (unfortunately), there also appears that there is no universal notational convention. For example, one popular notation is to use the subscript "c" followed by the station number to denote a corrected parameter and the reference condition. If we were considering the high spool speed N_2 corrected to the inlet (Station 2), we would denote its corrected form as N_{2c2}. If we were correcting this quantity to the inlet of the HPC (Station 25), we would use the notation, N_{2c25}. Relative to Equation (3.5), we would carry over the station identifier to the parameters θ and δ as well. For example, if we consider the traditional correction for HPC spool speed N_2, (which we will derive in the next section),

$$N_{2c2} = \frac{N_2}{\sqrt{\theta_2}} \quad , \quad N_{2c25} = \frac{N_2}{\sqrt{\theta_{25}}} \quad where \quad \theta_2 = \frac{T_2}{518.67} \quad and \quad \theta_{25} = \frac{T_{25}}{518.67} \tag{3.6}$$

Throughout this text, unless otherwise explicitly stated, *we will assume that the reference location is Station 2, the inlet of the engine.* Thus, within the context of this book, the appearance of $\boldsymbol{\theta}$ **without an explicit subscript will mean** $\boldsymbol{\theta}_{\mathbf{2}}$**.** The same convention applies to δ.

Note that, with our notational convention in mind, Equation (3.1) can be stated more generally as

$$X = f(X_{ck}, T_k, P_k) \tag{3.7}$$

wherein $X_{c\,k}$ is the *corrected* form of gas path parameter X, using engine Station k as the reference point. Repeating our manipulations presented in Equations (3.2), (3.3), and (3.4), we would arrive at an equivalent form of Equation (3.5), i.e.,

$$X_{ck} = \frac{X}{\theta_k^{\alpha}\ \delta_k^{\beta}} \quad where \quad \theta_k = \frac{T_k}{T_{ISA}}, \quad \delta = \frac{P_k}{P_{ISA}} \tag{3.8}$$

where

$$T_{ISA} = 518.67 \deg R\ ,\ P_{ISA} = 14.696\ psia \quad or \quad T_{ISA} = 288.15 \deg K\ ,\ P_{ISA} = 101325.353\ Pa.$$

Equation (3.8) is now consistent (in notation) with our example in Equation (3.6) with $X = N_2$ and $k = 2$ and 25, respectively, where $\alpha = 0.5$, and $\beta = 0$.

In the next section, we will list the most common parameter corrections that are used in performing a GPA and provide a motivation for a few of them that have simple derivations.[3] These include parameters such as spool speeds, air flow, (mass) fuel flow, temperatures, pressures, thrust, horsepower, torque, acceleration (rate of change in spool speed), metal temperature, and rate of change of metal temperature. All of these corrections will take the form of Equation (3.5) (or equivalently Equation (3.8)).

3.3 TRADITIONAL CORRECTIONS

The corrected parameters that we will introduce in this section are commonplace in the gas turbine industry and are used in a variety of engineering disciplines, most notably engine performance, control, and diagnostics, prognostics, and health management (DPHM). These have been tabulated in Table 3.1.

The experienced reader will no doubt recognize most, if not all, of these entries. The theta and delta *exponents* listed in this table are what we are referring to as

TABLE 3.1
Traditional Gas Path Parameter Corrections

Parameter	Symbol	α	β	Corrected Parameter
Temperature	T	1	0	$T_c = \frac{T}{\theta}$
Pressure	P	0	1	$P_c = \frac{P}{\delta}$
Speed	N	0.5	0	$N_c = \frac{N}{\sqrt{\theta}}$
Air flow	w_a	-0.5	1	$w_{ac} = \frac{w_a\sqrt{\theta}}{\delta}$
Fuel flow	w_f	0.5	1	$w_{fc} = \frac{w_f}{\sqrt{\theta}\,\delta}$
Thrust	F_n	0	1	$F_{nc} = \frac{Fn}{\delta}$
Horsepower	HP	0.5	1	$HP_c = \frac{HP}{\sqrt{\theta}\,\delta}$
Torque	Q	0	1	$Q_c = \frac{Q}{\delta}$
Metal temp.	T_m	1.24	-0.2	$T_{mc} = \frac{T_m\delta^{0.2}}{\theta^{1.24}}$
Acceleration	$\dot{N}$	0	1	$\dot{N}_c = \frac{\dot{N}}{\delta}$
Metal temp. rate	$\dot{T}_m$	0.74	0.8	$\dot{T}_{mc} = \frac{\dot{T}_m}{\theta^{0.74}\delta^{0.8}}$

traditional in the sense that they are those that would appear as reduced variables in dimensional analysis utilizing Buckingham's π theorem. Inherent in that analysis are some additional assumptions, the most important of which is that *specific heats (cp, cv)* and, hence, their ratio $\gamma = cp/cv$are *constant*. These approximate exponents are sufficient for most of what we discuss in this text, however, there are occasions when greater accuracy is warranted, such as in the construction of real-time models (which will be discussed in Chapter 7). Relaxation of this assumption would produce a refinement to the exponents given in Table 3.1. Data driven methods, which will be discussed briefly in the next section of this chapter, can also be employed to refine the exponents for a particular application. For a given parameter, the traditional correction applies to all instances of that parameter in the engine's gas path, so we have dropped any (subscript) reference to specific engine stations in the table.

We will now attempt to supply some rationale for the corrections of a few of these commonly used gas path parameters appearing in Table 3.1. Most will be motivated from simple thermodynamic relationships and do not require an extensive knowledge ofthermodynamics or gas turbine operation. We will begin with the first parameter listed, temperature *T*.

3.3.1 Temperature

As a general rule, it seems reasonable that engine inlet temperature (T_2) should have a direct effect on downstream temperatures in the engine's gas path. It seems natural, for example, that for a given operating condition, increasing T_2 would, in turn, increase T_{25}, which, in turn, would increase T_3, etc. To motivate this concept, we need to consider the T-S diagram. To be definite, let us consider the compression induced by the Fan (inner diameter) and the low-pressure compressor (LPC) from Station 2 to Station 25 of our sample engine appearing (figuratively) in Figure 1.1. Without loss of generality, an isentropic compression has been assumed. Figure 3.1 depicts a compression from P_2 to P_{25}, starting at different (inlet) temperatures (T_2 and T_2*).

At this point we will note two properties of the T-S diagram, which we will state without proof. (The reader can derive these for himself/herself, armed only with the elementary laws of thermodynamics.) The relationships are

1. The curves of constant pressure are monotonically increasing with entropy *S*
2. The constant pressure curves diverge from one another with increasing entropy *S*, in the sense that, at two arbitrary levels of entropy as depicted, the ΔT between the two pressure curves are increasing, i.e., (T_{25}* - T_2*) < (T_{25} - T_2).

Nevertheless, we will now show that the temperature *ratios* remain approximately constant, i.e., T_{25} over T_2 is approximately equal to T_{25}* over T_2*.

Without loss of generality, we can assume that * represents what would be observed on standard day conditions; therefore, $T_2^* = T_{ISA}$. To proceed, we need one

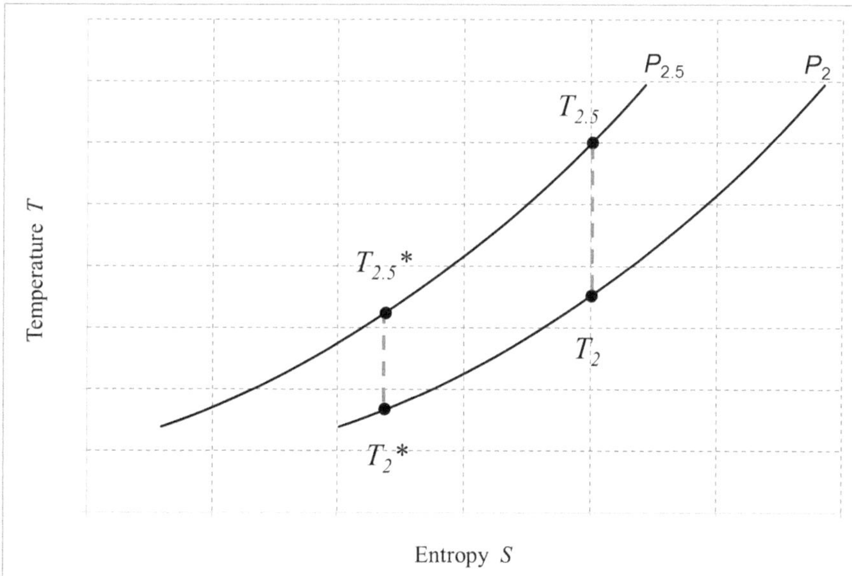

FIGURE 3.1 Sample compression T-S diagram

more relationship from basic thermodynamics (which follows from the definition of entropy and the perfect gas law).

$$dS = \frac{dh}{T} - R\frac{dP}{P} \tag{3.9}$$

where *S, h, T, P, R* denote entropy, enthalpy, temperature, pressure and the gas constant, respectively. Integrating Equation (3.9) between States 2 and 25, and *assuming constant specific heat*, yields

$$\begin{aligned} 0 &= \int_2^{25} dS = \int_2^{25} \frac{dh}{T} - R\int_2^{25} \frac{dP}{P} = cp\int_2^{25} \frac{dT}{T} - R\int_2^{25} \frac{dP}{P} \\ &= c_p \ln\left(\frac{T_{25}}{T_2}\right) - R\ln\left(\frac{P_{25}}{P_2}\right) \end{aligned} \tag{3.10}$$

Equation (3.10) implies that

$$\frac{T_{25}}{T_2} = \left(\frac{P_{25}}{P_2}\right)^{R/cp} \tag{3.11}$$

Likewise, integrating between 2* and 25*, with the same assumption on specific heats, provides

$$\frac{T_{25}^*}{T_2^*} = \left(\frac{P_{25}}{P_2}\right)^{R/cp} \tag{3.12}$$

Combining Equations (3.11) and (3.12) yields

$$\frac{T_{25}^*}{T_2^*} = \frac{T_{25}}{T_2} \Rightarrow T_{25}^* = \frac{T_{25}}{(T_2/T_2^*)} = \frac{T_{25}}{\theta} \tag{3.13}$$

Now, because* represents what we would observe on a standard day, we have that

$$T_{25}^* = T_{25\,c} = \frac{T_{25}}{\theta} \tag{3.14}$$

as required. By using the same arguments from Station 25 to 3, we have that

$$\frac{T_3^*}{T_{25}^*} = \frac{T_3}{T_{25}} \Rightarrow \frac{T_3^*}{T_3} = \frac{T_{25}^*}{T_{25}} = \frac{T_2^*}{T_2} \Rightarrow T_{3\,c} = T_3^* = \frac{T_3}{(T_2/T_2^*)} = \frac{T_3}{\theta} \tag{3.15}$$

and so forth. Therefore, we have in general

$$T_c = \frac{T}{\theta} \tag{3.16}$$

The foregoing establishes the first entry in Table 3.1.

3.3.2 Pressure

The next most common gas path parameter that immediately comes to mind is pressure, the next entry in our table. The pressure changes experienced at various stations throughout the engine's gas-path are the effect of either compressions or expansions resulting from the action of the engine's turbomachinery. For purposes of motivating the correction for pressure, we can consider again, without loss of generality, a compression process; the argument for expansion pressures would be similar. Let us consider, once more, the pressure P_{25} at the exit of the LPC. This pressure is related to temperature T_{25} and inlet temperature and pressure, T_2 and P_2 respectively, by virtue of the following relationship

$$\frac{T_{25}}{T_2} = \left(\frac{P_{25}}{P_2}\right)^{\frac{\gamma-1}{\gamma\,\eta}} \tag{3.17}$$

where η represents polytropic efficiency[4] and $\gamma = cp/cv$is our familiar ratio of specific heats. At a fixed engine operating point, the corrected parameter $T_{25\,c} = T_{25}/\theta$ is constant which implies that T_{25}/T_2 = constant. Thus, by virtue of Equation (3.17), we have that P_{25}/P_2 is also constant, which in turn implies that P_{25}/δ is constant, which establishes the correction[5] at station 25, namely,

$$P_{25c} = \frac{P_{25}}{\delta} \tag{3.18}$$

Note, that implicit in the preceding argument, is that the ratio of specific heats γ and efficiency η do not change. Proceeding downstream to station 3 we have that

$$\frac{P_3}{P_{25}} = \left(\frac{T_3}{T_{25}}\right)^{\frac{\gamma\eta}{\gamma-1}} = \left(\frac{T_3/\theta}{T_{25}/\theta}\right)^{\frac{\gamma\eta}{\gamma-1}} = const. \tag{3.19}$$

since we know from Equation (3.16) that T/θ is a corrected parameter and hence constant. If we now divide Equation (3.19) by δ, we obtain

$$\frac{P_3}{\delta} = \frac{P_3}{P_{25}}\frac{P_{25}}{\delta} = const \tag{3.20}$$

Thus, we proceed downstream and repeat our manipulation we establish in general that P/δ = constant as required. Therefore, we have in general

$$P_c = \frac{P}{\delta} \tag{3.21}$$

This establishes the second entry in Table 3.1.

3.3.3 Speed

Rotational spool speed N is another fundamental and important parameter that appears next in our tabular list. Tangential velocity is related to rotational speed (rpm) by the radius of the object (compressor or turbine blade) in question, as depicted in Figure 3.2. It is also related to *acoustic* velocity by Mach number and the square root of temperature.

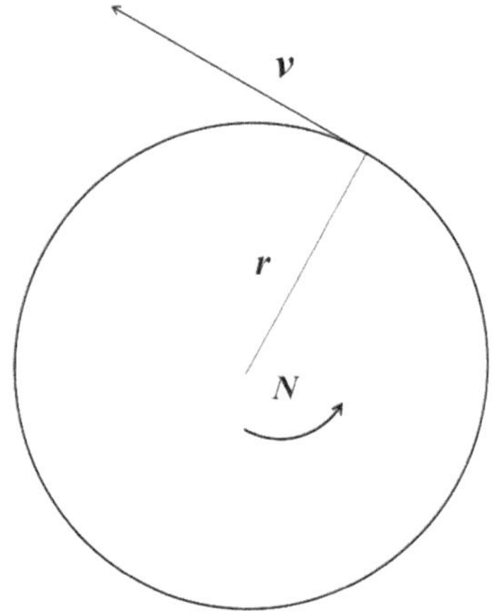

FIGURE 3.2 Rotational speed

This relationship can be written as follows:

$$v = r\,N = M_n\sqrt{\gamma RT} \tag{3.22}$$

where v is the velocity in m/s, r is the radius (m), N is the rotational speed (rad/s), γ is the ratio of specific heats, R the gas constant, and M_n is Mach number. We will, again, make our usual assumption that specific heats are constant (hence γ is constant) and that the local Mach number is constant, as mentioned in the previous section of this chapter.

We will employ our simple manipulation A2.1 of the Appendix, i.e., of taking logs and differentiating Equation (3.22), to obtain

$$\frac{dN}{N} = \underbrace{\frac{dM_n}{M_n}}_{zero} + \frac{1}{2}\left(\underbrace{\frac{d\gamma}{\gamma} + \frac{dR}{R}}_{zero} + \frac{dT}{T}\right) = \frac{1}{2}\frac{dT}{T} \tag{3.23}$$

This implies that

$$\frac{dN}{N} - \frac{1}{2}\frac{d\theta}{\theta} = 0 \Rightarrow \frac{N}{\sqrt{\theta}} = \text{constant} \tag{3.24}$$

Therefore,

$$N_c = \frac{N}{\sqrt{\theta}} \tag{3.25}$$

thus establishing the third entry in our table of traditional corrections.

Before moving on to derive one last correction from Table 3.1, the fourth entry, air flow, we would like to pause momentarily to take a closer look at speed and explore how Equation (3.25) can be refined by relaxing some assumptions made in its derivation. In particular, suppose we do *not* make the assumption of constant specific heats. In that case, Equation (3.23) becomes

$$\frac{dN}{N} = \frac{1}{2}\left(\frac{d\gamma}{\gamma} + \frac{dT}{T}\right) \tag{3.26}$$

Because $\gamma = cp/cv$, we have that

$$\frac{d\gamma}{\gamma} = \frac{dcp}{cp} - \frac{dcv}{cv} \tag{3.27}$$

We also recall that $R = cp - cv$ and, therefore,

$$0 = \frac{dR}{R} = \frac{dcp}{cp - cv} - \frac{dcv}{cp - cv} = \left(\frac{cp}{cp - cv}\right)\frac{dcp}{cp} - \left(\frac{cv}{cp - cv}\right)\frac{dcv}{cv} \Rightarrow$$
$$\frac{dcp}{cp} = \left(\frac{cv}{cp}\right)\frac{dcv}{cv} = \frac{1}{\gamma}\frac{dcv}{cv} < \frac{dcv}{cv} \quad \text{since } \gamma > 1 \tag{3.28}$$

Combining with Equation (3.27), we obtain

$$\frac{d\gamma}{\gamma} = \frac{dcp}{cp} - \frac{dcv}{cv} < 0 \tag{3.29}$$

If we go back to Equation (3.26) and make use of this last inequality, we see that

$$\frac{dN}{N} = \frac{1}{2}\left(\frac{d\gamma}{\gamma} + \frac{dT}{T}\right) = \frac{1}{2}\left[\frac{\left(\frac{d\gamma}{\gamma} + \frac{dT}{T}\right)}{\frac{dT}{T}}\frac{dT}{T}\right] = \frac{a}{2}\frac{dT}{T} \quad \text{where } a < 1 \tag{3.30}$$

Equation (3.30) implies that

$$\frac{N}{T^{a/2}} = \text{constant} \quad \Rightarrow \quad \frac{N}{\theta^{a/2}} = \text{constant, with } \alpha = a/2 < 1/2 \tag{3.31}$$

This simple analysis suggests that if we take into account the changes in specific heat (and hence the changes in the ratio of specific heats), the familiar traditional correction of rotor speed (Equation (3.25)) should be replaced by an exponent less than one-half, i.e., less than square root. To determine how much less, we need to approximate how the ratio of specific heats change (with temperature). We will approximate the ratio of specific heat (γ) as follows

$$\gamma \approx \alpha T_s^{\beta} \quad \Rightarrow \quad \ln(\gamma) = \ln(\alpha) + \beta \ln(T) \quad \Rightarrow \quad \frac{d\gamma}{\gamma} = \beta\frac{dT}{T} \tag{3.32}$$

Figure 3.3 depicts this relationship along with an exponential curve fit.

Hence, $\beta = -0.041$, and we have

$$\begin{aligned}\frac{dN}{N} &= \frac{1}{2}\left(\frac{d\gamma}{\gamma} + \frac{dT}{T}\right) \\ &= \frac{1}{2}\left(-0.041\frac{dT}{T} + \frac{dT}{T}\right) = \frac{0.959}{2}\frac{dT}{T} \\ &= 0.4795\frac{dT}{T} = (0.4795)\frac{d\theta}{\theta}\end{aligned} \tag{3.33}$$

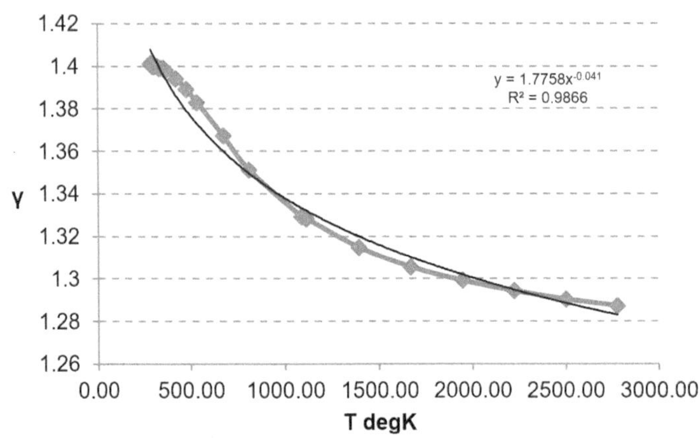

FIGURE 3.3 Ratio of specific heats (γ) as a function of temperature

We then arrive at a refined correction for rotor speed as

$$N_c = \frac{N}{\theta^{0.48}} \tag{3.34}$$

3.3.4 Air Flow

We will provide one more sample derivation before we move on to a discussion on data-based methods in the next section, namely the fourth entry in our table of traditional parameters, i.e., mass air flow rate. If we consider the air flow, w_a (in g/s), moving with uniform velocity, v, through a pipe with a cross-sectional area of A (m^2), then

$$w_a = \rho\ Av \quad = \quad \text{density}(\text{g/m}^3) \times \text{area}\ (\text{m}^2) \times \text{velocity}\ (\text{m/sec}) \tag{3.35}$$

For an ideal gas, we also know that $P = \rho RT$, from which it follows that

$$\frac{w_a}{A} = \rho\ v = \frac{P}{RT}v \tag{3.36}$$

From the definition of Mach number, M_n, we have

$$M_n = v / \sqrt{\gamma\, RT} \tag{3.37}$$

Joining Equations (3.36) and (3.37), we obtain the relationship

$$\frac{w_a}{A} = \frac{PM_n\sqrt{\gamma RT}}{RT} \quad \Rightarrow \quad \frac{w_a\sqrt{T}}{AP} = M_n\sqrt{\frac{\gamma}{R}} \tag{3.38}$$

Invoking our usual assumptions that Mach number and specific heats are constant, we see that

$$\frac{w_a\sqrt{T}}{AP} = const \tag{3.39}$$

If we now multiply both sides of Equation (3.39) by $P_{ISA}/T_{ISA} = const$, we arrive at

$$\frac{w_a\sqrt{T/T_{ISA}}}{AP/P_{ISA}} = \frac{w_a\sqrt{\theta}}{\delta} = const \;\Rightarrow\; w_{ac} = \frac{w_a\sqrt{\theta}}{\delta} \tag{3.40}$$

as specified in Table 3.1.

3.4 DATA-BASED METHODS

In the previous section, we concentrated on what we referred to as *traditional* corrections. These are approximations that can be deduced from simple principles, a little calculus, and algebraic manipulation, under several simplifying assumptions, namely constant local Mach number and constant specific heats. We took a short diversion, using engine rotor speed (N) as an example, as to what refinement might be possible if we relaxed the latter of these assumptions. A small change in theta exponent was suggested by this exercise, wherein the traditional rotor speed exponent of 0.5 was seen to be 0.48 instead. For most applications this may not matter much, but for other usages, the difference might be hugely beneficial. This leads us to a fundamental question. To what degree can we rely solely on theoretically derived relationships based on underlying approximating, and perhaps unrealistic, assumptions in a real-world application? A natural response to this philosophical question would be to apply it to real-world data and observe whether the metrics important to the application at hand are satisfactory. Within the context of *performance diagnostics and prognostics*, which is the primary focus of this book, reducing variability is a fundamental metric. Chapter 2 discusses this at great length. Parameter correction is one step in the diagnostic process to control and mitigate the natural variation inherent in the captured data and its subsequent reduction to reliable (hence meaningful) health information. In this section, we will (briefly) discuss an approach to determine the corrected parameter θ and δ exponents empirically from observed engine data. In the next section, we will apply traditional corrections to sample datasets to observe how well variability is reduced and then apply the empirical approach discussed in this section to note any further variability mitigation.

We know from Chapter 2 that for performance diagnostics, and in particular GPA, it is *change* that is important, and it is the variability in parameter change, i.e., parameter Δ variation, that is most important. With this in mind, our discussion will focus on minimizing variability in Δ parameters. Empirics notwithstanding, we still must make some assumptions. We will assume that Equation (3.1) is still in play and the assumptions needed to infer from it are the general form of a corrected parameter X_c, given by Equation (3.5). There is one subtle difference. While Equation (3.1) requires something to be held constant, we will not be making the outright assumption that it is local Mach numbers. In fact, the approach we are about to describe assumes the use of discrete data points, which implies that an engine steady-state condition is in effect, and thus, it is more likely that some (engine control) power setting parameter is being held constant. It does not much matter what it is because we will not be making direct use of it in our approach. It is only the assumption that something is being held constant to induce a steady-state condition so that Equation (3.1) can be used along with its consequent, Equation (3.5).

For simplicity, we will consider only temperature (θ) corrections; pressure (δ) corrections can be handled in an analogous manner. If we let *X* denote the gas path parameter under consideration (i.e., this could be a temperature, pressure, rotor speed, etc.), we will assume that we have available to us a set of observed data $\{X_i, \theta_i, \delta_i\}_{i=1}^n$ along with fundamental design information on what we would consider nominal levels for this parameter, $\{X_{ci}^{(NOM)}\}_{i=1}^n$ to which we could compare the observational data in order to calculate a series of parameter residuals $\{\Delta X_i\}_{i=1}^n$. *To simplify our discussion, we will make the assumption that the nominal levels are generated at sea level, static, standard day conditions* (temperature and pressure) and are typically calculated from a steady-state engine model of some sort.[6] Thus, *the nominal values are corrected quantities*. Before proceeding, we need to clarify some implicit *assumptions* that we will impose on this process we are about to describe. The data $\{X_i, \theta_i, \delta_i\}_{i=1}^n$ can be from a single engine or a family of engines of the same model type. In the latter case, where more than one engine is represented, each engine must be at *approximately* the same performance deterioration level. Because we most likely will not know a given engine's health state, *a-priori*, we can qualify this assumption by insisting that the family consists of engines whose operating cycles (or hours) fall within a reasonably small band. The importance of this assumption will become clear in the next section.

There are several empirical approaches possible, but for the sake of brevity, we will describe only one wherein we will determine the exponent that minimizes the standard deviation ($\sigma_{\Delta X}$) of the parameter deltas $\{\Delta X_i\}$. To keep this brief, we will assume that the parameter *X* has a corrected form of $X_c = X/(\theta^\alpha \delta^\beta)$(as mentioned previously). We can calculate the parameter deltas[7] as follows:

$$\Delta X_i = 100\left(\frac{\dfrac{X_i}{\theta_i^\alpha \delta_i^\beta} - X_{ci}^{(NOM)}}{X_{ci}^{(NOM)}}\right) \tag{3.41}$$

We can rewrite Equation (3.41) as

$$\frac{X_i}{\theta_i^\alpha \delta_i^\beta} = X_{ci}^{(NOM)}\left(1 + \frac{\Delta X_i}{100}\right) \tag{3.42}$$

Thus, ΔX_i represents the % deviation[8] (from nominal), and the *variation* we observe in these residuals are reflective of performance deterioration, sensor inaccuracies, and inappropriate theta correction due to using an incorrect exponent α. Now, if we plot these deltas against θ, and if α were the true correction exponent, we would expect to see a scatter plot that would have zero correlation, i.e., a linear regression line fit horizontal with the theta axis. An upward or downward trend, however, would indicate the extent to which the exponent α was deficient.

The methodology we will use can be described as follows:

1. Collect n data points $\left\{X_i\,,\ T_{2i}\,,P_{2i}\right\}_{i=1}^{n}$
2. Calculate $\theta_i = T_{2i}/T_{ISA}$, where it is assumed that temperature is in *deg R* or *deg K*, and $\delta_i = P_{2i}/P_{ISA}$
3. Pick a range of theta exponents to try, $\{\alpha_1, \alpha_2, \cdots, \alpha_K\}$, where this range is chosen, such that the true exponent would be covered by this selection. This requires a bit of engineering judgment. For example, if $X = N_2$, then it would be reasonable (perhaps generous) to choose the range between 0.4 and 0.6 with an increment of 0.01.
4. Iteratively, for each $\alpha_k \in \{\alpha_1, \alpha_2, \cdots, \alpha_K\}$
5. Calculate ΔX_i using Equation (3.41), $\forall i$=1, 2, …, n
6. Calculate the standard deviation of the $\{\Delta X_i\}_{i=1}^{n}$, which we will denote by σ_{α_k}
7. Determine the minimum standard deviation $\sigma_{\alpha_J} = \min\left\{\sigma_{\alpha_k}\right\}_{k=1}^{K}$
8. Choose α_J as the best exponent to correct parameter X

This minimization method carries a computational overhead in that the iterative nature of the process requires that a statistical analysis be performed at each newly chosen exponent, where the exponent is modulated in small increments over some range. After that is completed, the minimum statistic is isolated, and the corresponding exponent is selected as the winner. To illustrate this method, we use hypothetical data in Figure 3.4.

The assumed nominal (reference) relationship is superimposed as the solid line appearing in the plot. Each of the 100 data points depicted were generated at different ambient conditions (varying levels of θ) with the addition of varying (temperature) bias and random noise. A theta exponent $\alpha = 0.92$ was also assumed in the generation process. If we received this data with no knowledge of the true α, we might *correct* the temperature data using the traditional exponent of unity ($\alpha = 1$) and then calculate the % deltas using Equation (3.41). A plot of these deltas vs theta appears in Figure 3.5. A linear fit to the data has been superimposed in the plot, and it is clear that we do *not* have a zero correlation, indicating that unity is not the appropriate exponent to use for correction. In fact, the negative trend indicates that the true exponent must be smaller than $\alpha = 1$.

If we now apply the iterative procedure described above, wherein we use a range of sample exponents $\alpha = [0.80, 0.81, 0.82, \cdots, 1.00]$ to calculate the associated variation (standard deviation) of the data associated with each exponent, we see from

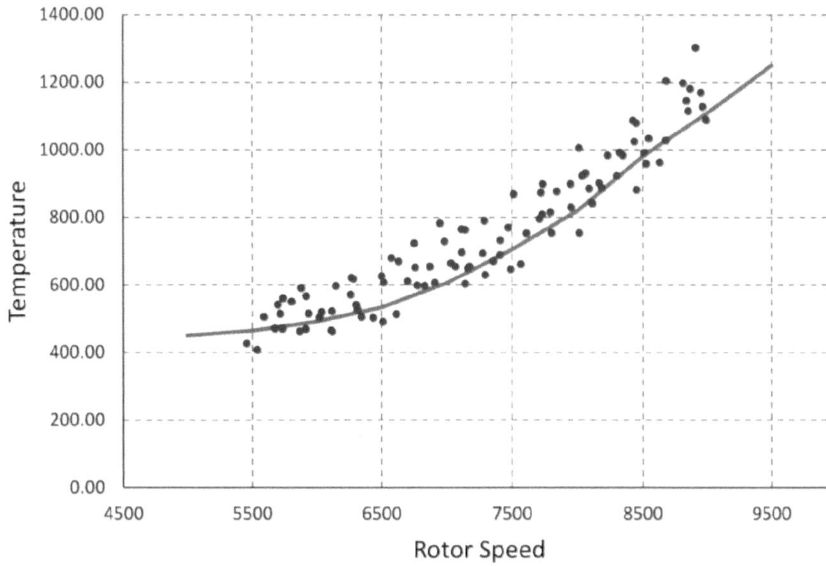

FIGURE 3.4 Hypothetical temperature vs rotor speed data

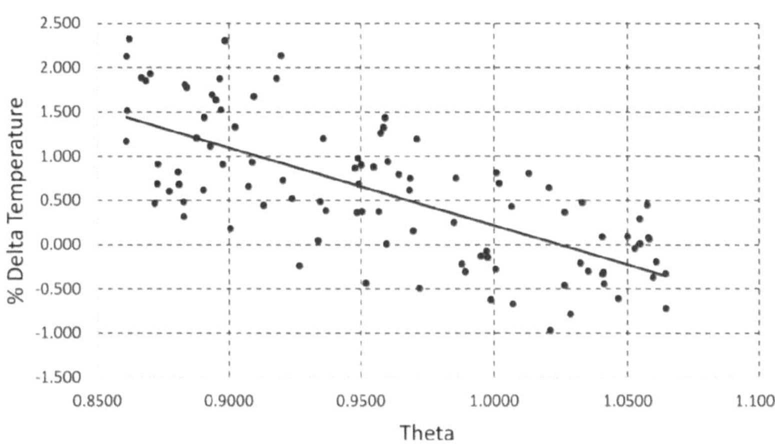

FIGURE 3.5 % Temperature deltas vs theta (using $\alpha = 1$)

Figure 3.6 that the minimum variability occurs around exponents 0.91 and 0.92, and while it is not discernible from the plot, the standard deviations for these two exponents are 0.5505 and 0.5502, respectively, making 0.92 the winner.

As engineers, we would be happy with this result, as if we use *either* 0.91 or 0.92 as the theta exponent to adjust our data, we would be significantly reducing the variability in the temperature Δ data over using the traditional unity exponent. What if we take a closer look and repeat our variability analysis using a refined range of exponents

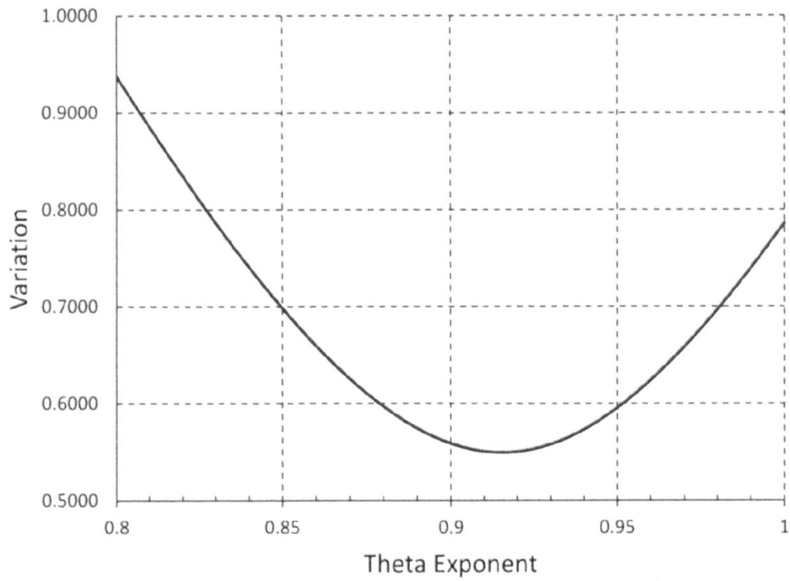

FIGURE 3.6 Effect of theta exponent α on parameter Δ variation

between 0.91 and 0.92 at 0.001 increments, i.e., $\alpha = [0.91, 0.911, 0.912, \cdots, 0.92]$. If we did so (using the same data), we would arrive at Figure 3.7.

We see from this *closer look* that the best exponent is 0.915 and not 0.92, which we know is the true exponent in this hypothetical exercise. What is going on here is (unfortunately) endemic to data driven methods. Results depend on the particular dataset being used. If we were to repeat this exercise with a different *sample* of data (using the same noise statistics) the result would, again, be slightly different. If we could repeat this an infinite number of times, the average best exponent would approach 0.92 in the limit. This quandary is not of any practical engineering consequence except to highlight the need to treat empirically derived results as an approximation and not physical fact.

Alpha	Std Dev
0.91	0.550485
0.911	0.550099
0.912	0.549793
0.913	0.549566
0.914	0.549419
0.915	0.549351
0.916	0.549363
0.917	0.549455
0.918	0.549626
0.919	0.549877
0.92	0.550208

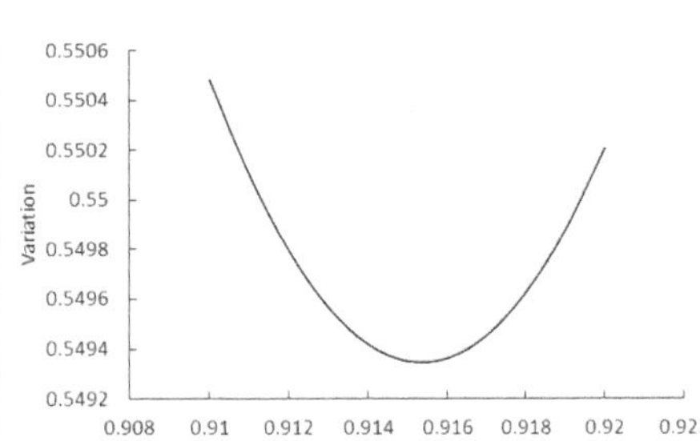

FIGURE 3.7 Refined effect of theta exponent α on parameter Δ variation

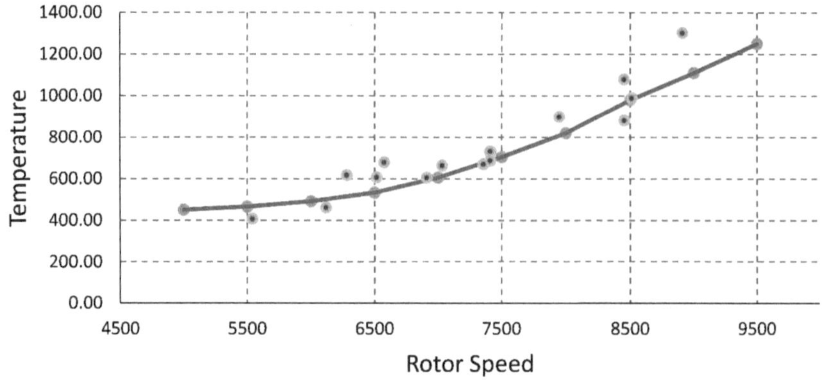

FIGURE 3.8 Reduced sample hypothetical temperature vs rotor speed data

To illustrate further, if we take our original dataset (depicted in Figure 3.4) and, instead of using all of the data, we use only 15 of the data points to create a small sample (as illustrated in Figure 3.8) and repeat the analysis using the same range of test exponents $\alpha = [0.80, 0.81, 0.82, \cdots, 1.00]$, we would obtain an optimal exponent close to 0.94, as suggested by Figure 3.9. This shift to 0.94 is solely the effect of

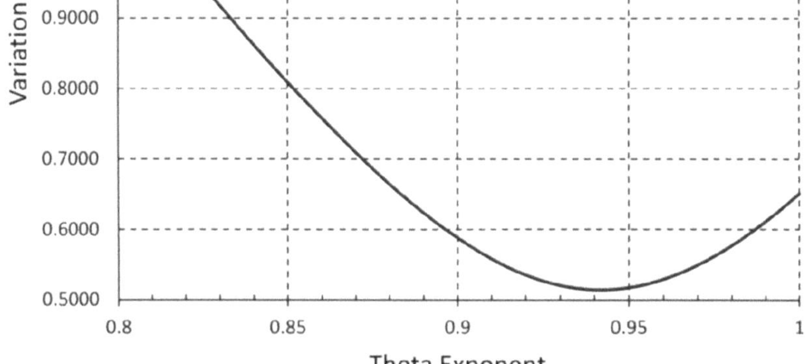

FIGURE 3.9 Reduced sample effect of theta exponent α on parameter Δ variation

sample size reduction, as the theta range represented in the small sample was reasonably extensive and appropriate for the analysis, i.e.,

	theta	*deg C*	*deg F*
min	0.8721	-21.86	-7.36
max	1.0597	32.20	89.95

What should be noted is that with all methods of an empirical nature, the characteristics of the data being utilized in support of the method are of critical importance. The number of data samples, the level of noise present in the data, and the degree to which the data represents the effects under scrutiny are pivotal to the outcome of a data-driven method. In particular, for this application, we need to ensure that our data represents a significant range of ambient conditions (i.e., θ). To do otherwise would severely compromise the analysis. Another potential problem is possible scarcity in appropriate data to use for any one engine in the fleet, driving the analyst to combine data taken from several engines that may unknowingly have differing performance levels, which would introduce bias and increased variability that could result in a corrupted analysis. Moving away from the ingenuous hypothetical data used in this section, the next offers some realistic high-fidelity data taken from a high bypass, twin spool, closed-loop engine simulation[9] to illustrate the material discussed in previous sections of this chapter.

3.5 EXAMPLES

To illustrate some of the ideas discussed in this chapter we will use some realistic engine data generated from a high bypass, twin spool, closed-loop engine simulation created by NASA called the commercial modular aero-propulsion system simulation (C-MAPSS) [3].

The first example is of a single engine, whose modular performance has deteriorated approximately 1.5% across all modules, running at sea level, static conditions. The throttle resolver angle (TRA) ranges from 80 to 100 degrees, and ambient temperature ranges between $[-29F, 89F]$, with a total sample size of 500 points being generated. For the purpose of illustration, we will only consider five output parameters; Fan speed (N_1), core speed (N_2), *EGT* (T_5), HPC inlet temperature (T_{25}), and mass fuel flow (w_f). Sensor non-repeatability (i.e., precision error) was also included. Table 3.2 lists the noise statistics used for this purpose. The nominal (non-deteriorated) C-MAPSS engine simulation running at sea level, standard day conditions will serve as the reference level from which % of point deltas will be computed. As an example, a graphic depiction of this data (for parameter N_1) is contained in Figure 3.10.[10]

As one can clearly see, the level of data scatter for N_1, relative to its reference (nominal) level (indicated in red), would be unacceptable in performing any sort of diagnostic analysis. Similar results hold for the other four parameters in question

TABLE 3.2
Sensor Noise Levels Used in C-MAPSS Simulation

Parameter	Noise Standard Deviation	Unit
N1	0.5	rpm
N2	1	rpm
T25	0.3	F
T5	1	F
Wf	50	pph
T2	0.1	F
P2	0.05	psia

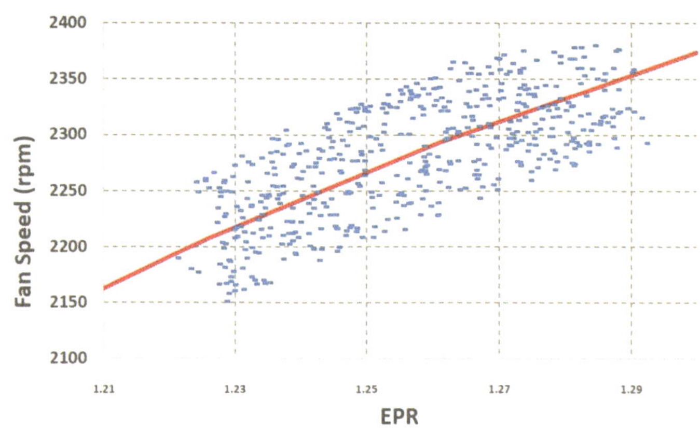

FIGURE 3.10 Sample C-MAPSS deteriorated engine data

(N_2, T_{25}, T_5, and w_f). Much of this variation is due to the effects of the changing ambient conditions. If we adjust this data by calculating *corrected* N_1 using the traditional relationship from Table 3.1, i.e., $N_{1c} = N_1/\sqrt{\theta}$, then we can observe a dramatic reduction in scatter, as seen in Figure 3.11.

This reduction in variability using a *corrected* parameter can be viewed more dramatically by calculating (and plotting) both the % deltas of the uncorrected data and the % deltas of the corrected data, relative to the reference, as shown in Figure 3.12. Here, the corrected delta variation (red) appears to be roughly an order of magnitude *less* than the uncorrected delta variation (blue). Figure 3.13 demonstrates the high correlation of the uncorrected deltas, with theta and a near zero correlation for the corrected deltas, which, of course, is what a corrected parameter is designed to achieve. A close examination of the corrected deltas in this graphic indicates a very slight negative slope (computed at -2.118), which implies that a slightly different exponent is needed to produce the desired zero slope.

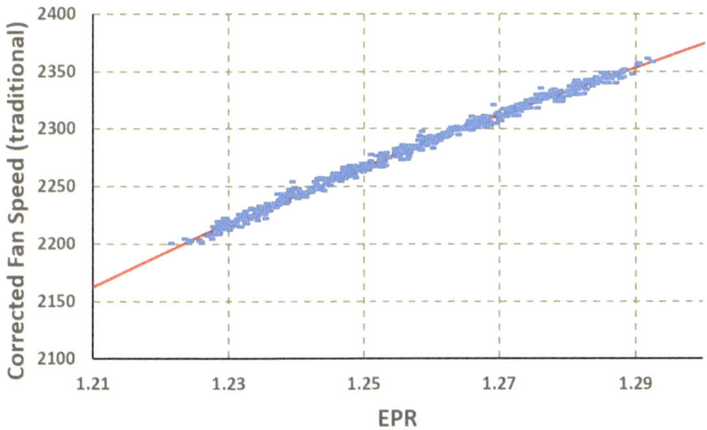

FIGURE 3.11 Adjusted C-MAPSS engine data using traditional correction

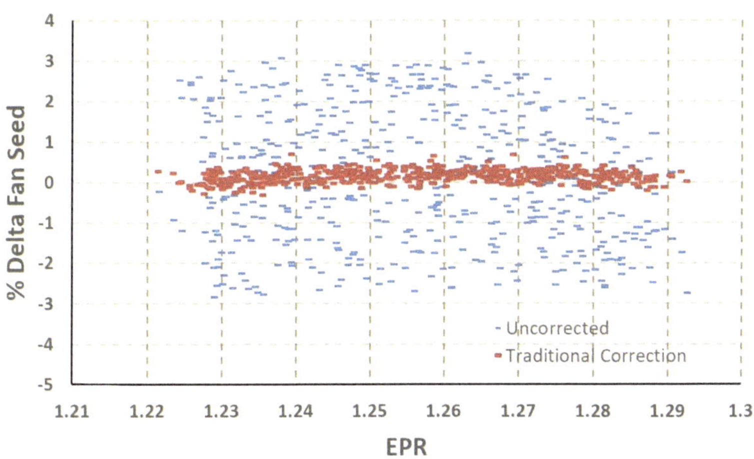

FIGURE 3.12 N_1 % delta variation reduction: corrected vs uncorrected

We might also ask if further reduction is possible using a different theta exponent than the traditional α = 0.5? Indeed, our theoretical exploration (in the previous section) of relaxing the assumption of constant specific heats led us to a refinement of the traditional correction (Equation (3.34)) that suggests that $\alpha \approx 0.48$ for rotor speeds. As an illustrative exercise, we will apply our sample N_1 data to the empirical process described in the previous section using a range of potential exponents $\alpha = [0.44, 0.45, \ldots, 0.52]$ to determine which produces the least variation. Figure 3.14 displays the results from which it can be seen that a minimum variability is achieved (interestingly enough) at $\alpha = 0.48$ in accordance with Equation (3.34).

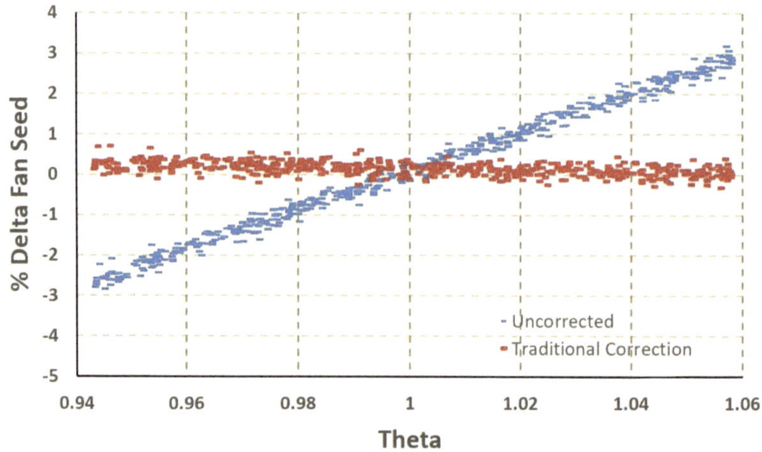

FIGURE 3.13 Theta dependence: corrected vs uncorrected

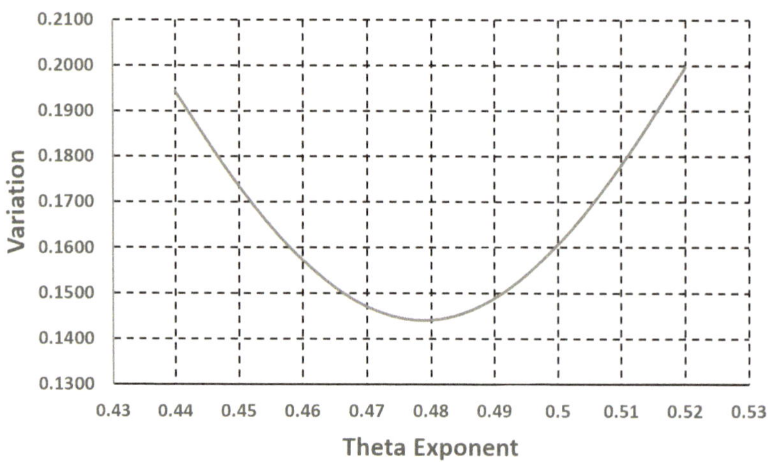

FIGURE 3.14 Effect of theta exponent α on Δ N_{1C} variability

Recalculating the slope of the corrected delta vs theta using this new exponent brings us closer to zero at -0.116.

Repeating this exercise with the remaining four parameters produces the exponents assembled in Table 3.3. Interestingly, T_{25} remained unchanged at unity.

Table 3.4 consolidates the effects that parameter correction had on parameter delta variability and theta (i.e., inlet temperature) correlation from the C-MAPSS simulations. The "None" column refers to uncorrected parameters. It is clear that the use-corrected quantities dramatically reduced variability (induced by the broad range of ambient temperatures) by an order of magnitude at the least. The correlation with theta was reduced considerably as well, which should come as no surprise,

TABLE 3.3
Theta Exponents Empirically Derived from C-MAPSS Simulation Data

	Theta Exponent	
Parameter	**Traditional**	**Empirical**
N1	0.5	0.48
N2	0.5	0.49
T25	1	1
T5	1	0.96
Wf	0.5	0.6

TABLE 3.4
Theta Exponent Impact on Parameter Delta Variability and Theta Correlation

	σ			θ Correlation		
Parameter	**None**	**Traditional**	**Empirical**	**None**	**Traditional**	**Empirical**
ΔN1	1.618	0.161	0.144	0.996	-0.443	-0.027
ΔN2	1.649	0.05	0.038	1	-0.638	0.046
ΔT25	3.355	0.134	0.134	0.999	-0.073	-0.073
ΔT5	3.245	0.173	0.124	0.999	-0.701	0.108
ΔWf	2.05	0.509	0.382	0.982	0.66	0.001

as a *corrected* parameter is, by its (colloquial) definition, intended to produce a zero correlation.

This exercise was intended to illustrate some of the concepts discussed in this chapter, and although the C-MAPSS can simulate realistic engine data, the exercise is arguably unrealistic in another sense, i.e., the enormous amount of data that was used. Five hundred data points at sea level, static conditions, from a mildly deteriorated,[11] but unchanging engine experiencing a very broad range of ambient temperatures is not a practical real-life occurrence. What if, instead, we only had 20 data points, how would that impact our data driven methodology? This is closer to a real-life situation of an engine that is not in commercial service but is used solely for test-stand calibration. Its performance level would essentially be unchanged and would operate in a reasonably broad range of ambient conditions due to seasonal fluctuations or from being shipped to test-stands in different parts of the world.

To explore this *in a simple fashion* we randomly extract 20 data points from our 500-point dataset and repeat the empirical exercise. A natural consequence of the reduced random sample of 20 points was, unfortunately, a reduced ambient temperature range, [+29F, 87F], approximately half the range of the original set. The analysis produced results that were different but reasonably close to those of the 500-point dataset. The minimum variance exponents are displayed in Table 3.5 along with those from the larger dataset for comparison.

The effects on variability and temperature correlation (see Table 3.6), likewise, are similar to those obtained with the large sample (Table 3.4).The largest deviation in exponents calculated from the small sample relative to the large sample is for Δw_f at 0.64 versus 0.60, respectively. Given that $\sigma_\theta = 0.033$, it is a simple matter to calculate (from information in Table 3.6) the slope of Δw_f corrected by 0.64 relative to θ as being -1.637. Moreover, this is the largest slope for any of the parameters in Table 3.6, corrected by their empirically derived exponents from Table 3.5 that minimize variability. What is interesting is that using the best exponent (derivable from the data) notwithstanding, the data additionally provides an implicit indication that the true exponent is actually lower (than 0.64) by virtue of the strong negative slope! We will leave the reader to ponder this conundrum and move on to our final example.

TABLE 3.5
Theta Exponents Empirically Derived from Small Sample C-MAPSS Simulation

	Theta Exponent		
Parameter	Traditional	Empirical (500 Data Points)	Empirical (20 Data Points)
N1	0.5	0.48	0.485
N2	0.5	0.49	0.495
T25	1	1	1.02
T5	1	0.96	0.97
Wf	0.5	0.6	0.64

TABLE 3.6
Theta Exponent Impact on Parameter Delta Variability and Theta Correlation for Small Data Sample

	σ			θ Correlation		
Parameter	None	Traditional	Empirical	None	Traditional	Empirical
ΔN1	1.601	0.157	0.151	0.995	-0.302	-0.097
ΔN2	1.66	0.042	0.042	1	-0.338	-0.338
ΔT25	3.41	0.167	0.157	0.999	0.354	-0.038
ΔT5	3.266	0.152	0.126	0.999	-0.572	0.153
ΔWf	2.099	0.59	0.4	0.984	0.742	-0.134

Our final example refers back to our statements regarding data driven methods and the type and quantity of data available in the real world. The previous example illustrated what might happen with a small sample of data, albeit from a single engine with non-changing performance, subjected to a healthy range of ambient temperatures. While possible, this does not represent mainstream data collected by operators. If we, again, restrict ourselves to sea level, static test data (for simplicity), it is more likely to have a handful of data points from a plurality of engines in the fleet, whose data, treated *collectively*, might span a healthy range of ambient conditions. In such a case, we would assume, but not necessarily know with certainty, that all of the engines are of approximately the same performance level. Not that this would ever happen, but what if our assumption turned out to be false. How would that affect the data driven methodology? To illustrate what could possibly happen, we will consider three cases. In each case, we will use two engines (instead of a fleet) for simplicity and the fact that the potential hazards can be adequately represented with this simplification. A total of *four engine conditions* is simulated using C-MAPSS and are referred to as Eng 1, Eng 2, Eng 3, and Eng 3 (cold). Within each case, two engines (from the four) are *combined* to create our small fleet of two engines. We describe the configurations below.

3.5.1 Performance

- Eng 1 is moderately deteriorated at -1.5% uniformly across all modules.[12]
- Eng 2 is the same as Eng 1, except -2% for LPC and HPC.
- Eng 3 is highly deteriorated at -2% on all modules except HPT at -3%.
- Eng 3 (cold) is the same as Eng 3, except for ambient conditions.

3.5.2 Cases

1) Combines Eng 1 and Eng 2
2) Combines Eng 1 and Eng 3
3) Combines Eng 1 and Eng 3 (cold)

3.5.3 Ambient Conditions

Inlet temperature, T_2, ranges for the three cases are depicted in Figure 3.15.

For *each* case, we will combine the data from two engines as described above and perform the empirical method to assess the best theta exponent for each of the five gas path parameters. We will also perform the empirical analysis again on each engine individually and note the differences in the exponents obtained.

3.5.3.1 Case 1

In this scenario, we are combining data (20 points each) from Eng 1 and Eng 2 to give us a 40-point sample. As described above, these engines are close in performance except for a 0.5% difference in HPC deterioration between them. Eng 1 has an inlet temperature range of 20 *deg F* more extensive than Eng 2, as can be seen

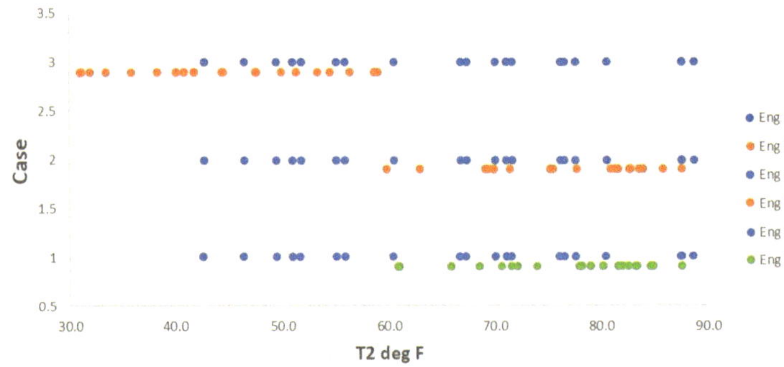

FIGURE 3.15 Inlet temperature ranges for Cases 1, 2, and 3

from Figure 3.15. Thus, our assumption of equal performance is approximately true. We can test this graphically by making a visual inspection of the raw (i.e., uncorrected) parameter deltas to see if any anomalies present themselves. Because w_f is among the most sensitive parameters to change with performance differences, we will save space and plot only its raw deltas in Figure 3.16. As can be seen, the two engines blend reasonably well, indicating a suitable match. Table 3.7 contains the results of the combined and individual empirical analyses, and Figures 3.17 and 3.18 contain the w_f deltas *corrected* using the *combined* and *individual* empirically derived exponents, respectively.

A quick glance at Table 3.7 shows results that are sensible, in the sense that they conform reasonably to those in Tables 3.3 and 3.5, which also used the Eng 1 deterioration configuration. Not surprisingly, the greatest departure is with the combined data w_f exponent at 0.69. Nevertheless, based on engineering experience, this seemingly inflated value still falls within the range that is possible for correcting w_f on some engine types. Likewise, Figures 3.17 and 3.18 show a near-zero correlation of

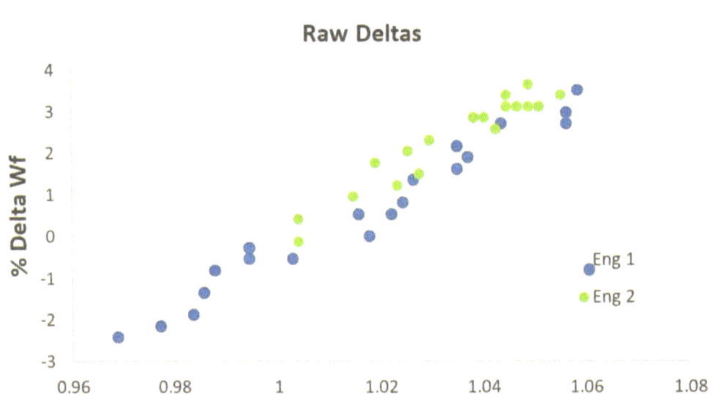

FIGURE 3.16 Case 1 combined data – raw delta w_f

TABLE 3.7
Case 1 Empirical Exponent Results

		Theta exponent		
		Empirical (20 Data Points)		
Parameter	Traditional	Eng 1	Eng 2	Combined
N1	0.5	0.47	0.49	0.47
N2	0.5	0.49	0.5	0.49
T25	1	1	1.05	1.01
T5	1	0.96	0.97	1.01
Wf	0.5	0.64	0.66	0.69
T2 (deg F)	Min	42.6	60.9	42.6
	Max	88.7	87.6	88.7

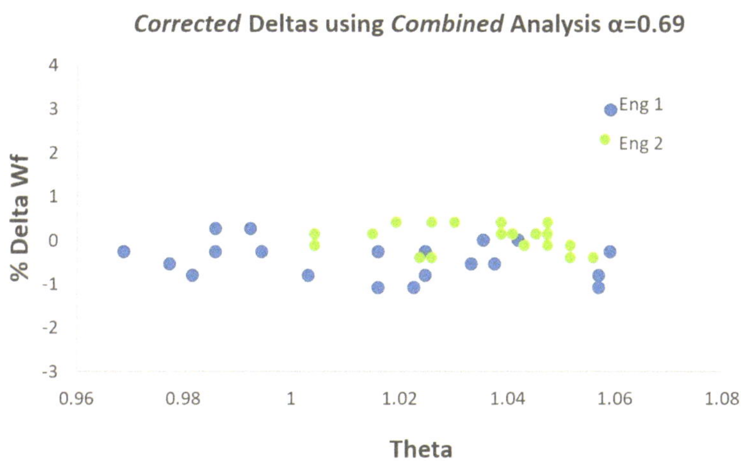

FIGURE 3.17 Case 1 combined data – corrected delta w_f

corrected deltas with theta as required, with little perceivable difference between them.

3.5.3.2 Case 2

Our second case combines and compares Eng 1 with Eng 2, both of which have radically different performance levels, violating our underlying assumption of approximate uniformity. Ambient temperature range is unchanged from Case 1. A quick plot (Figure 3.19) of the raw w_f deltas would suggest that the engines are not from the same performance population. For most, this should be enough evidence to abandon a blind application of the data driven method to this type of data (or attempt to "remove" engine performance discrepancy before mixing the data). If we forged

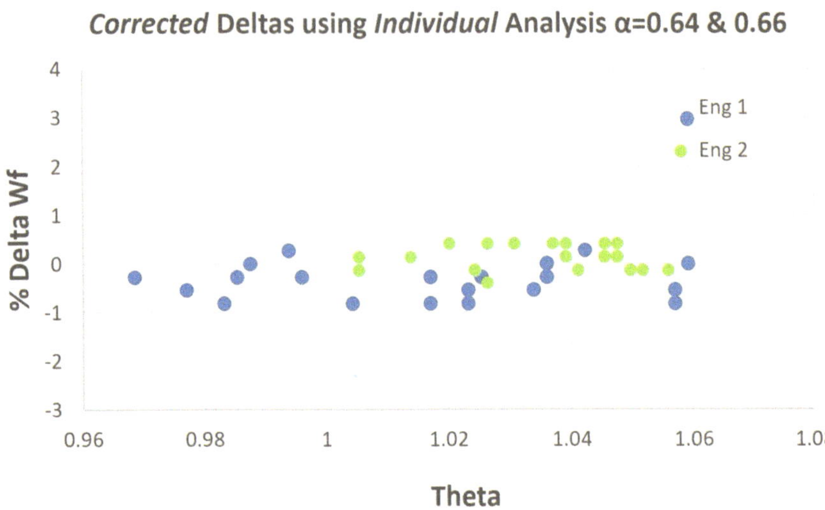

FIGURE 3.18 Case 1 individual data – corrected delta w_f

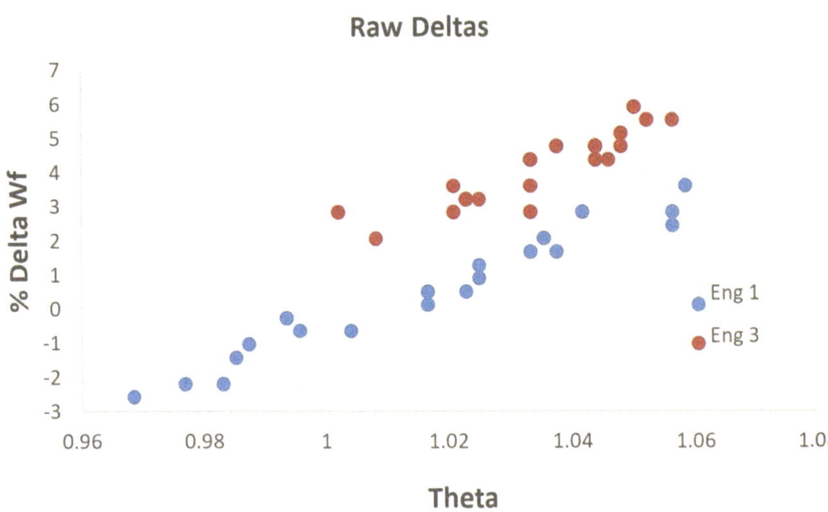

FIGURE 3.19 Case 2 combined data – raw delta w_f

ahead, notwithstanding these warning signs, we would obtain the correction exponents tabulated in Table 3.8. The rotor speeds and T_{25} for the combined data look reasonable; however, w_f and T_5 appear abnormally high. Interestingly, the individual analyses provide exponents that appear very reasonable. These observations are further substantiated by Figures 3.20 and 3.21.

TABLE 3.8
Case 2 Empirical Exponent Results

		Theta exponent		
		Empirical (20 Data Points)		
Parameter	Traditional	Eng 1	Eng 3	Combined
N1	0.5	0.47	0.48	0.47
N2	0.5	0.49	0.5	0.47
T25	1	1	1.03	1.01
T5	1	0.96	0.98	1.19
Wf	0.5	0.64	0.64	0.85
T2 (deg F)	Min	42.6	59.8	42.6
	Max	88.7	87.6	88.7

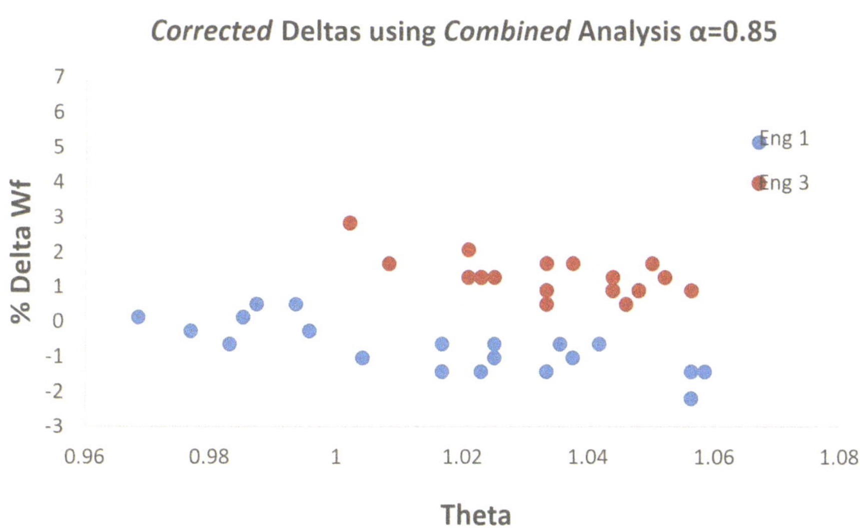

FIGURE 3.20 Case 2 combined data – corrected delta w_f

3.5.3.3 Case 3

Our final scenario combines and compares the same engines as in Case 2, with the difference being that Eng 3 has a different ambient range that is colder than what was used in Case 2. Additionally, about 40% of its ambient data does *not* overlap Eng 1, whose ambient conditions are unchanged from Case 2. This can be seen from Figure 3.15 and the usual plot of raw w_f deltas in Figure 3.22.

The same uniform population concern is still apparent, and we now have the additional concern that a large part of the data does not share a common inlet temperature range; in fact, less than 50% of the data overlap in ambient condition. With this

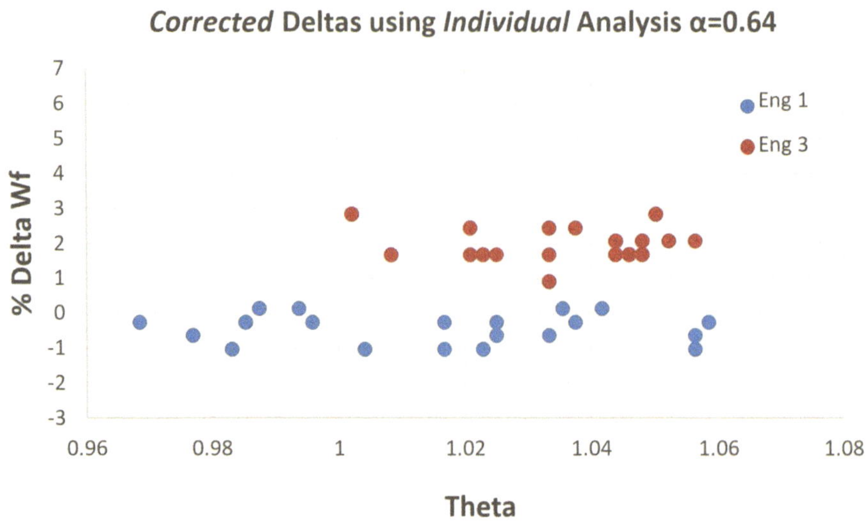

FIGURE 3.21 Case 2 individual data – corrected delta w_f

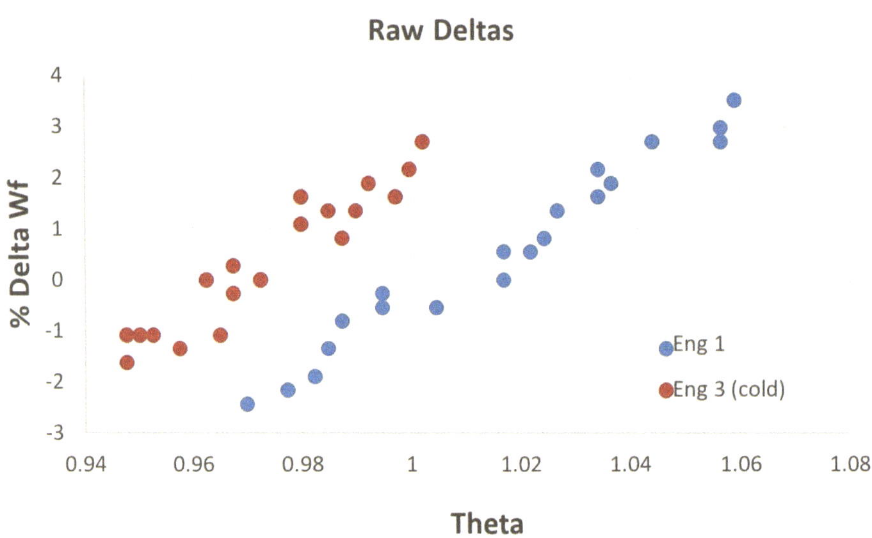

FIGURE 3.22 Case 3 combined data – raw delta w_f

added disparity, the data-driven methodology generates even more bizarre results than before, as can be seen in Table 3.9.

As the reader can observe, the exponents for w_f and T_5 using the combined data are beyond ludicrous. One might even muse whether using a random number generator to produce exponents would be more viable! Indeed, even N_2 produces an exponent of 0.52 in stark contradiction to our theoretical result (Equation (3.31)) that

TABLE 3.9
Case 3 Empirical Exponent Results

		Theta exponent		
		Empirical (20 Data Points)		
Parameter	Traditional	Eng 1	Eng 3 (cold)	Combined
N1	0.5	0.47	0.49	0.48
N2	0.5	0.49	0.49	0.52
T25	1	1	1.01	0.99
T5	1	0.96	0.96	0.68
Wf	0.5	0.64	0.64	0.36
T2 (deg F)	Min	42.6	31.0	42.6
	Max	88.7	59.0	88.7

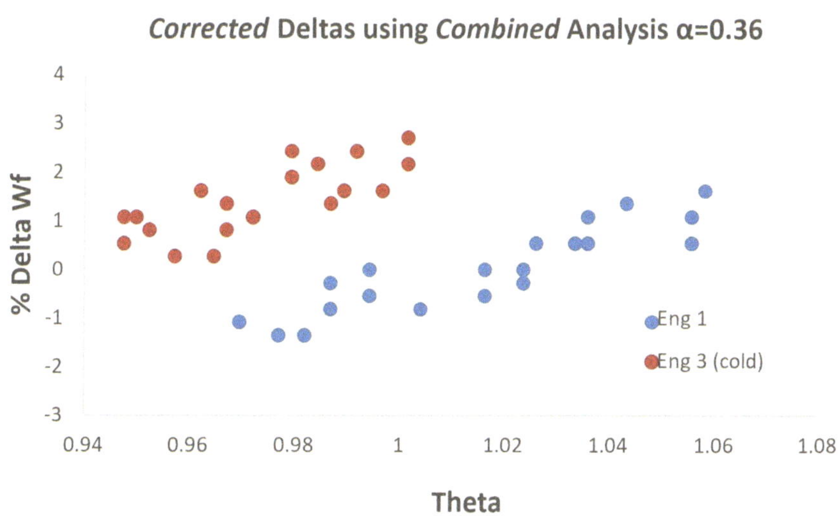

FIGURE 3.23 Case 3 combined data – corrected delta w_f

rotor speeds must admit a theta exponent strictly less than 0.5. As in the previous two cases, the individual analyses produce more reasonable results. We include Figures 3.23 and 3.24 as before, for completeness.

In summary, the examples presented in this section were chosen to both emphasize the effectiveness of the correction process and to illustrate (relative to the data-driven methodology) what can go right and what can go wrong. When employing empirical methods, it is imperative to know the suitability of the data being employed and to heed the warning of the ancient Romans: *purgamentum init, exit purgamentum.*[13]

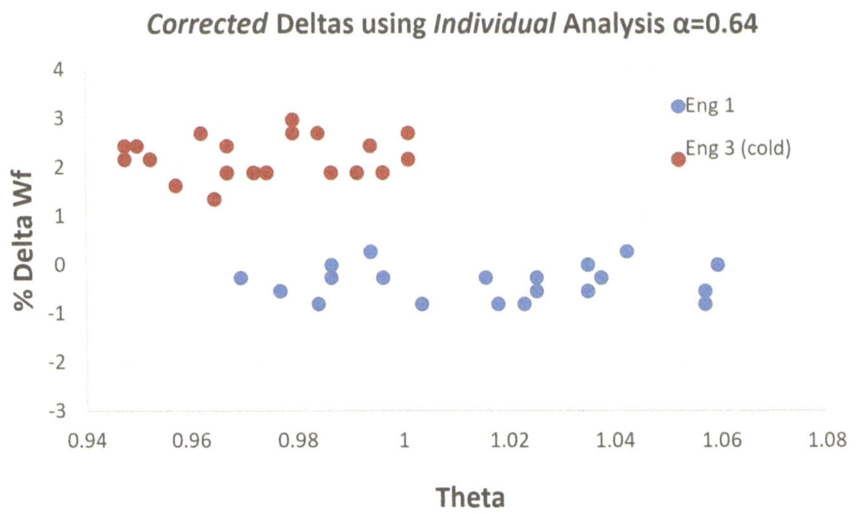

FIGURE 3.24 Case 3 individual data – corrected delta w_f

Before leaving this chapter, we will briefly discuss another adjustment that is important to ground-based applications: the effect of humidity on gas path parameters.

3.6 HUMIDITY CORRECTIONS

It is well known that high humidity levels in the ambient air can affect engine performance, [4–7]. Because measured gas path quantities are used in evaluating engine performance deterioration, the omission of the effect of humidity essentially introduces increased uncertainty in subsequent diagnostic calculations. Just as ambient (inlet) temperature and pressure are used to adjust gas path parameters before performance diagnostics are performed (standard day corrections), a similar correction process is possible if the level of humidity is known. For the most part, humidity levels are not measured in aero-engines during flight, so much of this discourse is most meaningful to ground testing or for land-based gas turbines. The following provides some insight into the magnitude of the effects encountered when humidity levels are present and measured. The approach we will offer differs from those already discussed in this chapter for standard day corrections and depend solely on the effect humidity has on *gas properties*, in particular on four parameters, namely, the specific heats at constant pressure and volume (cp and cv), their ratio (γ), and the gas constant R.

To facilitate our discussion, we require several definitions:

- Water to air ratio: $war = \dfrac{water\ mass}{dry\ air\ mass} = \dfrac{mass_w}{mass_{a,d}}$

- Specific humidity: $SH = \frac{mass_w}{total\ mass} = \frac{mass_w}{(1+war)mass_{a,d}} = \frac{war}{1+war}$

Additionally, we will define additional notation for specific heats as follows:

$$cp_{a,d} = \text{specific heat of dry air at constant pressure}$$
$$cp_w = \text{specific heat of water (vapor) at constant pressure}$$
$$cp_{a,w} = \text{specific heat of wet air at constant pressure}$$

These are associated by the following relationship given in Equation (3.43).

$$\begin{aligned} cp_{a,w} &= \left(\frac{mass_{a,d}}{total\ mass}\right)cp_{a,d} + \left(\frac{mass_w}{total\ mass}\right)cp_w \\ &= \left(\frac{1}{1+war}\right)cp_{a,d} + \left(\frac{war}{1+war}\right)cp_w \end{aligned} \tag{3.43}$$

Therefore, in terms of specific humidity, this becomes

$$cp_{a,w} = (1-SH)\,cp_{a,d} + SH\,cp_w \tag{3.44}$$

Additionally, we can write

$$\frac{dcp}{cp} = \frac{\Delta cp}{cp_{a,d}} \equiv \frac{cp_{a,w} - cp_{a,d}}{cp_{a,d}} = \frac{(1-SH)\,cp_{a,d} + SH\,cp_w - cp_{a,d}}{cp_{a,d}} = \frac{SH\,cp_w - SH\,cp_{a,d}}{cp_{a,d}}$$

With a little manipulation, we arrive at the expression

$$\frac{dcp}{cp} = SH\left(\frac{cp_w}{cp_{a,d}} - 1\right) \tag{3.45}$$

Equation (3.45) is one of the fundamental relationships that we will leverage in developing the association between the specific heat of wet air ($cp_{a,w}$) and the specific heat of dry air ($cp_{a,d}$). Likewise, we will need to develop analogous relationships for the gas constant, R, and the ratio of specific heats, γ. We define

$$R_{a,d} = \text{gas constant of dry air}$$
$$R_w = \text{gas constant of water (vapor)}$$
$$R_{a,w} = \text{gas constant of wet air}$$

These are associated by the following relationship given in Equation (3.46).

$$R_{a,w} = \left(\frac{mass_{a,d}}{total\ mass}\right)R_{a,d} + \left(\frac{mass_w}{total\ mass}\right)R_w = (1-SH)\,R_{a,d} + SH\,R_w \tag{3.46}$$

Additionally,

$$\frac{dR}{R}=\frac{\Delta R}{R_{a,d}}\equiv\frac{R_{a,w}-R_{a,d}}{R_{a,d}}=\frac{(1-SH)R_{a,d}+SH\,R_w-R_{a,d}}{R_{a,d}}=\frac{SH\,R_w-SH\,R_{a,d}}{R_{a,d}}$$

Arriving at the expression

$$\frac{dR}{R}=SH\left(\frac{R_w}{R_{a,d}}-1\right) \tag{3.47}$$

We need one more relationship involving the ratio of specific heats, $\gamma=cp/cv$. We start with a familiar relationship, i.e.,

$$\frac{R}{cp}=\frac{cp-cv}{cp}=1-\frac{cv}{cp}=1-\frac{1}{\gamma}=\frac{\gamma-1}{\gamma} \tag{3.48}$$

If we employ our favorite manipulation (outlined in A2.1 of the Appendix) by taking logs of both sides and differentiating, we obtain

$$\frac{dR}{R}-\frac{dcp}{cp}=\frac{d\gamma}{\gamma-1}-\frac{d\gamma}{\gamma}=\left(\frac{\gamma}{\gamma-1}\right)\frac{d\gamma}{\gamma}-\frac{d\gamma}{\gamma}=\left(\frac{\gamma}{\gamma-1}-\frac{\gamma-1}{\gamma-1}\right)\frac{d\gamma}{\gamma}=\left(\frac{1}{\gamma-1}\right)\frac{d\gamma}{\gamma}$$

(3.49)

Some rearranging of terms yields the final relationship

$$\frac{d\gamma}{\gamma}=(\gamma-1)\left(\frac{dR}{R}-\frac{dcp}{cp}\right) \tag{3.50}$$

The three expressions, Equations (3.45), (3.47), and (3.50) will provide the foundation upon which to build a suitable correction for the humidity effect on gas path parameters. We list them together in Equation (3.51).

$$\begin{aligned}\frac{dcp}{cp}&=SH\left(\frac{cp_w}{cp_{a,d}}-1\right)\\ \frac{dR}{R}&=SH\left(\frac{R_w}{R_{a,d}}-1\right)\\ \frac{d\gamma}{\gamma}&=(\gamma-1)\left(\frac{dR}{R}-\frac{dcp}{cp}\right)\end{aligned} \tag{3.51}$$

Furthermore, if we evaluate the expressions in Equation (3.51) using the following standard values for these gas parameters, i.e., $cp_{a,d}$= 1004 J/kg *deg K*, cp_w= 1859 J/kg *deg K*, $R_{a,d}$= 287 J/kg *deg K*, and R_w= 461 J/kg *deg K*, we would arrive at the approximate relationships given below.

$$\frac{dcp}{cp} = 0.852\,SH$$
$$\frac{dR}{R} = 0.61\,SH \qquad (3.52)$$
$$\frac{d\gamma}{\gamma} = -0.0968\,SH$$

Equation (3.52) quantifies the extent to which the basic gas properties of specific heat, gas constant, and ratio of specific heats vary once we know the specific humidity. Our ultimate goal is to determine a correction factor that will transform a sensed gas path parameter that was captured in a humid environment to the *equivalent value* it would have under *dry conditions*. To be specific, if X represents the gas path parameter in question, the correction will be multiplicative factor μ_X, such that

$$X_{dry} = X_{wet}\,\mu_X \qquad (3.53)$$

wherein, X_{wet} is the observed value under humid conditions, X_{dry} its equivalent value under dry conditions, and $\mu_X = f(cp, R, \gamma) = f(SH)$ by virtue of Equation (3.52). The actual form of μ_X can be determined by the following identity and a little manipulation using a generic symbol Z.

$$\frac{Z_{a,w}}{Z_{a,d}} = 1 + \frac{Z_{a,w}}{Z_{a,d}} - \frac{Z_{a,d}}{Z_{a,d}} = 1 + \frac{dZ}{Z} \quad \Rightarrow \quad \frac{Z_{a,d}}{Z_{a,w}} = \frac{1}{\left(1 + \frac{dZ}{Z}\right)} \qquad (3.54)$$

Now applying Equation (3.52) individually to Z in Equation (3.54) yields the following relationships.

$$\frac{\gamma_{a,d}}{\gamma_{a,w}} = \frac{1}{[1 - 0.0968\,SH]}$$
$$\frac{R_{a,d}}{R_{a,w}} = \frac{1}{[1 + 0.61\,SH]} \qquad (3.55)$$
$$\frac{cp_{a,d}}{cp_{a,w}} = \frac{1}{[1 + 0.852\,SH]}$$

The form and values for μ_X will be combinations of the terms appearing in Equation (3.55). As an illustration, consider engine rotor speed N. We know from Equation (3.22) that

$$v = r\,N = M_n\sqrt{\gamma\,RT}$$

After some manipulation and assuming $M_n = const$, we can assert

$$\left(\frac{N}{\sqrt{T}}\right)_{a,w} = \left(\frac{N}{\sqrt{T}}\right)_{a,d} \sqrt{\frac{\gamma_{a,w}}{\gamma_{a,d}} \frac{R_{a,w}}{R_{a,d}}} \tag{3.56}$$

Combining Equations (3.55) and (3.56) yields

$$\left(\frac{N}{\sqrt{T}}\right)_{a,d} = \left(\frac{N}{\sqrt{T}}\right)_{a,w} \left(\frac{1}{\sqrt{[1-0.0968\,SH][1+0.61\,SH]}}\right) \tag{3.57}$$

TABLE 3.10
Summary of Humidity Corrections

Parameter	Symbol	Humidity Correction	Relationship
Spool speed	$\frac{N}{\sqrt{\theta}}$	$\frac{1}{\sqrt{[1-0.0968\,SH][1+.61\,SH]}}$	μ_N
Air flow	$\frac{w_a\sqrt{\theta}}{A\delta}$	$\sqrt{\frac{1+0.61\,SH}{1-.0968\,SH}}$	μ_{w_a}
Enthalpy	$\frac{\Delta h}{T}$	$\frac{1}{(1-0.0968SH)(1+0.61SH)}$	$\mu_{\Delta h}$
Δ Temperature	$\frac{\Delta T}{T}$	$\frac{(1+0.852SH)}{(1-0.0968SH)(1+0.61SH)}$	$\frac{cp_{a,w}}{cp_{a,d}}\mu_{\Delta h}$
Horsepower	$\frac{SHP}{\sqrt{\theta}\,\delta}$	$\frac{1}{\sqrt{(1+0.61SH)}\,(1-0.0968SH)^{1.5}}$	$\frac{cp_{a,d}}{cp_{a,w}}\mu_w\,\mu_{\Delta T/T}$
Torque	$\frac{Q}{\delta}$	$\frac{1}{(1-0.0968SH)}$	$\frac{\mu_{SHP}}{\mu_N}$
Acceleration	$\frac{\dot{N}}{\delta}$	$\frac{1}{(1-0.0968SH)}$	μ_Q
Fuel flow	$\frac{w_f}{\sqrt{\theta}\,\delta}$	$\frac{1}{(1-0.0968SH)^{1.5}\sqrt{(1+0.61SH)}}$	$\mu_{w_f} = \mu_w\,\mu_{\Delta h}$
Thrust	$\frac{F_n}{\delta}$	$\frac{1}{(1-0.0968SH)}$	μ_{F_N}
Pressure	$\frac{P}{\delta}$	$\frac{1}{(1-0.0968SH)}$	$\mu_P = \mu_{F_N}$
Temperature	$\frac{T}{\theta}$	$\left(\frac{1}{(1-0.0968SH)}\right)^{1/3.5}$	μ_T

Thus, we have shown that the multiplicative factor to correct for humidity effects on rotor speed is

$$\mu_N = \left(\frac{1}{\sqrt{[1-0.0968\,SH\,][1+0.61\,SH\,]}} \right) = f(SH)$$

We can multiply both sides of Equation (3.57) by $\sqrt{T_{ISA}}$ to obtain the final form given in Equation (3.58).

$$\left(\frac{N}{\sqrt{\theta}} \right)_{a,d} = \left(\frac{N}{\sqrt{\theta}} \right)_{a,w} \left(\frac{1}{\sqrt{[1-0.0968\,SH\,][1+0.61\,SH\,]}} \right) \tag{3.58}$$

Because we will not be making direct use of humidity corrections for the remainder of this text, we will forego providing any further derivations and simply supply a list of known humidity corrections (borrowed from [1]) in Table 3.10.

As a final note, for aero-engine data obtained from ground runs or power generation land-based gas turbine data, the most *general form for a corrected parameter* (extending Equation (3.5)) for arbitrary gas path parameter, X, becomes

$$X_c = \frac{X}{\theta^{\alpha}\,\delta^{\beta}}\,\mu_X \quad where \quad \theta = \frac{T_2}{T_{ISA}}, \quad \delta = \frac{P_2}{P_{ISA}}, \quad \mu_X = f(SH) \tag{3.59}$$

The avid reader can find additional information of interest on parameter corrections in the following references [8–10].

NOTES

1. There is no clear evidence to suggest that this assumption is true except our strong desire to simplify the inter-relationships and arrive at a palatable approximation to use in calculations. It is more probable that these partials vary. The hope is that they do not vary enough to invalidate our usage of the resultant *corrected parameter* formulation. Fortunately, decades of usage offer a form of vindication to this engineering approximation.
2. See Appendix: Chapter 3 Addendum A3.1
3. A complete set of derivations and refinements can be found in [1].
4. For simplicity we will also assume η to be constant
5. Because ratio is constant $\forall\delta$, it has the same value as it does when δ =1 on a standard day.
6. The form of the model is not germane to our exposition and can take any form whatsoever. Its function here is merely to act as a reference from which parameter Δs can be calculated.
7. See Appendix A2.2
8. Parameter % deltas (ΔX) are *corrected* quantities by virtue of A3.6 of the Appendix.
9. Unfortunately, the use of real engine data for this purpose is not possible due to proprietary and data ownership issues.

10. For brevity, we omit plots of the other four parameters cited, as they are graphically similar.
11. The level of deterioration is inconsequential in this exercise, i.e., a highly deteriorated engine simulation would have produced essentially identical results.
12. Eng 1 is the same deterioration configuration used to produce Table 3.5 results.
13. Latin translation: "garbage in, garbage out."

REFERENCES

1. Volponi, A., *Gas Turbine Parameter Corrections*, Springer Publications, ISBN 978-3-030-41075-9, March 2020.
2. Buckingham, E., On physically similar systems: illustrations of the use of dimensional equations, *Physical Review*, Vol 4, pp. 345–376, 1914.
3. Liu, Y., Frederick, D. K., DeCastro, J. S., and Chan, W. W., *User's Guide for the Commercial Modular Aero-Propulsion System Simulation (C-MAPSS)*, NASA/TM—2012-217432, 2012.
4. Samuel, J.C., and Gale, B.M., Effect of humidity on performance of turbojet engines, *National Advisory Committee for Aeronautics*, Technical Note 2119, June 1950.
5. Mathioudakis, K., and Tsalavoutas, A., Uncertainty reduction in gas turban performance diagnostics by accounting for humidity effects, *Proceedings of 2003 ASME TURBO EXPO: Power for Land, Sea, & Air*, paper GT2001-0010, June 4–7, 2001, New Orleans, USA.
6. AGARD, Recommended practices for the assessment of humidity effects of atmospheric water ingestion on the performance and operability of gas turbine engines, *AGARD Advisory Report* No 332, September 1995.
7. Hanachi, Houman, et.al., Effects of the intake air humidity on the gas turbine performance monitoring, *Proceedings of 2015 ASME TURBO EXPO: Power for Land, Sea, & Air*, paper GT2015-43026, June 15–19, 2015, Montreal, Canada.
8. Sanders, N. D., Performance parameters for jet-propulsion engines, NACA Technical Note No. 1106, 1946.
9. Kurzke, J., 2003, Model based gas turbine parameter corrections, *Proceedings of 2003 ASME TURBO EXPO: Power for Land, Sea, & Air*, paper GT2003-38234, June 16–19, 2003, Atlanta, Georgia, USA
10. Assadi, M, et al., A novel correction technique for simple gas turbine parameters, *Proceedings of 2003 ASME TURBO EXPO: Power for Land, Sea, & Air*, paper GT2001-0009, June 4–7, 2001, New Orleans, USA.
11. Volponi, A.J., Gas turbine parameter corrections, *Transactions of the ASME, Journal of Engineering for Gas Turbines and Power*, Vol 121, No 4, pp. 613–621, October 1999.

4 Gas Path Analysis Fundamentals

4.1 OVERVIEW

Most performance diagnostic methods rely on discernable changes in observable parameters in order to detect physical faults. In general, physical faults consist of combinations of problems such as foreign object damage (FOD), blade erosion and corrosion, worn seals, excess clearances, or plugged nozzles, which result in changes in the thermodynamic performance of the engine as measured by adiabatic efficiencies, compressor flow capacities, and effective nozzle areas, etc. These thermodynamic changes, in turn, produce changes in observable engine parameters measured along the engine gas path, such as rotor speeds, temperatures, pressures, fuel flow, power output, and the like. The fundamental task of GPA is to infer from these observable shifts the level of deterioration in the various modules of the engine, thus allowing correction of the underlying physical faults through proper maintenance action.

In this chapter, we will formally introduce GPA and its mathematical framework. The presentation will provide some historical perspective and proceed in a chronological fashion following the methodology's advancement in the first few decades of its development. With nearly five decades of evolution, there have been numerous approaches proposed by researchers and practitioners from academic, government, and industrial sectors. Fortunately, the majority of these have a common thread that is congruent with the approach originally taken by Louis A. Urban,[1] which is the approach we will be taking and expanding upon in this book. The truly *alternative* methodologies will be discussed briefly in Chapter 5.

The focus of this chapter will be on the GPA of a single data point *snapshot* at steady-state conditions. This will serve as a foundation to address time series processing and ultimately non-steady-state analysis in the chapters to follow. This chapter will leverage the concepts and results presented in the preparatory material of the first three chapters and their associated addendums.

4.2 FUNDAMENTAL CONCEPTS

Amongst those intimately familiar with GPA, there would, no doubt, be little disagreement that its modern-day genesis began with Urban's pioneering work in the early 1970s [1–3]. In truth, one might argue that its roots can be traced to one of his earlier works [4], a privately published monograph, that (in Urban's own words) "was intended as a working tool to allow those who were not necessarily performance

DOI: 10.1201/9781003649274-4

specialists, to get good first order approximate solutions to most performance questions." The differential methods introduced in that monograph would later become essential ingredients in the modeling elements inherent in the GPA methodology. Before we launch into a mathematical description, we begin with the fundamental concept of GPA as proposed by Urban in the '70s.

As mentioned in the Overview section, this type of diagnostic method relies on discernable *changes* in observable parameters in order to detect physical faults occurring in the engine's major modules and sub-systems. The basic tenet underlying this approach is that physical faults occurring in the engine (such as blade erosion, corrosion, tip clearance, fouling, etc.) induce a *change* in component performance (as modeled by efficiencies, flow capacities, etc.), which, in turn, produce observable *changes* in measurable parameters along the engine's gas path (such as temperatures, pressures, speeds, etc.). Through inverse relationships, it should be possible to estimate the component shifts responsible for the measurement shifts observed that, in turn, provide information needed to address the underlying physical fault(s). This basic concept is depicted in Figure 4.1,which is a digital scan of Lou Urban's original overhead slide which he used in all of his GPA presentations. This figure has been redrawn and used by countless authors over the last 40 years in hundreds of GPA related papers.

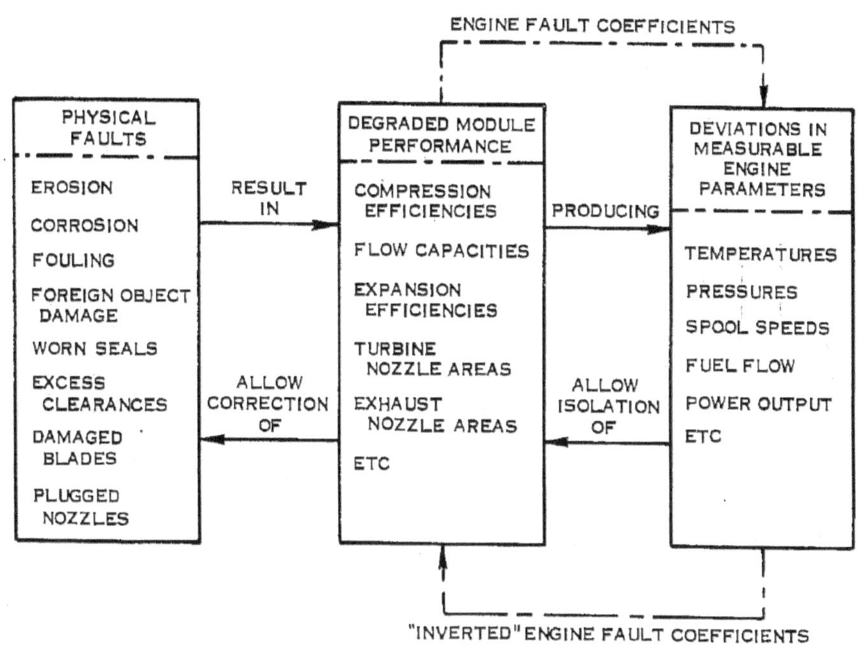

FIGURE 4.1 GPA principle.

This type of diagnostic can be referred to as *relative* engine performance diagnostics, in that it assesses *changes* in engine performance, *not absolute* performance levels. This makes it possible to track performance of a particular engine against itself (say from when the engine was installed on an aircraft) or how a particular engine has changed relative to some production level or fleet average. Whatever the base reference point, valuable information can be obtained that can directly impact the maintenance schedule, logistics, and cost.

Historically, the general approach taken was through the use of a linearized model approximation evaluated at a selected engine operating point. This provides a matrix relationship between changes in engine component performance (*independent* parameters) and the attendant changes in the typically measured engine parameters. The GPA problem then reduces to one of estimating the fault vector from knowledge of the measurement deltas from reference. While this might seem like a simple task, there are many difficulties encountered, not the least of which is that, in practice, the number of unknowns is far greater than the number of equations, as well as observability issues between certain faults amongst the measured quantities, poor signal to noise ratio, engine non-linearity effects, and a host of others. For this reason, the estimation problem has been the focus of considerable research over the past several decades, with methodologies incorporating a wide range of techniques ranging from Kalman filters, artificial neural networks, fuzzy logic, Bayesian belief networks, genetic algorithms, and combinations of the above. In the course of this chapter and the remaining chapters of the book, we will provide some detail on these methods, although some provide greater depth than others.

4.2.1 Calculating Measurement Δs

In order to infer changes in these performance characteristics, we must be able to observe discernible *changes* (from reference) in various engine parameters taken along the engine's gas path and have a mathematical model inter-relating the two sets of variables. Historically (in the '70s) the only gas path parameters available were those already resident on the engine for control purposes. That list would vary somewhat with engine model and type, and, as mentioned in previous chapters, the set of sensors would grow with time as EHM took hold and its benefits realized. For purposes of discussion, we can assume it to be some sub-set of sensors depicted in the Figure 1.1 schematic. We can generate, from m steady-state observations, an $(m \times 1)$ *measurement delta vector, Z*, as given in Equation (4.1).

$$Z = \begin{bmatrix} z_1 \\ z_2 \\ z_3 \\ z_4 \\ z_5 \\ \vdots \\ z_m \end{bmatrix} = \begin{bmatrix} \Delta N_1 \\ \Delta N_2 \\ \Delta w_f \\ \Delta T_3 \\ \Delta P_3 \\ \vdots \\ \Delta T_5 \end{bmatrix} = \begin{bmatrix} \Delta\ \textit{Low Spool Speed} \\ \Delta\ \textit{High Spool Speed} \\ \Delta\ \textit{Fuel Flow} \\ \Delta\ \textit{HPC Exit Temperature} \\ \Delta\ \textit{HPC Exit Pressure} \\ \vdots \\ \Delta\ \textit{Exhaust Gas Temperature} \end{bmatrix} = 100 \begin{bmatrix} (N_{1(meas)} - N_{1(ref)}) / N_{1(ref)} \\ (N_{2(meas)} - N_{2(ref)}) / N_{2(ref)} \\ (w_{f(meas)} - w_{f(ref)}) / w_{f(ref)} \\ (T_{3(meas)} - T_{3(ref)}) / T_{3(ref)} \\ (P_{3(meas)} - P_{3(ref)}) / P_{3(ref)} \\ \vdots \\ (T_{5(meas)} - T_{5(ref)}) / T_{5(ref)} \end{bmatrix} \tag{4.1}$$

Equation (4.1) clearly indicates a *% of point delta* calculation. These deltas are expressed as a percent (in lieu of engineering units) as a matter of convenience. Doing so creates uniformity in the numerical values that will contribute to the overall numerical stability of the (inverse) estimation process that will be described in this Chapter and Chapter 5. We refer the reader back to Chapter 2 and its associated Addendums for a complete explanation as to why this particular form of delta was chosen.

Although Equation (4.1) appears deceptively simple, some care and caution are advised. One must be mindful of measurement uncertainty and the degree to which it can be mitigated as well as which adjustments are appropriate. This depends in part on the type and form of the *reference* that is being used. Historically, the reference was a set of gas generator curves at standard day conditions. For the moment, let us assume that is the form of the reference. In that case, the gas path parameter observations would need to be adjusted by applying standard day corrections to the raw data prior to any delta calculation. Depending on where the data is being acquired, there are other physical effects that may need to be accounted for before a measurement delta is calculated. These consist of things such as (but not necessarily limited to) Reynold's effects, Mach number effects, scheduled bleed effects, and installation effects (for in-flight acquired data) and test cell effects for ground acquired data. In short, the difference between our observed value and the reference contains a myriad of possible contributors, many of which can be removed through adjustments. Failure to do so will introduce added uncertainty in the resultant measurement delta, which, in turn, will add uncertainty to our estimation of engine module performance change. Borrowing another of Urban's original GPA slides, Figure 4.2 depicts the general situation in calculating the measurement delta.[2]

To express this mathematically, we let $\{Raw_i\}_{i=1}^{m}$ represent the gas path parameter values in our (steady state) snapshot observation at ambient condition $\{\theta, \delta\}$ with $\{Corr_i\}_{i=1}^{m}$ being the associated standard day values $Corr_i = Raw_i / \theta^{\alpha_i}\delta^{\beta_i}$, (see Chapter 3 and accompanying Addendums), and $\{Base_i\}_{i=1}^{m}$ are the baseline reference values from our gas generator curves. Figure 4.2 and Equation (4.1) can be (alternatively) expressed as

$$\begin{aligned} Corr_i &= \underbrace{Base_i\left(1+\frac{Adjust\Delta_1}{100}\right)\cdots\left(1+\frac{Adjust\Delta_K}{100}\right)}_{Base_i^*}\left(1+\frac{z_i}{100}\right) \\ &= Base_i^*\left(1+\frac{z_i}{100}\right) \quad for\ \ i=1,2,\ldots,m \end{aligned} \tag{4.2}$$

In effect, we are adjusting the reference to match the *conditions* under which the data point was collected. Naturally, the type and number of adjustments needed depends on conditions associated with the data as well as the reference being employed. Using simple gas generator curves as the reference, as in the historical case, would require the most adjustments, whereas using a sophisticated non-linear aero-thermal engine model, such as numerical propulsion system simulation (NPSS),[3] for example, would

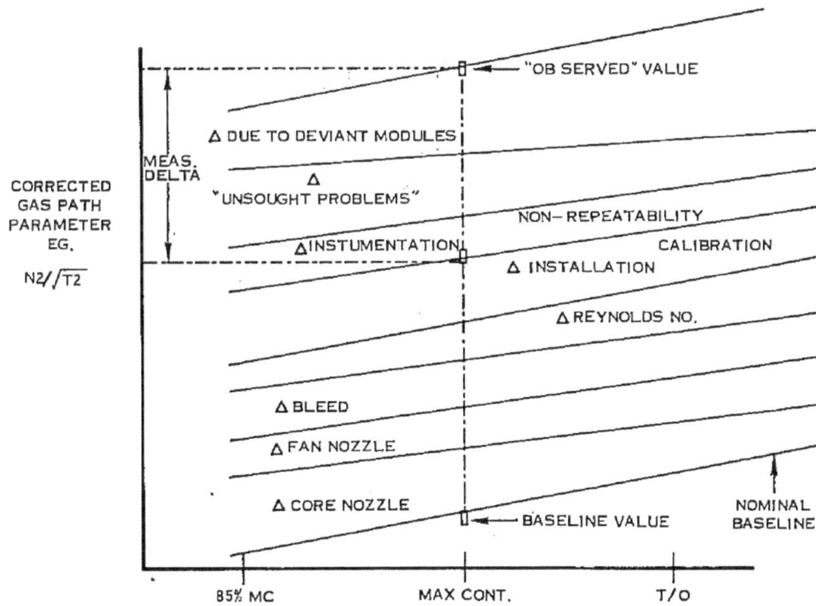

FIGURE 4.2 Calculating measurement deltas.

require far less. In this latter case the (NPSS) model can be run at the same ambient condition as the data collected, thereby negating the need to apply standard day correction to the data. If we assume, for simplicity, that only an installation calibration adjustment ($Inst\Delta_i$) is necessary, then Equation (4.2) would further simplify to the following:

$$Raw_i = Base_i^{(NPSS)}\left(1+\frac{Inst\Delta_i}{100}\right)\left(1+\frac{z_i}{100}\right) \quad for \ \ i=1,2,\ldots,m \tag{4.3}$$

The takeaway from this and Figure 4.2 is that:

1. The *conditions* under which the *data* are collected and the *conditions* that the chosen form of *reference* can represent dictate the adjustments that need to be applied before the measurement deltas are calculated.
2. The measurement deltas calculated from a properly adjusted reference contain the effects of not only the sought after module performance faults alluded to in Figure 4.1 but also a host of other uncertainties and errors, such as:
   - Instrumentation bias errors,
   - Measurement non-repeatability errors,
   - Unsought engine system problems, and
   - Modeling and adjustment errors.

This last point cannot be stressed enough. It is an affirmation of the potential estimation uncertainty that we will face in the GPA process. Before we can discuss these effects, their impact, and potential mitigation, we continue our presentation of the classical GPA mathematical model in the next section.

4.3 DEVELOPMENT OF THE MATHEMATICAL MODEL

Guided by Figure 4.1, we will evolve our basic model by defining the *independent* variables, which will constitute the *states* of the model, as the changes in thermodynamic performance of the major modules of the engine as characterized by *delta*[4] adiabatic efficiencies ($\Delta\eta$), flow capacities (ΔFC), and nozzle areas (ΔA).We will represent this collection as a $(n_e \times 1)$ vector x_e. For example, using our hypothetical engine configuration of Figure 1.1, we can choose the following engine fault parameters to model (where $n_e = 10$).

$$x_e = \begin{bmatrix} \Delta\eta_{FAN} \\ \Delta FC_{FAN} \\ \Delta\eta_{LPC} \\ \Delta FC_{LPC} \\ \Delta\eta_{HPC} \\ \Delta FC_{HPC} \\ \Delta\eta_{HPT} \\ \Delta A_4 \\ \Delta\eta_{LPT} \\ \Delta A_{45} \end{bmatrix} = \begin{bmatrix} \Delta\ \textit{Fan Efficiency} \\ \Delta\ \textit{Fan Flow Capacity} \\ \Delta\ \textit{Low Compressor Efficiency} \\ \Delta\ \textit{Low Compressor Flow Capacity} \\ \Delta\ \textit{High Compressor Efficiency} \\ \Delta\ \textit{High Compressor Flow Capacity} \\ \Delta\ \textit{High Turbine Efficiency} \\ \Delta\ \textit{High Turbine Nozzle Area} \\ \Delta\ \textit{Low Turbine Efficiency} \\ \Delta\ \textit{Low Turbine Nozzle Area} \end{bmatrix} \tag{4.4}$$

4.3.1 Engine Influence Coefficients

The observed changes in the m gas path parameters (*dependent* variables) have already been defined as the $(m \times 1)$ vector Z in Equation 4.1. Although the relationship between these quantities (x_e , Z) is, in general, a non-linear one, our basic model will be a *linear approximation* of this relationship. At a given engine operating point, this will take the form of an $(m \times n_e)$ matrix of *influence coefficients* (ICs).[5] The matrix of influence coefficients consists of partial derivatives inter-relating each *independent* parameter, with each *dependent* parameter holding some other parameter constant. Typically, the parameter held constant is a control power setting parameter such as corrected Fan speed or *engine pressure ratio* (*EPR*); however, there are occasions when other variables are used (consult Addendum A4.4 for details). As an example of an influence coefficient, the influence of Fan efficiency on N_1 at constant EPR would be the partial derivative

$$IC(N_1 \mid \eta_{FAN})_{EPR=const} \triangleq \left.\frac{\partial N_1 / N_1}{\partial\eta_{FAN} / \eta_{FAN}}\right|_{EPR=const}$$

	Fan Eff	Fan FC	LPC Eff	LPC FC	HPC Eff	HPC FC	HPT Eff	A4	LPT Eff	A45
N1	0.170	-0.339	0.050	-0.144	-0.019	-0.004	-0.027	0.007	0.304	-0.170
N2	-0.084	0.138	-0.110	-0.151	0.253	-0.277	0.362	-0.128	-0.063	0.223
T25	0.155	-0.283	-0.219	0.240	-0.171	-0.037	-0.242	0.061	0.223	-0.247
T5	-0.192	0.340	-0.219	-0.404	-0.587	-0.126	-0.830	0.209	-0.577	-0.127
Ps3	-0.015	-0.054	-0.004	0.052	0.049	-0.017	0.242	-1.117	0.167	0.046
T3	-0.038	0.057	-0.200	-0.087	-0.399	-0.057	0.200	-0.322	0.010	0.097
P25	0.575	-1.089	-0.012	1.061	-0.747	-0.167	-1.051	0.273	0.803	-0.958
Wf	-0.297	0.436	-0.128	-0.539	-0.534	-0.121	-0.765	0.207	-0.111	-0.326

FIGURE 4.3 Sample engine influence coefficient matrix H_e

This quantity expresses the effect (at constant EPR) of a percent change in Fan efficiency relative to a percent change in N_1, in terms of percent. As a linear approximation, we have the relationship

$$\Delta N_1 \equiv \frac{dN_1}{N_1} \approx \underbrace{\left(\left. \frac{\partial N_1 / N_1}{\partial \eta_{FAN} / \eta_{FAN}} \right|_{EPR=const} \right)}_{IC(N_1|\eta_{FAN})_{EPR=const}} \frac{d\eta_{FAN}}{\eta_{FAN}} \Rightarrow \tag{4.5}$$

$$\Delta N_1 \approx IC(N_1 \,|\, \eta_{FAN})_{EPR=const} \, \Delta\eta_{FAN}$$

If we were to evaluate these partial derivatives for each independent–dependent parameter combination at a specific engine operating condition, we would generate a matrix of engine influence coefficients, which we will denote by H_e. Figure 4.3 provides an illustrative example generated by the engine simulation with sea level, static, standard day conditions at constant EPR.

The colloquial interpretation of the matrix is that each column represents the effect that a +1% change in that independent variable has on the measurement deltas. For example, in the first column, it states that a +1% change in the Fan efficiency *map* (at the operating point that the matrix was generated) will (after rematch) result in a +0.170% change in ΔN_1, a -0.084% change in ΔN_2, a +0.155% change in ΔT_{25}, and so forth. As explained in Addendum A4.4, the perturbations to produce a column of this matrix are introduced

- At steady state conditions;
- In a closed loop high fidelity simulation, i.e., with the engine control in the loop so that both engine rematch and control effects are captured in the resultant measurement deltas; and
- At the *component performance map level* to model a change in engine component hardware and not just a change due to rematch.

The last bullet may seem like a nuance, but it is an important nuance. For example, if a handling bleed at Station 25 were to fail open (at steady state), the HPC operating line would shift down away from surge and create a new operating point on the HPC map (and, hence, would be at a different operating efficiency). The delta HPC efficiency realized is NOT what is represented by the column parameter *HPC Eff.* The reason is that a broken bleed valve does not implicate (necessarily) a performance fault in the HPC. Broken bleed valves are certainly an issue worthy of diagnostic scrutiny, and we will address that later in the GPA evolutionary development.

To illustrate this nuance, consider an artist rendition of a Fan outer diameter (OD) compressor map for our hypothetical engine of Figure 1.1, given below in Figure 4.4.

The (operational) adiabatic efficiency for this compressor (η_{Fan}), at a given operating point, is given (functionally) as

$$\eta_{Fan} = f(\ N_1 / \sqrt{\theta_2}\ ,\ P_{15} / P_2\ ,\ \eta_{Fan}|_D\ ) \tag{4.6}$$

where $\eta_F|_D$ represents the design efficiency, i.e., the efficiency of the *nominal* engine at that operating point given the compressor characteristic depicted in Figure 4.4. This (temporary) change of notation dictates that Equation (4.5) is more accurately stated as

$$\Delta N_1 \equiv \frac{dN_1}{N_1} \approx \underbrace{\left(\left. \frac{\partial N_1 / N_1}{\partial \eta_{FAN}|_D / \eta_{FAN}|_D} \right|_{EPR=const} \right)}_{IC(N_1|\eta_{FAN}|_D)_{EPR=const}} \frac{d\eta_{FAN}|_D}{\eta_{FAN}|_D} \Rightarrow$$
$$\Delta N_1 \approx IC(N_1 | \eta_{FAN}|_D)_{EPR=const}\ \Delta\eta_{FAN}|_D \tag{4.7}$$

With this understanding, the first column of H_e in Figure 4.3 represents $Fan\ Eff = \Delta\eta_{Fan}|_D$. Similar arguments can be applied to the other columns of H_e, and Addendum 4.4 provides some of the detail. With this in mind, it is (hopefully) clearer that what is being represented is the map change depicted (in red) in Figure 4.5 for a deteriorated compressor.

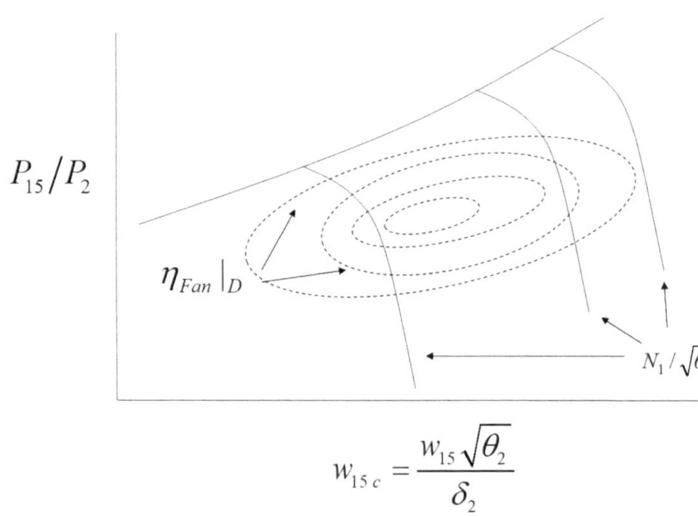

FIGURE 4.4 Hypothetical Fan OD compressor map.

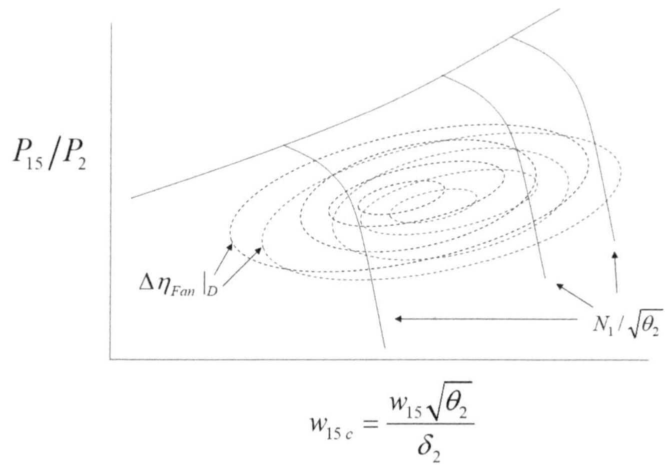

FIGURE 4.5 Hypothetical Fan OD compressor map change with deterioration.

The mathematical descriptions that follow in this text are complicated enough to digest without this change of notation. For the sake of clarity and simplification, we will simply refer to these changes as $\Delta\eta_{Fan}$ etc., now that the reader understands what, in truth, is being represented.

Before leaving this topic, we must mention that there is one final nuance that has been implicitly introduced in the foregoing discussion that the reader should be aware of. It has been implicitly assumed that the changes in (map) efficiency are *uniform across the entire map.* In truth, deterioration would, no doubt, be non-uniform. If we were making just one snapshot assessment, it would not matter; however, to trend a collection of *snapshots* performed at varying operating points or assess a time series of data points (next chapter), this assumption is required. Without this assumption, meaningful GPA could not be performed.

4.3.2 Basic Model

In its most *basic* form, we have the following (linear approximate) relationship for Figure 4.1.

$$Z = H_e x_e + \varphi \tag{4.8}$$

where Z is the $(m \times 1)$vector[6] of measured parameter deltas, x_e is an $(n_e \times 1)$ vector of (engine module) fault deltas, H_e is an $(m \times n_e)$ matrix of engine fault (influence) coefficients and φ is an $(m \times 1)$ random vector representing the non-repeatabilities inherent in the measurement Δ process. It is typically assumed that φ has a Normal distribution with zero mean and covariance *R*, i.e., $\varphi \in N(0,R)$. Note that this simple formulation assumes that only module performance shifts (plus noise) are responsible for the observed parameter shifts. Later we will amend this formulation to include other performance faults (e.g., sensor faults). In the next few paragraphs, we summarize an analysis already detailed in Chapter 2 and its Addendums.

For this formulation of the problem, we can consider three possibilities:

1. $n_e = m$ number of unknowns is equal to the number of equations
2. $n_e < m$ number of unknowns is less than the number of equations
3. $n_e > m$ number of unknowns is greater than the number of equations

For the *first* possibility $(n_e = m)$, the matrix H_e is square, and if it has proper rank, it can be inverted to provide an estimate of the fault Δs, $\hat{x}_e$, i.e.,

$$\hat{x}_e = H_e^{-1} Z \tag{4.9}$$

For the *second* possibility, $(n_e < m)$, we have an over-determined situation and can solve the problem with a *least squares* solution, i.e.,

$$\hat{x}_e = (H_e^T H_e)^{-1} H_e^T Z \tag{4.10}$$

where H_e^T denotes the matrix *transpose*. Now, if in addition, we have knowledge of the measurement Δ non-repeatabilities (*R*), we can easily generate a *weighted least squares* solution for the engine faults, i.e.,

$$\hat{x}_e = (H_e^T R^{-1} H_e)^{-1} H_e^T R^{-1} Z \tag{4.11}$$

In this latter solution, the fault assessments are being determined by *weighing* the influence of each measurement in accordance with the reciprocal of its non-repeatability. Thus, accurate measurements provide a greater weight in terms of their influence on the fault estimation than inaccurate measurements.

For these three solutions, each of the matrices

$$H_e^{-1} \quad , \quad (H_e^T H_e)^{-1} H_e^T \quad , \quad (H_e^T R^{-1} H_e)^{-1} H_e^T R^{-1} \tag{4.12}$$

forms a set of *inverse* influence coefficients, which illustrates how a % of point change in each measured parameter affects a % of point change in each of the engine fault parameters. We will, in general, refer to such matrices as *diagnostic* matrices for the remainder of this book.

From, an historical perspective, these were the solutions attempted in the early '70s. From Figure 4.1 and Equation (4.4) we actually have $n_e = 10$, as we are modeling only performance faults for five modules using two independent parameters (efficiency and a flow parameter[7]) per module. At that time, only a handful of measured parameters were available ($m \leq 5$) which actually places us firmly in Category 3 above, with $(n_e > m)$. Thus, in order to use one of the diagnostic matrices represented in Equation (4.12), the early investigators would attempt to reduce the fault set (from 10) to a value that satisfied $(n_e \leq m \leq 5)$ so that some form of solution could be obtained. One method used was to perform a combinatorial assessment of sub-sets of faults of size n_e, from the original 10, that satisfied $(n_e \le m)$. The assessment was performed, in simulation, in a manner to yield a fault sub-set that maximized module coverage with minimal estimation error. This was a time-consuming and arduous task executed on mainframe computers available at that time. Another method used was to conjoin the two performance measures of each module (efficiency and flow parameter) into a *single* representative performance measure, thereby reducing the fault size to $n_e = 5$, which would allow one of the diagnostic matrices to be used when $m = 5$. This process is often referred to as *coupling* and is outlined in Addendum 4.5.

Neither of these solutions are particularly ideal for many reasons, not the least of which was the limitation imposed on the diagnostic coverage of the system, which left little to no room for future expansion. Like it or not, any suitable form of GPA imposes the constraint that $(n_e > m)$. Following the chronological development, an estimation method emerged in the late '70s to provide an estimate solution to Equation (4.8), which was initially dubbed the *probabilistic method* and later became known as the *Kalman filter method*, a label that it still carries today.

4.4 KALMAN FILTER APPROACH – SNAPSHOT ANALYSIS

4.4.1 Basic Solution

The basic linear model given by Equation (4.8) with $(n_e > m)$, as already noted, has no unique solution for x_e but, in fact, infinitely many. If we cast this as a probabilistic system, where *both* x_e and Z are random vectors with Normal distributions and are linearly related, as in Equation (4.8), then what we seek for our solution is an estimate $\hat{x}_e$ that is *optimal* in some sense. As you might suspect, there are many ways to define optimality. We invite the reader to consult Addendum A4.2 where we have defined a number of optimal estimators, derived some properties and relationships, and indicated that if everything is Normal distributed and a linear measurement system is in play (our Equation (4.8)), then all of them produce the same estimate! With that knowledge in hand, we can choose one and proceed with a brief derivation to the final solution. To begin, we will state all of our assumptions in Equation (4.13) below.

$$\begin{aligned} & Z = H_e\, x_e + \varphi \\ \textit{where}\quad & x_e = (n_e \times 1)\ \textit{fault}\ \Delta\ \textit{vector},\quad x_e \in N(\mu_{x_e}, P_{x_e}) \\ & \varphi = (m \times 1)\ \textit{noise vector},\quad \varphi \in N(0, R) \\ & Z = (m \times 1)\ \textit{measurement}\ \Delta\ \textit{vector} \\ & H_e = (m \times n_e)\ \textit{matrix of IC's} \\ & x_e\ \textit{and}\ \varphi\ \textit{independent} \end{aligned} \tag{4.13}$$

From the Normal distribution assumptions, we can write the probability densities as

$$p(x_e) = const\ e^{-\frac{1}{2}\left\{\left(x-\mu_{x_e}\right)^T P_{x_e}^{-1}\left(x-\mu_{x_e}\right)\right\}} \tag{4.14}$$

$$p(\varphi) = const\ e^{-\frac{1}{2}\{\varphi^T R^{-1} \varphi\}} \tag{4.15}$$

Because Z is a linear function of two Gaussian vectors, it is itself Gaussian with mean μ_Z and covariance P_Z, where

$$\begin{aligned} \mu_Z &= E(Z) = E(H_e\, x_e + \varphi) = H_e\, E(x_e) + E(\varphi) = H_e\, \mu_{x_e} \\ P_Z &= Var(H_e x_e + \varphi) = Var(H_e x_e) + Var(\varphi) = H_e\, P_{x_e}\, H_e^T + R \end{aligned} \tag{4.16}$$

Therefore, we can write the probability density of Z as

$$p(Z) = const\ e^{-\frac{1}{2}\left\{\left(Z - H_e\, \mu_{x_e}\right)^T \left(H_e\, P_{x_e}\, H_e^T + R\right)^{-1} \left(Z - H_e\, \mu_{x_e}\right)\right\}} \tag{4.17}$$

The optimal estimate $\hat{x}_e$ that we will choose to derive is the so-called *maximum a-posteriori* estimate, defined as the value that maximizes the conditional probability function $p(x \mid Z)$, i.e.,

$$\hat{x}_e = \hat{x}_{MAP} = \arg\max_x \left\{ p(x \mid z) \right\} \tag{4.18}$$

By definition

$$p(x_e \mid Z) = \frac{p(x_e, Z)}{p(Z)} = \frac{p(Z \mid x_e)\, p(x_e)}{p(Z)} \tag{4.19}$$

Because x_e and Z are Gaussian, $p(Z \mid x_e)$ is a Normal density; therefore, all we need is mean and covariance to establish the probability density, keeping in mind that, in this conditional, x_e is being held *fixed* so that $E(x_e) = x_e$, and we have from definition

$$\begin{aligned} E(Z \mid x_e) &= E(H_e x_e + \varphi \mid x_e) = H_e\, E(x_e \mid x_e) + E(\varphi \mid x_e) = H_e\, x_e \\ P_{Z|x_e} &= E\left\{(Z - H_e\, x_e)(Z - H_e\, x_e)^T \mid x_e\right\} = E\{\varphi \varphi^T\} = R \end{aligned} \tag{4.20}$$

Therefore, its pdf becomes

$$p(Z \mid x_e) = const\ e^{-\frac{1}{2}\left\{\left(Z - H_e \mu_{x_e}\right)^T R^{-1} \left(Z - H_e \mu_{x_e}\right)\right\}} \tag{4.21}$$

Applying Equations (4.14), (4.17), and (4.21) to Equation (4.19) yields

$$p(x_e \mid Z) \propto e^{-\frac{1}{2}\left[(Z - H_e x_e)^T R^{-1} (Z - H_e x_e) + (x - \mu_{x_e})^T P_x^{-1} (x - \mu_{x_e}) - \underbrace{(Z - H_e \mu_{x_e})^T (H_e P_{x_e} H_e^T + R)^{-1} (z - H \mu_{x_e})}_{NOT\ dependent\ on\ x_e}\right]} \tag{4.22}$$

For our estimate, $\hat{x}_e = \hat{x}_{MAP}$, *maximizing* $p(x_e \mid Z)$ is the same as *minimizing* the exponent quadratic form J given below.

$$J = \frac{1}{2}\left\{(Z - H_e x_e)^T R^{-1} (Z - H_e x_e) + (x_e - \mu_{x_e})^T P_{x_e}^{-1} (x_e - \mu_{x_e})\right\} \tag{4.23}$$

IMPORTANT NOTE: Referring to Addendum A4.2, Equation (A4.2.17), we see that the first term in the quadratic form J is the same quadratic form that is minimized to obtain a weighted least squares estimate, $\hat{x}_{WLSQ}$, in which the weighting matrix is taken to be $W = R^{-1}$. In this light, our estimate $\hat{x}_e = \hat{x}_{MAP}$ becomes the solution to an *extended* or *generalized* WLSQ solution. What this means is that we could have started with the quadratic form J and seek a generalized WLSQ solution, wherein x_e *does not have to be a random variable*. This is a critical observation, and we will return to this later in this section.

To determine the value that minimizes J, we set its derivative to zero and solve for x_e as follows:

$$0 = \frac{dJ}{dx_e} = \left[-H_e^T R^{-1}(Z - H_e x_e) - (Z - H_e x_e)^T R^{-1} H_e\right] + \left[P_{x_e}^{-1}(x_e - \mu_{x_e}) + (x_e - \mu_{x_e})^T P_{x_e}^{-1}\right]$$

$$= \underbrace{(-H_e^T R^{-1} Z + (H_e^T R^{-1} H_e)x_e + P_{x_e}^{-1} x_e - P_{x_e}^{-1}\mu_{x_e})}_{\text{column vector}} + \underbrace{x_e^T (H_e^T R^{-1} H_e) - Z^T R^{-1} H_e + x_e^T P_{x_e}^{-1} - \mu_{x_e}^T P_{x_e}^{-1}}_{\text{row vector}}$$

Thus, both column vector and row vector portions must be zero. Setting the column portion to zero yields

$$(H_e^T R^{-1} H_e)x_e + P_{x_e}^{-1} x_e = P_{x_e}^{-1}\mu_{x_e} + H_e^T R^{-1} Z$$

$$\left(P_{x_e}^{-1} + H_e^T R^{-1} H_e\right)x_e = H_e^T R^{-1} Z + P_{x_e}^{-1}\mu_{x_e}$$

$$x_e = \left(P_{x_e}^{-1} + H_e^T R^{-1} H_e\right)^{-1}\left(H_e^T R^{-1} Z + P_{x_e}^{-1}\mu_{x_e}\right)$$

Note: The inverse of $\left(P_{x_e}^{-1} + H_e^T R^{-1} H_e\right)$ exists by Property 9 for positive definite matrices in Addendum A4.3. Setting the row vector portion to zero yields the same result; therefore, our solution becomes

$$\hat{x}_e = \hat{x}_{MAP} = \left[P_{x_e}^{-1} + H_e^T R^{-1} H_e\right]^{-1}\left\{H_e^T R^{-1} Z + P_{x_e}^{-1}\mu_{x_e}\right\} \tag{4.24}$$

Rearranging Equation (4.24)

$$\hat{x}_e = \left[P_{x_e}^{-1} + H_e^T R^{-1} H_e\right]^{-1}\left\{H_e^T R^{-1} Z + P_{x_e}^{-1}\mu_{x_e}\right\}$$

$$= \underbrace{\left(P_{x_e}^{-1} + H_e^T R^{-1} H_e\right)^{-1} H_e^T R^{-1}}_{D} Z + \left(P_{x_e}^{-1} + H_e^T R^{-1} H_e\right)^{-1} \underbrace{P_{x_e}^{-1}}_{P_{x_e}^{-1} + (H_e^T R^{-1} H_e) - (H_e^T R^{-1} H_e)} \mu_{x_e}$$

$$= DZ + \left(P_{x_e}^{-1} + H_e^T R^{-1} H_e\right)^{-1}\left[P_{x_e}^{-1} + (H_e^T R^{-1} H_e) - (H_e^T R^{-1} H_e)\right]\mu_{x_e}$$

$$= DZ + \left[I - \left(P_{x_e}^{-1} + H_e^T R^{-1} H_e\right)^{-1}(H_e^T R^{-1} H_e)\right]\mu_{x_e}$$

$$= DZ + \mu_{x_e} - DH_e\mu_{x_e} = \mu_{x_e} + D(Z - H_e\mu_{x_e})$$

$$= \mu_{x_e} + \left(P_{x_e}^{-1} + H_e^T R^{-1} H_e\right)^{-1} H_e^T R^{-1}(Z - H_e\mu_{x_e})$$

(4.25)

We see that our solution takes on the form of a predictor–corrector, as seen in Equation (4.26).

$$\hat{x}_e = \underbrace{\mu_{x_e}}_{\text{Predictor}} + \underbrace{\underbrace{\left(P_{x_e}^{-1} + H_e^T R^{-1} H_e\right)^{-1} H_e^T R^{-1}}_{\text{Gain Matrix}} \underbrace{(Z - H_e\,\mu_{x_e})}_{\text{Residual}}}_{\text{Corrector}} \tag{4.26}$$

The prediction is the mean value μ_{x_e}, and the correction is a gain (matrix) applied to the residual vector $Z - H_e\,\mu_{x_e}$. The vector $\hat{Z} = H_e\,\mu_{x_e}$ is what we would observe ideally,[8] given that our model, Equation (4.13), is faithful and the residual is the difference from what we actually observe , i.e., $Z - \hat{Z}$. This is a particularly useful form for the solution to possess and is perfectly suited for extending GPA from *snapshot* analysis to *time series* analysis, which we will explore in the next chapter.

At this juncture, the reader may be wondering why this has been referred to as the Kalman filter approach. This is best explained after we expand this snapshot analysis to time series data in which we will be discussing the discrete time Kalman filter where the reader will see that Equation (4.26) is an over-simplified degenerative form of the Kalman filter, given a time series of *one* data point.

Before resuming our discussion, let us revisit the question raised earlier as to whether we interpret Equation (4.26) as a *probabilistic* solution or a *generalized WLSQ* solution. This hinges on the assignment of whether we allow x_e to be random. The measurement deltas Z remain random in either event, due to the random variable φ that represents the measurement delta non-repeatability or precision error. Historically, x_e was modeled as random and followed the derivation that brought us to Equation (4.24). With this assumption, the quantities μ_{x_e} and P_{x_e} have statistical meaning. Because the nominal engine was thought to be at the center of the distribution, it was assumed that $\mu_{x_e} = 0$. The diagonal of the covariance matrix P_{x_e} represents the variances associated with each fault and was determined from historical data on the basis of how frequently those faults occurred.[9] As we learned very quickly, this created a philosophical problem. For the engines of the day,[10] LPT faults were extremely rare, which imposed a very small variance as opposed to other modules, in particular, the core modules HPC and HPT that had greater frequency and level of deterioration. The small variance had the effect that an LPT fault, should it ever occur, would be virtually invisible to the estimator. Philosophically, that was an unacceptable consequence, and the random assumption was abandoned and replaced with a generalized WLSQ interpretation, wherein the *weighting* matrix P_{x_e} could be specified by the user in a manner that would be more equitable amongst all of the module faults to ensure their visibility. However, as a matter of convenience, matrices, such as P_{x_e}, will still be referred to as covariance matrices in the remainder of this book.

4.4.2 Shortcomings of the Basic Solution

Returning to Equation (4.26), it is typical, but certainly not necessary, that the predictor is assumed to be zero, i.e., $\mu_{x_e} = 0$ in a snapshot analysis. In that case, Equation (4.6) becomes

$$\hat{x}_e = \left(P_{x_e}^{-1} + H_e^T R^{-1} H_e\right)^{-1} H_e^T R^{-1} Z = D Z \tag{4.27}$$

The matrix *D*, which we will refer to as the *diagnostic* matrix, is the *inverted engine fault coefficients* alluded to in Figure 4.1. This type of analysis can be referred to as *multiple fault analysis* (*MFA*), or *module performance analysis* (*MPA*). The term *fault*[11] being used here is a bit of a misnomer in that what is really being assessed is *deterioration* or degradation in the modules. As a diagnostic methodology, this basic solution has some shortcomings. The most critical include the following.

1. There is no *direct* accommodation for measurement Δ error except for precision error (non-repeatability) represented by the covariance matrix *R*.
2. There is no accommodation for unsought problems.
3. Because it is a linear solution, it does not accurately accommodate *large* perturbations that generally occur with a (true) fault that has been detected via a trend detection method (see Chapter 2)

We will discuss the first of these here and the others in Chapter 5.

Measurement delta uncertainty was discussed generically in Chapter 2. Within the context of MFA, this can take many forms, such as sensor drift, non-repeatability, model errors and inadequacies, and baseline adjustment errors, to name a few. We could label the set of all possible contributors as *apparent* sensor errors, as they affect the measurement deltas associated with individual sensor measurements; they are *apparent* because not all of these contributors are necessarily associated with the actual sensor hardware system. We will shorten the label further by referring to them simply as *sensor errors* to be consistent with much of the GPA literature. This deficiency is a notable one, as the presence of any of these sensor errors will be erroneously assigned to a module performance fault. As you can imagine, this could have serious economic consequences in terms of driving unneeded inspections, D&Cs, and possibly UERs. Fortunately, this was realized in the late '70s by Urban and addressed by expanding the basic model to include sensor errors.[12] (This also provides partial mitigation for Number 2 in our list above).

4.4.3 Expanding the Basic Model

Addressing *real* and *apparent* measurement error is a focal point of a performance diagnostic algorithm. The success of a given scheme depends on the extent to which it can isolate and compensate for errors occurring in the measurement and estimation process. Consequently, the fault vector given in the basic model in Equation (4.13) can be configured to contain components directly related to sensor error in addition to engine fault deltas. That is to say, we can denote by x the concatenation of an $(n_e \times 1)$ engine fault vector x_e (Equation (4.4)) and an $(n_s \times 1)$ sensor (error) fault vector x_s, as follows:

$$x = \left[x_e \,\vdots\, x_s\right]^T \quad (n \times 1) \text{ vector where } n = n_e + n_s \tag{4.28}$$

The x_s sensor fault vector must contain a component for each of the m measurement delta components appearing in Equation (4.1) plus a component for inlet temperature T_2, pressure P_2, and the power setting parameter (*PSP*) such as corrected Fan speed or engine pressure ratio (EPR). It should be clear that whether we use Equation (4.2) or (4.3) to calculate the measurement delta Z, any perturbation in any of these three measurements will corrupt the values of Z. Thus, $n_s = m + 3$ and x_s appear as follows:

$$x_s = \begin{bmatrix} \Delta N_1 error \\ \Delta N_2 error \\ \Delta w_f error \\ \Delta T_3 error \\ \Delta P_3 error \\ \vdots \\ \Delta T_5 error \\ \Delta T_2 error \\ \Delta P_2 error \\ \Delta PSP error \end{bmatrix} \tag{4.29}$$

As with the engine fault Δ vector, all components of the sensor fault Δ vector are % of point quantities. To amend our basic model, we form

$$Z = H_e x_e + H_s x_s + \varphi = \left[H_e \;\; \vdots \;\; H_s\right] \begin{bmatrix} x_e \\ \cdots \\ x_s \end{bmatrix} + \varphi \; = \; H\,x + \varphi \tag{4.30}$$

where H_s is an $(m \times n_s)$ matrix of sensor fault (influence) coefficients, and H is an $(m \times n)$matrix of influences formed by concatenating H_e and H_s. We can now rewrite Equation (4.13) as

$$\begin{aligned} & Z = H\,x + \varphi \\ \textit{where}\quad & x = (n \times 1) \textit{ fault } \Delta \textit{ vector}, \quad x \in N(\mu_x, P_x) \\ & \varphi = (m \times 1) \textit{ noise vector}, \quad \varphi \in N(0, R) \\ & Z = (m \times 1) \textit{ measurement } \Delta \textit{ vector} \\ & H = (m \times n) \textit{ matrix of IC's} \\ & \quad x \textit{ and } \varphi \textit{ independent} \end{aligned} \tag{4.31}$$

A quick inspection shows that Equation (4.31) is identical to Equation (4.13) except for the lack of subscripts and the larger vectors and matrix. Because we still have an under-determined system with $n = n_e + n_s > m$, we can perform the same derivation to yield our estimate

$$\begin{aligned} \hat{x} = \hat{x}_{MAP} &= \left[P_x^{-1} + H^T R^{-1} H \right]^{-1} \left\{ H^T R^{-1} Z + P_x^{-1} \mu_x \right\} \\ &= \underbrace{\mu_x}_{\text{Predictor}} + \underbrace{\underbrace{\left(P_x^{-1} + H^T R^{-1} H \right)^{-1} H^T R^{-1}}_{\text{Gain Matrix}} \underbrace{(Z - H\ \mu_x)}_{\text{Residual}}}_{\text{Corrector}} \end{aligned} \tag{4.32}$$

Naturally, all of the statements made regarding the previous solution in Equation (4.26) applies to Equation (4.32). What is different is that we now have $n_s = m + 3$ additional fault Δs in our solution. As we shall see, this addresses some of our problems but not all of them. Before discussing the improvements, as well as some consequences of this solution, we will outline the creation of two important numerical quantities central to the model, i.e., H_s and R.

4.4.4 Sensor Influence Coefficients

The sensor error influence coefficient matrix, H_S, relates how % errors in the raw measurement values, inlet conditions, and power setting parameter affect % changes in the (*corrected*) measurement residuals (from baseline). This set of relationships form the basic model from which sensor error deltas are estimated. As with most things, there is more than one way to generate these coefficients. For our purposes, there are essentially *two approaches*, and we will present and discuss both, as they produce slightly different coefficients, as well as the interpretation that accompanies them. We will start chronologically, with the classic approach.

4.4.4.1 Classical Approach

In this approach, the primary elements supporting the calculation of the sensor error ICs are the gas path parameter steady-state slopes (sss),[13] theta (θ), and delta (δ) exponents used in the standard day correction process.[14] The following description and derivation can be found in the literature, (see [5]) and is repeated here for completeness.

Recall Equation (3.5) from Chapter 3 from which we can describe the (standard day) *corrected* value X_c of an arbitrary gas path parameter *X* by

$$X_c = \frac{X}{\theta^{\alpha_X}\ \delta^{\beta_X}} \quad where \quad \theta = \frac{T_2}{T_{ISA}}, \quad \delta = \frac{P_2}{P_{ISA}} \tag{4.33}$$

where α_X and β_X are suitably chosen exponents. This also applies to the power setting parameter (PSP) which, itself, is typically a corrected parameter,[15] $X_c^{(PSP)}$, i.e.,

$$X_c^{(PSP)} = \frac{X^{(PSP)}}{\theta^{\alpha_{X^{(PSP)}}}\ \delta^{\beta_{X^{(PSP)}}}} \tag{4.34}$$

The % delta ΔX from the baseline reference level X_{Base} is (by virtue of Equation (A3.6.1) of the Appendix) given by

$$1+\frac{\Delta X}{100}=\frac{X_c}{X_{Base}} \tag{4.35}$$

Taking natural logarithms of Equations (4.33) and (4.34), and differentiating, provides the following relationships

$$\frac{dX_c}{X_c}=\frac{dX}{X}-\alpha_X\left(\frac{d\theta}{\theta}\right)-\beta_X\left(\frac{d\delta}{\delta}\right) \tag{4.36}$$

$$\frac{dX_c^{(PSP)}}{X_c^{(PSP)}}=\frac{dX^{(PSP)}}{X^{(PSP)}}-\alpha_{X^{(PSP)}}\left(\frac{d\theta}{\theta}\right)-\beta_{X^{(PSP)}}\left(\frac{d\delta}{\delta}\right) \tag{4.37}$$

Taking logs of Equation (4.35), we have

$$\begin{aligned}\ln\left(1+\frac{\Delta X}{100}\right)&\approx\frac{\Delta X}{100}\quad \textit{since}\ \frac{\Delta X}{100}\ \textit{is small}\\ &=\ln(X_c)-\ln(X_{Base})\end{aligned} \tag{4.38}$$

Differentiating Equation (4.38) yields

$$\frac{d(\Delta X)}{100}=\frac{dX_c}{X_c}-\frac{dX_{Base}}{X_{Base}} \tag{4.39}$$

It should be clear that $X_{Base}=f(X_c^{(PSP)})$; therefore,

$$\frac{dX_{Base}}{X_{Base}}=\left(\underbrace{\frac{dX_{Base}/X_{Base}}{dX_c^{(PSP)}/X_c^{(PSP)}}}_{sss(X_{Base})_{X_c^{(PSP)}}}\right)\left(\frac{dX_c^{(PSP)}}{X_c^{(PSP)}}\right)=sss(X_{Base})_{X_c^{(PSP)}}\left(\frac{dX_c^{(PSP)}}{X_c^{(PSP)}}\right) \tag{4.40}$$

where $sss(X_{Base})_{X_c^{(PSP)}}$ is the sss of the gas generator curve at the operating point under consideration. Combining Equations (4.36), (4.38), (4.39), and (4.40) yields the following relationship,

$$\begin{aligned}\frac{d(\Delta X)}{100}=\frac{dX}{X}&+\left[\alpha_{X^{(PSP)}}\,sss(X_{Base})_{X_c^{(PSP)}}-\alpha_X\right]\frac{d\theta}{\theta}\\ &+\left[\beta_{X^{(PSP)}}\,sss(X_{Base})_{X_c^{(PSP)}}-\beta_X\right]\frac{d\delta}{\delta}-sss(X_{Base})_{X_c^{(PSP)}}\frac{dX^{(PSP)}}{X^{(PSP)}}\end{aligned} \tag{4.41}$$

If we have m gas path parameters under consideration, then Equation (4.41) will generate an $m\times(m+3)$ matrix as X and ΔX range independently over the m measured gas path parameters for an arbitrary power setting parameter, $X_c^{(PSP)}$. If we consider the high bypass twin spool turbofan figuratively presented in Figure 1.1, there are two PSPs in common usage: corrected Fan speed (N_{1c2}) and EPR ($EPR = P_5/P_2$).[16] For these two cases, we have the following (traditional) exponents:

$$\begin{aligned} X_c^{(PSP)} &= N_{1c2} \quad \therefore \quad \alpha_{N_1} = 0.5 \quad , \quad \beta_{N_1} = 0 \\ X_c^{(PSP)} &= EPR \quad \therefore \quad \alpha_{EPR} = 0 \quad , \quad \beta_{EPR} = 1 \end{aligned} \tag{4.42}$$

For these two common cases, the general Equation (4.41) reduces to

$$\frac{d(\Delta X)}{100} = \frac{dX}{X} + [0.5\, sss(X_{Base})_{N_{1c2}} - \alpha_X]\frac{d\theta}{\theta} - \beta_X \frac{d\delta}{\delta} - sss(X_{Base})_{N_{1c2}} \frac{dN_1}{N_1} \tag{4.42}$$

And

$$\frac{d(\Delta X)}{100} = \frac{dX}{X} - \alpha_X \frac{d\theta}{\theta} + [\, sss(X_{Base})_{EPR} - \beta_X]\frac{d\delta}{\delta} - sss(X_{Base})_{EPR} \frac{dP_5}{P_5} \tag{4.44}$$

for N_{1c2} and EPR respectively. It is evident from these equations that the sensor error *Influence Coefficients* consist of an ($m \times m$) *Identity matrix* plus three columns representing the effect of the inlet and power setting parameters. Using Equation (4.44), where EPR is the PSP, we can produce a sample H_s matrix as depicted below in Figure 4.6.

4.4.4.2 Model Generation Approach

This method requires a non-linear aero-thermal engine model such as NPSS, running closed loop at a fixed steady-state condition. This is essentially the same procedure as when engine influence coefficients are generated, except that the perturbations are introduced to the sensor signals, one at a time, across the m main gas path sensors plus inlet temperature, pressure, and the power setting parameter. As with engine ICs, something is being held constant, most typically the PSP. The major difference between coefficients generated in this manner, compared to the classical method, is the control rematch effect. For example, many engines have variable stator vanes in the HPC for stall mitigation and performance enhancement. The control of the vanes runs off a basic schedule[17] to determine the set point (i.e., vane angle) and then a separate control for the actuation of the vane movement to achieve that set point. Because we are running at steady state, this latter control is, generally, not of concern in generating these ICs. The basic schedule is typically a function of a corrected

	N1 ER	N2 ER	Wf ER	T5 ER	T25 ER	T3 ER	P25 ER	P3 ER	P2 ER	T2 ER	P5 ER
N1	1	0	0	0	0	0	0	0	0.71	-0.49	-0.71
N2	0	1	0	0	0	0	0	0	0.45	-0.49	-0.45
Wf	0	0	1	0	0	0	0	0	1.80	-0.64	-2.80
T5	0	0	0	1	0	0	0	0	0.82	-1.04	-0.82
T25	0	0	0	0	1	0	0	0	0.16	-1	-0.16
T3	0	0	0	0	0	1	0	0	0.56	-0.91	-0.56
P25	0	0	0	0	0	0	1	0	-0.64	0	-0.36
P3	0	0	0	0	0	0	0	1	0.75	0	-1.75

FIGURE 4.6 Sample sensor influence coefficient matrix – classical.

	N1 ER	N2 ER	Wf ER	T5 ER	T25 ER	T3 ER	P25 ER	P3 ER	P2 ER	T2 ER	P5 ER
N1	1	0.00	0	0	0.01	0	0	0	0.73	-0.49	-0.71
N2	0	0.98	0	0	0.15	0	0	0	0.44	-0.49	-0.44
Wf	0	-0.11	1	0	0.13	0	0	0	1.74	-0.64	-2.77
T5	0	-0.12	0	1	0.15	0	0	0	0.78	-1.03	-0.81
T25	0	-0.04	0	0	1.04	0	0	0	0.18	-0.99	-0.17
T3	0	-0.08	0	0	0.08	1	0	0	0.55	-0.90	-0.57
P25	0	-0.16	0	0	0.19	0	1	0	-0.56	0	-0.39
P3	0	0.01	0	0	0.00	0	0	1	0.74	0	-1.74

FIGURE 4.7 Sample sensor influence coefficient matrix with control effects.

gas path parameter, usually core speed corrected to HPC inlet, i.e., $N_{2c25} \approx N_2/\sqrt{\theta_{25}}$. Therefore, if either N_2 or T_{25} is perturbed, then the resultant N_{2c25} is altered causing a different (than nominal) vane angle to be specified. This change in geometry within the engine will affect all the gas path parameters (to varying degree),[18] and the N_2 and T_{25} sensor error columns in the H_s matrix will no longer be all zeros with unity on the diagonal as they are in the classical IC matrix. Similar effects are possible for perturbations made to inlet temperature, pressure, and PSP. A sample IC matrix, generated using the engine simulation, is depicted in Figure 4.7.

4.4.4.3 Interpretation and Application

In the foregoing sections, we introduced two methods to derive the sensor fault (influence) coefficient matrix H_s. As the reader can see, they produce matrices that are not identical but not terribly different either. The initial intent of adding "sensor faults," x_s, in the expanded model was to capture instrumentation drift, some of the non-repeatability associated with the measurement, and portions of other errors present in the measurement delta process as alluded to in Figure 4.2. In that sense, it helps mitigate (not remove) error effects that can corrupt the engine module performance fault estimates, $\hat{x}_e$. There are arguments on both sides that would favor one of these matrices over the other for various applications and fault assumptions; however, both forms have been used effectively for many decades, and one could conclude that it really does not matter a great deal given the small differences between the two matrices. That being said, we will leverage the closed form mathematical expressions of the classical form of H_s to help in motivating the derivation of other numerical quantities, such as the covariance matrix R in the next section.

It should be noted that some real-world applications, inlet temperature, and/or inlet pressure are not available (for a variety of reasons), and an alternate procedure is necessitated. See Addendum A4.8 for an example solution. Another situation that can occur is when we have a gas path parameter that is *not directly measured* but can be indirectly specified as a function of other parameters. A common example presents itself in ground testing when a bellmouth is used and fitted with both total and static pressure probes. In this case, total inlet airflow can be indirectly inferred as a $f(P_{Total}/P_{Static})$. Addendum A4.9 provides a derivation for the total airflow "sensor" influence coefficient so that total airflow can be used as one of the m "measured" gas path parameters in a GPA analysis (see [5]).

4.4.5 Covariance Matrix R

In both the basic and expanded models presented in the preceding sections, R represents the covariance matrix for the random variable $\varphi \in N(0,R)$. By virtue of the linear model, it also represents the non-repeatability of Z under the *generalized WLSQ* assumption discussed earlier. In this case, the fault delta variable x is *deterministic* and not random; therefore, its variance is zero. By Addendum Equation (A4.1.7) we have

$$Var(Z) = H\ Var(x)\ H^T + Var(\varphi) = R \tag{4.45}$$

There are essentially two situations to consider: using *corrected* quantities or *raw* quantities to calculate the measurement deltas.[19] Interestingly, both situations yield the same results. For brevity, we will describe the calculation in the corrected measurement delta case. In this situation. we gather our measurements (N_1, N_2, w_f, etc.) as well as T_2, P_2, and a *power setting* parameter, which, for definiteness, we will assume to be $EPR = P_5/P_2$. Parameters are corrected to standard day conditions to form (N_{1c}, N_{2c}, w_{fc}, etc.) by applying the appropriate θ and δ corrections and then compared to a *reference* (or base) level *corrected* quantity determined as a function of the *target* parameter, e.g., ($N_{1c\_ref} = f(EPR)$, $N_{2c\_ref} = f(EPR)$, etc). The % difference produces the corrected measurement deltas:

$$\Delta N_{1c} = 100\left(\frac{N_{1c} - N_{1c\_ref}}{N_{1c\_ref}}\right), \Delta N_{2c} = 100\left(\frac{N_{2c} - N_{2c\_ref}}{N_{2c\_ref}}\right), \Delta w_{fc} = 100\left(\frac{w_{fc} - w_{fc\_ref}}{w_{fc\_ref}}\right), etc$$

In order to calculate the $(m \times m)$ matrix, $R = [R_{ij}]_{i,j=1,2,\dots,m}$, we will borrow Equation (4.41).

$$\frac{d(\Delta X)}{100} = \frac{dX}{X} + \left[\alpha_{X^{(PSP)}}\ sss(X_{Base})_{X_c^{(PSP)}} - \alpha_X\right]\frac{d\theta}{\theta}$$
$$+\left[\beta_{X^{(PSP)}}\ sss(X_{Base})_{X_c^{(PSP)}} - \beta_X\right]\frac{d\delta}{\delta} - sss(X_{Base})_{X_c^{(PSP)}}\frac{dX^{(PSP)}}{X^{(PSP)}}$$

To preserve our eyesight (and sanity) we will make a temporary change in notation and rewrite the above equation, for the ith gas path parameter, as

$$\frac{d(\Delta X_i)}{100} = \frac{dX_i}{X_i} + \left[\alpha_{PSP}\ sss_i - \alpha_i\right]\frac{d\theta}{\theta} + \left[\beta_{PSP}\ sss_i - \beta_i\right]\frac{d\delta}{\delta} - sss_i\frac{d\,PSP}{PSP} \tag{4.46}$$

The diagonal of R is the variance of Equation (4.46) and off-diagonal elements the covariance applied using Equation (4.46) (assuming that the noise elements of the ith and j^{th} parameters are uncorrelated) to obtain:

$$R_{ii} = \sigma_i^2 + (\alpha_{PSP}\ sss_i - \alpha_i)^2\sigma_{T_2}^2 + (\beta_{PSP}\ sss_i - \beta_i)^2\sigma_{P_2}^2 + sss_i^2\sigma_{PSP}^2$$

and for $i \neq j$

$$R_{ij} = (\alpha_{PSP}\ sss_i - \alpha_i)(\alpha_{PSP}\ sss_j - \alpha_j)\sigma_{T_2}^2$$
$$+(\beta_{PSP}\ sss_i - \beta_i)(\beta_{PSP}\ sss_j - \beta_j)\sigma_{P_2}^2 + sss_i^2 sss_j^2\sigma_{PSP}^2 \tag{4.47}$$

In this expression, the terms $(\sigma_1, \sigma_2, \ldots, \sigma_m)$ denote the standard deviations representing the measurement delta non-repeatabilities and

$$\sigma_{T_2} = T_2 \ \textit{non-repeatability}$$
$$\sigma_{P_2} = P_2 \ \textit{non-repeatability}$$
$$\sigma_{P_5} = P_5 \ \textit{non-repeatability}$$

If we form the $(m+3) \times (m+3)$ diagonal matrix $\sum$, as the matrix formed by the $m+3$ standard deviations squared (variances), i.e.,

$$\sum = \begin{bmatrix} \sigma_1^2 & & & & & & \\ & \sigma_2^2 & & & & & \\ & & \ddots & & & & \\ & & & \sigma_m^2 & & & \\ & & & & \sigma_{T_2}^2 & & \\ & & & & & \sigma_{P_2}^2 & \\ & & & & & & \sigma_{P_5}^2 \end{bmatrix}$$

then, R can be more succinctly expressed as

$$R = H_s \Sigma H_s^T \tag{4.48}$$

An example of a measurement covariance matrix is given in Figure 4.8.

The diagonal elements of this matrix are the variances for each measured parameter Δ, and the off-diagonal elements are the covariances relating the pairs of measurement Δ parameters. The covariances contain the effects of the inlet and power setting parameter uncertainties. If the uncertainties of these latter parameters were zero, i.e., $\sigma_{T_2} = \sigma_{P_2} = \sigma_{P_5} = 0$, then the R matrix would be diagonal. In any event, R is a symmetric, positive definite[20] matrix.

4.4.6 Diagnostic Matrix D and Properties

From our expanded model (Equation (4.31)) and snapshot solution (Equation (4.32)), we have

	ΔN1	ΔN2	ΔWf	ΔT5	ΔT25	ΔT3	ΔP25	ΔP3
ΔN1	0.062	0.015	0.073	0.029	0.013	0.022	0.008	0.042
ΔN2	0.015	0.040	0.045	0.019	0.019	0.015	0.002	0.026
ΔWf	0.073	0.045	0.636	0.089	0.035	0.064	0.035	0.160
ΔT5	0.029	0.019	0.089	0.273	0.032	0.033	0.011	0.047
ΔT25	0.013	0.019	0.035	0.032	0.086	0.024	0.014	0.010
ΔT3	0.022	0.015	0.064	0.033	0.024	0.065	0.008	0.033
ΔP25	0.008	0.002	0.035	0.011	0.014	0.008	0.111	0.021
ΔP3	0.042	0.026	0.160	0.047	0.010	0.033	0.021	0.162

FIGURE 4.8 Sample measurement covariance matrix R.

$$\begin{aligned}\hat{x} &= \mu_x + \left(P_x^{-1} + H^T R^{-1} H\right)^{-1} H^T R^{-1}(Z - H\ \mu_x) \\ &= \mu_x + D(Z - H\ \mu_x)\end{aligned} \tag{4.49}$$

The matrix $D = \left(P_x^{-1} + H^T R^{-1} H\right)^{-1} H^T R^{-1}$ is referred to as the *diagnostic* matrix as well as the *Kalman gain* matrix, (because it has the same form as the gain matrix in the discrete time Kalman filter). This matrix provides (colloquially speaking) a set of inverse *IC*s. There are several forms for this matrix (see Addendum A4.11). The two most often seen in the literature are the following.

$$\begin{aligned}&1)\quad D = \left(P_x^{-1} + H^T R^{-1} H\right)^{-1} H^T R^{-1} \\ &2)\quad D = P_x H^T (H P_x H^T + R)^{-1}\end{aligned} \tag{4.50}$$

We will use both forms of the diagnostic matrix in various derivations throughout the book; however, from a computational perspective, the second expression of Equation (4.50) is preferred. as it has only one matrix inversion of size $m \times m$, whereas the other has three inversions – two of which are of a much larger size $n \times n$.

4.4.6.1 Estimate Properties

The estimate given in Equation (4.49) is *unbiased*, i.e., $E(\hat{x}) = E(x)$. This follows easily from the model, as shown in Equation (4.51).

$$\begin{aligned}Z &= Hx + \varphi \quad \Rightarrow \quad E(Z) = H\mu_x \\ \hat{x} &= \mu_x + D(Z - H\mu_x) \Rightarrow \\ \mu_{\hat{x}} &= E(\hat{x}) = \mu_x + D(\underbrace{E(Z) - H\mu_x}_{zero}) \\ &= \mu_x = E(x)\end{aligned} \tag{4.51}$$

To obtain an expression for the estimate covariance matrix, $P_{\hat{x}}$, we start with a little manipulation as follows:

$$\begin{aligned}\hat{x} - \mu_{\hat{x}} &= \hat{x} - \mu_x \\ &= D(Z - H\mu_x) \\ &= D\big(\varphi + H(x - \mu_x)\big) \\ &= D\varphi + DH(x - \mu_x) \quad \Rightarrow \\ (\hat{x} - \mu_{\hat{x}})(\hat{x} - \mu_{\hat{x}})^T &= D\varphi\varphi^T D + DH(x - \mu_x)(x - \mu_x)^T H^T D^T \\ &\quad + \ \textit{cross terms in } \varphi \textit{ and } x\end{aligned} \tag{4.52}$$

By definition, we have

$$
\begin{aligned}
P_{\hat{x}} &= E\left((\hat{x}-\mu_{\hat{x}})(\hat{x}-\mu_{\hat{x}})^T\right) \\
&= DRD^T + DH\,P_xH^TD^T + \underbrace{E(cross\ terms)}_{zero} \\
&= DRD^T + DH\,P_xH^TD^T \\
&= D\left(H\,P_xH^T + R\right)D^T \\
&= \left[P_xH^T\left(H\,P_xH^T+R\right)^{-1}\right]\left(H\,P_xH^T+R\right)\left[P_xH^T\left(H\,P_xH^T+R\right)^{-T}\right] \\
&= P_xH^T\left(H\,P_xH^T+R\right)^{-1}HP_x \\
&= DHP_x
\end{aligned}
\tag{4.53}
$$

In summary,

$$
\begin{aligned}
\mu_{\hat{x}} &= \mu_x \\
P_{\hat{x}} &= DHP_x
\end{aligned}
\tag{4.54}
$$

4.4.6.2 Estimation Error Properties

The estimation error is defined simply as the difference between the fault delta value and our analyzed estimate, i.e., $e_x \triangleq x-\hat{x}$. Thus,

$$
\mu_{e_x} = E(e_x) = \mu_x - \mu_{\hat{x}} = 0 \tag{4.55}
$$

To determine the estimation error covariance, $P_e \equiv P_{e_x}$, we have another simple manipulation as follows:

$$
\begin{aligned}
P_e = P_{e_x} &= E\left((x-\hat{x})(x-\hat{x})^T\right) \\
&= E\left\{\left((x-\mu_x)-(\hat{x}-\mu_{\hat{x}})\right)\left((x-\mu_x)-(\hat{x}-\mu_{\hat{x}})\right)^T\right\} \\
&= \underbrace{E\left\{(x-\mu_x)(x-\mu_x)^T\right\}}_{P_x} + \underbrace{E\left\{(\hat{x}-\mu_{\hat{x}})(\hat{x}-\mu_{\hat{x}})^T\right\}}_{P_{\hat{x}}} - E\left\{(\hat{x}-\mu_{\hat{x}})(x-\mu_x)^T\right\} \\
&\quad - E\left\{(x-\mu_x)(\hat{x}-\mu_{\hat{x}})^T\right\}
\end{aligned}
$$

Therefore,

$$
\begin{aligned}
P_e &= P_x + P_{\hat{x}} - \underbrace{E\left\{(\hat{x}-\mu_{\hat{x}})(x-\mu_x)^T\right\}}_{DHP_x} - \underbrace{E\left\{(\hat{x}-\mu_{\hat{x}})(x-\mu_x)^T\right\}^T}_{(DHP_x)^T = DHP_x} \\
&= P_x + DHP_x - DHP_x - DHP_x \\
&= (I-DH)P_x
\end{aligned}
\tag{4.56}
$$

In summary,

$$\begin{aligned} \mu_e &\equiv \mu_{e_x} = 0 \\ P_e &\equiv P_{e_x} = \left(I - DH\right)P_x \end{aligned} \tag{4.57}$$

If we use the form $D = (P_x^{-1} + H^T R^{-1} H)^{-1} H^T R^{-1}$ in Equation (4.57), we obtain

$$P_e = (I - DH)P_x = \left[I - (P_x^{-1} + H^T R^{-1} H)^{-1} H^T R^{-1} H\right] P_x \tag{4.58}$$

From Addendum A4.11 Equation (A4.11.6), we have

$$\underbrace{P_x H^T (H P_x H^T + R)^{-1}}_{D} = \left[I - (P_x^{-1} + H^T R^{-1} H)^{-1} H^T R^{-1} H\right] P_x H^T R^{-1} \tag{4.59}$$

Multiplying Equation (4.58) by $H^T R^{-1}$and combining with Equation (4.59) yields

$$\begin{aligned} D = P_e H^T R^{-1} \quad &\Rightarrow \quad P_e = \left(P_x^{-1} + H^T R^{-1} H\right)^{-1} \quad \Rightarrow \\ P_e^{-1} - P_x^{-1} = H^T R^{-1} H \geq 0 \quad &\Rightarrow \quad P_e \leq P_x \end{aligned} \tag{4.60}$$

From Equation (4.60), we have the three results:

$$\begin{aligned} &1)\ \ P_e = \left(P_x^{-1} + H^T R^{-1} H\right)^{-1} \\ &2)\ \ P_e \leq P_x \\ &3)\ \ P_e^{-1} \geq P_e^{-1} - P_x^{-1} = H^T R^{-1} H \end{aligned} \tag{4.61}$$

Result Number 3 from Equation (4.61) supports our intuitive notion that more reliable measurements should produce more accurate estimates in that, as the non-repeatability of the measurement deltas, R, drops, the covariance of the error drops.

Before leaving this section, we state three interesting, but far from obvious, relationships involving the estimation error e_x. Mathematical details can be found in Addendum A4.12.

$$\begin{aligned} &1)\ \ \textit{If}\ \ \mu_x = 0\ \ \textit{then}\ \ \sqrt{\lambda_{\min}^{(I-DH)^T(I-DH)}} \leq \frac{\|e_x\|}{\|x\|} \leq \sqrt{\lambda_{\max}^{(I-DH)^T(I-DH)}} \\ &2)\ \ \textit{If}\ \ \mu_x = 0\ \ \textit{then}\ \ E\left(\frac{\|e_x\|^2}{\|x\|^2}\right) = \frac{1}{n}\sum_{i=1}^{n} \lambda_i^{(I-DH)^T(I-DH)} = \frac{1}{n}\textit{Trace}[(I-DH)^T(I-DH)] \\ &3)\ \ E\left(\frac{\|e_x\|^2}{\|x\|^2}\right) > 1 - \frac{m}{n} \quad \textit{where}\ \ m < n \end{aligned} \tag{4.62}$$

It is clear from Number 1 of Equation (4.62) that the *faults* most accurately determined by the diagnostic system with no *a-priori* information are those contained in the space spanned by the eigenvector having the *minimum eigenvalue*. Thus, knowing the eigenvectors (and values) of $(I - DH)^T(I - DH)$ yields information on the strengths and weakness of the diagnostic system, i.e., which faults are the most and least identifiable from the measurement suite at hand under the constraints of the assumed measurement and state covariances (R and P_x). The simple calculation provided by Number 2 yields a quantitative measure of the *overall goodness* of the diagnostic system in question and can be used to compare the merits of several systems. The result in Number 3 provides a very broad bound on the average squared error for an arbitrary diagnostic system in much the same spirit as Chebyshev's inequality in probability theory. As suspected, the average squared error becomes *worse* as the number of faults become progressively larger for a fixed measurement set, thereby introducing a limiting factor for the number of faults a practical diagnostic system should carry.

If we have no *a-priori* knowledge, i.e., $\mu_x = 0$, then the magnitude of the estimate is *attenuated* from the magnitude of the underlying faults. Generally speaking, each component of the $\hat{x}$ vector will receive some non-zero value, and this "property" has been referred to, in the literature, as the *smearing* effect. As an example, if we take the *coupled* engine fault IC matrix (Addendum A4 Figure A4.3) and concatenate the classical sensor fault IC matrix (Figure 4.6), we obtain the H matrix depicted in Figure 4.9 below.

If we take P_x to be the identity matrix and compute the diagnostic matrix D as per Equation (4.5), we obtain a D matrix, as shown in Figure 4.10.

	FAN	LPC	HPC	HPT	LPT	N1 ER	N2 ER	Wf ER	T5 ER	T25 ER	T3 ER	P25 ER	P3 ER	P2 ER	T2 ER	P5 ER
N1	-0.270	-0.137	-0.023	-0.032	0.406	1	0.00	0	0	0.01	0	0	0	0.73	-0.49	-0.71
N2	0.095	-0.307	-0.024	0.452	-0.197	0	0.98	0	0	0.15	0	0	0	0.44	-0.49	-0.44
Wf	0.269	-0.829	-0.654	-0.910	0.085	0	-0.11	1	0	0.13	0	0	0	1.74	-0.64	-2.77
T5	0.250	-0.744	-0.713	-0.977	-0.501	0	-0.12	0	1	0.15	0	0	0	0.78	-1.03	-0.81
T25	-0.213	0.093	-0.208	-0.285	0.371	0	-0.04	0	0	1.04	0	0	0	0.18	-0.99	-0.17
T3	0.036	-0.313	-0.455	0.426	-0.048	0	-0.08	0	0	0.08	1	0	0	0.55	-0.90	-0.57
P25	-0.841	1.367	-0.914	-1.242	1.378	0	-0.16	0	0	0.19	0	1	0	-0.56	0	-0.39
P3	-0.086	0.064	0.032	1.024	0.140	0	0.01	0	0	0.00	0	0	1	0.74	0	-1.74

FIGURE 4.9 Example H matrix.

	N1	N2	WF	T5	T25	T3	P25	P3
Fan	-0.161	-0.008	0.073	0.015	-0.031	0.003	-0.093	-0.044
LPC	-0.046	0.004	-0.071	-0.051	0.040	-0.015	0.169	0.124
HPC	0.141	-0.016	0.015	-0.127	0.104	-0.219	-0.158	0.016
HPT	0.004	0.107	-0.136	-0.138	-0.008	0.188	-0.088	0.222
LPT	0.175	-0.027	0.030	-0.178	0.098	0.019	0.126	-0.016
N1 ER	0.724	-0.055	-0.113	0.014	-0.142	-0.072	-0.037	-0.091
N2 ER	-0.055	0.858	-0.014	-0.041	-0.079	-0.162	0.133	-0.120
Wf ER	-0.113	-0.014	0.237	-0.212	0.034	0.003	-0.032	-0.152
T5 ER	0.014	-0.041	-0.212	0.436	-0.094	-0.114	0.036	0.175
T25 ER	-0.142	-0.079	0.034	-0.094	0.637	-0.168	-0.118	0.067
T3 ER	-0.072	-0.162	0.003	-0.114	-0.168	0.641	0.049	-0.107
P25 ER	-0.037	0.133	-0.032	0.036	-0.118	0.049	0.168	0.011
P3 ER	-0.091	-0.120	-0.152	0.175	0.067	-0.107	0.011	0.376
P2 ER	0.190	0.009	0.082	-0.004	-0.021	0.011	-0.086	-0.024
T2 ER	-0.063	-0.115	0.094	-0.107	-0.300	-0.184	0.011	0.049
PSP ER	0.050	-0.008	-0.136	0.005	0.038	0.045	-0.062	-0.211

FIGURE 4.10 Example D matrix.

Now we introduce a -3% HPT fault (x) and obtain the following estimate:

Component	Fault	Estimate
FAN	0	-0.0033
LPC	0	-0.0511
HPC	0	-0.5644
HPT	-3	-2.1762
LPT	0	0.1911
N_1 ER	0	-0.0097
N_2 ER	0	-0.32
w_f ER	0	0.4074
T_5 ER	0	0.4152
T_{25} ER	0	0.0221
T_3 ER	0	-0.5648
P_{25} ER	0	0.2633
P_3 ER	0	-0.6658
P_2 ER	0	-0.059
T_2 ER	0	-0.0392
P_5 ER	0	0.054

The *smearing* is evident. One consequence of the *smearing* effect is that, without having any *a-priori* knowledge of the underlying fault, x, the quantitative value represented by $\hat{x}$, can be underestimated and should, therefore, be interpreted somewhat broadly. By this we mean that the estimated values are best viewed as relative magnitudes that point to the faulty module(s), sensor(s), or combinations of these entities. By doing so, $\hat{x}$ becomes more of a fault *indicator* than a fault *estimator*. Add to this quandary the fact that not all faults are equally visible for a given fixed measurement suite. Some will *estimate* (i.e., *indicate*) better than others. Admittedly, this is not what we had hoped for, but it is the situation at hand for a *single snapshot* of data with no *a-priori* information. As we shall see in the examples to follow and Chapter 5, all is not lost. There are several things that will improve our situation:

- More data points (multiple snapshots, or time series data collected over many days, months, etc.),
- A-priori information that could be in the form of assumptions,
- Redundant sensors to allow an independent assessment of sensor faults, and
- Additional gas path sensors to admit greater fault visibility.

Our discussion of these potential remedies will start in Chapter 5 and continue throughout the remainder of the book.

TABLE 4.1
Delta Parameters for the Simulated HPC Fault

	N1	N2	Wf	T5	T25	T3	P25	P3
Delta (%)	0.024	0.080	0.643	0.728	0.212	0.460	0.914	-0.048

4.5 EXAMPLES

Before we conclude this chapter, we will give a few examples illustrating the various results one could find by applying different assumptions (or *a-priori* information) to a given snapshot data. *These are purely illustrative and are not intended to reflect real case scenarios but, rather, to demonstrate the mathematical inter-relationships given in this chapter.* The snapshot data used here was generated from the engine simulation program at sea level, static conditions (*EPR* at 1.494) with a simulated HPC fault that has resulted in 1% efficiency loss, as well as 1.2% flow capacity loss. The standard gas path sensor suite consists of N_1, N_2, w_f, T_5, T_{25}, T_3, P_{25}, and P_3. Note that P_5 is used to calculate the power setting parameter (*EPR*); therefore, it is not used as an additional measurement in the analysis. The delta parameters are calculated based on Equation (4.1) using a steady-state model as reference and are given in Table 4.1.

The H matrix is based on the influence coefficients as shown in Figures 4.3 and 4.7. The measurement covariance matrix R adopts the values as shown in Figure 4.8. The diagnostic matrix (D) is calculated using Equation (4.50) but is different in each of the following cases in order to incorporate different assumptions that are reflected in the design of the P_x matrix. As explained earlier in this chapter, assumptions (or *a-priori* information) are needed to solve the under-determined problem, i.e., solving 21 unknowns (both module health parameters and sensor biases) based on eight observations.

4.5.1 Case 1: Assuming Sensors Are Accurate and Using the "Correct" Fault Coupling Factors

The sensor accuracy assumption is not entirely unrealistic for snapshot analysis when the sensor measurements have been verified and rectified, which may be possible in a test cell environment. This helps to "remove" or "discount" sensor bias estimates by using a particular P_x matrix, as shown in Table 4.2. This P_x matrix has adopted the following coupling factors (CF) for Fan (1.3), LPC (1.3), HPC (1.2), HPT (-0.7) and LPT (-0.6). Apparently, the CF for HPC happens to be the same as the simulated fault, which we will relax in the next case.

The diagnostic results listed in Table 4.3 are calculated using Equation (4.49) with a predictor $\mu_x = 0$ (i.e., no *a-priori* information available, the same assumptions adopted for the other cases below).

TABLE 4.2
P_x Matrix for Case 1

100	130	0	0	0	0	0	0	0	0	0	0	0	0	0	0	0	0	0	0
130	169	0	0	0	0	0	0	0	0	0	0	0	0	0	0	0	0	0	0
0	0	100	130	0	0	0	0	0	0	0	0	0	0	0	0	0	0	0	0
0	0	130	169	0	0	0	0	0	0	0	0	0	0	0	0	0	0	0	0
0	0	0	0	100	120	0	0	0	0	0	0	0	0	0	0	0	0	0	0
0	0	0	0	120	100	0	0	0	0	0	0	0	0	0	0	0	0	0	0
0	0	0	0	0	0	100	-70	0	0	0	0	0	0	0	0	0	0	0	0
0	0	0	0	0	0	-70	49	0	0	0	0	0	0	0	0	0	0	0	0
0	0	0	0	0	0	0	0	100	-60	0	0	0	0	0	0	0	0	0	0
0	0	0	0	0	0	0	0	-60	36	0	0	0	0	0	0	0	0	0	0
0	0	0	0	0	0	0	0	0	0	.1	0	0	0	0	0	0	0	0	0
0	0	0	0	0	0	0	0	0	0	0	.1	0	0	0	0	0	0	0	0
0	0	0	0	0	0	0	0	0	0	0	0	.1	0	0	0	0	0	0	0
0	0	0	0	0	0	0	0	0	0	0	0	0	.1	0	0	0	0	0	0
0	0	0	0	0	0	0	0	0	0	0	0	0	0	.1	0	0	0	0	0
0	0	0	0	0	0	0	0	0	0	0	0	0	0	0	.1	0	0	0	0
0	0	0	0	0	0	0	0	0	0	0	0	0	0	0	0	.1	0	0	0
0	0	0	0	0	0	0	0	0	0	0	0	0	0	0	0	0	.1	0	0
0	0	0	0	0	0	0	0	0	0	0	0	0	0	0	0	0	0	.1	0
0	0	0	0	0	0	0	0	0	0	0	0	0	0	0	0	0	0	0	.1

TABLE 4.3
Diagnostic Results for Case 1

	Fan Eff	Fan Flow	LPC Eff	LPC Flow	HPC Eff	HPC Flow	HPT Eff	A4	LPT Eff	A45	
Truth	0	0	0	0	-1	-1.2	0	0	0	0	
Estimate	-0.04	-0.06	-0.02	-0.02	-0.97	-1.18	0	0	-0.02	0.01	
	N1 ER	N2 ER	Wf ER	T50 ER	T25 ER	T3 ER	P25 ER	P3 ER	P2 ER	T2 ER	P5 ER
Truth	0	0	0	0	0	0	0	0	0	0	0
Estimate	0	0	0	0	0	0	0	0	0	0	0

As expected, the estimated HPC efficiency and flow capacity changes are rather close to the simulated "truth," while all other fault estimates are close to zero (with slight smearing effects on other modules). Obviously, this result is achieved by ignoring sensor faults and using the "correct" coupling factors. What if a different HPC coupling factor is utilized? That leads us to the next case.

4.5.2 Case 2: Assuming Sensors Are Accurate and Using a Different Set of Coupling Factors

If we adopt a coupling factor of 1 for HPC, while still assuming the sensors are accurate, the P_x matrix changes to the form illustrated in Table 4.4 (with differences highlighted in red).

With this change, the diagnostic results are shown in Table 4.5. It can be noticed that the estimate for HPC flow capacity loss becomes less accurate. The smearing effects on Fan and LPT also become more visible.

4.5.3 Case 3: Assuming Sensors Are Accurate and Adding More Gas Path Measurements

For interested readers who have been following the theories and examples in this chapter, the following question may naturally arise: what if we had two more gas path measurements available? Can we better estimate the fault without assuming coupling factors? In this case, if we are confident with the sensor accuracy (again, an assumption), we are dealing with 10 unknown parameters with 10 measurements. The answer is positive.

We reconfigured the engine simulation program and made T48 (LPT inlet temperature) and P48 (LPT inlet pressure) measurable. For the same HPC fault used in the first two cases, the ΔT48 and ΔP48 calculated are 0.65% and -0.09%, respectively. In addition, because new sensors were added, the R and H matrices also had to be augmented. To relax the assumptions on coupling factors, the P_x matrix in Table 4.6 is utilized. Note that two columns/rows were added for the additional sensors (with the same diagonal values 0.1). As a result, the augmented P_x matrix is 23 × 23 as compared with the dimension of 21 × 21 in the first two cases.

A new diagnostic matrix D was calculated based on the updated matrices. For the same simulated fault, the diagnostic results are shown in Table 4.7. It is obvious that, with the addition of two sensors, the estimates of HPC efficiency and flow capacity changes are rather accurate without assuming any coupling factors, and the smearing effects are also minimal.

As pleasing as this result may suggest, it has to be noted that the contributions of additional sensors in the diagnosis process vary significantly depending on the location of the sensors, hence the influence of the module fault as observed by the sensors. Moreover, adding sensors is often prohibitive due to increased production and maintenance costs.

TABLE 4.4

P_x Matrix for Case 2

100	130	0	0	0	0	0	0	0	0	0	0	0	0	0	0	0	0	0	0	0
130	169	0	0	0	0	0	0	0	0	0	0	0	0	0	0	0	0	0	0	0
0	0	100	130	0	0	0	0	0	0	0	0	0	0	0	0	0	0	0	0	0
0	0	130	169	0	0	0	0	0	0	0	0	0	0	0	0	0	0	0	0	0
0	0	0	0	100	100	0	0	0	0	0	0	0	0	0	0	0	0	0	0	0
0	0	0	0	100	100	0	0	0	0	0	0	0	0	0	0	0	0	0	0	0
0	0	0	0	0	0	100	-70	0	0	0	0	0	0	0	0	0	0	0	0	0
0	0	0	0	0	0	-70	49	0	0	0	0	0	0	0	0	0	0	0	0	0
0	0	0	0	0	0	0	0	100	-60	0	0	0	0	0	0	0	0	0	0	0
0	0	0	0	0	0	0	0	-60	36	0	0	0	0	0	0	0	0	0	0	0
0	0	0	0	0	0	0	0	0	0	.1	0	0	0	0	0	0	0	0	0	0
0	0	0	0	0	0	0	0	0	0	0	.1	0	0	0	0	0	0	0	0	0
0	0	0	0	0	0	0	0	0	0	0	0	.1	0	0	0	0	0	0	0	0
0	0	0	0	0	0	0	0	0	0	0	0	0	.1	0	0	0	0	0	0	0
0	0	0	0	0	0	0	0	0	0	0	0	0	0	.1	0	0	0	0	0	0
0	0	0	0	0	0	0	0	0	0	0	0	0	0	0	.1	0	0	0	0	0
0	0	0	0	0	0	0	0	0	0	0	0	0	0	0	0	.1	0	0	0	0
0	0	0	0	0	0	0	0	0	0	0	0	0	0	0	0	0	.1	0	0	0
0	0	0	0	0	0	0	0	0	0	0	0	0	0	0	0	0	0	.1	0	0
0	0	0	0	0	0	0	0	0	0	0	0	0	0	0	0	0	0	0	.1	0
0	0	0	0	0	0	0	0	0	0	0	0	0	0	0	0	0	0	0	0	.1

TABLE 4.5
Diagnostic Results for Case 2

	Fan Eff	Fan Flow	LPC Eff	LPC Flow	HPC Eff	HPC Flow	HPT Eff	A4	LPT Eff	A45	
Truth	0	0	0	0	-1	-1.2	0	0	0	0	
Estimate	-0.1	-0.14	-0.01	-0.02	-1.02	-1.02	0	0	-0.07	0.04	
	N1 ER	N2 ER	Wf ER	T50 ER	T25 ER	T3 ER	P25 ER	P3 ER	P2 ER	T2 ER	P5 ER
Truth	0	0	0	0	0	0	0	0	0	0	0
Estimate	0	0.04	0	0	0	-0.01	0.01	-0.01	0	0	0

TABLE 4.6
P_x Matrix for Case 3

100	0	0	0	0	0	0	0	0	0	0	0	0	0	0	0	0	0	0	0	0	0	0
0	100	0	0	0	0	0	0	0	0	0	0	0	0	0	0	0	0	0	0	0	0	0
0	0	100	0	0	0	0	0	0	0	0	0	0	0	0	0	0	0	0	0	0	0	0
0	0	0	100	0	0	0	0	0	0	0	0	0	0	0	0	0	0	0	0	0	0	0
0	0	0	0	100	0	0	0	0	0	0	0	0	0	0	0	0	0	0	0	0	0	0
0	0	0	0	0	100	0	0	0	0	0	0	0	0	0	0	0	0	0	0	0	0	0
0	0	0	0	0	0	100	0	0	0	0	0	0	0	0	0	0	0	0	0	0	0	0
0	0	0	0	0	0	0	100	0	0	0	0	0	0	0	0	0	0	0	0	0	0	0
0	0	0	0	0	0	0	0	100	0	0	0	0	0	0	0	0	0	0	0	0	0	0
0	0	0	0	0	0	0	0	0	100	0	0	0	0	0	0	0	0	0	0	0	0	0
0	0	0	0	0	0	0	0	0	0	.1	0	0	0	0	0	0	0	0	0	0	0	0
0	0	0	0	0	0	0	0	0	0	0	.1	0	0	0	0	0	0	0	0	0	0	0
0	0	0	0	0	0	0	0	0	0	0	0	.1	0	0	0	0	0	0	0	0	0	0
0	0	0	0	0	0	0	0	0	0	0	0	0	.1	0	0	0	0	0	0	0	0	0
0	0	0	0	0	0	0	0	0	0	0	0	0	0	.1	0	0	0	0	0	0	0	0
0	0	0	0	0	0	0	0	0	0	0	0	0	0	0	.1	0	0	0	0	0	0	0
0	0	0	0	0	0	0	0	0	0	0	0	0	0	0	0	.1	0	0	0	0	0	0
0	0	0	0	0	0	0	0	0	0	0	0	0	0	0	0	0	.1	0	0	0	0	0
0	0	0	0	0	0	0	0	0	0	0	0	0	0	0	0	0	0	.1	0	0	0	0
0	0	0	0	0	0	0	0	0	0	0	0	0	0	0	0	0	0	0	.1	0	0	0
0	0	0	0	0	0	0	0	0	0	0	0	0	0	0	0	0	0	0	0	.1	0	0
0	0	0	0	0	0	0	0	0	0	0	0	0	0	0	0	0	0	0	0	0	.1	0
0	0	0	0	0	0	0	0	0	0	0	0	0	0	0	0	0	0	0	0	0	0	.1

TABLE 4.7
Diagnostic Results for Case 3

	Fan Eff	Fan Flow	LPC Eff	LPC Flow	HPC Eff	HPC Flow	HPT Eff	A4	LPT Eff	A45	
Truth	0	0	0	0	-1	-1.2	0	0	0	0	
Estimate	-0.01	-0.01	0	-0.01	-1.02	-1.21	-0.01	0	0	0	
	N1 ER	N2 ER	Wf ER	T50 ER	T25 ER	T3 ER	P25 ER	P3 ER	P2 ER	T2 ER	P5 ER
Truth	0	0	0	0	0	0	0	0	0	0	0
Estimate	0	0	0	0	0	0	0	0	0	0	0
	T48 ER	P48 ER									
Truth	0	0									
Estimate	0	0									

TABLE 4.8
P_x Matrix for Case 4

100	130	0	0	0	0	0	0	0	0	0	0	0	0	0	0	0	0	0	0	0
130	169	0	0	0	0	0	0	0	0	0	0	0	0	0	0	0	0	0	0	0
0	0	100	130	0	0	0	0	0	0	0	0	0	0	0	0	0	0	0	0	0
0	0	130	169	0	0	0	0	0	0	0	0	0	0	0	0	0	0	0	0	0
0	0	0	0	100	120	0	0	0	0	0	0	0	0	0	0	0	0	0	0	0
0	0	0	0	120	100	0	0	0	0	0	0	0	0	0	0	0	0	0	0	0
0	0	0	0	0	0	100	-70	0	0	0	0	0	0	0	0	0	0	0	0	0
0	0	0	0	0	0	-70	49	0	0	0	0	0	0	0	0	0	0	0	0	0
0	0	0	0	0	0	0	0	100	-60	0	0	0	0	0	0	0	0	0	0	0
0	0	0	0	0	0	0	0	-60	36	0	0	0	0	0	0	0	0	0	0	0
0	0	0	0	0	0	0	0	0	0	10	0	0	0	0	0	0	0	0	0	0
0	0	0	0	0	0	0	0	0	0	0	10	0	0	0	0	0	0	0	0	0
0	0	0	0	0	0	0	0	0	0	0	0	10	0	0	0	0	0	0	0	0
0	0	0	0	0	0	0	0	0	0	0	0	0	10	0	0	0	0	0	0	0
0	0	0	0	0	0	0	0	0	0	0	0	0	0	10	0	0	0	0	0	0
0	0	0	0	0	0	0	0	0	0	0	0	0	0	0	10	0	0	0	0	0
0	0	0	0	0	0	0	0	0	0	0	0	0	0	0	0	10	0	0	0	0
0	0	0	0	0	0	0	0	0	0	0	0	0	0	0	0	0	10	0	0	0
0	0	0	0	0	0	0	0	0	0	0	0	0	0	0	0	0	0	10	0	0
0	0	0	0	0	0	0	0	0	0	0	0	0	0	0	0	0	0	0	10	0
0	0	0	0	0	0	0	0	0	0	0	0	0	0	0	0	0	0	0	0	10

TABLE 4.9
Diagnostic Results for Case 4

	Fan Eff	Fan Flow	LPC Eff	LPC Flow	HPC Eff	HPC Flow	HPT Eff	A4	LPT Eff	A45	
Truth	0	0	0	0	-1	-1.2	0	0	0	0	
Estimate	-0.04	-0.05	0.05	0.07	-0.79	-0.84	-0.07	0.05	-0.04	0.02	
	N1 ER	N2 ER	Wf ER	T50 ER	T25 ER	T3 ER	P25 ER	P3 ER	P2 ER	T2 ER	P5 ER
Truth	0	0	0	0	0	0	0	0	0	0	0
Estimate	-0.04	0.05	-0.01	0.03	-0.04	0.07	0.04	0	0.01	-0.06	-0.03

TABLE 4.10
Diagnostic Results for Case 5

	Fan Eff	Fan Flow	LPC Eff	LPC Flow	HPC Eff	HPC Flow	HPT Eff	A4	LPT Eff	A45	
Truth	0	0	0	0	-1	-1.2	0	0	0	0	
Estimate	0.02	-0.04	-0.06	0.09	-0.26	-0.09	-0.21	0.07	-0.02	-0.1	
	N1 ER	N2 ER	Wf ER	T50 ER	T25 ER	T3 ER	P25 ER	P3 ER	P2 ER	T2 ER	P5 ER
Truth	0	0	0	0	0	0	0	0	0	0	0
Estimate	-0.14	0.12	0	0.14	-0.15	0.22	0.19	-0.07	-0.02	-0.18	-0.1

4.5.4 Case 4: Relaxed Sensor Accuracy Assumptions

Now let's go back to the configuration setup in Case 1 (without having T48 and P48). While keeping the same coupling factor assumptions, we relax the sensor accuracy assumption with the P_x matrix shown in Table 4.8, i.e., we acknowledge that the sensors would be (equally) noisy. This would produce the results shown in Table 4.9. As compared to results from Case 1, the HPC fault estimates turn out less accurate due to the smearing effects on sensor bias estimates.

4.5.5 Case 5: No A-Priori Assumptions

In this last case study, we will relax all assumptions, i.e., assuming no *a-priori* information is available. The P_x matrix becomes a 21×21 diagonal matrix of 100 and the diagnostic results are shown in Table 4.10.

While HPC fault may, arguably, still appear to be the likely fault diagnosis, the estimates of efficiency and flow capacity changes are significantly attenuated. This is the result of attempting to solve an under-determined problem without the use of appropriate additional information or assumptions. In the following chapters, the readers will be introduced to the various forms of a-priori information and reasonable assumptions that assist in resolving these challenges.

There is a large number of publications available in the general literature on various aspects of GPA. A grossly abbreviated list appears below. Some of the earlier references [6–27] discuss fundamental concepts covered in this chapter and may provide further insight for the interested reader.

REFERENCES

1. Urban, L.A., 1972, *Gas Path Analysis Applied to Turbine Engine Conditioning Monitoring*, AIAA/SAE Paper 72-1082.
2. Urban, L.A., *Gas Path Analyzer*, US Patent 3731070, May 1, 1973.
3. Urban, L.A., 1974, Parameter Selection for Multiple Fault Diagnostics of Gas Turbine Engines, *AGARD Conference Proceedings* No. 165.
4. Urban, L.A., 1969, *Gas Turbine Engine Parameter Interrelationships,* HSD UTC, Windsor Locks, CT. (1st ed. 1967, 2nd ed. 1969).
5. Volponi, A.J., 1994, Sensor Error Compensation in Engine Performance Diagnostics, *ASME Turbine and Aero-Engine Congress.* The Hague, Netherlands, June 13–16, ASME Paper No. 94-GT-58.
6. Smetana, F., 1974, Turbojet Engine Gas Path Analysis – A Review, AGARD-CP-165.
7. Volponi, A.J., 1983, Gas Path Analysis: An Approach to Engine Diagnostic, *Time-Dependent Failure Mechanisms and Assessment Methodologies*, Cambridge University Press.
8. Volponi, A.J., 2003, *Foundations of Gas Path Analysis I&II*, (Von Karman Lecture Series VKI LS 2003-01, Gas Turbine Condition Monitoring and Fault Diagnosis), Von Karman Institute, Rhode-Saint-Genese, Belgium.
9. Doel, D.L., 1992, TEMPER - A Gas Path Analysis Tool for Commercial Jet Engines, *Proceedings of the American Society of Mechanical Engineers Turbine and Aeroengine Congress.* Cologne, Germany, June 1–4, ASME Paper 92-GT-315.

10. Doel, D.L., 1993, An Assessment of Weighted-Least-Squares Based Gas Path Analysis, *Proceedings of the American Society of Mechanical Engineers Turbine and Aeroengine Congress*. Cincinnati, Ohio, May 24–27, ASME Paper 93-GT-119.
11. Stamatis, A., et al, 1991, Jet Engine Fault Detection with Discrete Operating Points Gas Path Analysis, *Journal of Propulsion*, Vol 7(6), pp 1043–1048.
12. Urban, L.A., Volponi, A.J., 1992, Mathematical Methods of Relative Engine Performance Diagnostics, SAE 1992 Transactions, *Journal of Aerospace*, Vol 101, pp 2025–2050. Paper # 922048.
13. Provost, M.J., 1994, *The Use of Optimal Estimation Techniques in the Analysis of Gas Turbines*, Ph.D. thesis, Cranfield University, 1994.
14. Dienger, G., Rheilander, H., *Economic Review on ATLAS JT9D Gas Path Analysis Application*, 17th ETMT Working Paper, Hamburg IF26, Rel 604, 1979.
15. Danielsson, S.G., *Gas Path Analysis Applied to Pre and Post Overhaul Testing of the JT9D Turbofan Engine*, SAE Paper 770993, 1977.
16. Stroobach, J., *GPA Evaluation of On-Wing Recorded Data with a Limited Number of Sensors on the Engines Installed PH-DTA*, KLM CED Report G-218, 1980.
17. Stroobach, J., *GPA Testcell Evaluation Performed on CF6-50C Engines*, KLM CED Report G-239, 1981.
18. Glenny, D.E. Gas Path Analysis and Engine Performance Monitoring in a Chinook Helicopter, *AGARD: Engine Condition Monitoring*, 1988.
19. Kwon, Oh-Kyu, et al., 1991, Fault Detection and Diagnosis in Gas Turbines, *ASME Journal of Engineering for Gas Turbines and Power*, Vol 276, pp 113.
20. Simon, D., et al., 2002, Kalman Filtering with State Equality Constraints, *IEEE Transaction on Aerospace and Electronic Systems*, Vol 39(1), pp 128–136.
21. Simon, D., et al., 2003, Aircraft Turbofan Engine Health Estimation using Constrained Kalman Filtering, *Proceedings of the American Society of Mechanical Engineers Turbine and Aeroengine Congress*. Atlanta, GA, June 16–19, Paper No. GT2003-38584.
22. Simon, D.L., et al., 2004, *Sensor Needs for Control and Health Management of Intelligent Aircraft Engines*, NASA/TM—2004-213202.
23. Litt, J.S., et al., 2005, *A Survey of Intelligent Control and Health Management Technologies for Aircraft Propulsion Systems*, NASA/TM—2005-213622.
24. Merrington, G.L., 1993, *Fault Diagnosis in Gas Turbines Using a Model Based Technique,* ASME Paper 93-GT-13.
25. Provost, M.J., 1988, *COMPASS: A Generalized Ground Based Engine Monitoring System,* AGARD-CP-448.
26. Mathioudakis, K., et al., Computer Modelling and Data Processing Methods. An Essential Part of Jet Engine Condition Monitoring and Fault Diagnosis. *Proceedings of the 15th International Symposium on Aircraft Integrated Monitoring Systems*, RWTH, Aachen, Germany, Sept. 12–14, 1989, pp. 185–213.
27. Volponi, A.J., 2003, *Basic Fault Model and Measurement Error Handling*, (Von Karman Lecture Series VKI LS 2003-01, Gas Turbine Condition Monitoring and Fault Diagnosis), Von Karman Institute, Rhode-Saint-Genese, Belgium.

NOTES

1. See Introduction chapter
2. See Volponi, A.J., 1983, *Gas Path Analysis: An Approach to Engine Diagnostic,* Time-Dependent Failure Mechanisms and Assessment Methodologies, Cambridge University Press.
3. Numerical propulsion system simulation.

4. These will be % of point deltas to be consistent with our measurement deltas.
5. See Addendum A4.4.
6. See Addendum A4.1.
7. Assumes turbine nozzles are choked.
8. With no noise.
9. Although gas turbines are extremely complicated machines, they are very reliable, and faulted datasets tend to be very small compared to non-faulted datasets.
10. In the late 1970s, the primary engines being analyzed were commercial engines such as the P&W JT9D and the GE CF6-50.
11. This usage is very common in the GPA literature, and we will continue to use it throughout the book.
12. See Urban, L.A., *Gas Path Analysis – A Tool for Engine Condition Monitoring,* 33rd Annual International Air Safety Seminar, Flight Safety Foundation, Inc., Christchurch, New Zealand, 1980.
13. See Addendum A4.6.
14. See Chapter 3.
15. For example, corrected Fan rotor speed or *EPR*.
16. Note: $EPR = (P5 / \delta)\,(1 / PISA) = \text{const} \times P_{5c}$; therefore, it is a corrected parameter.
17. There may be additional schedules as well, used to *trim* the basic schedule for flight conditions, such as Mach and altitude.
18. No pun intended.
19. Equations (4.2) or (4.3), respectively.
20. Positive definite matrix properties can be found in Addendum A4.3.

5 Discrete Time Gas Path Analysis

5.1 OVERVIEW

In the previous chapter, we established most of the fundamental relationships and properties that we will need in our succeeding discussions on GPA. Our emphasis has been on a model-based approach that lends itself to a generalized LSQ estimation strategy, which has often been referred to as the Kalman filter approach. In this chapter, we will continue this discussion and extend the analysis from single point snapshot to multi-point and time series data, which is where almost all GPA analyses are performed today. In doing so, we will leverage other information and assumptions that arise with time series data to mitigate the shortcomings noted in the snapshot case and provide some illustrative examples for clarification.

In the process of extending our approach to time series data, we will introduce the discrete time Kalman filter, from which it will be evident as to why the methodology has been linked to this construct. We have also alluded to the fact that over the past five decades, other approaches have been proposed. Some of these are closely aligned to the methodology we have adopted here; however, others have been truly different in some fundamental way. For the latter, we will provide a brief discussion, along with references, in the final section of this chapter. Because most of these have been singular publications with little evidence of sustained research, they will not be pursued in the remainder of the book.

5.2 SUMMARY OF FUNDAMENTAL RELATIONSHIPS

In Chapter 4, we established a basic model and numerous parameter inter-relationships in our description of GPA for the single data point (snapshot) analysis. We rewrite the most important of these here as a convenient reference before considering the extension to multiple data points.

5.2.1 Model

$$
\begin{aligned}
& Z = H\,x + \varphi \\
\textit{where}\quad & x = (n\times 1)\ \textit{fault}\ \Delta\ \textit{vector},\quad x \in N(\mu_x, P_x) \\
& \varphi = (m\times 1)\ \textit{noise vector},\quad \varphi \in N(0, R) \\
& Z = (m\times 1)\ \textit{measurement}\ \Delta\ \textit{vector} \\
& H = (m\times n)\ \textit{matrix of IC's} \\
& \quad x\ \textit{and}\ \varphi\ \textit{independent}
\end{aligned}
\tag{5.1}
$$

DOI: 10.1201/9781003649274-5

Moreover,

$$Z = H_e x_e + H_s x_s + \varphi = \begin{bmatrix} H_e & \vdots & H_s \end{bmatrix} \begin{bmatrix} x_e \\ \cdots \\ x_s \end{bmatrix} + \varphi = H\,x + \varphi \tag{5.2}$$

where H_e is an $(m \times n_e)$ matrix of engine fault (influence) coefficients, H_s is an $(m \times n_s)$ matrix of sensor fault (influence) coefficients, and H is an $(m \times n)$matrix of influences formed by concatenating H_e and H_s.

5.2.2 Estimate

$$\hat{x} = \underbrace{\mu_x}_{\text{Predictor}} + \underbrace{\underbrace{D}_{\text{Gain Matrix}} \underbrace{(Z - H\,\mu_x)}_{\text{Residual}}}_{\text{Corrector}} \tag{5.3}$$

The diagnostic matrix *D*, can be computed by either formulation given below.

$$\begin{aligned} &1)\quad D = \left(P_x^{-1} + H^T R^{-1} H\right)^{-1} H^T R^{-1} \\ &2)\quad D = P_x H^T (H\,P_x H^T + R)^{-1} \end{aligned} \tag{5.4}$$

From a computational perspective, the second expression of Equation (5.4) is preferred, as it has only one matrix inversion of size $m \times m$, whereas the other has three inversions, two of which are much larger being of size $n \times n$.

The estimation error, $e_x = x - \hat{x}$, has covariance P_e given by

$$P_e = (I - DH)P_x = \left(P_x^{-1} + H^T R^{-1} H\right)^{-1} \tag{5.5}$$

Therefore, we can rewrite Equation (5.3) as

$$\hat{x} = \mu_x + \underbrace{P_e H^T R^{-1}}_{D} (Z - H\,\mu_x) \tag{5.6}$$

We are now in a position to consider the practical situation of having multiple data points and how to analyze them to produce a single result.[1] For the most part, engine monitoring is a long-term endeavor spanning many months, perhaps many years. This is certainly the case for flight acquired data for aero-engines and power generation plant data for stationary gas turbines as part of any engine heath management program. Before attacking the general problem, we will begin with the simplest case, that of multiple snapshots.

5.3 MULTIPLE SNAPSHOTS

Multiple steady-state data point snapshots are a common occurrence found in test stand performance assessments of aero-engines before they are installed on an aircraft. These engines are typically either new or recently overhauled and, to a lesser degree, test as received (TAR) after removal from an aircraft. It is not uncommon to have four or five steady-state data points captured at different power conditions to determine whether the engine meets certain performance criteria before returning to service. These requirements differ from engine type and model but generally include items such as hot day *EGT* margin, thrust available at takeoff, acceptable specific fuel consumption at cruise, compressor stall margin, and the like. If an engine does not pass the required standards, it will be returned to the shop for additional rework. In these instances, a GPA analysis can offer some insight into the potential modules responsible for the performance measure deficiency and impact the subsequent overhaul rework. It was in this setting that the early GPA programs were utilized with reasonable success.

To begin our discourse, we will assume the following:

1) N measurement deltas, $\{Z_1, Z_2, \ldots, Z_N\}$ at power conditions $\{PSP_1, PSP_2, \ldots, PSP_N\}$ are available.
2) Corresponding IC matrices $\{H_1, H_2, \ldots, H_N\}$ have been generated.
3) Instrumentation non-repeatabilities are known, and corresponding measurement covariance matrices $\{R_1, R_2, \ldots, R_N\}$ have been generated.
4) The time span between the data points collected is negligible (total time 1–2 hours) so that we can assume nothing has changed in the engine's performance level during that time interval.

The last assumption is very reasonable for this type of dataset, as engine deterioration is a slow process (months, not hours). Accordingly, we can assume the same for the test stand instrumentation. In short, we are assuming that *during* the testing, our *fault state* (engine and sensor), represented by x, *has not changed.*

The strategy is to perform individual snapshot analyses using Equation (5.3), wherein the a-priori estimate (predictor) for the kth data point is the estimate from the preceding $(k\text{-}1)^{th}$ data point analysis. In addition, by virtue of Assumption 4, we can use a fixed design state covariance matrix P_x when calculating the diagnostic matrices $\{D_1, D_2, \ldots, D_N\}$, i.e.,

$$D_k = P_x H_k^T (H_k P_x H_k^T + R_k)^{-1} \tag{5.7}$$

Thus, the analysis proceeds as follows:[2]

$$\begin{aligned}
\hat{x}_1 &= D_1(Z_1) \\
\hat{x}_2 &= \hat{x}_1 + D_2(Z_2 - H_2\hat{x}_1) \\
\hat{x}_3 &= \hat{x}_2 + D_3(Z_3 - H_3\hat{x}_2) \\
&\vdots \\
\hat{x}_N &= \hat{x}_{N-1} + D_N(Z_N - H_N\hat{x}_{N-1})
\end{aligned} \tag{5.8}$$

As we have seen from Chapter 4, Equation (4.62), the normalized error, at a given operating point, is bounded by the eigenvalues of $(I-DH)^T(I-DH)$, i.e.,

$$\sqrt{\lambda_{\min}^{(I-DH)^T(I-DH)}} \le \frac{\|e_x\|}{\|x\|} \le \sqrt{\lambda_{\max}^{(I-DH)^T(I-DH)}}$$

Numerically, this can be quite large, for example

$$0.0026 \le \frac{\|e_x\|}{\|x\|} \le 1.8736$$

The actual bounds will vary as the matrix D varies. The matrices H and R are fixed at a particular engine operating point, leaving P_x as the sole vehicle for *tuning* the diagnostic system to perform in some optimal fashion. Optimality for such systems is defined as the ability to assess with acceptable accuracy those faults or combination of faults that are considered reasonable or likely to occur in the given operational environment under consideration. Test stand applications, in which we have multiple snapshots, may have pre- and post-overhaul engines to analyze, which would offer engine fault combination possibilities different from those that would be anticipated in the long-term degradation of an in-flight analysis. This subjectivity has made the exercise of tuning the system more of a black art than an exact science. The designer typically strives for some level of uniformity among the class of faults to be assessed with engine fault assessment as one concern and sensor error detection as another. Since the inception of this technique in the late '70s, the clarion call has been to *tune* the system to provide adequate accuracy in *both* engine and sensor faults alike. Unfortunately, with unknowns outnumbering equations roughly 2:1, the system behaves like an n dimensional seesaw.[3] When emphasis is directed toward one fault, it is accompanied by an attendant loss in another. If all single engine and sensor faults are considered equally important, the best the designer can achieve is a balanced seesaw that assesses all $n_e + n_s$ singletons at the same approximate level of accuracy. What, then, would be a reasonable level to expect?

A partial answer can be obtained by the following rationale. If we assume that the average level for our $n_e + n_s$ faults is approximately the same as the average error among all of the infinitely possible combinations occurring under a uniform distribution law, then an approximate lower bound for the normalized error is given by $\sqrt{1-(m/n)}$. In our examples, $m = 8$ and $n = 21$ ($n = 16$ for coupled engine faults), which gives an average error ≥ 0.79 (0.71 for coupled engine faults), tuning efforts notwithstanding. In either case, this would not cause the diagnostician to be imbued with overwhelming optimism. Nevertheless, the situation is not as bleak as it might appear, as tradeoffs between fault assumptions can be made and better accuracy achieved for a selected sub-set.

The desire for increased accuracy in engine fault assessments, combined with the need to remain robust with respect to measurement errors, can be addressed, at least partially, through additional analyses. These are somewhat dependent on the application and data source. We will explore some of these enhancements as we proceed through this chapter. Relative to the multi-snapshot scenario, there are two that are worthy of mention. The first deals with the use of average deterioration models to provide a-priori information, and the second is a methodology to accommodate a large measurement error.

5.3.1 Deterioration Models

In the mid-1970s, various studies were conducted to determine the long-term deterioration effects on the major modules of high bypass turbofan engines in commercial revenue service [1–3]. These investigations provided information that allowed the construction of deterioration models, wherein performance measures, such as changes in component efficiency, flow capacity and nozzle area, could be estimated as a function of flight cycle or hours of operation. This practice has been sustained through the decades by the engine manufacturers and provides an approximate estimation of these parameters based on engine build and mission analysis. We cannot attest to the accuracy of these models; however, it is clear that the estimates generated are at least of the appropriate polarity that reflect the fact that with time, engines deteriorate, and do not improve. The utilization of such models to provide an *a-priori* guess for GPA began in the early '80s for TAR engine tests. As a generality, the closer the model is to the actual degradation of the engine under scrutiny, the greater the accuracy of the estimates. Even a rough guess in the *proper direction* (polarity) can be beneficial.

5.3.2 Large Measurement Error Recovery

As we have seen in our previous examples, some improvement can be made in our assessment of engine faults if we change our perspective toward measurement errors. If we view sensor faults as a detection mechanism whose purpose is to absorb small measurement perturbations and flag large sensor excursions, then the *tuning* can be slanted toward the engine faults. If only small measurement errors are present, such as analysis would provide greater accuracy of the engine faults present. The foregoing examples already illustrated this principle. For *post-overhaul* (or *new installation*) engines, we are somewhat forced into this *tuning* bias, as the only feasible a-priori information to use in this scenario is zero deterioration present. Nevertheless, if a large sensor error was present, the linear gain (D) would smear the signal across the engine fault estimates as well, rendering them untenable, even if the large sensor fault had been *detected*. If we are willing to restrict ourselves to the possibility of dealing with only *one large measurement error,* the situation can be made manageable.

Unlike engine faults that generally affect changes in a plurality of measurements, the sensor error (for the most part) affects only ***one*** measurement and, thus, has the potential for greater detection visibility and correction. The exception to this statement involves errors in $\{T_2, P_2, PSP\}$, which are utilized in the data normalization and baseline generation of all parameters and, thus, must be handled separately. The methodology is derived in [4] and is repeated in Addendum A5.1 as a matter of convenience. *This method is appropriate for either single or multiple snapshot analysis.* For time series data, as we shall see in the sections to follow (in particular sections 5.5 and 5.8), there are other methods to employ for this purpose.

Before moving on, we provide a few simple examples to illustrate the effect that some of these assumptions and manipulations have on the analysis accuracy.

5.3.3 Examples

The following hypothetical examples for discrete GPA have been simulated at sea level, static condition at five different power levels to steady state wherein data is captured to analyze major module performance. The analysis is typically done to provide module health assessment to help decide the maintenance work scope for each module. Using the built-in deterioration model in the simulation program, the level of module health deteriorations, as shown in Table 5.1, is introduced in our examples.

Using the built-in *EPR*-based controller, the simulated engine is run to steady state at five different PLAs as shown in Table 5.2 below. The delta parameters are calculated using a simple steady state model representing a healthy, nominal engine.

ICs (H) derived for PLAs at 80, 70, and 60 are interpolated to generate the ICs for PLAs at 75 and 65. With the measurement Δ vector and ICs (H) matrix for each power level generated, we can now apply Equations (5.3) and (5.4) to estimate the faults (a 21 x 1 vector including 10 module health parameters and 11 sensor faults).

In the first attempt, we will *assume* all sensor measurements are accurate by using a P_x matrix that heavily discounts sensor faults, as used in Table 4.6. Equation (5.3) is applied sequentially, as shown in Equation (5.8), to each measurement Δ vector from PLA 80 to 60, with the initial predictor μ_x being assigned 0 and subsequently taking the $\hat{x}$ from the previous data point. The fault estimates are shown in Table 5.3 along with the simulated faults. While the final estimates (in column "PLA 60") appear reasonably close to the simulated faults (except for Fan efficiency and flow capacity due to the smearing effect), it should be noted that the sensor accuracy assumption is often unrealistic, as BOM sensors (instead of calibrated test cell sensors) are commonly used in the test cells.

It is also worth noting that, although it sounds fair to expect the fault estimates to approach the simulated faults in simulation studies, it is, indeed, a false anticipation to expect a perfect match due to the nonlinearity of engine dynamics with multiple-fault presence. The performance deviation (Δs) induced by the presence of multiple faults is not a linear combination (sum) of each fault's influence due to the rebalancing of engine aero-thermodynamics. Moreover, the "ground truth" is never known in actual practice. Assumptions or any forms of *a-priori* information have to be introduced one way or another based on the user's experience to arrive at an acceptable solution.

In the next case, we will relax the sensor accuracy assumption by using an identity P0 matrix (i.e., $P_0 = I_{21\times21}$, and the D matrix for each power level is updated accordingly). The updated fault estimates are shown in Table 5.4 along with the simulated fault. The results still correctly point to HPC as the major contributor to performance deterioration; however, the overall smearing effects could mislead the user to HPT and T3 and T5 sensors.

While the smearing effect is bad, all is not hopeless. The key is to utilize reliable *a-priori* information to improve the estimates. This process may involve multiple trial-and-error attempts, and eventually, it is up to an expert to make a thoughtful decision. For example, if a nominal engine deterioration model is available, even just pointing roughly in the right direction, the results can be significantly improved. To illustrate that, we set the initial μ_x to *half* of the simulated values and reprocess the five data points, the results of which are shown in Table 5.5. The updated estimates clearly point to the HPC, and the sensor fault estimates are also much more attenuated. Note again that, in reality, the "ground truth" is not available to help making adjustments to the parameters used in this process; one can only rely on experience and knowledge to help.

For the examples given so far, ignoring sensor faults always appears to give better estimates. That's simply because we haven't simulated any sensor faults. The authors certainly don't want the readers to have the impression that it's always a good idea to assume sensors are accurate (unless they are diligently checked and calibrated before the test, which is costly). In the following example, T5 sensor faults are simulated to demonstrate the smearing effect, as well as the large sensor error recovery process (as detailed in Chapter 5 Addendum) that can be utilized for compensation.

To *simplify this illustration*, we will use only the data captured at PLA 80. The same process can be applied to all data points sequentially, as illustrated in the previous examples. Two test cases are generated with T5 sensor bias of 2% and 5%, respectively. Using the same P0 matrix and the same initial assignment of μ_x as in the above example, the fault estimates are shown in Table 5.6. As compared with the previous case, with no T5 sensor fault (included in column "No T5 ER"), the T5 sensor fault estimates (1.06 and 2.32) are attenuated, and all module health parameter estimates are affected.

The question then arises as to how one could be alerted to this (unknown) sensor fault and potentially recover from it. In this case, the large T5 error (ER) estimates (1.06% and 2.32%) should be noticed for potential sensor fault. Following the large measurement error recovery method described in Chapter 5 Addendum A5.1, each sensor fault estimate is compared to a *threshold* value, as described in Equation (A5.1.17), and the one with the maximal ratio is identified as the faulty sensor. To achieve that, the threshold value for each sensor has to be established using Equations (A5.1.19)–(A5.1.21). In this example, the lower limit matrix *L*, calculated using Equation (A5.1.19), is shown in Table 5.7. The upper limit matrix *U*, calculated using Equation (A5.1.20), is shown in Table 5.8.

Based on the lower and upper limit matrices and Equation (A.5.1.21), the following thresholds are picked for each sensor as shown in Table 5.9 column "*Ti.*" The ratios between sensor fault estimates (columns "T5 ER = 2%" and "T5 ER = 5%") and thresholds are shown in columns "Ratio_1" and "Ratio_2," respectively. In both cases, the T5 sensor is identified as the faulty sensor.

TABLE 5.1
Simulated Module Health Deteriorations

	Fan Eff	Fan FC	LPC Eff	LPC FC	HPC Eff	HPC FC	HPT Eff	HPT A4	LPT Eff	LPT A45
Change (%)	-1.3	-0.23	-0.75	-0.68	-2.78	-0.43	-0.15	0.11	-0.13	0.07

TABLE 5.2
Gas Path Data Captured at Different Power Levels

PLA (deg)	80	75	70	65	60
ΔN1c (%)	-0.08	-0.06	-0.05	-0.06	-0.10
ΔN2c (%)	-0.38	-0.21	-0.16	-0.15	-0.17
Δwfc (%)	2.36	2.55	2.64	2.73	2.78
ΔT5c (%)	2.56	2.67	2.70	2.72	2.76
ΔT25c (%)	0.38	0.38	0.35	0.28	0.24
ΔT3c (%)	1.34	1.40	1.39	1.37	1.36
ΔP25c (%)	0.94	0.88	0.70	0.54	0.42
ΔP3c (%)	-0.37	-0.36	-0.36	-0.36	-0.36

TABLE 5.3
Fault Estimates Assuming Accurate Sensors

	PLA 80	PLA 75	PLA 70	PLA 65	PLA 60	Simulated Fault
Fan Eff	-0.18	-0.23	-0.28	-0.28	-0.24	-1.30
Fan FC	0.31	0.36	0.44	0.44	0.41	-0.23
LPC Eff	-0.83	-0.83	-0.76	-0.76	-0.81	-0.75
LPC FC	-0.70	-0.70	-0.69	-0.73	-0.64	-0.68
HPC Eff	-2.80	-2.83	-2.81	-2.81	-2.83	-2.78
HPC FC	-0.39	-0.60	-0.38	-0.40	-0.31	-0.43
HPT Eff	-0.12	-0.12	-0.12	-0.15	-0.18	-0.15
HPT A4	0.12	0.13	0.13	0.13	0.13	0.11
LPT Eff	-0.21	-0.20	-0.22	-0.23	-0.31	-0.13
LPT A45	0.04	0.05	0.05	0.04	0.06	0.07
N1 ER	0.00	0.00	0.00	0.00	0.00	0.00
N2 ER	0.00	0.00	0.00	0.00	0.00	0.00
wf ER	0.00	0.00	0.00	0.00	0.00	0.00
T5 ER	0.00	0.00	0.00	0.00	0.00	0.00
T25 ER	0.00	0.00	0.00	0.00	0.00	0.00
T3 ER	0.01	0.01	0.01	0.01	0.01	0.00
P25 ER	0.00	0.00	0.00	0.00	0.00	0.00
P3 ER	0.00	0.00	0.00	0.00	0.00	0.00
P2 ER	0.00	0.00	0.00	0.00	0.00	0.00
T2 ER	0.00	0.00	0.00	0.00	0.00	0.00
P5 ER	0.00	0.00	0.00	0.00	0.00	0.00

TABLE 5.4
Fault Estimates without Assuming Accurate Sensors

	PLA 80	PLA 75	PLA 70	PLA 65	PLA 60	Simulated Fault
Fan Eff	-0.11	-0.14	-0.17	-0.18	-0.19	-1.30
Fan FC	0.21	0.25	0.29	0.31	0.31	-0.23
LPC Eff	-0.15	-0.16	-0.16	-0.15	-0.16	-0.75
LPC FC	-0.07	-0.10	-0.14	-0.16	-0.17	-0.68
HPC Eff	-0.73	-0.74	-0.71	-0.71	-0.72	-2.78
HPC FC	-0.02	-0.04	-0.02	-0.02	-0.01	-0.43
HPT Eff	-0.60	-0.62	-0.61	-0.62	-0.63	-0.15
HPT A4	0.25	0.28	0.29	0.31	0.31	0.11
LPT Eff	-0.31	-0.35	-0.40	-0.42	-0.44	-0.13
LPT A45	-0.15	-0.14	-0.11	-0.09	-0.09	0.07
N1 ER	-0.38	-0.42	-0.41	-0.41	-0.42	0.00
N2 ER	-0.35	-0.30	-0.33	-0.33	-0.36	0.00
Wf ER	0.11	0.12	0.14	0.15	0.14	0.00
T5 ER	0.42	0.46	0.46	0.47	0.49	0.00
T25 ER	-0.28	-0.31	-0.31	-0.32	-0.32	0.00
T3 ER	0.62	0.61	0.58	0.57	0.57	0.00
P25 ER	0.17	0.17	0.12	0.10	0.09	0.00
P3 ER	-0.24	-0.26	-0.27	-0.28	-0.27	0.00
P2 ER	0.13	0.15	0.16	0.15	0.13	0.00
T2 ER	-0.43	-0.45	-0.42	-0.41	-0.40	0.00
P5 ER	-0.18	-0.18	-0.16	-0.12	-0.10	0.00

Once the faulty sensor is identified, its estimate can be updated accordingly. A simple calculation following Equation (A5.1.16) produces T5 (a-priori) error estimates of 2.55 and 5.55 for the two cases, respectively. After placing these updated estimates for the T5 sensor in the predictor μ_x and repeat the calculations for each case, the new results are shown in Table 5.10. A quick comparison with the case with no sensor fault (listed in Column "No T5 ER") reveals that the module health parameter estimates are no longer affected by the presence of T5 measurement errors, illustrating the effectiveness of the large measurement error recovery method.

These few examples demonstrate how additional *information* can be leveraged to increase the accuracy of a given analysis. Information can come in many forms: models, additional analytics, assumptions, additional data, and other sources. This will be a reoccurring theme throughout the remainder of our GPA discussions and will be summarized in Chapter 8 when we discuss the topic of *information fusion*. Next, we consider time series data and the possibilities it affords relative to reducing some of the limitations in the analysis imposed by the underdetermined nature of our GPA diagnostic system.

TABLE 5.5
Fault Estimates with Rough Initial Assignments

	PLA 80	PLA 75	PLA 70	PLA 65	PLA 60	Simulated Fault
Fan Eff	-0.71	-0.72	-0.73	-0.74	-0.74	-1.30
Fan FC	-0.01	0.01	0.03	0.03	0.03	-0.23
LPC Eff	-0.45	-0.46	-0.45	-0.45	-0.45	-0.75
LPC FC	-0.38	-0.39	-0.41	-0.42	-0.42	-0.68
HPC Eff	-1.76	-1.77	-1.75	-1.75	-1.77	-2.78
HPC FC	-0.22	-0.24	-0.22	-0.22	-0.22	-0.43
HPT Eff	-0.38	-0.39	-0.38	-0.39	-0.41	-0.15
HPT A4	0.18	0.20	0.20	0.21	0.21	0.11
LPT Eff	-0.22	-0.24	-0.27	-0.28	-0.28	-0.13
LPT A45	-0.04	-0.03	-0.02	-0.02	-0.02	0.07
N1 ER	-0.19	-0.21	-0.20	-0.20	-0.20	0.00
N2 ER	-0.18	-0.13	-0.17	-0.17	-0.20	0.00
Wf ER	0.05	0.06	0.07	0.07	0.06	0.00
T50 ER	0.22	0.24	0.24	0.24	0.26	0.00
T25 ER	-0.14	-0.16	-0.15	-0.16	-0.16	0.00
T3 ER	0.32	0.32	0.29	0.28	0.28	0.00
P25 ER	0.08	0.10	0.06	0.05	0.06	0.00
P3 ER	-0.13	-0.14	-0.14	-0.14	-0.14	0.00
P2 ER	0.07	0.07	0.08	0.08	0.06	0.00
T2 ER	-0.23	-0.24	-0.21	-0.21	-0.21	0.00
P5 ER	-0.08	-0.08	-0.07	-0.06	-0.04	0.00

5.4 TIME SERIES DATA

As mentioned previously,[4] the appearance of onboard engine monitoring hardware in the late ’70s ushered in the age of EHM. Data became available for analysis with every flight and allowed long-term trending to ensue. The time series data that was collected for trending was typically steady-state cruise data. When stable conditions were realized, 10–60 seconds of data would be collected, and the sample was time-averaged, thereby producing a vector of steady state gas path parameters that were stored on the QAR (or transmitted during flight through ACARS) for subsequent gas path analysis. Typically, this was done once or twice during flight, and from these *snapshots*, a series of (computed) *delta* vectors would be generated in the GPA process and trended over time. The early form of these trend charts, affectionately referred to as *worm charts*, were generated by the ground processing station and *visually* inspected for anomalies, [69–71] An early example is given in Figure 5.1; these were later replaced by more modern data plots as computerized processing evolved over time. Before the parameter deltas were plotted, some form of data

TABLE 5.6
Fault Estimates with T5 Error

	T5 ER=2%	T5 ER=5%	No T5 ER	Simulated Fault
Fan Eff	-0.75	-0.81	-0.71	-1.30
Fan FC	0.08	0.21	-0.01	-0.23
LPC Eff	-0.50	-0.56	-0.45	-0.75
LPC FC	-0.49	-0.66	-0.38	-0.68
HPC Eff	-1.95	-2.24	-1.76	-2.78
HPC FC	-0.28	-0.36	-0.22	-0.43
HPT Eff	-0.78	-1.38	-0.38	-0.15
HPT A4	0.11	0.00	0.18	0.11
LPT Eff	-0.62	-1.23	-0.22	-0.13
LPT A45	-0.04	-0.03	-0.04	0.07
N1 ER	-0.14	-0.06	-0.19	0.00
N2 ER	-0.17	-0.15	-0.18	0.00
Wf ER	-0.32	-0.89	0.05	0.00
T5 ER	1.06	2.32	0.22	[2, 5, 0]
T25 ER	-0.35	-0.68	-0.14	0.00
T3 ER	0.09	-0.25	0.32	0.00
P25 ER	0.13	0.21	0.08	0.00
P3 ER	0.09	0.42	-0.13	0.00
P2 ER	0.09	0.12	0.07	0.00
T2 ER	-0.47	-0.84	-0.23	0.00
P5 ER	0.01	0.13	-0.08	0.00

smoothing was performed to reduce the overall data scatter and enhance the analyst's ability to visually identify parameter changes. This typically required some form of rolling average and statistical manipulation. The parameters appearing in Figure 5.1 represent processed, smoothed data.

When a suspect trend was identified by the analyst, a visual delta–delta excursion from the main trend line for each parameter was noted (as illustrated in Figure 5.2). These parameter ΔΔs were then compared to a pre-assembled list of parameter shifts (referred to as a *fingerprint chart*) to determine a closest pattern match. Figure 5.3[5] illustrates an abbreviated fingerprint chart that might be used in the analysis. In the case depicted in Figure 5.2, it would be deemed that the HPT module had deviant performance, and a hot section borescope probably would have been ordered. Depending on the damage observed during the borescope inspection, a decision would be made as to whether maintenance was required before the engine could return to service.

It should be evident that the process described in the foregoing paragraphs was labor intensive and could be quite subjective depending on the analyst's experience

TABLE 5.7
Sensor Fault Lower Limit Matrix

1.00	0.11	0.49	0.06	0.19	0.11	0.24	0.18	0.56	0.10	0.04
0.12	1.00	0.13	0.01	0.15	0.17	0.77	0.23	0.02	0.20	0.07
0.16	0.04	1.00	0.45	0.07	0.01	0.09	0.41	0.23	0.13	0.23
0.04	0.01	0.83	1.00	0.17	0.18	0.13	0.34	0.03	0.20	0.06
0.18	0.13	0.19	0.25	1.00	0.29	0.72	0.18	0.03	0.45	0.06
0.11	0.14	0.03	0.27	0.29	1.00	0.27	0.31	0.06	0.31	0.05
0.07	0.19	0.07	0.06	0.22	0.08	1.00	0.03	0.28	0.04	0.10
0.09	0.10	0.59	0.26	0.09	0.16	0.06	1.00	0.07	0.11	0.24
0.28	0.01	0.34	0.03	0.02	0.03	0.51	0.07	1.00	0.27	0.44
0.09	0.16	0.36	0.29	0.44	0.30	0.13	0.21	0.49	1.00	0.01
0.04	0.06	0.68	0.10	0.07	0.05	0.34	0.49	0.88	0.02	1.00

TABLE 5.8
Sensor Fault Upper Limit Matrix

1.00	8.45	6.14	26.07	5.66	9.24	15.02	11.29	3.56	10.80	27.21
9.37	1.00	25.48	127.7	7.91	6.90	5.18	9.95	114.5	6.16	16.10
2.06	7.69	1.00	1.20	5.22	34.64	13.98	1.70	2.95	2.77	1.47
16.22	71.65	2.23	1.00	3.93	3.72	16.72	3.83	37.26	3.43	10.00
5.29	6.66	14.54	5.89	1.00	3.45	4.64	10.79	63.96	2.29	14.37
8.75	5.90	98.03	5.67	3.50	1.00	12.53	6.43	31.55	3.39	18.43
4.18	1.30	11.61	7.48	1.38	3.68	1.00	17.69	1.94	7.61	2.92
5.44	4.33	2.45	2.97	5.57	3.27	30.66	1.00	14.52	4.70	2.04
1.79	52.03	4.44	30.18	34.54	16.78	3.52	15.18	1.00	2.05	1.14
9.80	5.04	7.49	5.01	2.22	3.24	24.81	8.85	3.70	1.00	61.08
27.24	14.53	4.39	16.08	15.41	19.47	10.51	4.23	2.27	67.38	1.00

TABLE 5.9
Sensor Fault Test Thresholds (Ti) & Ratios

Sensor ER	*Ti*	T5 ER = 2%	Ratio_1	T5 ER = 5%	Ratio_2
N1 ER	0.82	-0.14	0.17	-0.06	0.07
N2 ER	1.32	-0.17	0.13	-0.15	0.11
wf ER	0.56	-0.32	0.58	-0.89	1.59
T5 ER	0.77	1.06	1.37	2.32	3.01
T25 ER	1.18	-0.35	0.30	-0.68	0.57
T3 ER	1.60	0.09	0.06	-0.25	0.15
P25 ER	0.49	0.13	0.27	0.21	0.42
P3 ER	0.71	0.09	0.13	0.42	0.60
P2 ER	0.44	0.09	0.20	0.12	0.28
T2 ER	0.95	-0.47	0.49	-0.84	0.88
P5 ER	0.61	0.01	0.01	0.13	0.22

TABLE 5.10
Fault Estimates with Large Measurement Error Recovery

	EGT ER = 2% with Recovery Value 2.55	*EGT* ER = 5% with Recovery Value 5.55	No T5 ER
Fan Eff	-0.70	-0.70	-0.71
Fan FC	-0.03	-0.03	-0.01
LPC Eff	-0.44	-0.44	-0.45
LPC FC	-0.35	-0.35	-0.38
HPC Eff	-1.71	-1.71	-1.76
HPC FC	-0.21	-0.21	-0.22
HPT Eff	-0.27	-0.27	-0.38
HPT A4	0.21	0.21	0.18
LPT Eff	-0.11	-0.11	-0.22
LPT A45	-0.04	-0.04	-0.04
N1 ER	-0.20	-0.20	-0.19
N2 ER	-0.18	-0.18	-0.18
Wf ER	0.16	0.16	0.05
T5 ER	**2.53**	**5.53**	0.22
T25 ER	-0.08	-0.08	-0.14
T3 ER	0.38	0.38	0.32
P25 ER	0.07	0.07	0.08
P3 ER	-0.19	-0.19	-0.13
P2 ER	0.06	0.06	0.07
T2 ER	-0.16	-0.16	-0.23
P5 ER	-0.10	-0.10	-0.08

and competence. These concerns were the catalyst for an attempt to automate the process and remove the subjective uncertainty.

One solution that emerged involved the use of a discrete time Kalman filter (see Addendum A5.2). The *predictor–corrector* nature of the estimation algorithm inherent in the Kalman filter approach can provide *one* mechanism for incorporating additional information in the sense that a *good* prediction supports the estimation accuracy by reducing the (diagnostic information) burden on the limited gas path measurements (*corrector*). For example, the incorporation of (assumed) average engine component deterioration levels that were alluded to in the preceding discussion on deterioration models is a step in this direction. The general form of the Kalman filter as applied to time series data (at time *k*) is given below (from Addendum A5.2).

$$\begin{aligned}
\hat{x}(k|k-1) &= \Phi(k)\hat{x}_{k-1} \\
P(k|k-1) &= \Phi(k)\, P_{k-1}\, \Phi^T(k) + Q_k \\
D_k &= P(k|k-1)\, H_k^T \left(H_k\, P(k|k-1)\, H_k^T + R_k\right)^{-1} \\
\hat{x}_k &= \hat{x}(k|k-1) + D_k\left[Z_k - H_k\, \hat{x}(k|k-1)\right] \\
P_k &= \left[I - D_k H_k\right] P(k|k-1)
\end{aligned} \tag{5.9}$$

where

$$\begin{aligned}
\Phi(k) &= \textit{State Transition Matrix} \\
H_k &= \textit{IC matrix} \\
D_k &= \textit{Kalman Gain (Diagnostic matrix)} \\
R_k &= \textit{Measurement covariance} \\
P_k &= \textit{State covariance} \\
Z_k &= \textit{Measurement } \Delta \textit{ vector} \\
\hat{x}_k &= \textit{Fault } \Delta \textit{ estimate vector}
\end{aligned}$$

Equations (5.9) provide a direct means to incorporate deterioration model information directly in the form of the state transition matrix, Φ, as described in Addendum A5.3. The use of Equation (5.9) directly as written, however, can introduce some numerical instabilities. These difficulties are well known [5–7]. These problems stem from the propagation of error introduced by the finite word length of the digital computer and can cause the covariance update to produce a P_k that is no longer positive definite. This, in turn, will corrupt the diagnostic matrix D_k and, hence, the fault estimates $\hat{x}_k$. As can be found in the cited references, there are a number of reformulations possible that will mitigate these difficulties but will also add some additional complexity to the implementation. We propose a more ad hoc approach

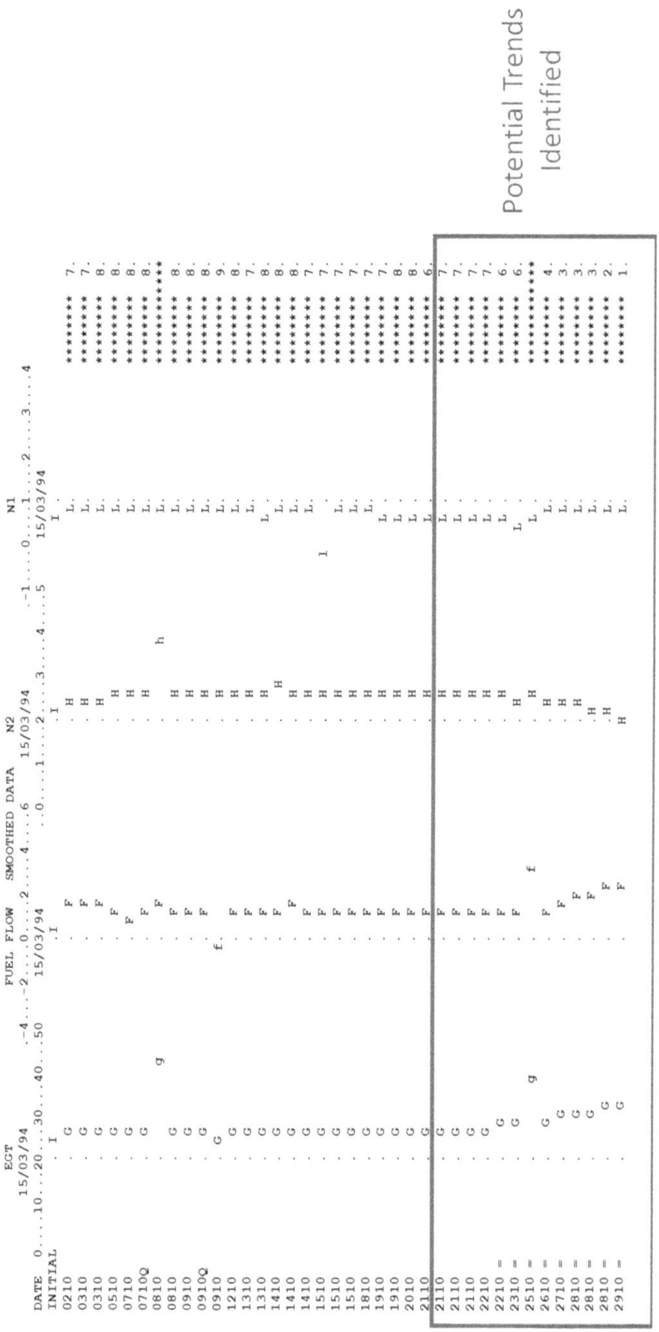

FIGURE 5.1 Worm chart

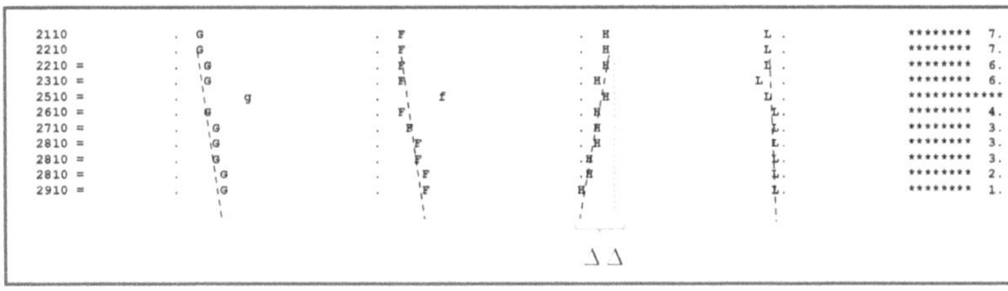

- EGT = + 6 ° C
- Fuel Flow = + 1.2 % Δ
- N2 = - 0.6 % rpm
- N1 = + 0.2 % rpm

FIGURE 5.2 Determine magnitude and direction of parameter ΔΔs

that maintains simplicity, solves the problem, and, while not (mathematically) optimal, still provides reasonable estimation accuracy.

The approach we will use is, in effect, a "fixed" Kalman gain approach in the sense that the state covariance will *not* be extrapolated or updated for $k \geq 1$. In doing so, the Kalman equations become the following for $k \geq 1$.

$$\begin{aligned} \hat{x}(k|k-1) &= \Phi(k)\hat{x}_{k-1} \\ D_k &= P_0 H_k^T \left(H_k \, P_0 \, H_k^T + R_k \right)^{-1} \\ \hat{x}_k &= \hat{x}(k|k-1) + D_k \left[Z_k - H_k \; \hat{x}(k|k-1) \right] \end{aligned} \tag{5.10}$$

This is essentially the same implementation discussed in the previous section for multiple snapshots, where $\Phi = I$. The relevant difference for time series is that we have essentially an open-ended stream of discrete data points taken (predominantly) at steady-state cruise conditions. This fact presents an opportunity to capture additional information that we can apply to our analysis methodology. The characteristic of this general information-leveraging potential comes by way of *inspecting* a time series of data points rather than just a single point in isolation. It is concerned with

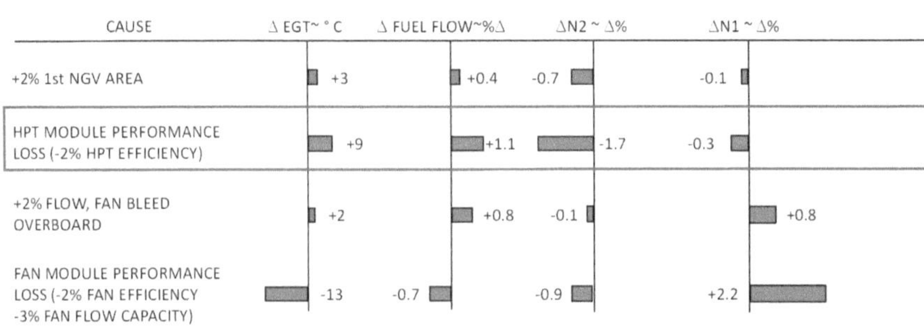

FIGURE 5.3 Abbreviated fingerprint chart

the difference between *gradual* deterioration and a *rapid* deterioration, in the temporal sense, and has provided a partial solution to the *smearing* effect discussed previously.

5.5 SEPARATING GRADUAL AND RAPID DETERIORATION

Engine performance changes can manifest themselves in one of two ways: 1) gradual (*long*-term) deterioration or 2) rapid (*short*-term) deterioration [71]. These may affect component (module) performance changes that can be characterized in terms of estimated parameters, such as efficiencies, flow capacities, and effective nozzle areas, or can be changes in performance of engine sub-systems, such as bleeds, cooling flows, and variable geometry mechanisms. Similarly, sensor faults can also admit *gradual* drift or *rapid* drift or failure. The methodology that we will introduce involves reconfiguring the general GPA solution to include fault detection to annunciate rapid fault conditions when they occur and apply separate analysis techniques more suitable to these conditions. In short, we have the following multi-step procedure.

1. Estimating module performance and sensor error shifts via Kalman filter GPA method.
2. Perusing the measurement delta time series to *detect* any *rapid* and *persistent* shifts.[6]
3. Making a *single fault* assumption and a separate analysis consistent with this assumption to identify the underlying fault if a *rapid* and *persistent* shift is detected in the time series.
4. Applying appropriate accommodation logic to reconcile the analysis made in Step 1 with the results from Step 3.

The *single fault* assumption made in Step 3 is a concession in the form of heuristic information based on experience, i.e., it is improbable that more than one failure occurs at the same time. The methodology that is outlined above combines a multiple fault analysis (MFA) with a single fault analysis (SFA) approach defined in such a manner as to allow for both of these processes to operate in concert with one another, automatically, and without corruptive interaction. It might be considered an example of algorithm fusion driven by observational information.

5.5.1 Partitioning the Measurement Δ

The success of a practical diagnostic system depends heavily on its ability to partition this total deviation into constituent parts representing the different fault scenarios to allow proper fault identification.

Without much loss of generality, we can consider two different fault mechanisms

1) Gradual deterioration
2) Rapid deterioration

The *former* concerns itself with the normal process of component degradation over time and would produce slowly, increasing and/or decreasing (average) changes in the measurement deviations, whereas the *latter* is the result of a system fault event that would produce a rapid (temporal) trend shift in the measurement deviations. Figure 5.4 illustrates this concept for an arbitrary measurement deviation.

In our notional depiction of Figure 5.4, we have introduced some new notation. $Z_0^{(INIT)}$ represents the average % delta offset of the installed engine from the nominal baseline reference, and Z_k denotes the measurement delta from the installed level at time k. If a fault event is detected at time k, then $Z_k = Z_k^{(MFA)} + \Delta\Delta_k$, where $\Delta\Delta_k$ is the temporal % measurement $\Delta\Delta$ due to the fault event. The remaining measurement delta, $Z_k^{(MFA)}$, is the quantity used in Equation (5.9), as Z_k performs the normal GPA. The superscript annotation MFA designates MFA to distinguish it from $\Delta\Delta_k$, which will be the quantity used in the SFA described in the next sub-section. Inherent in all of this are the following two general assumptions.

1) Gradual deterioration exists for all components under consideration, which is assumed to be characterized by slow performance shifts in *all* of the major modules of the gas turbine (for example, FAN, LPC, HPC, HPT, LPT) as well as the sensors for each of the measurements. The analysis methodology for this type of fault identification will be referred to as MFA.
2) Rapid degradation occurs as the result of a *single* component failure (full or partial). The analysis methodology for this type of temporal *event* fault identification will be referred to as SFA.

Thus, the overall isolation strategy at time k requires that we

1) Recognize which fault mechanism (*slow* or *rapid*) is occurring.
2) Partition the current total measurement deviation into its appropriate constituent parts.
3) Apply the appropriate analysis methodology to isolate the fault at time k

5.6 SFA

As we have already discussed, engine (and sensor) performance changes can manifest themselves in one of two categories: 1) gradual (*long*-term) deterioration or 2) rapid (*short*-term) deterioration. In this section, we will concern ourselves with the latter and consider the problem of fault *isolation*, given the premise that a fault event has been detected. The problem of detection becomes one of recognizing a step or rate change in a gas path parameter or a collection of parameters. Some of the problems associated with fault detection and the mechanisms that can be applied to accomplish this task have already been discussed in Chapter 2 and will not be directly addressed in this section. We will assume that a method exists to *detect* the fault event by virtue of detecting that a (persistent) shift (or rapid change) in a set of measured engine parameters has occurred at some discrete time k.

It is important to recognize that there can exist (or should exist) a certain time *latency* inherent in the detection process. If we denote by k_{start} and k_{detect} the *start* and *detection* (discrete) times, respectively, of a temporal event, then it is generally true that $k_{detect} \neq k_{start}$. The reason for this is that it would not be prudent to declare an event detection on the basis of one data point, as it might be a statistical outlier, and a practical architecture should support this philosophy. The notion of a *lateny period* is pivotal to constructing an architecture that will support both MFA and SFA processing. In addition to *knowing* detection time k_{detect}, it is also important to know when the event has concluded or stabilized. We will denote this time by k_{end}, and it is assumed that the trend detection algorithm, which is monitoring the measured parameters, estimates this time along with k_{start}. For the purpose of this discussion, we will define the period between the *start* and *end* of the event as the *event period*. Figure 5.5 provides a hypothetical graphical illustration of a *short*-term event, *start*, *detection*, and *end* time defining the latency and event periods.

Before we outline a particular methodology for the SFA, there are a couple of matters to note.

An assumption we shall make is that processing (fault detection and isolation) is taking place at each discrete time point. The *default* isolation process is taken to be the MFA, wherein performance changes (relative to reference) in the modules and sensors are inferred from the collection of measurement parameter changes (i.e., Δs from reference) at time *k* using the Kalman filter GPA. Once a trend detection is confirmed, and the start time of the event is established, we will switch to a different isolation strategy (SFA) beginning in the *latency period*. This necessitates that we *reprocess* the data within the *latency period*, as prior to trend annunciation, this data was analyzed by MFA methods with a possible attendant corruption of performance results. For example, if an engine bleed leak was responsible for the observed trend shift, the MFA processing would have erroneously assigned the problem to

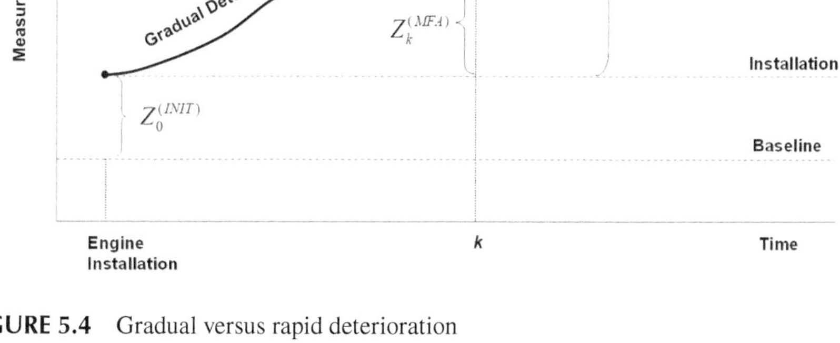

FIGURE 5.4 Gradual versus rapid deterioration

a combination of component performance and or sensor error faults (as the bleed problem is not among the $n_e + n_s$ faults addressed by the MFA algorithm). Another important point is that it is essential that the SFA operate on *temporal* changes, i.e., measurement parameter $\Delta\Delta$s, as indicated in Figure 5.5. This provides a performance $\Delta\Delta$ from the initiation of the event. The reference in this case is whatever the performance was at discrete time $k_{start}-1$. This is what distinguishes a *short*-term (event) deterioration from *long*-term (gradual) deterioration.

Before we can describe an illustrative SFA methodology, we need to consider what types of *short*-term (event) deterioration faults we can estimate. In general, these can be engine module performance faults, sensor faults, and engine sub-system faults. The latter would consist of things such as engine bleed system faults, variable geometry faults (e.g., compressor variable stator vane problems), turbine case cooling faults, and the like. The vector of possible faults might look something like the following.

$$\Delta x_k^{(SFA)} = \begin{bmatrix} \Delta FAN \\ \Delta LPC \\ \vdots \\ \Delta LPT \\ \\ \Delta N1_{error} \\ \Delta N2_{error} \\ \vdots \\ \Delta Bleed_{25} \\ \Delta Bleed_{3} \\ \Delta SVA \\ \vdots \\ \Delta TCC \end{bmatrix} = \left.\begin{array}{l} \left.\begin{array}{c} \textit{coupled FAN fault} \\ \textit{coupled LPC fault} \\ \vdots \\ \textit{coupled LPT fault} \\ \\ \end{array}\right\} n_e \\ \left.\begin{array}{c} N_1 \textit{ sensor error} \\ N_2 \textit{ sensor error} \\ \vdots \end{array}\right\} n_s \\ \left.\begin{array}{c} \textit{station 25 bleed fault} \\ \textit{station 3 bleed fault} \\ \textit{HPC stator vane fault} \\ \vdots \\ \textit{Turbine Case Cooling fault} \end{array}\right\} n_{sys} \end{array}\right\} n \tag{5.11}$$

where

$$\begin{aligned} n_e &= \textit{number of engine module faults} \\ n_s &= \textit{number of sensor error faults} \\ n_{sys} &= \textit{number of engine sub-system faults} \\ n &= n_e + n_s + n_{sys} \end{aligned}$$

Once a trend shift in one or more measurement delta parameters is confirmed, and the (discrete) time, k_{start}, of the underlying fault is determined, temporal $\Delta\Delta$s can be calculated, as indicated in Figure 5.5. Typically, the (temporal) reference level, from which the $\Delta\Delta$ is calculated, is taken to be a *rolling* or *exponential* average of previous points before the start of the fault event. The calculation applies to the entire *event period*, including the reprocessed *latency period*. The $\Delta\Delta_k$ values are computed as the difference from the current measurement Δ and the average level (prior to the event), i.e.,

$$\Delta\Delta_k(i) = Z_k(i) - A_{kstart-1}(i) \quad , \quad i = 1,2,\ldots,m \qquad (5.12)$$
$$for \;\; k \geq k_{start}$$

where (if we use exponential averages)

$$A_k = \alpha \; A_{k-1} + (1-\alpha)\, Z_k \qquad (5.13)$$
$$0 < \alpha < 1$$

Following these steps, we have at current time k within an *event period* (i.e., $k_{start} \leq k \leq k_{end}$), the $\Delta\Delta_k$ vector quantities, which provide a *signature* in the measurement $\Delta\Delta$ space representative of the underlying *event fault* $\Delta x_k^{(SFA)}$. The important factor is that *reprocessing* within the *latency* period allows a separate method to be applied to calculate the incremental change in performance $\Delta\hat{x}_k^{(SFA)}$ due to the *event* that can then be added back to the *known* (estimated) performance state $\hat{x}_{k_{start-1}}$ (before the *event* initiated) to form a more accurate estimate, i.e.,

$$\hat{x}_k = \hat{x}_{k_{start-1}} + \Delta\hat{x}_k^{(SFA)} \quad for \;\; k_{start} \leq k \leq k_{end} \qquad (5.14)$$

This provides some mitigation for the well-documented problem of "fault smearing" in an estimation process of an underdetermined system in which the number of faults is larger than the number of measurements, i.e., $n>m$.

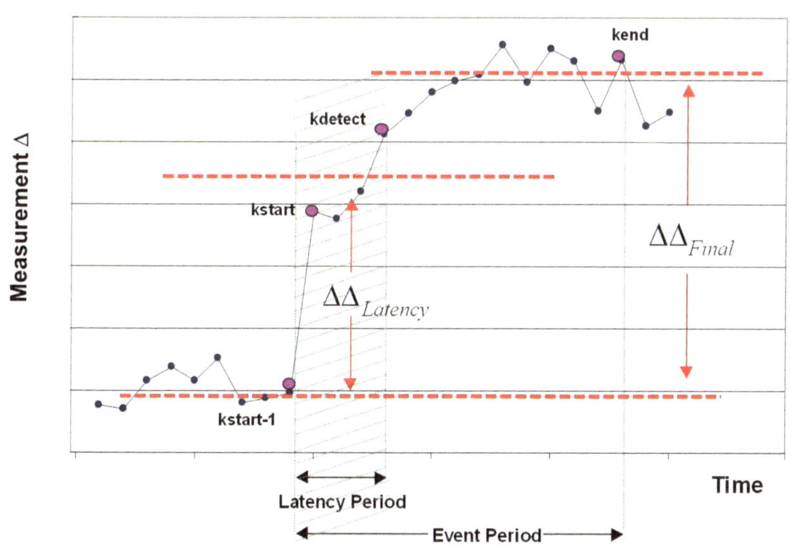

FIGURE 5.5 Latency and event periods

To address the task of isolating the (single) fault in question, various algorithms exist. These range from methods employing statistical, fuzzy logic, Bayesian belief networks and artificial neural network (ANN) elements.[7] For simplicity, what we will describe in the remainder of this section is an approach that draws upon the methods we have already introduced (for the MFA) utilizing a Kalman filter type of processing. To set up the model, we need a set of ICs that describe the potential single faults in question. Using Equation (5.11) as a guide, we already have the first two of the three IC components, i.e., H_e and H_s. The third component, which we will denote by H_{sys}, deals with the engine sub-system faults. An example of a full SFA IC matrix is illustrated in Figure 5.6 for the engine simulation.[8] This set of ICS was generated at cruise condition (35,000 feet, Mach number 0.75, *EPR* = 1.452). Coupling factors for Fan (1.3), LPC (1.3), HPC (1.2), HPT (-0.7), and LPT (-0.6) are applied to create the major module performance fault ICs, wherein each column represents 1% performance increase (a polarity change from 1% performance loss). The variable bleed valve (VBV) column represents a VBV actuation system fault (or bleed leakage) that results in 1% core flow loss. The variable stator vane (VSV) column represents an actuation system fault that results in 1 degree bias of the stator vanes on average. Each of the sensor columns represent 1% bias on the particular sensor. *PSP* used in this case is *EPR*.

Thus, our SFA model can be expressed as

$$
\begin{aligned}
\Delta\Delta_k &= H_k^{(SFA)} \Delta x_k^{(SFA)} + \varphi \\
H^{(SFA)} &= \left[H_e \vdots H_s \vdots H_{sys} \right] \quad , \quad m \times (n_e + n_s + n_{sys}) \; matrix \\
&\varphi = (m \times 1) \; noise \; vector, \quad \varphi \in N(0, R)
\end{aligned}
\tag{5.15}
$$

Equation (5.15) is a familiar format, and we can apply our Kalman filter (KF) solution as we have done previously; however, we will need to make a slight alteration. The usual KF solution would take the form given in Equation (5.16).

$$
\Delta\hat{x}_k^{(SFA)} = P_0 (H_k^{(SFA)})^T \left[H_k^{(SFA)} P_0 (H_k^{(SFA)})^T + R_k \right]^{-1} \Delta\Delta_k \tag{5.16}
$$

The “alteration” will be in how we choose the P_0 matrix. The *single fault* isolation is obtained by processing Equation (5.16) *iteratively* to provide a *snapshot* analysis for *each* of the $n = n_e + n_s + n_{sys}$ single fault causes under consideration. Each *iteration* of Equation (5.16) will be made with a different P_0 matrix chosen to accentuate the jth single fault cause and zeroing out all others. Because these are snapshot analyses, the covariance update calculation is not required. The a-priori state estimate is also assumed to be zero. This process will yield estimates for *each* single fault under consideration. It is necessary to *rank* each of these estimates and determine the top two or three single faults. A measure, which could be used to compare estimates for this purpose of ranking, is a normalized measurement error norm, which we will describe below. The single fault admitting the minimum error is deemed the most likely, the second smallest error, the next likely, and so forth.

The P_0 matrix in Equation (5.16) is an $n \times n$ *diagonal* matrix, where $n = n_e + n_s + n_{sys}$. In the Jth iteration, we set the Jth element to a large number, say 1000, i.e.,[9]

$$P_0 = diag\left\{0, 0, \cdots, 0, \underbrace{1000}_{J^{th}}, 0, \cdots, 0\right\} \Rightarrow \Delta\hat{x}_k^{(SFA)}(i) = 0 \ \forall i \neq J \tag{5.17}$$

Thus, after the nth iteration, we have at discrete time k, $(k_{start} \leq k \leq k_{end})$, n potential single fault solutions, $\{\Delta\hat{x}_k^{(SFA)}(1), \Delta\hat{x}_k^{(SFA)}(2), \cdots, \Delta\hat{x}_k^{(SFA)}(n)\}$. To determine which of these is the best solution, we need a measure of some sort to rank the solutions. One possibility is to apply the IC matrix $H_k^{(SFA)}$ to each of these to obtain the measurement $\Delta\Delta$ vector that would be anticipated if, indeed, the Jth fault estimated, $\Delta\hat{x}_k^{(SFA)}(J)$, was the true underlying fault, and then compare this to the observed measurement $\Delta\Delta$ vector for *closeness*. Assume, for the moment, that we have a metric that reflects a level of closeness that is denoted by $\Delta\Delta error_k$.

$$\begin{aligned} \Delta\Delta_k^{(j)} &= H_k^{(SFA)}\Delta\hat{x}_k^{(SFA)}(j) \quad , \quad 1 \leq j \leq n_{sys} \\ \Delta\Delta error_k^{(j)} &= closeness(\Delta\Delta_k \, , \Delta\Delta_k^{(j)}) \end{aligned} \tag{5.18}$$

In Equation (5.18), $\Delta\Delta_k^{(j)}$ is an $m \times 1$ vector of measurement shifts we would expect to see with the estimated *single* fault $\Delta\hat{x}_k^{(SFA)}(j)$, and $\Delta\Delta_k$ is the $m \times 1$ vector of measurement shifts actually observed. For our closeness metric, we will take a normalized Euclidean distance,[10] as defined in Equation (5.19).

$$\Delta\Delta error_k^{(j)} = \frac{\sqrt{\sum_{i=1}^{m}\left(\frac{\Delta\Delta_k(i) - \Delta\Delta_k^{(j)}(i)}{\sigma_i}\right)^2}}{\sqrt{\sum_{i=1}^{m}\left(\frac{\Delta\Delta_k(i)}{\sigma_i}\right)^2}} \tag{5.19}$$

The σ_i , $i = 1,2,...,m$ represent the familiar measurement Δ parameter's non-repeatabilities used in calculating the measurement covariance R. Ordering the errors from *min* to *max* we obtain

$$\Delta\Delta error_k^{(j_1)} \leq \Delta\Delta error_k^{(j_2)} \leq \cdots \leq \Delta\Delta error_k^{(j_n)} \tag{5.20}$$

corresponding to the single fault estimates

$$\left\{\Delta\hat{x}_k^{(SFA)}(j_1), \Delta\hat{x}_k^{(SFA)}(j_2), \cdots, \Delta\hat{x}_k^{(SFA)}(j_n)\right\} \tag{5.21}$$

in order of likelihood. In most cases, the first of these ranked single faults is deemed to be the underlying fault and is reported to the user. In some instances, however, it may be prudent to report the first and second single faults, as there exists the possibility for erroneous fault identification, especially if the associated error norms are "close." Depending on the circumstances, the SFA may experience confusion between two single faults. The aliasing of single faults depends on the single faults themselves, their relative magnitude with respect to the non-repeatability of the measurements, and the number of measurements available for the analysis. All these factors can contribute to a *confounding* or *ambiguity* of the underlying single fault with another single fault. To mitigate the possibility of false identifications, some post-processing of the SFA results may be required. This will improve the overall accuracy of the system. We will defer discussion of this important topic to Chapter 9 when we discuss ambiguity and its resolution.

Before we present a few examples of the SFA method, it is worth mentioning that the technique can be easily extended to isolating *double* faults, *triple* faults, and so forth, up to m number of faults. To enable this for *double* faults, for example, we would have to consider all combinations of n faults taken two at a time, i.e.,

$$nC2 \equiv \binom{n}{2} = \frac{n!}{(n-2)!2!} = \frac{n(n-1)}{2}$$

cases.[11] For our sample system depicted in Figure 5.6, with $n = 18$, this would produce 153 double fault cases to consider. The procedure is essentially the same as described above, except the P_0 matrix would have two non-zero diagonal entries. For example, to estimate the double fault, $\left\{\Delta x_k^{(SFA)}(I), \Delta x_k^{(SFA)}(J)\right\}$, where $I \neq J$, we would set P_0 as indicated in Equation (5.22).

$$P_0 = diag\left\{0, \cdots, 0, \underbrace{1000}_{I^{th}}, 0, \cdots, 0, \underbrace{1000}_{J^{th}}, 0, \cdots, 0\right\} \tag{5.22}$$

If we let this double fault estimate be denoted by $\Delta\hat{x}_k^{(SFA)}(I,J)$, then we calculate the anticipated measurement $\Delta\Delta$ and attendant error, mimicking Equations (5.18) and (5.19).

$$\Delta\Delta_k^{(I,J)} = H_k^{(SFA)} \Delta\hat{x}_k^{(SFA)}(I,J) \tag{5.23}$$

$$\Delta\Delta error_k^{(I,J)} = \frac{\sqrt{\sum_{i=1}^{m}\left(\frac{\Delta\Delta_k(i) - \Delta\Delta_k^{(I,J)}(i)}{\sigma_i}\right)^2}}{\sqrt{\sum_{i=1}^{m}\left(\frac{\Delta\Delta_k(i)}{\sigma_i}\right)^2}}$$

	Fan	LPC	HPC	HPT	LPT	VBV	VSV	N1	N2	Wf	T5	T25	T3	P25	P3	P2	T2	PSP
ΔN1	-0.35	-0.18	0.01	0.01	0.50	0.08	0.00	1.00	0.00	0.00	0.00	0.00	0.00	0.00	0.00	0.55	-0.49	-0.48
ΔN2	0.24	-0.39	0.03	0.64	-0.36	0.31	-0.09	0.00	0.98	0.00	0.00	0.17	0.00	0.00	0.00	0.33	-0.49	-0.36
ΔWf	0.54	-0.92	-0.47	-0.56	0.01	0.84	-0.09	0.00	-0.07	1.00	0.00	0.13	0.00	0.00	0.00	0.98	-0.63	-1.99
ΔT5	0.48	-0.87	-0.60	-0.69	-0.76	0.68	-0.11	0.00	-0.09	0.00	1.00	0.16	0.00	0.00	0.00	0.71	-1.03	-0.75
ΔT25	-0.30	0.11	-0.24	-0.27	0.47	-0.34	-0.05	0.00	-0.03	0.00	0.00	1.06	0.00	0.00	0.00	0.18	-0.99	-0.12
ΔT3	0.13	-0.38	-0.43	0.58	-0.14	0.19	-0.07	0.00	-0.06	0.00	0.00	0.10	1.00	0.00	0.00	0.43	-0.90	-0.43
ΔP25	-1.08	1.39	-0.94	-1.08	1.66	-1.35	-0.18	0.00	-0.14	0.00	0.00	0.25	0.00	1.00	0.00	-0.49	0.00	-0.30
ΔP3	-0.09	0.11	0.02	1.03	0.28	-0.10	0.00	0.00	0.00	0.00	0.00	0.01	0.00	0.00	1.00	0.21	0.00	-1.16

FIGURE 5.6 Example SFA influence coefficients at cruise condition

As before, the errors are ranked

$$\Delta\Delta error_k^{(I_1,J_1)} \leq \Delta\Delta error_k^{(I_2,J_2)} \leq \cdots \leq \Delta\Delta error_k^{(I_{nC2},J_{nC2})} \tag{5.24}$$

which determines the ranking across all $nC2$ *double* faults. This naturally increases the computational burden and would only be feasible for ground-based processing. In theory, this procedure could be repeated for *triple* faults and so forth; however, it rapidly becomes impractical from a computational perspective as $nC3 = 816$, $nC4 = 3060$, etc. These values are assuming the total number of possible faults to be $n = 18$ for the hypothetical engine simulation. In a real application, a value of $n = 30$ would not be unusual, giving rise to values of 435 *double* faults, 4,060 *triple* faults, 27,405 *quadruple* faults, and so on. Fortunately, the *single fault hypothesis* works well in practice when trend shifts are detected and the *latency period is short*, before collateral damage can occur.

5.6.1 SFA Examples

In this section, we will use three fault scenarios to illustrate the SFA method using time series data obtained from the engine simulations:

- Case 1: major *module fault* (1% HPT efficiency loss, 0.7% A4 increase)
- Case 2: *sensor fault* (1.5% T5 measurement bias, which is about 8 *deg C*)
- Case 3: *subsystem fault* (1.9% core flow leakage due to VBV actuation system fault)

For all three cases, module health parameter adjustments, as shown in Figure 5.7, are introduced to simulate a deteriorated engine. The simulations are run to various cruise conditions, and an averaged steady-state data point is captured for each simulation. A total of 35 simulations are carried out, resulting in a time series of 35 data points (each representing one flight). The fault is injected in the 31st data point in the time series data.

5.6.1.1 Case 1: Major Module Fault

In this example, the fault being injected is a -1% coupled HPT performance fault (efficiency loss of 1% and A4 increase of 0.7%) and is added to the simulations starting from data point (or Flight) 31 and onward. The steady-state data points are normalized by a nominal cruise model to calculate the delta parameters, as shown in Figure 5.8. It can be observed that the delta parameters don't initialize near zero due to the simulated deteriorations. While the parameter shifts, at the 31st data point, are apparent on several parameters, such as Delta N2 and Delta T5, it may not be as visible on other parameters, such as Delta N1 and Delta T3, due to apparent sensor noise. Trending plots like these can be thought of as a more modern form of the *worm charts* mentioned earlier to monitor engine performance in service.

In practice, each incoming data point triggers the trend detection process (as described in Chapter 2). A fault is detected when a trend shift is identified in one or more parameters. For this dataset, a simple trend detector is able to detect a persistent shift, in multiple parameters, at Flight 33 (i.e., $k_{dectect} = 33$) and determine the start of this event (i.e., $k_{start} = 31$). To calculate the temporal ΔΔs using Equation (5.12), a (temporal) reference level for each delta parameter is taken to be an *exponential* average of previous points before the start of the fault event ($A_{kstart-1}$). The reference levels ($A_{kstart-1}$) and ΔΔs calculated (by the trend detector) for the subsequent data points in the latency period are listed in Table 5.11. The last row ($\Delta\Delta_{(avg)}$) is the average of the ΔΔs calculated manually and added to the table to illustrate the effect of apparent sensor noise.

The ΔΔ parameters in Table 5.11 are subsequently processed by an SFA algorithm, as described by Equations (5.15)–(5.18), the top three fault isolation results, together with the normalized errors as defined by Equation (5.19), are listed in Table 5.12 below. With the presence of realistic sensor noises in this case, the SFA is able to correctly identify the fault and provide rather accurate estimates as compared to the "simulated truth (-1% HPT performance fault)." The normalized errors of the top candidate are also relatively small as compared to the second and third candidates.

The detection and isolation of an abrupt fault (by the SFA) provides additional evidence to guide the MFA process as explained earlier. Therefore, when reprocessing the data within the latency period, the SFA estimates are utilized to adjust the MFA estimates using Equation (5.14). This allows some mitigation of the smearing effect caused by the abrupt fault. The MFA outputs are shown in Figures 5.9 and 5.10, wherein the orange curve indicates the results when considering SFA estimates and the blue without. It can be observed that the smearing of the HPT fault on HPC and other components has been improved by the use of SFA estimates.

5.6.1.2 Case 2: Sensor Fault

For an easy comparison with the previous case, the first 30 data points of Case 2 are identical to the data used in Case 1. A T5 sensor bias of 8 *deg C* (or 1.5%) is added to Data Point 31 and onward. The delta parameter trending plots for this case are shown in Figure 5.11. The only parameter that exhibits a noticeable (and detectable) trend shift is Delta T5.

Similarly, the trend detection process would recognize a persistent shift at Flight 33 (i.e., $k_{dectect} = 33$) on Delta T5 and determine the start of this event at $k_{start} = 31$. The reference levels ($A_{kstart-1}$) and Δ Δs calculated for the subsequent data points in the latency period are listed in Table 5.13.

Following the same procedure, the delta–delta parameters from Table 5.13 are processed through the SFA. The top three fault isolation results, as well as the normalized errors, are shown in Table 5.14. For all the faulty data points, the SFA correctly identifies the fault and provides rather accurate estimates as compared to the "simulated truth (1.5% T5 measurement bias)."

Similar to Case 1, the data within the latency period are reprocessed to remove the smearing effect of the T5 sensor fault. The MFA outputs are shown in Figures 5.12 and 5.13, wherein the orange curve indicates the results when considering SFA estimates and the blue without. It is clear that the smearing induced by the T5 sensor fault has been well compensated by the use of the SFA estimates.

5.6.1.3 Case 3: Sub-system Fault

In this case, we introduce a VBV actuator fault starting from Point 31. The first 30 data points are identical to the previous cases. The delta parameter trending plots are shown in Figure 5.14. Trend shifts are apparent on multiple parameters, which can be easily detected by a trend detector. Similar to the first two cases, ΔΔs are calculated and shown in Table 5.15; SFA fault isolation results are shown in Table 5.16. The SFA correctly identifies the fault and provides rather accurate estimates as compared to the "simulated truth (1.9% core flow leakage due to VBV fault)."

There are a few ways to compensate for the VBV fault in the MFA. One simple method is to remove the identified fault influence from the ΔΔs in the latency period, i.e., subtracting the VBV ICs multiplied by the estimated fault size in this case. The resulting MFA outputs are shown in Figures 5.15 and 5.16. It can be observed that the smearing effect of the VBV fault has been well compensated.

5.7 OTHER CONSIDERATIONS

In the foregoing sections, we have outlined a general methodology for the analysis of time series data based on a KF formulation developed in Chapter 4. Before we move on to a short discussion on alternative approaches, there are several topics that are worthy of some consideration. The discussions to follow are intentionally brief and intended only as an introductory overview to these topics. Where applicable, references are provided for the interested reader.

5.7.1 Fault Selection & Sensor Selection

Relative to the MFA, the more faults that are selected to be included in the model, the more difficult the estimation problem, as $n \gg m$. There are a number of ways that one can attempt to address this basic difficulty. One approach is to introduce additional information. We have seen several ways to employ this tactic in the examples in this chapter as well as those in Chapter 4, e.g., the use of deterioration models to provide *a-priori* estimates, *coupling* engine module faults, assumptions regarding the health of the sensors and underweighting those faults through the P_0 matrix, the use of trending and single fault assumption when a measurement delta trend shift is detected, etc. The underlying theme for these types of accommodations is one of adding information (independent of the m measurements) to lessen the discrepancy between n and m. Another possibility is to select the faults to be estimated based on the informational content of the m sensors in our measurement suite. The visibility of any given fault depends on the measurement suite at hand. Some measured parameters contain greater thermodynamic information for a given engine module fault than others. A quick glance at Figure 1.1 suggests, for example, that most of the measured parameters we have been using are concentrated in the core of the engine, i.e., the high spool. Thus, it should not be surprising that high spool faults are better represented and easier to estimate. This raises the obvious question as to what might be the best or optimal choice of engine faults to model for a given set of measurements. Optimality might be defined in terms of minimizing an estimation error.

In practice, we may be interested in a set of ancillary parameters,[12] y, that are dependent on our independent fault parameters, x. In general, we have

$$\begin{aligned} Z &= H\,x + \varphi \\ y &= G\,x \end{aligned} \tag{5.25}$$

and we estimate

$$\begin{aligned} \hat{x} = DZ &= P_0 H^T \left(H P_0 H^T + R \right)^{-1} Z \\ \hat{y} &= G\hat{x} \end{aligned} \tag{5.26}$$

The ancillary parameters represented by y might contain elements such as operating line shifts for the compression elements (FAN, LPC, HPC), takeoff thrust, *EGT* margin change at tropical day conditions, cruise SFC, and the like. One interesting approach to performing an optimal fault selection was introduced by Litt [12] and later generalized by Simon and Garg [13], [14]. Considering only engine performance faults, x_e, where we have $n_e > m$, the essential theme underlying these three papers is to determine a linear transformation V^* in which the linear combination of engine faults, $q = V^* x_e$, become the new (virtual) set of faults to be estimated by the KF, producing $\hat{q}$, that minimizes the estimation error of the ancillary parameters of interest, $\hat{y} = G(V^*)^{\dagger}\hat{q}$. The matrix V^* is $\ell \times n_e$ where $\ell \le m$ and (V^*) represents the pseudo-inverse of V^*. It should be noted that the module faults, x_e, can be represented in the vector y, in Equation (5.28) by proper choice of the G matrix. This approach has utility in exploring, for a given measurement suite, the potential fault and/or ancillary parameter estimation accuracy in design studies. It does, however, have some limitations, one of which is that it requires that the data being analyzed be at a *fixed steady state* condition.[13] The calculation of V^* cannot be made in closed form, thereby requiring a computationally intensive search algorithm to approximate its value, which would only be possible off-board.

Health management goals can often dictate what fault visibility is required in a diagnostic system. We might then consider the converse question, i.e., given a fixed set of faults to be estimated, what would be the best sensor measurement suite to estimate them? This is not a new question and, in fact, was explored by Urban in the early ’70s. He was not alone, and many other researchers have explored this question, and some publications on this issue have emerged in recent years [15–19]. The methods are varied and, to some degree, depend on the diagnostic estimation methodology being utilized. However, as a very general statement, the approaches

Fan Eff	Fan Flow	LPC Eff	LPC Flow	HPC Eff	HPC Flow	HPT Eff	A4	LPT Eff	A45
-1%	-1.20%	-0.70%	-0.90%	-2%	-2.20%	-2.50%	1.70%	-0.80%	0.60%

FIGURE 5.7 Simulated engine deteriorations

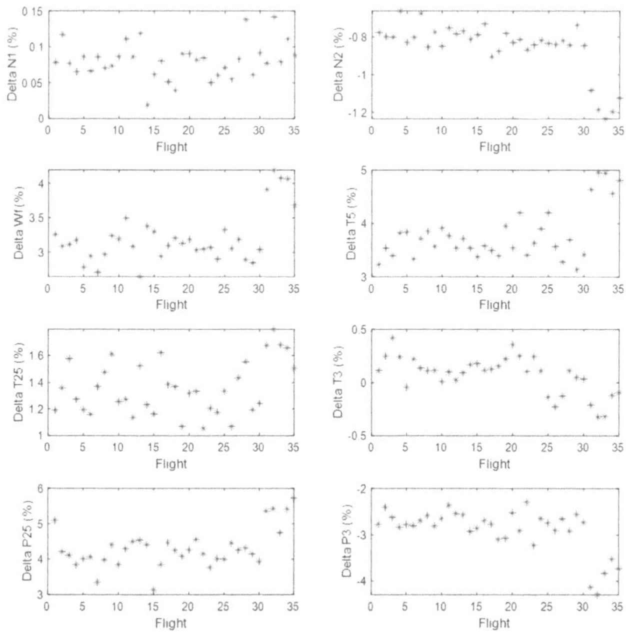

FIGURE 5.8 Simulated time series data with HPT fault

usually begin with the assumption that a *basic* set of sensors are already in use (typically for control purpose) and the question becomes one of selecting additional sensors to add to this *basic* sensor suite. Therefore, given a set of n possible sensors to add, what would be the *best* sub-set of r sensors, $1 \le r \le n-1$, that would minimize *fault* estimation error and maximize accuracy. Different authors have used different error metrics for this purpose, but many typically rely on massive simulation studies in which each of the sub-sets of n sensors taken r at a time are evaluated across all possible fault scenarios.

TABLE 5.11
Delta–Delta Parameters and Reference Levels for Case 1

	N1	N2	wf	T5	T25	T3	P25	P3
$A_{kstart-1}$	0.08	-0.82	3.03	3.55	1.28	0.04	4.14	-2.75
$\Delta\Delta_{(31)}$	0.00	-0.26	0.88	1.09	0.39	-0.25	1.22	-1.37
$\Delta\Delta_{(32)}$	0.06	-0.37	1.16	1.41	0.51	-0.36	1.29	-1.54
$\Delta\Delta_{(33)}$	0.00	-0.41	1.05	1.40	0.40	-0.36	0.61	-1.08
$\Delta\Delta_{(34)}$	0.03	-0.38	1.03	1.02	0.37	-0.16	1.27	-0.77
$\Delta\Delta_{(35)}$	0.01	-0.30	0.65	1.26	0.22	-0.13	1.58	-0.98
$\Delta\Delta_{(avg)}$	0.02	-0.34	0.95	1.24	0.38	-0.25	1.20	-1.15

Although fault and sensor selection are important elements for a diagnostic system, an in-depth discussion would take us too far afield and are outside the scope of this book. At the end of the day, economics will influence many of these decisions. Sensors introduce additional cost, from initial procurement, added engine weight, maintenance, and so forth. Sensors, quite literally, must buy their way onto the engine in terms of increased safety, reduced maintenance cost, and enhanced controllability.

5.7.2 Constrained Solutions

We conclude this section with a brief mention of another tactic to employ for MFA that falls within the realm of additional information. The additional information is simply that engines *deteriorate* over time, i.e., they do not improve unless maintenance actions are performed. This is akin to the use of deterioration models, which indicate a negative tendency for module efficiencies and flow capacities, and a positive trend for turbine nozzle areas. Outside of using such models as *a-priori* estimates in the KF solution, is there a means to *constrain* the solution to conform to these tendencies? A positive answer to this question was provided by Daniel Simon [20] and explored together with Donald Simon (NASA GRC) [21], specifically for the MFA application of GPA. The latter paper introduces two solutions. The first imposes *hard* constraints to administer inequalities on the fault estimates (essentially providing a range within which the estimate must reside) and requires a quadratic programming solution in addition to the base KF. The second solution imposes *soft* constraints that differ from the former in that they are only required to be satisfied *approximately*. The results (from simulation studies performed by Simon and Simon) indicate some marginal improvement but at the cost of increased computational burden and complexity.

As with all of the methods we have discussed, the individual practitioner needs to decide whether the potential improvement in analysis outweighs the burden to implement for the particular application at hand.

5.8 ALTERNATIVE APPROACHES

5.8.1 Overview

At this juncture, we would like to provide a brief synopsis of other approaches that have emerged in the literature over the past few decades that are fundamentally different from the KF methodology that we have described in this and the preceding chapters. Having said that, we also want to emphasize that there has also been much written on the KF approach, and model-based methods similar to it, whose details differ, in some ways, to the particular formulation we have presented in this text. Some notable examples can be found in the works of Doel [22–25]; Provost [26]; Simon [21], [27] and, [28]; Mathioudakis [29–31]; Li [32]; and Borguet [33].

In terms of other approaches that involve technical devices other than KFs or WLSQ, the large majority can be partitioned into the following general categories:

TABLE 5.12
SFA Fault Isolation Results for Case 1

	1st Rank			2nd Rank			3rd Rank		
	Fault	Estimate	Error	Fault	Estimate	Error	Fault	Estimate	Error
$\Delta\Delta_{(31)}$	HPT	-0.95	0.43	P3ER	-1.48	0.76	VSV	-17.35	0.81
$\Delta\Delta_{(32)}$	HPT	-1.14	0.4	P3ER	-1.67	0.78	VSV	-18.81	0.83
$\Delta\Delta_{(33)}$	HPT	-0.88	0.41	P3ER	-1.13	0.81	T5ER	1.54	0.88
$\Delta\Delta_{(34)}$	HPT	-0.85	0.41	VSV	-17.87	0.69	HPC	-1.17	0.73
$\Delta\Delta_{(35)}$	HPT	-0.9	0.48	P25ER	1.57	0.68	VSV	-21.38	0.68
$\Delta\Delta_{(avg)}$	HPT	-0.95	0.38	VSV	-17.44	0.78	HPC	-1.15	0.81

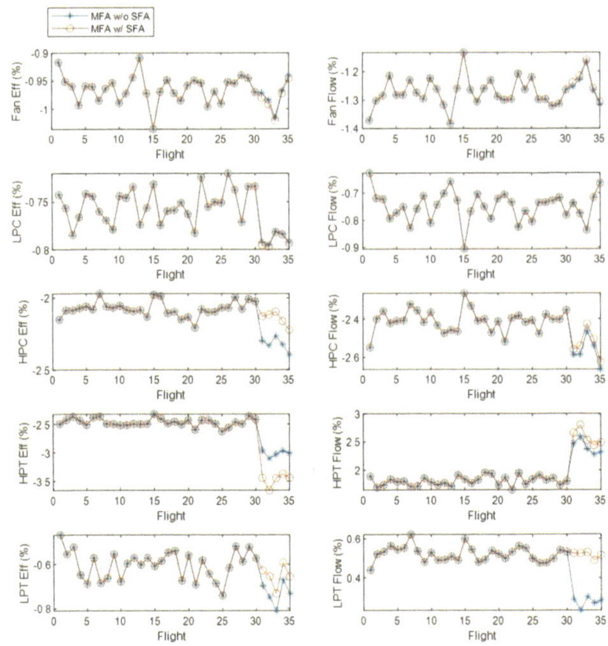

FIGURE 5.9 Adjusted MFA major module fault estimates using SFA result (Case 1)

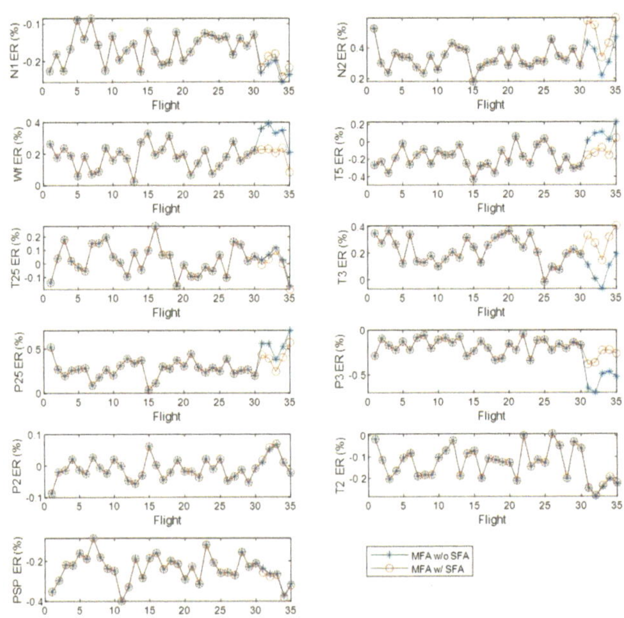

FIGURE 5.10 Adjusted MFA sensor fault estimates using SFA result (Case 1)

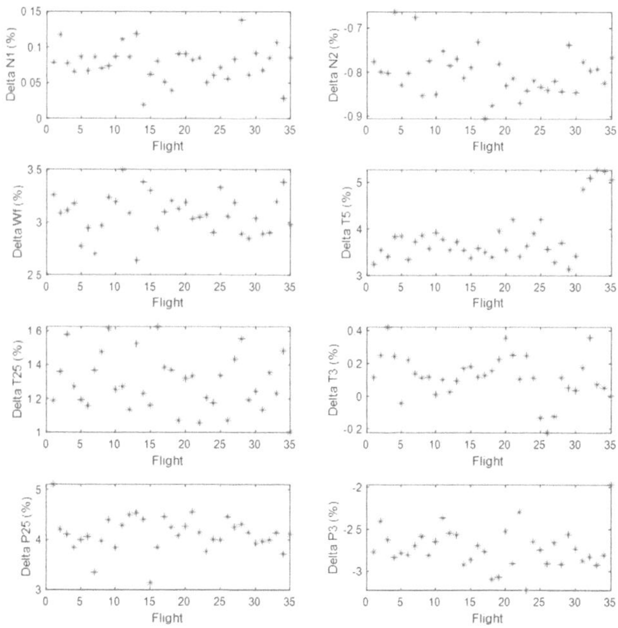

FIGURE 5.11 Simulated time series data with T5 sensor fault

- Genetic algorithms
- Fuzzy logic
- ANNs
- Bayesian belief networks

We have chosen not to discuss these in-depth because these methods are, for the most part, not among the historical diagnostic methodologies that appear in the literature. This may be due, in part, to complexity, computational burden, lack of appropriate

TABLE 5.13
Delta–Delta Parameters and Reference Levels for Case 2

	N1	N2	Wf	T5	T25	T3	P25	P3
$A_{kstart-1}$	0.08	-0.82	3.03	3.55	1.28	0.04	4.14	-2.75
$\Delta\Delta_{(31)}$	-0.01	0.04	-0.14	1.31	-0.15	0.13	-0.17	-0.13
$\Delta\Delta_{(32)}$	0.00	0.02	-0.13	1.55	0.07	0.32	-0.14	-0.08
$\Delta\Delta_{(33)}$	0.03	0.03	0.16	1.72	-0.05	0.03	0.00	-0.18
$\Delta\Delta_{(34)}$	-0.05	-0.01	0.35	1.70	0.20	0.01	-0.43	-0.06
$\Delta\Delta_{(35)}$	0.00	0.05	-0.05	1.51	-0.28	-0.04	-0.02	0.78
$\Delta\Delta_{(avg)}$	-0.01	0.03	0.04	1.56	-0.04	0.09	-0.15	0.07

TABLE 5.14
SFA Fault Isolation Results for Case 2

	1st Rank			2nd Rank			3rd Rank		
	Fault	Estimate	Error	Fault	Estimate	Error	Fault	Estimate	Error
$\Delta\Delta_{(31)}$	T5ER	1.33	0.38	LPC	-0.28	0.85	VBV	0.32	0.87
$\Delta\Delta_{(32)}$	T5ER	1.47	0.43	T2ER	-0.31	0.83	LPC	-0.28	0.87
$\Delta\Delta_{(33)}$	T5ER	1.74	0.19	LPC	-0.23	0.93	P2ER	0.28	0.94
$\Delta\Delta_{(34)}$	T5ER	1.65	0.43	LPC	-0.36	0.82	VBV	0.40	0.85
$\Delta\Delta_{(35)}$	T5ER	1.47	0.66	P3ER	0.67	0.80	PSPER	-0.44	0.85
$\Delta\Delta_{(avg)}$	T5ER	1.53	0.21	LPC	-0.25	0.88	P2ER	0.29	0.90

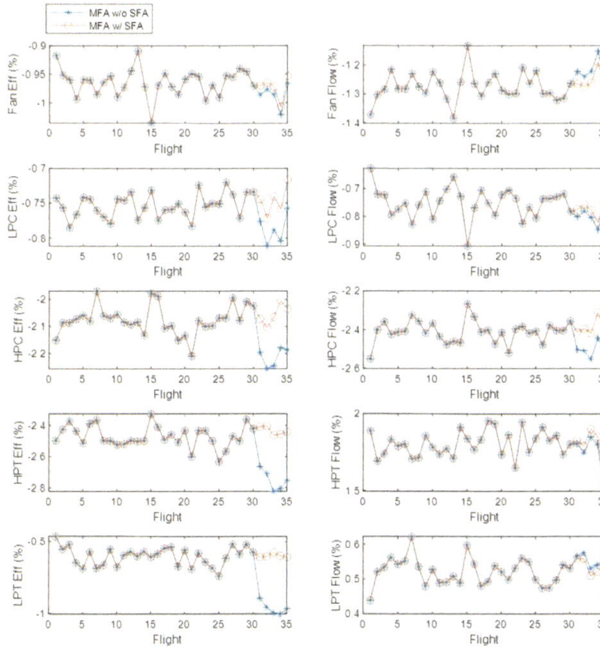

FIGURE 5.12 Adjusted MFA major module fault estimates using SFA result (Case 2)

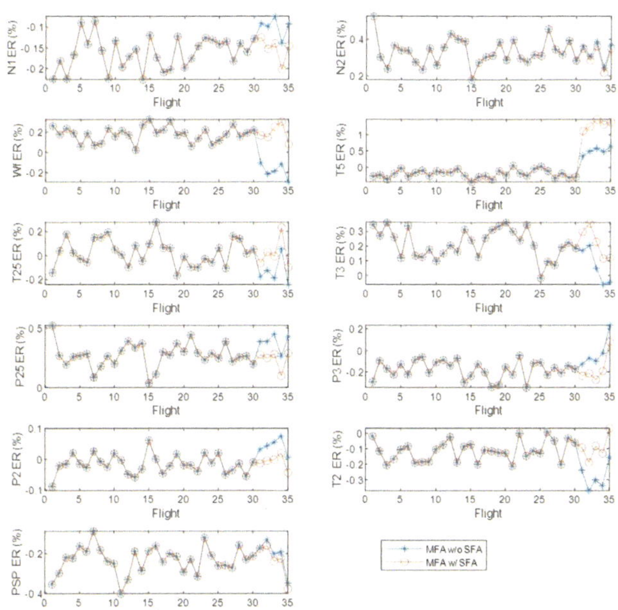

FIGURE 5.13 Adjusted MFA sensor fault estimates using SFA result (Case 2)

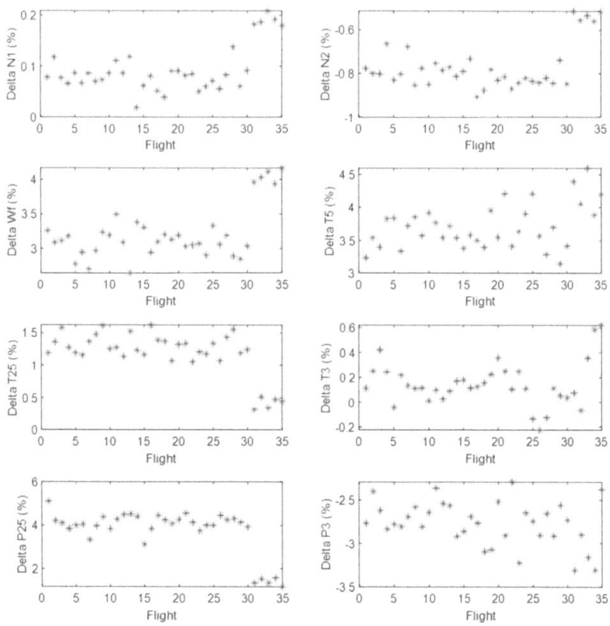

FIGURE 5.14 Simulated time series data with VBV fault

(fault) data,[14] and perceived level of benefit relative to the traditional algorithms. We will mention a few selected approaches briefly and provide references for the interested reader.

5.8.2 Genetic Algorithms

Genetic algorithms (GAs) are commonly used to find solutions to optimization and search problems by using biologically inspired operators such as mutation, crossover,

TABLE 5.15
Delta–Delta Parameters and Reference Levels for Case 3

	N1	N2	wf	T5	T25	T3	P25	P3
$A_{kstart-1}$	0.08	-0.82	3.03	3.55	1.28	0.04	4.14	-2.75
$\Delta\Delta_{(31)}$	0.10	0.31	0.93	0.84	-0.97	0.03	-2.79	-0.56
$\Delta\Delta_{(32)}$	0.11	0.26	1.00	0.50	-0.77	-0.11	-2.61	-0.15
$\Delta\Delta_{(33)}$	0.13	0.29	1.08	1.05	-0.94	0.31	-2.78	-0.41
$\Delta\Delta_{(34)}$	0.11	0.26	0.91	0.34	-0.81	0.54	-2.56	-0.56
$\Delta\Delta_{(35)}$	0.10	0.30	1.13	0.65	-0.84	0.58	-2.97	0.37
$\Delta\Delta_{(avg)}$	0.11	0.28	1.01	0.68	-0.87	0.27	-2.74	-0.26

TABLE 5.16
SFA Fault Isolation Results for Case 3

	1st Rank			2nd Rank			3rd Rank		
	Fault	Estimate	Error	Fault	Estimate	Error	Fault	Estimate	Error
$\Delta\Delta_{(31)}$	VBV	2.05	0.31	LPT	-1.29	0.46	P25ER	-2.51	0.48
$\Delta\Delta_{(32)}$	VBV	1.78	0.34	P25ER	-2.42	0.41	LPT	-1.11	0.50
$\Delta\Delta_{(33)}$	VBV	2.08	0.24	LPC	-1.74	0.42	LPT	-1.32	0.46
$\Delta\Delta_{(34)}$	VBV	1.88	0.33	LPC	-1.60	0.45	LPT	-1.19	0.50
$\Delta\Delta_{(35)}$	VBV	2.05	0.31	LPC	-1.73	0.43	P25ER	-2.80	0.46
$\Delta\Delta_{(avg)}$	VBV	1.97	0.26	LPC	-1.65	0.44	P25ER	-2.51	0.45

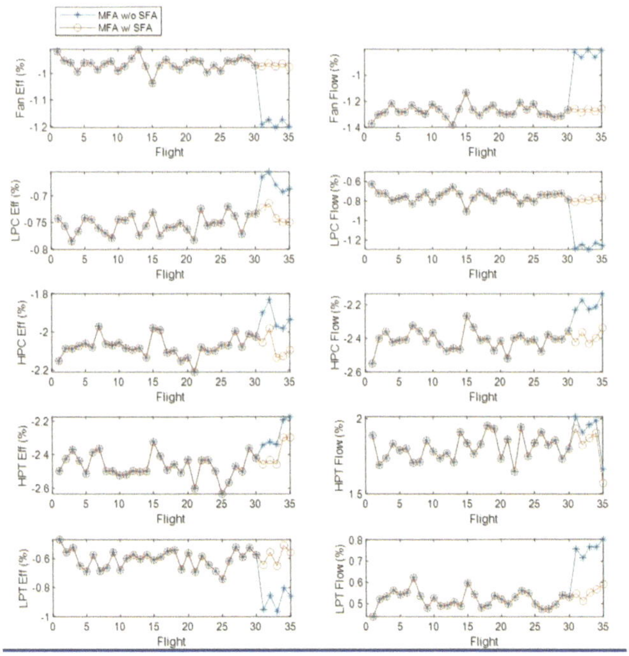

FIGURE 5.15 Adjusted MFA major module fault estimates using SFA result (Case 3)

and selection [36]. If we recall our fundamental modeling and estimation formulation as described in Equations (5.1)–(5.3), the GPA problem can essentially be formulated as an optimization problem. This creates opportunities for sophisticated (albeit computationally expensive) optimization algorithms such as GAs. The necessity arises when using high-fidelity non-linear engine models, as well as imposing parameter restrictions to guide the optimization (or search) process. More specifically, instead of using a linear model, as in Equation (5.1), a non-linear model was defined in [37] and [38] as follows,

$$Z = h(x, w) + b + v \tag{5.30}$$

where h is a non-linear function (as represented by a non-linear engine model) of the performance parameter vector x, and the vector of the environment and power setting parameter w. Vectors b and v are the measurement bias and noise parameters, respectively. The optimization process is set to minimize the discrepancy between actual and predicted measurements with considerations of measurement noise and biases across multiple power settings and ambient conditions. As depicted in Figure 5.17, the approach implants candidate fault scenarios (from a *population* generated by the GAs) in a non-linear engine simulation, runs the simulation, selects top candidates that produce smaller discrepancies, and evolves to the next generation using the *mutation* and *crossover* operators until a global minimum is achieved (or reaching

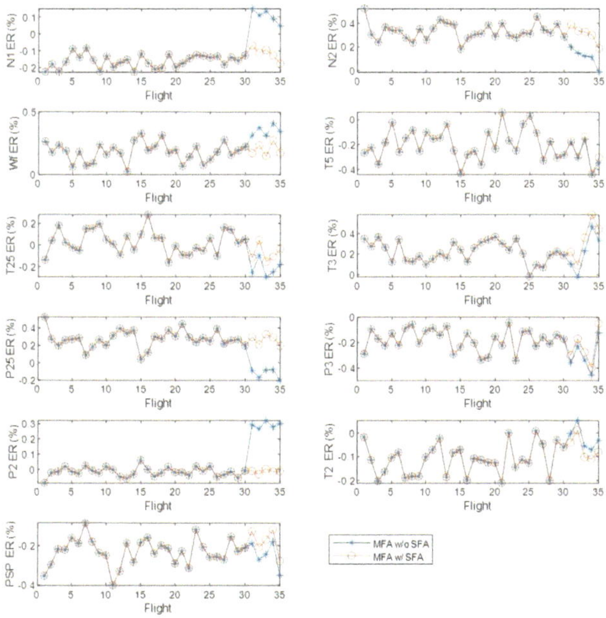

FIGURE 5.16 Adjusted MFA sensor fault estimates using SFA result (Case 3)

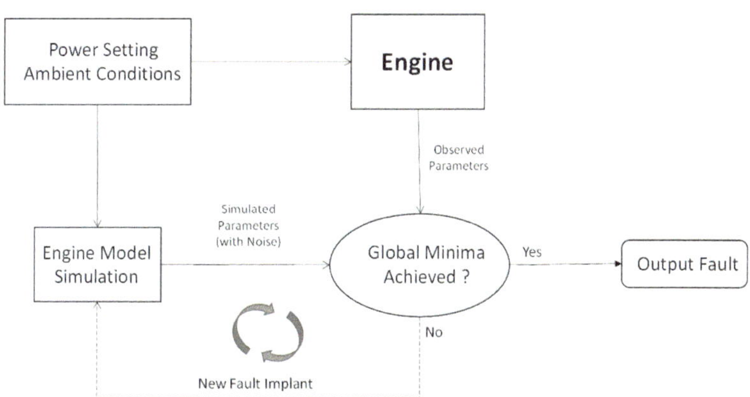

FIGURE 5.17 Genetic algorithms based engine and sensor diagnosis approach

a satisfactory criterion). Notably, this approach *does not involve the use of delta parameters*; instead, it seeks to match the simulation outputs to the measurements directly by searching for the optimal performance parameters.

This method has been reportedly applied to engine OEM test cell data with satisfactory results [37]. The drawbacks of the method, as the authors also pointed out, include the computational burden and the lack of diagnostics confidence measure.

5.8.3 Fuzzy Logic

Fuzzy logic (FL) is one of the most widely used artificial intelligence (AI) methods to approximate the relationship between dependent and independent parameters based on a set of IF-THEN statements.[15] As explained in Addendum A5.6, the general FL approach consists of four basic components: fuzzy rules (sets of IF-THEN statements), fuzzifier (maps input signals into the fuzzy set), inference engine (combines fuzzysets with each other), and defuzzifier (calculates the output values). The application of FL to GPA is rather straightforward due to its rule-based architecture, which can be designed to perform pattern recognition of engine/sensor fault signatures. The fuzzy approach (i.e., fuzzification and defuzzification) is advantageous for dealing with the uncertainties both in modeling and measurement errors. For example, the ICs for HPT fault in Figure 5.6 can be fuzzified to construct a (simplified) SFA rule, such as

IF (deltaWf is *Increased*) AND (deltaT5 is *Increased*)
  AND (deltaN2 is *Decreased*) AND (deltaT3 is *Decreased*)
THEN (HPTFaultLikely is *High,* HPCFaultLikely is *Medium, …*).

The belief associated with each input (e.g., deltaWf) being *increased* or *decreased* is calculated from the fuzzy membership functions. Alternatively, the problem can be formulated as an *estimation* problem (instead of the *classification* formulation in the above example) by defining the outputs as a vector of health parameters, as shown in Figure 5.18.

While most fuzzy logic applications to GPA are focused on SFA [39–41], it has also been applied (partially) to MFA by defuzzifying the output to achieve a quantitative solution [42]. In that approach, an additional step was taken to tune the rules of the fuzzy diagnostics system over a known deterioration model (baseline) that is

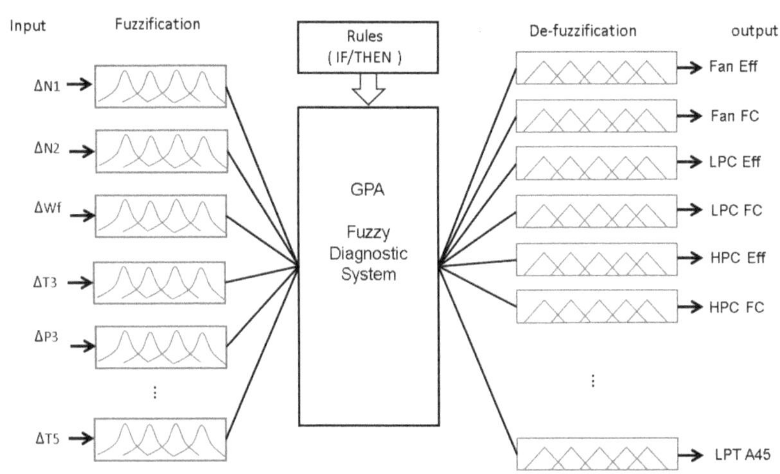

FIGURE 5.18 An example of a fuzzy logic diagnostic system

calculated at the previous flight with an MFA method. This feature enables SFA and MFA processes to operate in concert with one another using FL.

5.8.4 ANNs

The discussion of ANNs and their applications in EHM is a big topic by itself. We do not intend to get into the algorithmic details of the various types of ANNs in this book. Interested readers can find a plethora of books and publications on ANNs, as machine learning has been a popular topic over the past decade. Generally speaking, ANNs are a series of algorithms that mimic the operations of the human brain to recognize relationships amongst parameters in a (potentially large) dataset. Among the variety of capabilities that ANNs can offer, *classification* (or *pattern recognition*) and *regression* (or *function approximation*) are the two most commonly used. Therefore, as long as the reader has a fair understanding of *classification* and *regression*, the rest of this section should be easy to follow. As a matter of fact, the ANNs in these applications can essentially be replaced by other *classification* or *regression* tools as machine learning techniques evolve. In addition to the quick introduction herein, we have provided Addendum A6.3 that briefly describes a particular form of ANNs, the multi-layer perceptron (MLP), in conjunction with its application to model alignment in Chapter 6.

5.8.4.1 Multi-Layer Perceptron (MLP)

The idea here is to *train* a set of MLP ANNs to learn the functional relationship between measured gas turbine parameters and the health parameters that are indicative of faulty conditions, i.e., to perform a *regression* between these two sets of parameters. If we consider the measured (steady state) gas turbine parameters Z as a (non-linear) function of health parameter x and operating conditions w, i.e., $Z = f(x,w)+v$, then the ANNs are *trained* to learn the inverse function $x = f^{-1}(Z,w)$ subject to measurement noise v. This is a typical regression problem in which the parameters of the ANNs (e.g., the weights and biases of the MLP) are estimated by a particular training algorithm (e.g., the back propagation algorithm for MLP) that minimizes the mean square errors between ANN outputs and the training target data.

When applied to GPA, the process, as shown in Figure 5.19, often starts from generating the training data by running a gas turbine simulation program at various operating conditions with simulated faults. The simulation program can be an NPSS[16] or Simulink model or simply use the ICs. Simulation data is often pre-processed (e.g., denoising and normalization) and divided into training and evaluation datasets. A set of ANNs is then trained to learn the inverse function (i.e., Z and w being the ANNs *inputs* and x being the ANN *output*) often using off-the-shelf neural network tools. The number of neurons in the input and output layers is determined by the number of input measurements and output parameters. The number of hidden layer nodes, as well as the number of hidden layers, is selected based on convergence criterion and the characteristic of the input–output mapping relationship.

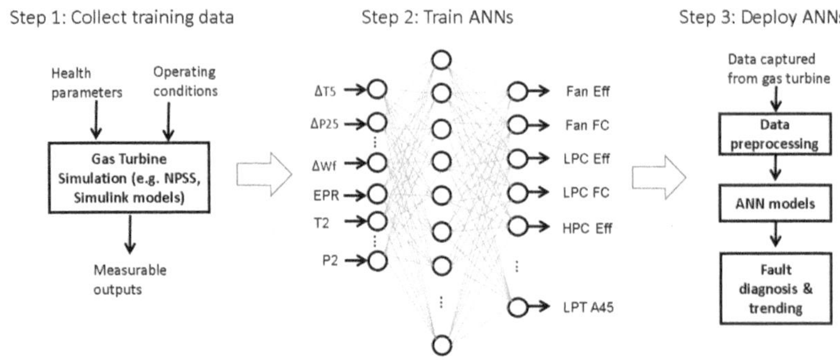

FIGURE 5.19 An illustrative ANN training and deployment process for GPA applications

Once the ANNs are trained, they can be deployed in a diagnostic system to support downstream trending, anomaly detection, and fault isolation functions. Clearly, the three-step process portrayed in Figure 5.19 is simplified. Real-world practice would involve an extensive verification and validation effort before and after deployment. MLP is commonly chosen in such applications due to its simplicity and maturity. Most earlier ANN applications in gas turbine fault diagnosis fall into this category [43–45]. While most published work only included major module health parameters, as shown in Figure 5.19, the ANNs can be expanded to include sub-system and sensor faults [35], [46].

5.8.4.2. Probabilistic Neural Network (PNN)

Unlike the MLP that performs *regression*, the PNN is a *classifier*, which, in mathematical form, is a discrete-valued function that assigns a (categorical) class label to the particular input data [47]. For example, given a set of measured gas turbine parameters, indicative of an HPT fault, a well-designed MLP will output the estimates for all faults (with HPT health parameter shifts being the highest). In contrast, a PNN sets the value for HPT fault class as 1 and the others as 0 without giving any fault estimates (or outputting the class label for an HPT fault).

PNN is an extension to the (also popular) radial basis function (RBF) neural network with the addition of a decision layer that discretizes the output. The architecture of a typical PNN, as shown in Figure 5.20, consists of three layers: an *input* layer, a *pattern* layer, and a *summation/decision* layer (also considered two separate layers by many). When an input is presented, the pattern layer computes distances from the input vector to the pattern vectors and produces a vector whose elements indicate how close the input is to a pattern. This is essentially done using a multivariate Gaussian probability density function (centered on the pattern vector) that takes the form of a radial basis function. The summation layer sums these contributions for each *class* to produce a vector of probabilities. Finally, a transfer function on the output of the summation/decision layer picks the maximum of these probabilities and produces a 1 for that class and a 0 for all other classes.

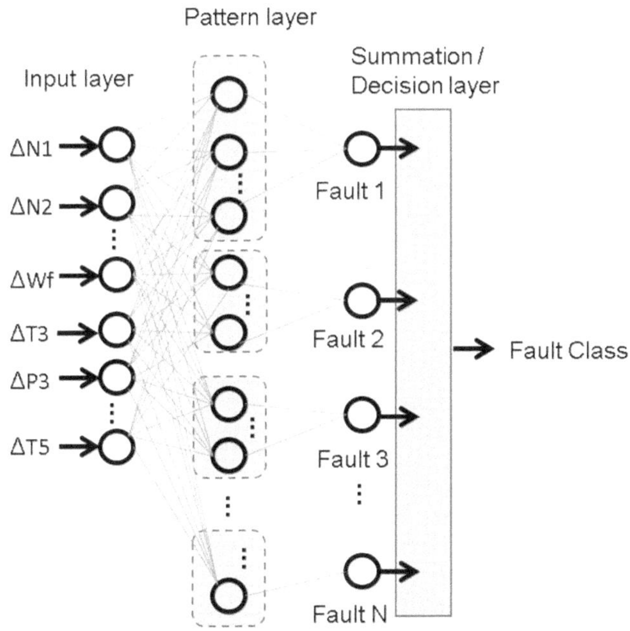

FIGURE 5.20 Structure of a probabilistic neural network

The classification capability of the PNN lends itself naturally to SFA applications [48–50]. In these applications, typically an engine performance model (or ICs in some cases) is used to support the construction of a network. Similar to the MLP applications, training data is generated using the model to simulate a comprehensive set of healthy and faulty situations. The network performance in detecting and quantifying sensor faults (using the outputs from the RBFs upstream of the decision layer) is validated on a large number of fault cases, often generated by the model as well [49]. This method has been applied to address not only performance faults but also sensor faults with the presence of component faults [50].

5.8.4.3. Auto Associative Neural Network (AANN)

A distinctive feature of the AANN is that the input and output layers have identical parameters (as its name implies). Its purpose is to reconstruct its own inputs (by learning the correlations amongst the inputs). Therefore, *unlike* MLP or PNN, AANN is an *unsupervised* learning[17] model.

AANNs have three hidden layers, namely the *mapping* layer, the *bottleneck* layer, and the *demapping* layer, and can be divided into two sub-networks, i.e., *compression* and *decompression* (Figure 5.21). The compression part reduces high dimensional data (from input layer) to a lower dimension (in the bottleneck layer) with an insignificant information loss based on the concept of (non-linear) principal component analysis (PCA). Therefore, the bottleneck layer nodes are representative of the principal components of the inputs. The decompression part tries to reconstruct

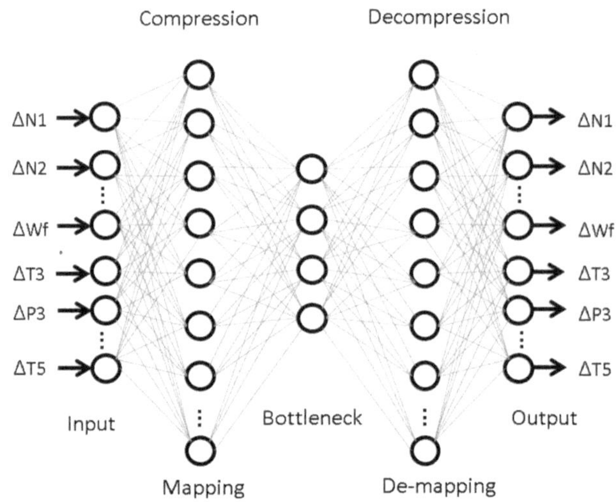

FIGURE 5.21 Structure of an auto associative neural network

the original data from the compressed version with minimum information loss [51]. In machine learning terms, the bottleneck layer describes a *code* used to represent the input. Therefore, the compression part becomes an *encoder* that maps the input into the *code*, and the decompression part becomes a *decoder* that maps the *code* to a reconstruction of the input. Hence, the AANN is also commonly known as *autoencoder.*

Due to the auto associative nature of its network architecture, AANNs are primarily used for sensor data validation and sensor fault diagnosis applications. AANNs also have certain advantages over conventional filtering techniques at handling outlier removal and noise reduction. Therefore, they can be used as virtual sensors (or analytic redundancy for physical sensors) in various applications.

One of the earlier applications of AANN-based sensor validation technique for a turbofan engine was proposed by Guo et al. [52]. The proposed approach utilized the AANN to perform dimensionality reduction to create a model for sensor failure detection and data recovery and applied to an F100 turbofan engine. Lu et al. [53] proposed a genetic AANN algorithm for offline sensor data validation with noise filtering and bias detection and correction. Studies also showed that the AANNs were capable of successful sensor failure diagnosis in the presence of component faults. The methodology proposed in [54] consisted of a bank of AANNs and provided a solution to the problem of both sensor and component fault detection and isolation.

5.8.4.4. Other Types of ANNs and Hybrid Solutions

As mentioned earlier, any ANN that performs either *regression* or *classification* can potentially be applied to gas path performance diagnosis. In addition to the aforementioned three popular types, other forms of ANNs have also found their way to the gas path analysis arena, e.g., radial basis function (RBF) neural networks [55], self-organizing map [56], dynamic neural network (DNN) [57], and cascaded network [58]. Studies are also abundant in hybrid solutions that combine different forms of ANNs [59], as well as combining ANNs with other approaches, such as GA [60], fuzzy logic (neuro-fuzzy) [61], and physics-based methods [35].

Before concluding our brief overview of ANN-based approaches, we will offer some observations and remarks. A well-versed reader might have noticed that, in most of the ANNs applications, the ANNs are trained with simulation data. This is due to the fact that faulty gas turbine data are hard to come by and are often incomplete in terms of fault coverage. This tends to raise the fair question that, if a physics-based model has to be used to train the ANN, why not use model-based methods instead? While one could argue that a compact (or *shallow*) ANN is often more computationally efficient than a high-fidelity physics-based model, this observation does point to a dilemma within many data-driven approaches, i.e., the lack of faulty data. Another important limitation to keep in mind is to avoid extrapolating the ANN outside the range of values in the dataset used for training. The further the extrapolation goes outside the range, the greater chance there is for the ANN to fail, as the modeling assumptions (baked-in implicitly by the training data) may no longer hold.

In the past few years, deep learning (DL) has attracted remarkable research attention particularly in the context of big data analytics. DL is part of a broader family of machine learning methods based on ANNs with representation learning, i.e., recognizing patterns in the data. The adjective "deep" in deep learning refers to the use of multiple layers in the network. Despite its success in other domains, few studies have been published on the application of DL for gas turbine fault diagnostics pertaining to GPA. This is an area to see more studies in the future. The value of this new technique needs yet to be proven both technically and economically to buy its way into real-world applications.

5.8.5 Bayesian Belief Networks

Bayesian belief networks (BBNs) are probabilistic expert systems that combine the principles of probability theory and graph theory [62]. A BBN consists of *nodes* representing discrete variables, as defined in probability theory. All possible discrete values of a variable, compose the set of *states* of the corresponding node. In the context of GPA applications, for example, a node can be ΔT5, Δwf, or HPT. The discrete states for the ΔT5 node can be "Decrease (D)", "Normal (N)", and "Increase (I)" (defined based on the designer's choice of parameter variation ranges). Similarly, the discrete states for the HPT node can be "Healthy (H)" and "Faulty (F)." The inter-relation among the variables is expressed through the *links* of the network. The quantitative relationship among the nodes is given through the conditional probability table (CPT) of each node. Continuing with the HPT example for an engine with four basic gas path measurements N1, N2, wf, and T5, a simple BBN can be created to show the HPT fault effect on the measured parameters (assuming no other faults are present), as shown in Figure 5.22. For example, the CPT of the ΔT5 node can be defined based on the polarity of the parameter shifts as indicated in the ICs (e.g., Figure 5.6). For instance, the number 0.9 in the first row of the CPT indicates

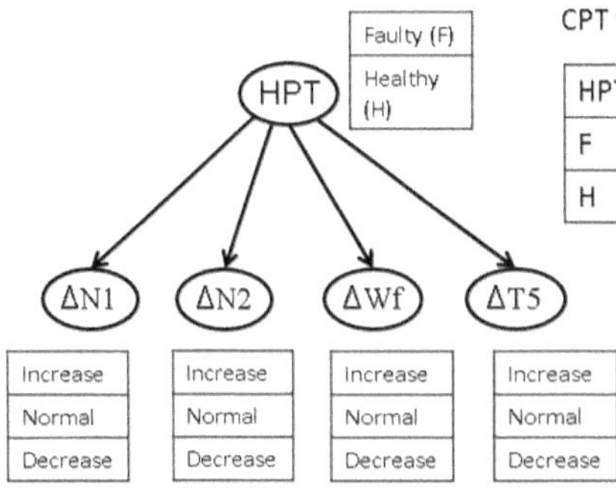

CPT for ΔT5:

HPT	P(ΔT5=I)	P(ΔT5=N)	P(ΔT5=D)
F	0.9	0.09	0.01
H	0.01	0.98	0.01

FIGURE 5.22 Nodes, links, and CPTs in a BBN

the conditional probability that ΔT5 will be in "Increase" state provided that HPT is "Faulty," while the conditional probability for ΔT5 to remain in "Normal" or to "Decrease" is 0.09 or 0.01, respectively. These conditional probabilities depend on the significance of the fault effect, the discretization of the nodes, and often expert opinions.

This example is clearly overly simplified. In a GPA application, additional health parameter nodes and (delta) measurement nodes have to be included, which will result in a BBN more like what is shown in Figure 5.23. Note that all the links lead from each health parameter node to each (delta) measurement node (indicating the causal relationship implicitly). Meanwhile, the number of discrete states of the health parameter nodes is often increased to support higher accuracy in diagnostic reasoning. For example, the (delta) HPT efficiency node can define five discrete states based on designated parameter ranges, e.g. "LargeDrop": [-3, -1.5]%; "SlightDrop": [-1.5, 0.5]%; "Healthy": [-0.5, 0.5]%; "SlightIncrease": [0.5, 1.5]; "LargeIncrease": [1.5, 3]%. As compared to the above simple example, the CPT for a (delta) measurement node becomes the set of probabilities for *all* possible combination of states of *all* parent nodes (i.e., health parameter nodes). The CPT can be derived by running engine

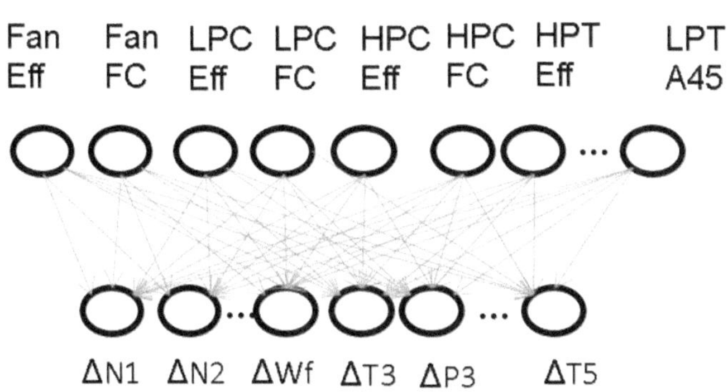

FIGURE 5.23 Structure of a Bayesian belief network

simulations with various levels of deteriorations and faults. A detailed procedure for specifying the CPTs of the diagnostic BBN for a GPA application is proposed in [65].

For this type of architecture consisting of a set of measurement nodes {z} and a set of health parameter nodes {h}, the joint probability of network nodes is given below.

$$P(z,h) = P(h)P(z \mid h) = P(z)P(h \mid z) \tag{5.31}$$

where $P(h)$ is the *a-priori* probability for health parameter h, $P(z \mid h)$ is the conditional probability for measurement parameter z, given health states h, i.e., the CPTs, $P(z)$ is the probability for measurement parameter, and $P(h \mid z)$ is the conditional probability for health parameter given measurement states. During diagnostic inference, the (conditional) probabilities of all states of health parameter nodes {h} are estimated, from the given states of measurement nodes {z} through measurements using Bayes' theorem as shown in Equation (5.31). Commercial off-the-shelf software tools are available for BBN system development on various platforms such as *Matlab* and *Python*.

Earlier applications of BBN in GPA were reported by Palmer [64] and Romessis [63]. A more comprehensive study was presented in [65] in which an engine performance model was used to aid the construction of CPTs. Lee et al. [66] proposed hierarchically arranged multiple BBN models based on an offline fault diagnosis method for industrial gas turbine engines that provided both qualitative and quantitative diagnostic reasoning.

Developing a BNN based on expert knowledge is highly complex, time consuming, and error prone. If large data samples (either from simulation or field data) are available, BBN can be trained in a supervised [67] or unsupervised [68] learning manner, but adjustments based on expert opinion are still needed on many occasions. Moreover, the estimation of probabilities with BBN has proven to be *NP-hard*, i.e., the required computational time increases exponentially with the size of the network. These factors limit the BNN's capability to handle a larger scale of fault modes such as sensor and sub-system faults.

NOTES

1. One of the earliest publications that explored this multi-point extension can be found in: Stamatis, A., et al, 1991, *Jet Engine Fault Detection with Discrete Operating Points Gas Path Analysis*, J. Propulsion, Vol. 7, No. 6.
2. We are assuming an *a-priori* estimate for $k = 1$.
3. See Urban, L.A., Volponi, A.J., 1992, *Mathematical Methods of Relative Engine Performance Diagnostics*, SAE 1992 Transactions, Vol 101, *Journal of Aerospace*, Paper # 922048.
4. See Chapter 1.
5. Volponi, A. J., 2014, *Gas Turbine Engine Health Management: Past, Present, and Future Trends. ASME. J. Eng. Gas Turbines Power.* May 2014; 136(5):051201.
6. See Chapter 2, section on Fault Detection & Isolation for a short discussion on trend detection.
7. We will provide a short discussion on alternative methods at the end of this chapter.

8. The C-MAPSS simulation has only two engine system faults (VBV and VSV). For a real application this list would contain other sub-systems, which will be illustrated in our discussions in Chapter 9.
9. This approximates an LSQ solution. In fact, the limiting construction, i.e., as the number in the Jth position approaches ∞, is a simple inverse relationship. See Addendum A5.4 for details.
10. See Addendum 5.5 for details.
11. In general,
12. See section *Ancillary Parameters Generated via ICs* in Addendum A4.4.
13. This is approximately accommodated through an averaging assumption in [14].
14. For methods using ANNs where training is required.
15. For those readers who are unfamiliar with fuzzy logic, see Addendum A5.6 for the basics. For further reading consult references [8–11] and [34].
16. Numerical propulsion system simulation (NASA)
17. The goal of unsupervised learning is to find hidden patterns in unlabeled data, i.e., no target or labels are provided in the training data.

REFERENCES

1. Sallee, G.P., 1973, *Performance Deterioration Based on In-Service Engine Data: JT9D Jet Engine Diagnostics Program*, NASA CR-159525.
2. Wulf, R.H., 1980, *CF6-50 Engine Performance Deterioration: Engine Diagnostic Program,* NASA CR-159867.
3. Ziemianski, J.A., Mehalic, C.M., 1980, *Investigation of Performance Deterioration of the CF6/JT9D High-Bypass Ratio Turbofan Engines,* NASA Technical Memorandum 81552.
4. Volponi, A.J., 1994, Sensor Error Compensation in Engine Performance Diagnostics, *Proceedings of the American Society of Mechanical Engineers Turbine and Aeroengine Congress.* The Hague, Netherlands, June 13–16, Paper No. 94-GT-58.
5. Grewal, M., Andrews, A., 1993, *Kalman Filtering: Theory & Practice,* Prentice Hall, NJ, ISBN 0-13-211335-X, Chapter 6.
6. Jazwinsky, A., 1970, Stochastic Processes and Filtering Theory*, Mathematics in Science and Engineering*, Vol. 64, New York: Academic Press, Chapter 8, Section 8.
7. Lewis, F., 1986, *Optimal Estimation,* John Wiley & Sons, New York, ISBN 0-471-83741-5, Chapter 4.
8. Zadeh, L.A., 1965, Fuzzy Sets, *Inform Control*, 8, 338–353.
9. Zadeh, L.A., 1973, Outline of a New Approach to the Analysis of Complex Systems and Decision Processes, *IEE Transactions on Systems, Man, and Cybernetics*, SMC-3,(1), pp. 28–44.
10. Ross, Timothy J., 1995, *Fuzzy Logic with Engineering Applications*, McGraw-Hill, Inc., ISBN 0-07-053917-0
11. Volponi, A., Solomonides, J., 1994, Advanced Engine Control Development Utilizing Fuzzy Logic and Rapid Prototyping, *Royal Aeronautical Society Proceedings* ISBN 1 85768 021 9.
12. Litt, Jonathan, 2005, *An Optimal Orthogonal Decomposition Method for Kalman Filter-Based Turbofan Engine Thrust Estimation,* ASME GT2005-68808.
13. Simon, D.L., Garg, S., 2009, *Optimal Tuner Selection for Kalman Filter-Based Aircraft Engine Performance Estimation*, ASME GT2009-59684.
14. Simon, D.L., Armstrong, J.B., 2011, *Application of an Optimal Tuner Selection Approach for On-Board Self-Tuning Engine Models*, ASME GT2011-46408.

15. Simon, D.L., Garg, S., 2009, *A Systematic Approach to Sensor Selection for Aircraft Engine Health Estimation,* NASA/TM-2009-215839.
16. Kamboukos, Ph., et.al., 2001, Optimizing Diagnostic Effectiveness of Mixed Turbofans by Means of Adaptive Modeling and Choice of Appropriate Monitoring Parameters, *AVT Symposium on Monitoring and Management of Gas Turbine Fleets for Extended Life and Reduced Costs*, Manchester, UK, 8–11 October 2001, Defense Technical Information Center Compilation Pat Notice ADP014126.
17. Simon, D., Rinehart, A., 2016, Sensor Selection for Aircraft Engine Performance Estimation and Gas Path Fault Diagnostics, *Journal of Engineering for Gas Turbines and Power, Transactions of the ASME* 138 (2016), 071201.
18. Sowers, T.S., et.al., 2008, *Application of the Systematic Sensor Selection Strategy for Turbofan Engine Diagnostics,* ASME paper GT2008-50525, Berlin, Germany.
19. Sowers, T.S., et.al., 2009, *Expanded Application of the Systematic Sensor Selection Strategy for Turbofan Engine Diagnostics,* ASME paper GT2009-59251, Orlando, FL.
20. Simon, D., Chia, T., 2002, Kalman Filtering with State Equality Constraints, *IEEE Transactions on Aerospace and Electronic Systems*, 39, 128–136.
21. Simon, D., Simon, D.L., 2005, Aircraft Turbofan Engine Health Estimation Using Constrained Kalman Filtering, *ASME Journal of Engineering for Gas Turbines and Power*, 127, 323–328.
22. Doel, D.L., 1992, TEMPER - A Gas Path Analysis Tool for Commercial Jet Engines, *Proceedings of the American Society of Mechanical Engineers Turbine and Aeroengine Congress*. Cologne, Germany, June 1–4, ASME Paper 92-GT-315.
23. Doel, D.L., 1993, An Assessment of Weighted-Least-Squares Based Gas Path Analysis, *Proceedings of the American Society of Mechanical Engineers Turbine and Aeroengine Congress*. Cincinnati, Ohio, May 24–27, ASME Paper 93-GT-119.
24. Doel, D.L., 2002, *Interpretation of Weighted Least Squares Gas Path Analysis Results*, ASME paper GT-2002-30025.
25. Doel, D.L., 2003, *Development of Baselines, Influence Coefficients and Statistical Inputs for Gas Path Analysis*, Von Karman Lecture Notes.
26. Provost, M.J., 1994, *The Use of Optimal Estimation Techniques in the Analysis of Gas Turbines*, Ph.D. Thesis, Cranfield University, 1994.
27. T. Kobayashi, T., Simon, D.L., 2003, Application of a Bank of Kalman Filters for Aircraft Engine Fault Diagnostics, *Proceedings of the American Society of Mechanical Engineers Turbine and Aeroengine Congress*. Atlanta, GA, June 16–19, Paper No. GT2003-38550.
28. T. Kobayashi, T., Simon, D.L., 2005, Evaluation of a Bank of Kalman Filters for Aircraft Engine Sensor Fault Diagnostics, *Transactions of the ASME, Journal of Engineering for Gas Turbines and Power*, 127, 497–504.
29. Kamboukos, Ph., Mathioudakis, K., 2007, Turbofan Engine Health Assessment by Combining Steady and Transient State Aerothermal Data, *7th European Turbomachinery Conference*, paper 137. Athens, Greece, March 5–9.
30. Stamatis, A., et al, 1991, Jet Engine Fault Detection with Discrete Operating Points Gas Path Analysis, *Journal of Propulsion*, 7(6), 1043–1048.
31. Aretakis, N., Mathioudakis, K., Stamatis, A., 2002, *Non-Linear Engine Component Fault Diagnosis from a Limited Number of Measurements Using a Combinatorial Approach*, ASME paper GT-2002-30031
32. Li, Y.G., 2003, Gas Turbine Diagnostic Approach with Transient Measurements, *Proceedings of the Institution of Mechanical Engineers, Part A, (Journal of Power and Energy)*, 217(A2), 169–177.
33. Borguet S., Leonard, O., 2010, A Sparse Estimation Approach to Fault Isolation, *ASME Journal of Engineering for Gas Turbines and Power*, 132(2), 513–522.

34. Ganguli, R., 2001, *Application of Fuzzy Logic for Fault Isolation of Jet Engines*, ASME paper 2001-GT-13.
35. Volponi, A.J., et al., The Use of Kalman Filter and Neural Network Methodologies in Gas Turbine Performance Diagnostics: A Comparative Study, *ASME 2000-GT-547, IGTI Turbo Expo 2000, Munich; Transactions of the ASME, Journal of Engineering for Gas Turbines and Power*, 125, 917–924.
36. Mitchell, M., 1996, *An Introduction to Genetic Algorithms*, Cambridge, MA: MIT Press. ISBN 9780585030944.
37. Zedda, M., 1999, *Gas Turbine Engine and Sensor Fault Diagnosis*, Ph.D. dissertation, Cranfield University. URI: http://dspace.lib.cranfield.ac.uk/handle/1826/9117
38. Singh, R., 2003, Fault Diagnostics Using Genetic Algorithm for Advanced Cycle Gas Turbine, *Gas Turbine Condition Monitoring and Fault Diagnosis*, Von Karman Lecture Series 2003-01, January 2003.
39. Ganguli, R., 2003, Application of Fuzzy Logic for Fault Isolation of Jet Engines, *Journal of Engineering for Gas Turbines and Power*, 125, 617–623.
40. Ogaji, S.O.T., et al., 2005, Gas-Turbine Fault Diagnostics: A Fuzzy-logic Approach, *Applied Energy*, 82, 81–89.
41. Kyriazis, A., Mathioudakis, K., 2009, Gas Turbine Fault Diagnosis Using Fuzzy-Based Decision Fusion, *Journal of Propulsion and Power*, 25, 335–343.
42. Marinai, L., 2004, *Gas-Path Diagnostics and Prognostics for Aero-Engines Using Fuzzy Logic and Time Series Analysis*, Ph.D. Thesis, Bedford, UK: Cranfield University.
43. Bettocchi, R., Pinelli, M., Spina, P.R., Venturini M., 2007, Artificial Intelligence for the Diagnostics of Gas Turbines - Part I: Neural Network Approach, *Journal of Engineering for Gas Turbines and Power*, 129(3), 711–719.
44. DePold, H.R., Gass, F.D., 1999, The Application of Expert Systems and Neural Networks to Gas Turbine Prognostics and Diagnostics, *Journal of Engineering for Gas Turbines and Power*, 121(4), 607–612.
45. Zedda, M., Singh, R., 1998, Fault Diagnosis of a Turbine Engine Using Neural Networks: A Quantitative Approach, *34th AIAA/ASME/SAE/ASEE Joint Propulsion Conference and Exhibit*, Cleveland, OH, USA, July 13–15.
46. Lu, P.J., et al., 2001, An Evaluation of Engine Faults Diagnostics Using Artificial Neural Networks, *Journal of Engineering for Gas Turbines and Power*, 123(2), 340–346.
47. Specht, D.F., 1990, *Probabilistic Neural Networks*, Neural Network, 3, 109–118.
48. Eustace, R., Merrington, G., 1995, Fault Diagnosis of Fleet Engines Using Neural Networks, *Proceedings of the Twelfth International Symposium on Air Breathing Engines*, Melbourne, Australia, September 10–15, paper no. ISABE 95-7085
49. Mathioudakis, K., Romessis, C., 2004, Probabilistic Neural Networks for Validation of On-board Jet Engine Data, *Proceedings of the Institution of Mechanical Engineers*, 218, 59–72.
50. Romesis, C., Mathioudakis, K., 2003, Setting Up of a Probabilistic Neural Network for Sensor Fault Detection Including Operation with Component Faults, *Journal of Engineering for Gas Turbines and Power*, 125(3), 634–641.
51. Kramer, M.A. 1991, Nonlinear Principal Component Analysis Using Autoassociative Neural Networks, *AIChE Journal*, 37, 233.
52. Guo, T.H., et al., 1996, Sensor Validation for Turbofan Engines Using an Autoassociative Neural Network. *Proceedings of the Guidance, Navigation, and Control Conference*, San Diego, CA, USA, 29–31 July, p. 3926.
53. Lu, P.J., Hsu, T.C., 2002, Application of Autoassociative Neural Network on Gas-path Sensor Data Validation, *Journal of Propulsion and Power*, 18, 879–888.
54. Sadough Vanini, Z.N., Meskin, N., Khorasani, k., 2014, Multiple-Model Sensor and Components Fault Diagnosis in Gas Turbine Engines Using Autoassociative Neural Networks, *Journal of Engineering for Gas Turbines and Power*, 136(9), 091603.

55. Ogaji, S.O.T., 2003, *Advanced Gas-path Fault Diagnostics for Stationary Gas Turbines*, Ph.D. Thesis, Bedford, UK: Cranfield University.
56. Cottrell, M., et al., 2009, Fault Prediction in Aircraft Engines Using Self-Organizing Maps, *International Workshop on Self-Organizing Maps*, Berlin/Heidelberg, Germany: Springer, pp. 37–44.
57. Sadough Vanini, Z.N., Khorasani, K., Meskin, N., 2014, Fault Detection and Isolation of a Dual Spool Gas Turbine Engine Using Dynamic Neural Networks and Multiple Model Approach, *Information Science*, 259, 234–251.
58. Ogaji, S.O.T., Singh, R. 2003, Advanced engine diagnostics using artificial neural networks, *Applied Soft Computing Journal*, 3, 259–271.
59. Zedda, M., Singh, R., 2002, Gas Turbine Engine and Sensor Fault Diagnosis Using Optimization Techniques, *Journal of Propulsion and Power*, 18, 1019–1025.
60. Roy, N., Ganguli, R., 2006, Filter Design Using Radial Basis Function Neural Network and Genetic Algorithm for Improved Operational Health Monitoring, *Applied Soft Computing*, 6, 154–169.
61. Palade, V., et al., 2002, Fault Diagnosis of an Industrial Gas Turbine Using Neuro-Fuzzy Methods, *IFAC Proceedings Volumes*, 35(1), 471–476.
62. Pearl, J., 1988, *Probabilistic Reasoning in Intelligent Systems: Networks of Plausible Inference*, 2nd ed., San Mateo, CA: Morgan Kaufmann.
63. Romessis., C., Mathioudakis, K., 2006, Bayesian Network Approach for Gas Path Fault Diagnosis, *Journal of Engineering for Gas Turbines and Power*, 128(1), 64–72.
64. Palmer, C.A., 1998, Combining Bayesian Belief Networks with Gas Path Analysis for Test Cell Diagnostics and Overhaul, *Proceedings of the ASME 1998 International Gas Turbine and Aeroengine Congress and Exhibition*, Stockholm, Sweden, 2–5 June 1998.
65. Romessis, C., Stamatis, A., Mathioudakis, K., 2001, *Setting up a Belief Network for Turbofan Diagnosis with the Aid of an Engine Performance Model*, ISABE paper 2001, Nr. 1032.
66. Lee, Y.K., et al., 2010, A Fault Diagnosis Method for Industrial Gas Turbines Using Bayesian Data Analysis. *Journal of Engineering for Gas Turbines and Power*, 132, 041602.
67. Wettig, H., et al., 2002, Supervised Naive Bayes Parameters, In P. Ala-Siuru, & S. Kaski (Eds.), *Proceedings of the Tenth Finnish Artificial Intelligence Conference*, Oulu, Finland: Finnish Artificial Intelligence Society, 71–83.
68. Pham, D.T., Ruz, G.A., 2009, Unsupervised Training of Bayesian Networks for Data Clustering, *Proceedings of the Royal Society A*, 465, 2927–2948.
69. Algun, F, 1988, F110 Engine Monitoring and Maintenance Management Systems for F-16 C/D, AGARD-CP-448, *Engine Condition Monitoring- Technology and Experience*, 448, 11.1–11.12. https://pascal-francis.inist.fr/vibad/index.php?action=getRecordDetail&idt=7117362
70. Chetail, P., 1988, LE CFM56-5 SUR A320 A Air France, AGARD-CP-448, *Engine Condition Monitoring- Technology and Experience*, 448, 15.1–15.19. https://pascal-francis.inist.fr/vibad/index.php?action=getRecordDetail&idt=7117364
71. Volponi, A.J., 2003, Extending Gas Path Analysis Coverage for Other Fault Conditions, (*von Karman Lecture Series VKI LS 2003-01, Gas Turbine Condition Monitoring and Fault Diagnosis*), von Karman Institute, Rhode-Saint-Genese, Belgium.

6 Transient Modeling

6.1 OVERVIEW

In the previous chapters, we discussed the fundamental methodologies for executing performance diagnostic assessments that emerged from engine condition monitoring activities that began in the '70s and are still common practice today. Increased capabilities of onboard monitoring hardware and processing throughput during the '90s and the ensuing decades opened the door for acquiring and processing higher bandwidth data that extended past steady-state operation of the engine.

In this chapter, we will discuss some basic modeling approaches that allow tracking engine operation during transient maneuvers. This will serve as a foundation to begin expanding our diagnostic methods beyond discrete time series to a more continuous time assessment with the ultimate goal of achieving onboard, real-time diagnostic evaluation. The content will be a general overview and is not intended as a guide to engine model construction. Onboard real-time modeling of gas turbine engines is widely used for controls and diagnostics in both industry and academia [40–47]. It is also a fundamental component of NASA's *intelligent engine* concept [48], [49].

6.2 INTRODUCTION

A practical consideration for implementing a real-time, onboard engine performance tracking system is the development of high-fidelity engine models capable of providing a reference level from which performance changes can be trended. Real-time engine models made their advent as state variable models (SVM) in the mid-1980s, which utilized a piece-wise linear model that granted a reasonable representation of the engine during steady-state operation and mild transients. Increased processor speeds over the next decade allowed more complex models to be considered that were a combination of linear and non-linear physics-based elements. While the latter provided good fidelity for control purposes over both transient operation and the engine operational flight envelope, they are often found insufficient to provide the high level of fidelity required for long-term performance tracking as well as not addressing the issue of engine-to-engine variation. Over time, these models may deviate enough from the actual engine being monitored as a result of improvements made during an engine's life cycle such as hardware modifications, bleed and stator vane schedules alterations, cooling flow adjustments, and the like. These deviations often result in inaccurate (and often misleading) module performance estimations. Thus, the

DOI: 10.1201/9781003649274-6

challenge is to find a modeling approach that will address all of these shortcomings while maintaining the execution speed required for real-time implementation.

The first attempts at constructing onboard engine models utilized a simple real-time engine model that took the form of an SVM.

6.3 SVMS

Increased emphasis on aircraft propulsion control beginning in the late '70s and '80s with the introduction of the FADEC, which spawned an interest in creating higher fidelity dynamic engine models capable of executing in real-time to realize the potential capabilities of these new controls [1–3], [36–38]. These ranged from simple additions such as analytic redundancy for sensor validation and emerging control philosophies leveraging model simulated non-measurable parameters to multi-variable model-based control for future variable cycle engines being proposed and studied. Regardless of the driving force, engine health monitoring and diagnostics would, again, be a natural beneficiary of these modeling endeavors.

The basic formulation of the SVM that emerged had two primary constituents: a *steady-state* model element and a *dynamic* model element. The *steady-state* model contains a set of basepoint values (aka *trim* values) tabulated for a series of points along the engine operating line from idle to full power at a given reference flight condition. Throughout the remainder of this section, $\boldsymbol{x}$ will refer to the engine state vector, $\boldsymbol{u}$ the deterministic engine *input* vector, $\boldsymbol{y}$ the measured engine *output* vector, and z the *uninstrumented engine outputs*. The table lookup procedure for the base-point model requires a bivariate linear interpolation algorithm. The independent variables in this procedure were typically corrected low rotor speed (N_{1c}) and ram pressure ratio (P_2/P_{amb}). Corrected low rotor speed is a *state* in the SVM; however, the ram pressure ratio is based on filtered measurements from the gas turbine instrumentation suite and is directly correlated with Mach number. The *dynamic* model reflects perturbation effects (i.e., linear dynamics) about a steady-state equilibrium condition. The complete piece-wise linear model is a collection of points generated along the nominal operating line from idle to full power. Hence, the perturbation points that characterize the linearized model coincide with the steady-state model at the tabulated flight conditions.

The basic formulation begins with the following characterization:

$$\begin{aligned} \dot{x} &= f(x,u) \\ y &= g(x,u) \\ z &= h(x,u) \end{aligned} \tag{6.1}$$

where f, g, and h are non-linear functions. Note that all the variables are a function of time (t), i.e., $\{\dot{x}(t), x(t), u(t), y(t), z(t)\}$; however, we will use a simplified symbolic representation in the sequel as a matter of notational convenience.

If we formally differentiate Equation (6.1), we obtain

$$
\begin{aligned}
d\dot{x} &= \underbrace{\left(\frac{\partial \dot{x}}{\partial x}\right)}_{A} dx + \underbrace{\left(\frac{\partial \dot{x}}{\partial u}\right)}_{B} du \\
dy &= \underbrace{\left(\frac{\partial y}{\partial u}\right)}_{C_1} du + \underbrace{\left(\frac{\partial y}{\partial u}\right)}_{D_1} du \\
dz &= \underbrace{\left(\frac{\partial z}{\partial u}\right)}_{C_2} du + \underbrace{\left(\frac{\partial z}{\partial u}\right)}_{D_2} du
\end{aligned}
\tag{6.2}
$$

This provides

$$
\begin{aligned}
\Delta\dot{x} &= A\ \Delta x + B\ \Delta u \\
\Delta y &= C_1\ \Delta x + D_1\ \Delta u \\
\Delta z &= C_2\ \Delta x + D_2\ \Delta u
\end{aligned}
\tag{6.3}
$$

where $\{A, B, C_1, C_2, D_1, D_2\}$ are matrices of partial derivatives at the equilibrium condition, and the Δs are changes realized from the small signal perturbation from equilibrium, i.e.,

$$
\begin{aligned}
\Delta x &= x - x_B \\
\Delta u &= u - u_B \\
\Delta y &= y - y_B \\
\Delta z &= z - z_B
\end{aligned}
\tag{6.4}
$$

The basepoints, $\{x_B, u_B, y_B, z_B\}$ and partials $\{A, B, C_1, C_2, D_1, D_2\}$ are generated in advance at a number of steady state points on the engine's operating line, covering idle to max power, and stored in tables that can later be interpolated for values in-between as the engine engages in transient maneuvers. The number of points generated and stored is a design choice that compromises between *computer memory and throughput* and *model accuracy*. It would not be unusual to have 20 to 30 steady state conditions represented. Indeed, there are many choices to be made that affect these competing attributes, not the least of which is the choice of the state variables represented in the vector x. Typically, some combination of rotor speeds and metal temperatures are used.[1] The input vector, u, contains the appropriate control variables, such as fuel flow, variable geometry commands, engine bleed commands, and the like. The output vector, y, contains the measured parameters that we have been dealing with throughout our discussions, and the vector z contains output parameters that are not measured, such as *TIT*, thrust, compressor stall margins, and other parameters of interest for control and diagnostics. In order to minimize the size and quantity of the numerical tables that would need to be stored for both the *steady-state* and *dynamic* model elements, *corrected parameters* are used in all of the above equations. An overview depiction of a simplified SVM architecture appears in Figure 6.1.

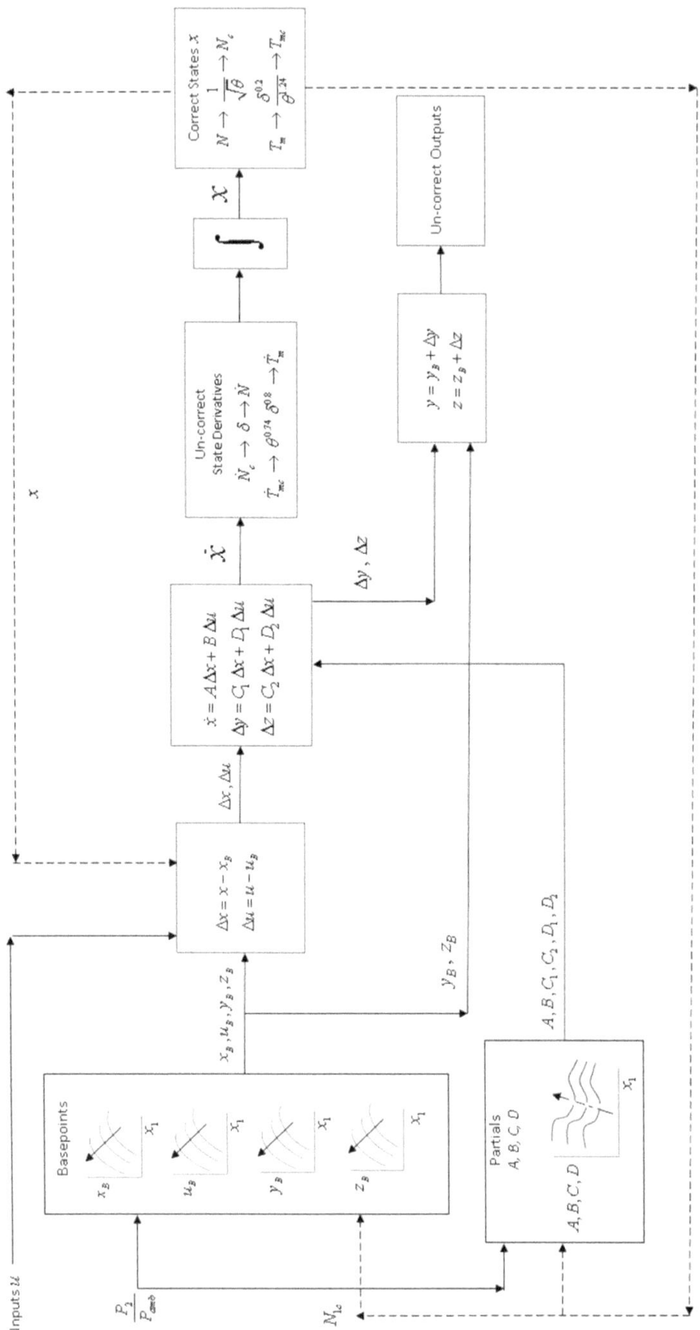

FIGURE 6.1 Basic state variable model architecture overview

The SVM provides a reference model for what would be considered a nominal engine. In theory, this could be used for diagnostics in much the same manner as the nominal steady-state models were used in the preceding chapters. One major difference is that the measured data stream that would use the model to calculate measurement deltas would be streaming at a much higher bandwidth[2] in order to track transient maneuvers. This would impose a considerable real-time computational burden if we factor in the calculation of the deltas, the calculation of the diagnostic matrix, along with its numerical constituents, and so forth. As fate would have it, the development of these models in the late '80s was driven primarily for control applications, with analytic redundancy being the first application. In this case, the requirement was to have the engine model reflect the state of the engine being monitored as it deteriorated over time. This very naturally imposed a requirement for modeling the deterioration of the major modules, not for diagnostic purposes so as much as for capturing the effects that the deterioration had on the observed measured gas path parameter values. In short, the model needed to be self-tuning.

The narrative is not a new one. Up until the '90s, EHM had not yet become entrenched, and diagnostics and prognostics proceeded slowly, leveraging what was available and put in place for control purposes. This was certainly true regarding engine instrumentation. As the new millennium approached, self-tuning models were already in use for engine control purposes. As a result, these models were adopted by the DPHM community for diagnostic purposes in an attempt to expand methods from discrete time series steady-state analysis to broader coverage within the flight envelope. In doing so, a slightly different methodology emerged, which will be the main topic of the following chapters. To set the foundation, we begin our discussion of a self-tuning model.

6.4 SELF-TUNING MODELS

The modeling approach that we will outline in this chapter was originally introduced in the mid-1980s and initially tested in flight trials by NASA in their *F-15 performance seeking control* program, [4–11], [33]. A real-time SVM was combined with an engine performance tracking filter (e.g., a Kalman filter observer) to form what was called a self-tuning onboard real-time model (STORM). To enable this ability, the SVM described in the previous section can be expanded to contain a vector of module performance elements, which we will denote by η. These elements will be referred to as *tuners* and are estimated in real time by the Kalman filter observer. To begin, we rewrite Equation (6.1) as seen below in Equation (6.5).

$$\begin{aligned} \dot{x} &= f(x,u,\eta) \\ y &= g(x,u,\eta) \\ z &= h(x,u,\eta) \end{aligned} \tag{6.5}$$

Mimicking our previous manipulations, we have, by formal differentiation, the following relationships

$$
\begin{aligned}
d\dot{x} &= \underbrace{\left(\frac{\partial \dot{x}}{\partial x}\right)}_{A} dx + \underbrace{\left(\frac{\partial \dot{x}}{\partial u}\right)}_{B_1} du + \underbrace{\left(\frac{\partial \dot{x}}{\partial \eta}\right)}_{B_2} d\eta \\
dy &= \underbrace{\left(\frac{\partial y}{\partial x}\right)}_{C_1} dx + \underbrace{\left(\frac{\partial y}{\partial u}\right)}_{D_{11}} du + \underbrace{\left(\frac{\partial y}{\partial \eta}\right)}_{D_{12}} d\eta \\
dz &= \underbrace{\left(\frac{\partial z}{\partial x}\right)}_{C_2} dx + \underbrace{\left(\frac{\partial z}{\partial u}\right)}_{D_{21}} du + \underbrace{\left(\frac{\partial z}{\partial \eta}\right)}_{D_{22}} d\eta
\end{aligned}
\tag{6.6}
$$

This results in the following set of relations.

$$
\begin{aligned}
\Delta \dot{x} &= A\,\Delta x + B_1\,\Delta u + B_2\,\Delta \eta \\
\Delta y &= C_1\,\Delta x + D_{11}\,\Delta u + D_{12}\,\Delta \eta \\
\Delta z &= C_2\,\Delta x + D_{21}\,\Delta u + D_{22}\,\Delta \eta
\end{aligned}
\tag{6.7}
$$

where $\{A, B, C_1, C_2, D_{11}, D_{12}, D_{21}, D_{22}\}$ are matrices of partial derivatives at the equilibrium condition, and the Δs are changes realized from the small signal perturbation from equilibrium, i.e.,

$$
\begin{aligned}
\Delta x &= x - x_B \\
\Delta u &= u - u_B \\
\Delta y &= y - y_B \\
\Delta z &= z - z_B \\
\Delta \eta &= \eta - \eta_B
\end{aligned}
\tag{6.8}
$$

The set of engine partials generated for such a STORM application differs somewhat from those used in a conventional control system design process. For the Kalman filter design model, we have included the impact that component *deterioration* has upon the engine *states* and *outputs*. The approach *treats component deterioration as a system input*; however, unlike the externally supplied engine inputs, we cannot directly measure component deterioration; therefore, these *inputs are unknown*. We need to formulate a methodology for dealing with the unknown input η. There are two approaches to solving this problem. In one case, we can formulate a state estimation problem that decouples the unknown inputs from the state estimation error, i.e., the so-called unknown input observer (UIO) problem [12]. This approach attempts to preserve estimation accuracy without directly calculating the unknown input vector. However, knowing η has intrinsic value in an engine diagnostic sense, the second approach of directly estimating the level of off-nominal engine performance is preferable for diagnostic purposes.

To solve this estimation problem, we need to assume a model that fits the deterioration process and can also be used effectively within a Kalman filter design framework. If we temporarily exclude abrupt engine fault occurrences (e.g., foreign object damage (FOD)) from the deterioration model, then an accurate depiction of the aging engine's life cycle is the so-called *slowly varying constant model* [13]. To properly formulate this model requires a stochastic process framework. In this framework, integrating a white noise source produces a slowly varying, random constant, i.e., the so-called *random walk* or *Wiener process*. Expressing this model in terms of a state of space realization results in the following constraints:

$$\begin{aligned} \dot{x}_\eta(t) &= w_\eta(t) \\ \eta(t) &= x_\eta(t) \\ E\left\{w_\eta(t_1)\, w_\eta^T(t_2)\right\} &= Q_\eta\, \delta(t_1 - t_2) \end{aligned} \tag{6.9}$$

where $w_\eta(t)$ denotes the white noise vector process with a covariance matrix of Q_η, the symbol $E\{\bullet\}$ represents the statistical expectation operator, and $\delta(\bullet)$ is the Dirac delta function. Because the function $\delta(t_1 - t_2)$ is 0 unless $t_1 = t_2$, the white noise process is uncorrelated in time. Moreover, a white noise process is a Gaussian, zero-mean process by definition. Therefore, specifying the Q_η covariance matrix completely characterizes the random process for our model of component deterioration. Deterioration is modeled in the same manner described in previous chapters, i.e., some combination of adiabatic efficiency and flow capacity changes for the compressor modules and adiabatic efficiency and nozzle area changes for the turbine components.

Each of the measurements in y contains, as before, a non-deterministic component, i.e., noise. Hence, the engine *output* measurement relationship given in Equation (6.7) needs to be modified to include this term as follows.

$$\Delta y = C_1\, \Delta x + D_{11}\, \Delta u + D_{12}\, \eta + \theta \tag{6.10}$$

where $E\left\{\theta(t_1)\theta^T(t_2)\right\} = R\,\delta(t_1 - t_2)$ and $E\left\{w_\eta(t)\,\theta^T(t)\right\} = 0 \quad \forall t$. The measurement noise is denoted by θ and is assumed to be a white, Gaussian, vector process with a covariance matrix of R. Furthermore, the random process driving the deterioration input, w_η, and the measurement noise, θ, are assumed to be uncorrelated in this problem formulation.

Before proceeding, we will expand on our equations by introducing a white noise variable (process noise) into the state equation as follows:

$$\dot{x} = A\,\Delta x + B_1\,\Delta u + B_2\,\eta + w_x \tag{6.11}$$

The process noise variable w_x is Gaussian with zero mean and covariance of $Q_x > 0$. Because Q_x is *positive definite,* a square root factorization exists, [14], i.e. $Q_x = Q_x^{1/2}\, Q_x^{T/2} = Q_x^{1/2}\, Q_x^{1/2}$. Likewise, because $Q_\eta > 0$, we have that $Q_\eta = Q_\eta^{1/2}\, Q_\eta^{T/2} = Q_\eta^{1/2}\, Q_\eta^{1/2}$.

To formulate the Kalman filter design model, we *merge* the engine and deterioration models to form *one model.* The resultant *augmented state vector* contains the normal engine states as well as the deterioration inputs, η .

$$\begin{bmatrix} \Delta\dot{x} \\ \dot{\eta} \end{bmatrix} = \underbrace{\begin{bmatrix} A & B_2 \\ 0 & 0 \end{bmatrix}}_{A_{KF}} \begin{bmatrix} \Delta x \\ \eta \end{bmatrix} + \underbrace{\begin{bmatrix} B_1 \\ 0 \end{bmatrix}}_{B_{KF}} \Delta u + \underbrace{\begin{bmatrix} Q_x^{1/2} \\ Q_\eta^{1/2} \end{bmatrix}}_{Q} \underbrace{\begin{bmatrix} w_x \\ w_\eta \end{bmatrix}}_{w}$$

$$\Delta y = \underbrace{\begin{bmatrix} C_1 & D_{12} \end{bmatrix}}_{C_{KF}} \begin{bmatrix} \Delta x \\ \eta \end{bmatrix} + D_{11}\,\Delta u + R^{1/2}\,\theta \tag{6.12}$$

$$\Delta z = \underbrace{\begin{bmatrix} C_2 & D_{22} \end{bmatrix}}_{M_{KF}} \begin{bmatrix} \Delta x \\ \eta \end{bmatrix} + D_{21}\,\Delta u$$

Thus,

$$\begin{bmatrix} \Delta\dot{x} \\ \dot{\eta} \end{bmatrix} = A_{KF} \begin{bmatrix} \Delta x \\ \eta \end{bmatrix} + B_{KF}\,\Delta u + Q\,w$$

$$\Delta y = C_{KF} \begin{bmatrix} \Delta x \\ \eta \end{bmatrix} + D_{11}\,\Delta u + R^{1/2}\,\theta \tag{6.13}$$

$$\Delta z = M_{KF} \begin{bmatrix} \Delta x \\ \eta \end{bmatrix} + D_{21}\,\Delta u$$

Using the augmented state equations in (6.13), we now have a system model that the Kalman filter design machinery can use for producing a recursive state estimator. The objective is to estimate x and η, and then to compute y and z using the output equations. The Kalman filter formulation minimizes the mean-square state estimation error induced by random disturbances and measurement noise in linear systems. In more generic terms, the Kalman filter provides a systematic framework for establishing a trade-off[3] between measured information and a process model so that an optimal linear estimate of the process outputs can be produced [15–20].

The Kalman filter generates an output error, i.e., the so-called filter *residual* vector, which is defined by

$$\begin{aligned} r &= y - \hat{y} \\ &= y - \left(C_{KF} \begin{bmatrix} \Delta\hat{x} \\ \hat{\eta} \end{bmatrix} + D_{11}\,\Delta u + y_B \right) \\ &= y - \left[C_1\,\Delta\hat{x} + D_{11}\,\Delta u + D_{12}\,\hat{\eta} + y_B \right] \\ &= \Delta y - \left[C_1\,\Delta\hat{x} + D_{11}\,\Delta u + D_{12}\,\hat{\eta} \right] \end{aligned} \tag{6.14}$$

where the "*hat*" symbol above y, x, and η denotes the Kalman filter *estimate* of the *output* measurement, engine *state*, and *tuner* vectors, respectively. Note that the second equality in Equation (6.14) expresses the output measurement estimate in terms of the augmented state estimates, the deterministic input vector, and the output base point, y_B. The Kalman filter applies a weighting factor to the residual vector and then feeds back this error signal to the state equation, i.e.,

$$\begin{aligned}\begin{bmatrix}\Delta\dot{\hat{x}}\\ \dot{\hat{\eta}}\end{bmatrix} &= A_{KF}\begin{bmatrix}\Delta\hat{x}\\ \hat{\eta}\end{bmatrix} + B_{KF}\,\Delta u + K\,r\\ &= \begin{bmatrix}A & B_2\\ 0 & 0\end{bmatrix}\begin{bmatrix}\Delta\hat{x}\\ \hat{\eta}\end{bmatrix} + \begin{bmatrix}B_1\\ 0\end{bmatrix}\Delta u + \begin{bmatrix}K_1\\ K_2\end{bmatrix}\{r\}\end{aligned} \tag{6.15}$$

This implies

$$\begin{aligned}\Delta\dot{\hat{x}} &= A\,\Delta\hat{x} + B_2\,\hat{\eta} + B_1\,\Delta u + K_1\,r\\ \dot{\hat{\eta}} &= K_2\,r\\ &\Rightarrow\\ \hat{\eta} &= \int_0^t K_2\,r(\tau)d\tau\\ \Delta\hat{x} &= \hat{x} - x_B\\ \Delta u &= u - u_B\end{aligned} \tag{6.16}$$

Hence, the state derivative prediction will be based not only on the *current state* and *input* but also on the weighted *output estimation error* from the previous state update. The weighting factor, $K = \begin{bmatrix}K_1\\ K_2\end{bmatrix}$, is the so-called *Kalman Filter gain matrix.* The K_1 matrix weights the residual error effect on the engine state estimates, whereas K_2 indicates the portion of the residual error that should be allocated to the model *tuners*. The integrator in Equation (6.16) guarantees that in the linear combination of $r(t)$ (i.e., $K_2\,r$) will be nulled as the Kalman filter approaches steady state. To compute the K-matrix, one needs to solve an algebraic *Riccati* equation given in Equation (6.17).

$$\begin{aligned}&AP + PA^T + Q - KRK^T = 0 \quad \textit{where} \quad K = PC^T R^{-1}\\ &\Rightarrow\\ &AP + PA^T + Q - PC^T R^{-1} CP = 0\end{aligned} \tag{6.17}$$

where P represents the *state covariance* matrix. Solving the *Riccati* equation in real-time would be computationally prohibitive; therefore, it is typical to generate the Kalman gain matrix in advance (offline) at steady state points on the engine's operating line covering idle to max power and stored in tables along with the other matrices used in the SVM. If we summarize our calculations, we have the following formulation.

$$
\begin{aligned}
&\begin{bmatrix} \Delta\dot{\hat{x}} \\ \dot{\hat{\eta}} \end{bmatrix} = (A_{KF} - K\,C_{KF})\begin{bmatrix} \Delta\hat{x} \\ \hat{\eta} \end{bmatrix} + K\Delta y + (B_{KF} - K\,D_{11})\Delta u \\
&\Delta\hat{y} = C_{KF}\begin{bmatrix} \Delta\hat{x} \\ \hat{\eta} \end{bmatrix} + D_{11}\,\Delta u \\
&\Delta\hat{z} = M_{KF}\begin{bmatrix} \Delta\hat{x} \\ \hat{\eta} \end{bmatrix} + D_{21}\,\Delta u \\
&\Delta\hat{x} = \hat{x} - x_B \quad , \quad \Delta\hat{y} = \hat{y} - y_B \quad , \quad \Delta\hat{z} = \hat{z} - z_B \quad , \quad \Delta y = y - y_B \\
&A_{KF} = \begin{bmatrix} A & B_2 \\ 0 & 0 \end{bmatrix} \quad , \quad B_{KF} = \begin{bmatrix} B_1 \\ 0 \end{bmatrix} \quad , \quad C_{KF} = \begin{bmatrix} C_1 & D_{12} \end{bmatrix} \quad , \quad M_{KF} = \begin{bmatrix} C_2 & D_{22} \end{bmatrix}
\end{aligned}
\tag{6.18}
$$

The relationships depicted in Equation (6.18) generalize our SVM representation to track engine module performance shifts in a real-time fashion. This is essentially the same representation as that given by Simon,[4] wherein a discrete time form is given that is more amenable for computer application. A simplified overview depiction of this system is illustrated in Figure 6.2.

In summary, the measured gas path parameters (temperatures, pressures, speeds, etc.) are compared to the estimated values from the SVM to generate *residual* differences that are used by the Kalman filter to estimate *tuner* values (module efficiency and flow capacity changes) that update the SVM through the state equations. This closed loop operation updates the tuner values to minimize the residuals. In this manner, the self-tuning model outputs estimated values to approach the observed measured values. It also provides non-measured parameter values of interest, z, that can be used for control and diagnostic purposes. These latter values can be whatever

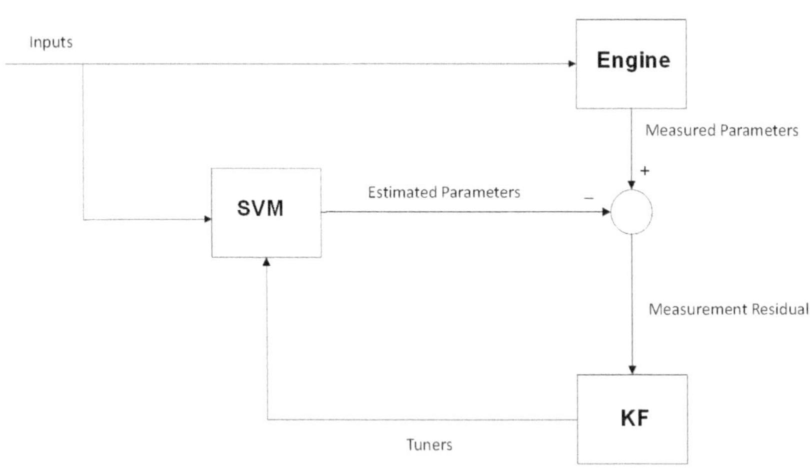

FIGURE 6.2 Self-tuning onboard real-time model overview

is of interest for the application at hand, such as thrust, stall margins, turbine inlet temperature, cruise specific fuel consumption, and the like.

Among the issues that must be addressed when constructing this dynamic system is the matter of system *observability*. In most cases, the practitioner is constrained to use the BOM engine instrumentation suite. Unfortunately, the selection of these sensors in some instances has been primarily to meet engine control system requirements rather than engine diagnostics performance metrics. From a strict mathematical point-of-view, a test for system *observability* produces a binary yes or no result. The mathematical definition for *observability* requires that the matrix

$$\begin{bmatrix} C \\ CA \\ \vdots \\ CA^{n-1} \end{bmatrix}$$

has maximal rank, i.e., rank n, where n is the dimension of the system state space. This is unsatisfactory from a practical (robust) engineering design point-of-view. A more satisfying *observability* criterion would be a measure of *how close an observable system was to being unobservable.* The parametric model data that determines system *observability* are the $\{A, C\}$ SVM matrices. Both the A and C matrices contain engine and tuner partials; hence, they represent the design matrices used to solve the matrix *Riccati* equation for the Kalman filter gain matrix K. To ensure that a unique and stable solution is generated by the *Riccati* equation for the STORM design, A and C must be fully observable. Moreover, the number of STORM *tuners* must be less than or equal to the number of linearly independent output measurements (see Addendum A6.2).

The key expression here is linearly independent output measurements. The Kalman filter analysis should contain a metric that indicates the degree to which the mathematical *observability* criterion is satisfied. One technique that can achieve this objective is to incrementally form the *observability* matrix and compute its rank using the singular value decomposition (SVD) technique (see [18], [19], [21], [22]). It is possible that the system *observability* computation defined above is numerically ill-conditioned. However, in most cases, we do not need to form the entire *observability* matrix. By forming this matrix in an iterative manner (i.e., one term in the composite *observability* matrix stack at each iteration) and computing its rank after each iteration avoids the nasty numerical problems. Nevertheless, passing this *observability* condition can still yield a system that has undesirable properties with regard to sensitivity.

One method to address these concerns is to consider the following analysis of the Kalman filter specified by the first expression in Equation (6.18). We will focus our attention on the relationship between Δy and $\Delta \hat{x}$ at steady state, i.e.,

$$\Delta u = 0$$

$$\Delta\dot{\hat{x}} = 0$$

Under these assumptions, we can rewrite this expression as

$$0 = (A - KC)\Delta\hat{x} + K\Delta y$$

Solving for $\Delta\hat{x}$ in terms of the system matrices and Δy, i.e.,

$$\Delta\hat{x} = -\underbrace{(A - KC)^{-1}K}_{G-Matrix}\Delta y \tag{6.19}$$

When expressed in this form, we can analyze some of the structural properties of the Kalman filter using multi-input/multi-output system tools such as the singular value decomposition (SVD) technique. The SVD of the $n \times m$ G-matrix can be expressed as the following product

$$G = U\Sigma V^T \tag{6.20}$$

where U is an *orthonormal* $n \times n$ matrix, V is an *orthonormal* $m \times m$ matrix, and Σ is an $n \times m$ matrix, with $k = \min\{n, m\}$ non-negative singular values, σ_i arranged in descending order along the main diagonal. All other elements of Σ are zero. The singular values are defined as the positive square roots of the eigenvalues of $G^T G$. *The ith column of U and V denote the output and input singular vectors, respectively, of the Kalman filter.* Note that in most Kalman filter applications, the number of outputs (n) is greater than the number of filter inputs (m), i.e., there are more states to estimate than output measurements.

The column vectors of U (which we denote as u_i) represent the *output* directions of the Kalman filter estimator. These column vectors are orthogonal and of unit length, that is

$$\|u_i\|_2 = \sqrt{\sum_{k=1}^{n} |u_{ik}|^2} = 1 \tag{6.21}$$
$$u_i^T u_i = 1 \quad , \quad u_i^T u_j = 0 \quad , \quad i \neq j$$

Likewise, the column vectors of V (which we denote as v_i) are orthogonal, of unit length, and represent the *input* directions of the Kalman filter. These *input* and *output* directions are related through the singular values of Σ. To derive this relationship, we can manipulate Equation (6.20) by post-multiplying both sides of this equation by V and exploiting the *orthonormal property* of V, i.e., $V^T V = I$. Performing this operation produces the relationship, $GV = U\Sigma$, which for the ith singular value, becomes

$$Gv_i = \sigma_i u_i \tag{6.22}$$

Note that both u_i and v_i are vectors, whereas σ_i is a scalar. Equation (6.22) indicates that if we consider the set of filter *inputs* in the direction of v_i, then the filter *outputs* will be in the direction of u_i. Furthermore, because both u_i and v_i are unit vectors

$$\frac{\| Gv_i \|}{\| v_i \|} = \| Gv_i \| = \| \sigma_i u_i \| = |\sigma_i| \| u_i \| = \sigma_i \tag{6.23}$$

which directly ties the gain of the matrix G to that input direction.

To interpret these results in a Kalman filter context, we need to analyze the *input* and *output* directions associated with the m-singular values defined for the G-matrix in Equation (6.20). One technique that can be used is to exploit information regarding the relative comparisons of the singular values of G. Therefore, all of the matrices that appear in Equation (6.20) must be normalized in some manner. In gas turbine applications the normalization process is equivalent to non-dimensionalizing the engine SVM.

As we begin the filter analysis process, we should note the ratio between the maximum and minimum singular values of G, i.e., (σ_1 / σ_m), where $\sigma_1 \leq \sigma_2 \leq ... \leq \sigma_m$. This ratio defines the condition number of the G-matrix, which will have ramifications with regard to the estimation performance of the filter. Ideally, this ratio should be relatively small, e.g., less than 25. This implies that the instrumentation suite was well designed for the STORM application and that all the sensors are making significant contributions to the state estimates. When this condition does not exist, either one particular input/output direction dominates the Kalman filter response, or one or more input/output directions are not making significant contributions to the Kalman filter estimates. As a rule, the latter condition is more probable than the former. A large condition number can also be symptomatic of a situation in which some of the sensors in the instrumentation suite are not linearly independent. The interested reader can consult [18] for a more in-depth discussion of the Kalman filter and its various forms and formulations in order to address various limitations and issues.[5]

6.5 HYBRID MODELS & MACHINE LEARNING

A practical consideration for implementing a real-time onboard engine component performance tracking system is the development of high-fidelity engine models capable of providing a reference level from which performance changes can be trended. SVMs coupled with a Kalman filter observer to form a STORM system are a step in this direction, but for purposes of *diagnostics* and EHM, there is still a requirement to *initialize* the system to accommodate the differences between the newly installed engine being monitored and the SVM. Initialization was briefly discussed in Chapter 2, and for discrete point data analysis, it is a means to align the reference model to a level that makes sense for the application at hand. In some instances (e.g., test stand snapshots), this is typically taken to be an average level of overhauled engines for a given facility, in others (e.g., long-term flight acquired data), it is usually tailored to the specific engine being monitored. Because these nominal reference models are steady-state models, the initialization is generally achieved through statistical analysis and culminates in an adjustment to the nominal baseline model, as described by Equation (4.2). This procedure essentially removes any inherent sensor bias, base model inaccuracy, and other effects that would otherwise corrupt the calculated measurement delta and the subsequent performance deltas resulting from them. The need for initialization for the self-tuning model is just as critical but, unfortunately, much more difficult, as it must address non-steady-state behavior as well. One approach that can be taken is to extend the STORM design to include an *empirical* element to capture the *initial* differences between the SVM and the actual engine being monitored as the engine operates. This strategy has been referred to as *enhanced* STORM (eSTORM) and has been described in the literature [23–30]. This methodology is essentially a basic form of *machine learning* and is well suited for our purposes. We provide an outline of this approach in the sections to follow.

6.5.1 Introduction to the Hybrid Model

The self-tuning models offer two main benefits, 1) engine parameter synthesis, both measured and non-measured, and 2) engine module performance tracking as provided by the *tuners*. The former being useful for FADEC fault detection and accommodation (FDA) logic and control law implementation, while the latter has the *potential* to provide valuable performance deterioration information to support engine maintenance operations and logistics as well as providing a means to detect performance anomalies onboard to enhance flight safety. This potential is in jeopardy if the physics-based model (SVM) being used is not a *faithful* representation of the actual engine being monitored. When the engine model does not provide an accurate representation of the monitored engine, then the residual difference will be interpreted (for the most part, erroneously) as changes in module performance by the *tuners*. In this instance, the tuners become a mathematical artifact to close between the model and engine, and as a result, have little to no value for true performance tracking purposes. A visual illustration of this problem is depicted in Figure 6.3, which portrays *tuner* value fluctuations due to SVM inaccuracy, engine performance variation (from paper engine model nominal), sensor biases, control schedule differences, and a host of other possible factors.

It is clear from this example that real changes in engine module performance would be difficult to discern and track over time. What is needed is a means to align the model to the data (in an initialization phase) to collapse all of these performance deltas represented by the tuners to a *consistent* zero level as illustrated in Figure 6.4. If this can be accomplished, then deviations in the tuners over time due to natural engine module deterioration or abrupt changes due to a more significant engine/sensor event, would be more easily discernible, and detectable, to allow appropriate maintenance action to ensue.

To help mitigate these effects, a *hybrid* engine model approach can be adopted, wherein the differences (or mismatch) between the physics-based model and the monitored engine are empirically modeled. One can think of this as a model adjustment process in which the resultant model is adjusted to represent the particular

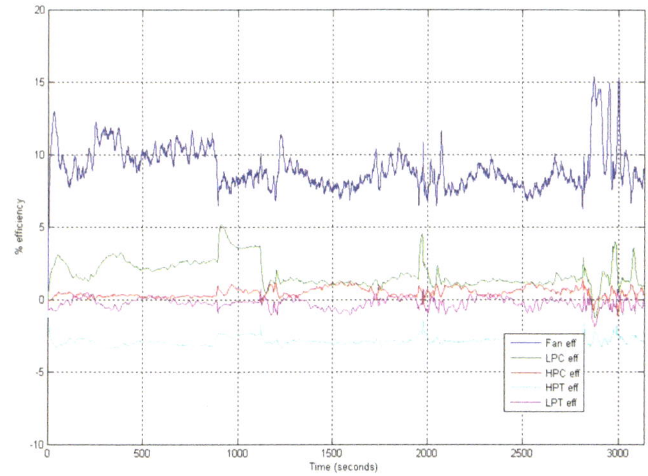

FIGURE 6.3 Possible STORM tuner fluctuation without initialization

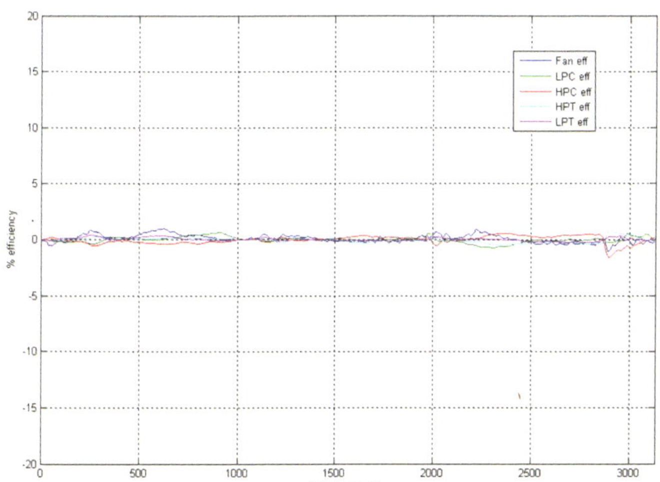

FIGURE 6.4 STORM tuner fluctuation with initialization

engine being monitored at installation. This will lessen the model error effects due to engine-to-engine variation as well as deficiencies in the physics model due to its simplicity (for real-time operation). A generic hybrid model configuration is depicted in Figure 6.5.

The empirical element can take many forms, including, but not limited to, regression models, ARMA[6] models, and ANN models. The inclusion of an engine performance estimation process in this architecture is not essential to the present discussion

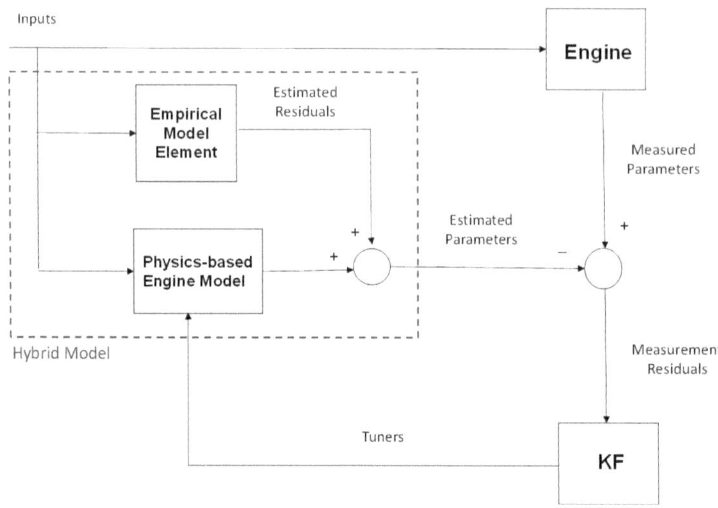

FIGURE 6.5 Generic hybrid model configuration

but is included to depict a typical application for which this hybrid structure is particularly helpful. When the empirical model is complete, the hybrid structure would take the general form shown in Figure 6.5. The combination of the empirical element and the physics-based engine model provides a more faithful representation for the particular engine being monitored. This provides more meaningful residual information from which an engine performance change assessment can be performed, as potential (physics-based) model inaccuracies and shortcomings have been effectively removed by virtue of the empirical element. In our present discussion, the physics-based model is the SVM; however, as a generality it can be any physics-based model that makes sense for the application at hand. We will refer to such a hybrid representation as or eSTORM.

Relative to the eSTORM concept, the empirical element was taken to be an ANN, as reported in [24–26], [28]. For simplicity, a garden variety MLP ANN was used because of its compact form and computational efficiency (see Addendum 6.3). During the initialization period, the ANN would capture the residual difference between the SVM and the engine being monitored, as figuratively depicted in Figure 6.6.

After the initialization period concludes, the ANN training ceases, and it is then used as depicted in Figure 6.5 to align the SVM to a closer match of the monitored engine, thereby producing residuals close to zero with minimal variation. This provides a more stable *reference* to track degradation over time as the engine deteriorates. As a figurative illustration, we refer the reader to Figure 6.7. This plot depicts the *tuner* response to simulated data in which the SVM is mismatched with the engine (as can be seen from the high variation in the first 1,000 seconds). An adjustment had been performed, and the ANN is activated around 1,100 seconds. As can been seen from the plot, the *tuners* are effectively nullified when the ANN is

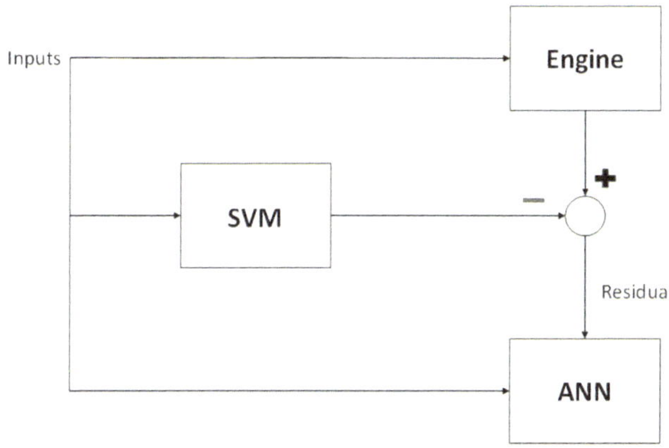

FIGURE 6.6 Learning residual difference

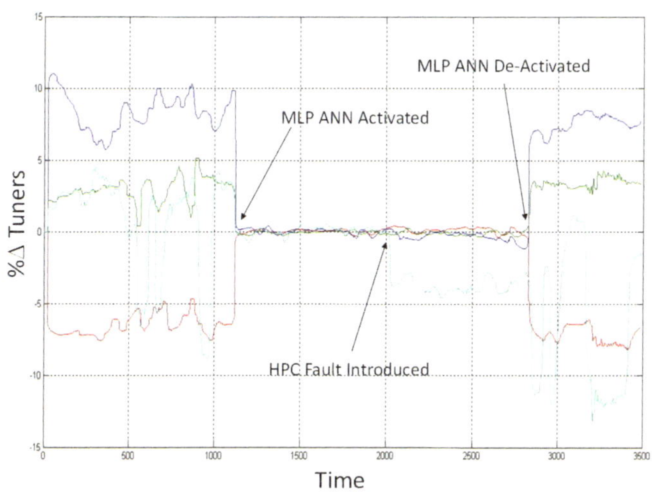

FIGURE 6.7 Illustration of fault visibility with hybrid model

engaged, as the hybrid model more effectively tracks the engine. At 2,000 seconds, an HPC fault was introduced into the engine simulation, and the shift in the tuner is easily discernible. When the ANN is disengaged, the tuner variation returns, and although a shift caused by the implanted fault is still visible, it would be much more difficult to detect.

Although the hybrid model concept is simple enough, its successful deployment is not. There is a plethora of practical considerations that need to be addressed, and, in doing so, some assumptions and compromises will be required. Among the issues that require attention are the following:

- Empirical model building considerations; onboard vs off-board
- Sequential model building strategy
- Single or multiple empirical elements
- Data compression

These concerns are *not* entirely independent of each other, and we will briefly discuss them together in the sections that follow.

6.5.2 Empirical Model Building Considerations

The construction of the hybrid model empirical element, for an onboard implementation, challenges the two primary commodities of time and space, i.e., computational throughput and memory. In addition, there is the issue of how long it will take to assemble enough flight data to adequately complete the initialization to begin using the hybrid model as the diagnostic reference.

Due to the high bandwidth of the data being processed by the STORM system (15–20Hz), the storage and retention of engine and flight input data over a series of flights to properly effect a reasonable initialization could be problematic. This could impose an unrealistic requirement in terms of storage capacity for an onboard system. To circumvent this constraint, a *sequential* building (and storing) strategy that uses a compact statistical and parametric summary of the data collected over time will be used.

One consideration is that measurement residuals are likely to vary with flight condition (Mach and altitude) for the same engine power condition. As such, it would be reasonable to partition the flight envelope to allow *individual* empirical representations to be derived in lieu of using one empirical model for the entire flight regime. This would support an *incremental* approach to empirical modeling in that it would not be expected that the engine would experience the entire flight regime in a single flight. There are two approaches to this that are worthy of mention. One approach would be to partition the *flight envelope* into sub-regions as a function of pertinent independent flight parameters. The other approach would enact a partition by *flight phase*, i.e., takeoff, climb, cruise, etc. We outline both tactics below.

6.5.2.1 Flight Envelope Partitioning

It is natural to consider partitioning a typical flight envelope into contiguous regions and developing individual empirical models for each region in order to provide adequate coverage. The resultant configuration would require a means to smoothly transition between models as the aircraft's flight trajectory progressed across the flight envelope. A suitable partition might entail having n_p regions, as depicted in Figure 6.8. The colored dashed lines offer an artist rendering of individual flight profiles. Each region (or *cell*) is defined by its center point (red dot) and a $\pm$ distance in altitude and Mach (about that point). Empirical sub-models would be derived for *each* measured parameter residual and *each* region (*cell*) in the partition. For the situation depicted in Figure 6.8, $n_p = 19$, hence the number of sub-models required would be 152 (= 8×19). This may seem like a large number; however, the use of small ANNs

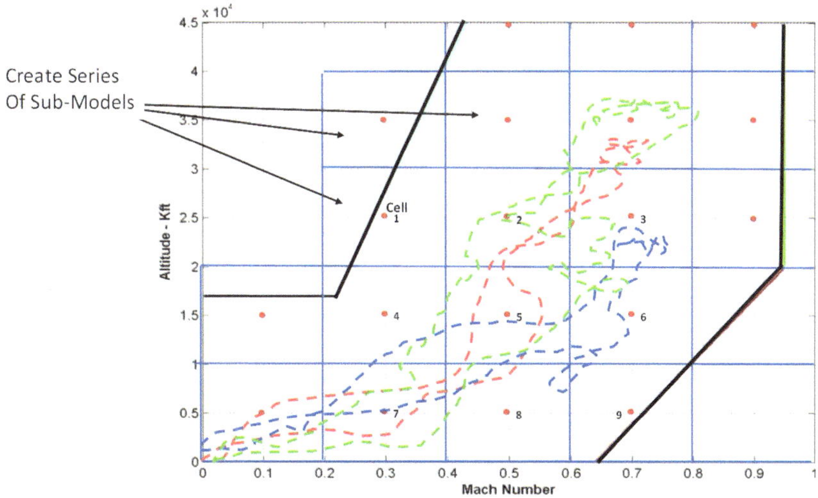

FIGURE 6.8 Notional flight envelope partition into cells

such as *MLPs* that are computationally efficient and compact, real-time implementation is easily achievable. The more difficult task to address is how to collect sufficient data and train the networks to provide sufficient flight coverage.

The creation of *cells* and sub-models reduces the empirical model problem to coping with many small models versus a single large model to cover the entire flight regime. There are advantages and disadvantages to this approach. Because the empirical (*difference*) model can only be invoked when it is complete, a clear advantage to the partition approach is that the time to complete a sub-model over a small area of the flight envelope is certainly shorter than the time needed to model the entire flight envelope. Thus, (completed) sub-models could be invoked for the purpose of anomaly detection and performance tracking when data is collected in those regions, while training the empirical sub-model in other region could progress when data become available from those regions. This allows a sequential modeling effort to ensue, with minimum time for invocation of the system (when data is collected in completed sub-model cells). The disadvantage is additional complexity. This takes several forms: 1) the integration of a system of sub-models and the seamless transition from one to another, and 2) as time accrues for the completion of sub-models in cells that are visited infrequently, engine deterioration effects must be separated from the observed residual differences so that the degradation is not absorbed into the difference model. Fortunately, there are processes to address both of these issues and will be discussed in the sections that follow.

6.5.2.2 Flight Phase Partitioning

As an alternative to partitioning the *flight envelope*, as described above, it is possible to create a partitioning by *flight phase*. The basic flight phases for a commercial revenue flight are given below.

- Start-up
- Taxi
- Takeoff
- Climb
- Cruise
- Descent
- Landing
- Shutdown

These phases can be notionally thought of as engine operational states and can be described by a finite state machine, as illustrated in Figure 6.9. The arrows connecting the states control the transition between states, and there would be a set of criteria involving aircraft and engine operational parameters to trigger the transitions.

For purposes of generating the empirical difference model for the hybrid representation, each flight phase would contribute its own empirical sub-model in the same manner that *cells* contributed sub-models in the flight envelope partitioning approach.

No matter which partitioning approach is utilized, the analyst is still faced with the basic problems of data capture, model building (i.e., training the ANN), what processes need to be resident onboard, and what needs to be accomplished off-board. Central to all of these is the potential impact on available processing throughput and memory. How we collect data and process the data will be a key factor in mitigating these constraints.

6.5.3 Data Compression

The most difficult task in the empirical modeling process is how to collect sufficient data and train the networks to provide sufficient flight envelope coverage. Traditionally, ANNs are trained *offline* in batch mode once sufficient data has been collected for that purpose. This would become a burdensome proposition to apply

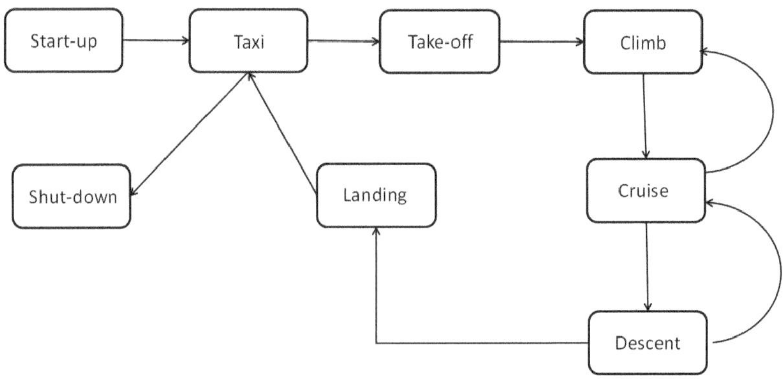

FIGURE 6.9 Notional commercial engine flight phase representation

across a fleet of engines. It would also defeat the goal of providing the level of autonomy we seek in an onboard diagnostic system. Because we are tracking the degree to which component performance may change on a given engine, a methodology for training the system onboard without the need to download engine and flight data to a ground station for batch processing is needed. It is also evident that this methodology must support sequential building of the empirical model (ANN) over many flights, as it is unlikely that sufficient data would be collected on any single flight to capture the entire flight envelope or flight phases. Such a method would need to be performed within the computational bandwidth of present-day microprocessors, not require excessive amounts of memory for storing data of the interim models, and have a means for self-determining when a given model is complete (in order to end the model building phase).

The partitioning of the flight regime contributes to reducing the complexity of the model by having a collection of smaller and, hence, simpler models, thereby reducing the computational load and data storage requirements. That statement is only true if a process can be defined to sequentially build each (simple) model in such a way as not to require the archival storage of past raw data. This seems to be a frivolous requirement, as an empirical model, regardless of its representation, is a function of the data collected! However, this requirement can be fulfilled by maintaining (sequentially) a compact statistical model representation of the data that can be used for the purpose of building the empirical model. This representation will take the form of Gaussian mixture models (GMMs).

Simply stated, a GMM is a collection of data *clusters*, each of which consists of a pair of n-dimensional vectors, one being a vector of means, the other a vector of standard deviations. In essence, the input and residual data are partitioned into clusters, and the clusters are statistically represented by the mean and standard deviation of each parameter under consideration (this forms the n-dimensional vectors). The GMM clusters are formed *on-the-fly* in real-time as data is collected. These will be stored (and archived) for subsequent training of an empirical difference model to be described below. This eliminates the need for the storage of the actual raw data. Pictorially, an example of a set of GMM clusters is depicted in the Figure 6.10. In this figure, we have a small area of a particular flight envelope partition (cell), which represents a fairly stable cruise condition at approximately Mn = 0.78 and 36k feet altitude. In this application, there are eight input parameters that cover flight condition, engine power level, and control effectors (e.g., variable geometry, handling and service bleeds, etc.). Depicted here are the first two dimensions of this eight-dimensional space. The green points represent the raw data (collected at 10Hz). There are approximately 32k raw data points depicted[7] in this figure. The pink points depict (in two dimensions) the GMM cluster centers representing this collection of data along with a 1-sigma standard deviation ellipse drawn around it. There are 23 GMM clusters depicted in Figure 6.10. The GMM clusters consist of the eight-dimensional input parameters along with the m-dimensional residual difference parameters. The assertion is that these 23 GMM clusters contain the requisite information to adequately model the residual difference of the original set of 36k (green) raw data

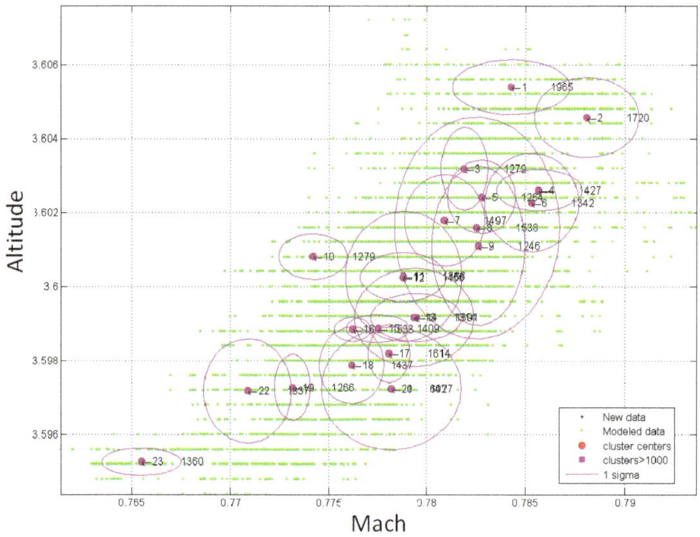

FIGURE 6.10 GMM data compression

points. That is to say, that an empirical model created from the GMM clusters would be *essentially* the same as one created from the original raw data itself.

6.5.4 Empirical Modeling Architecture

To accomplish the empirical model building task that would be suitable for a real-time, onboard environment, operating in a totally autonomous operational mode, a two-stage model building strategy can be adopted. This strategy incorporates both a real-time process and a non-real-time (offline) process. *Both processes are envisioned to be accomplished onboard*, automatically, without manual intervention.

The first stage of the process is performed in real-time. It is the process of forming the GMMs alluded to in the preceding section. In essence, this can be thought of as a data compression phase, wherein clusters of input and residual parameter data are formed, in real-time during flight and temporarily stored for subsequent processing in the second (offline, but onboard) stage of the model building procedure. In addition to forming the (compressed) data clusters, the first stage of this process also performs

1) regime recognition, i.e., knowing what cell partition (or flight phase) and, hence, sub-model, is being built.
2) whether the current data point should be part of a new cluster or is, indeed, already covered by an existing cluster from a previous flight (in which case, the current data point is already modeled, and the eSTORM configuration in Figure 6.5 can be executed using existing MLPs to track performance changes).

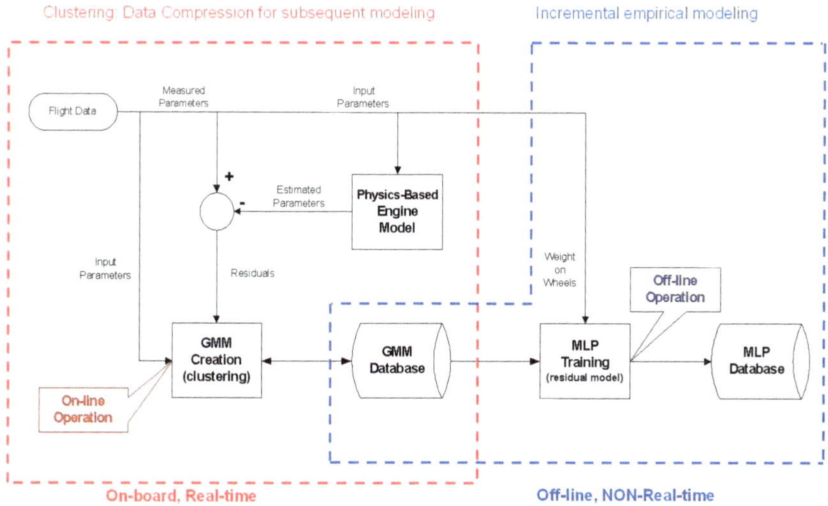

FIGURE 6.11 Two-stage empirical modeling approach

The *overview* structure of this two-stage process is depicted in Figure 6.11.

Relative to Figure 6.11, the steps enclosed within the red dashed lines form the first stage of the process, which is performed onboard in real-time. As each data point is received, (typically at 10–20Hz), its *cell* (or *flight phase*) classification is determined. This defines where in the flight regime the data resides and what sub-model is in effect. Using the input parameters (of this data point), a determination is made as to whether the data point should be processed by an existing MLP sub-model (as depicted in Figure 6.5) or whether the current data point represents an area in the flight regime that has not been previously modeled. This determination is made by establishing whether or not the n-dimensional input vector falls within a *neighborhood* of an already existing GMM cluster residing in the GMM database. A nearest neighbor criterion can be used for this determination. If the data point falls outside any already established GMM cluster, it is a candidate for the formation of a new GMM cluster. The formation of a new GMM element is essentially a data clustering process that is performed in real-time.

The steps enclosed in the blue dashed lines in Figure 6.11 are performed offline (i.e., non-real-time) and are performed onboard at the end of the flight. This could, for example, be launched by a weight-on-wheels (WOW) signal. This is the second stage of the incremental model building process. The GMM database is perused to determine whether new clusters have been added to the database. In this case, on a cell-by-cell basis, new MLPs are generated for each *residual* parameter for each cell (or sub-model) that has new GMMs and becomes the current incremental model in the MLP database. The training of the MLPs is done individually. Relative to a traditional MLP batch training methodology that would train directly on stored flight data, it is possible to train on the GMM *clusters* directly.[8] The effect on the *residual* model is essentially the same. Of course, this last statement is true only if certain basic assumptions are met. There is an implicit assumption that the (residual) difference (during the initialization phase) between the monitored engine and the model (SVM) is essentially a continuous function. If the difference is due to collective module performance, sensor bias, and model inaccuracies, this is a fair assumption. In this case, we would not expect dramatic shifts in residual value as the GMMs are being formed; therefore, the GMM average would be a fair approximation for the actual residual difference. On the other hand, if our SVM model was terribly deficient, for example, if it were missing an essential control effector such as a handling bleed, then the activation or deactivation of the bleed, unbeknownst to the model, would cause discontinuities in the residual difference. In this latter case, the GMM averaging would be corrupted, and their use would be less than marginal. For more information see [33–35], [39].

The hybrid model approach described above is a step toward producing a *digital twin* for the engine being monitored in order to enable performance health assessments. It is, of course, not the only approach possible and the reader can consult [31], [32] for some other perspectives.

NOTES

1. See [27].
2. Typically, between 15–20 Hz.
3. Essentially a form of information fusion.
4. Simon, D.L., Garg, S., 2009, *Optimal Tuner Selection for Kalman Filter-Based Aircraft Engine Performance Estimation*, ASME GT2009-59684.
5. Many of which we have chosen not to discuss for brevity.
6. Autoregressive–moving-average model
7. Note: A single green point depicted in this two-dimensional plot may represent many points that are unique in the other six dimensions not appearing on this plot.
8. See Addendum A6.4.

REFERENCES

1. Mihaloew, J., Roth, S., 1982, *A Piecewise Linear State Variable Technique for Real Time Propulsion System Simulation*, NASA, Technical Memorandum 82851.
2. French, M., 1982, Development of a Compact Real-Time Turbofan Engine Dynamic Simulation, *SAE Aerospace Congress & Exposition*, Anaheim, CA, paper 821401.
3. Weinberg, M., Adams, G., 1975, Low Order Linearized Models of Turbine Engines, *Directorate of Propulsion & Power Engineering*, Wright-Patterson AFP, Ohio, Technical Report ASD-TR-75-24.
4. Behbahani, A., et.al., 2009, *Integrated Model-Based Controls and PHM for Improving Turbine Engine Performance, Reliability, and Cost*, AFRL-RZ-WP-TP-2009-2178.
5. Armstrong, J.B., Simon, D.L., 2012, Constructing an Efficient Self-Tuning Aircraft Engine Model for Control and Health Management Applications, *Annual Conference of Prognostics and Health Management Society*, Minneapolis, Minnesota, USA.
6. Luppold, R.H., et.al., 1989, *Estimating In-Flight Engine Performance Variations Using Kalman Filter Concepts*, AIAA Paper 89-2584.

7. Gallops, G.W., et.al., 1992, *In-Flight Performance Diagnostic Capability of An Adaptive Engine Model,* AIAA Paper 92-3746.
8. Kerr, L.J., et.al., 1991, *Real-Time Estimation of Gas Turbine Engine Damage Using a Control Based Kalman Filter Algorithm*, ASME Paper 91-GT-216.
9. Lambert, H., et.al., 1991, *Preliminary Flight Evaluation of an Engine Optimization Algorithm*, AIAA Paper, AIAA-91-1998.
10. Maine, T., et.al., 1990, *A Preliminary Evaluation of an F100 Engine Parameter Estimation Process Using Flight Data*, AIAA Paper, AIAA-90-1921.
11. Meyers, L, Burcham, F., 1984, *Preliminary Flight Test Results of the F100 EMD Engine in an F-15 Airplane*, NASA TM-85902.
12. Chen, J., et al., 1996, Design of Unknown Input Observers and Robust Fault Detection Filters, *International Journal of Control*, 63(1), 10, 85–105.
13. Kwakernaak, H., Sivan, R., 1972, *Linear Optimal Control Systems*, Wiley-Interscience, Ch. 4.
14. Lewis, F., 1986, *Optimal Estimation*, John Wiley & Sons. Hoboken, New Jersey, USA.
15. Gelb, A., et al., *Applied Optimal Estimation*, Cambridge, MA: MIT Press.
16. Jaw, L., Mattingly, J., 2009, *Aircraft Engine Controls: Design, System Analysis, and Health Monitoring,* AIAA Education Series, ISBN 978-1-60086-705-7.
17. Lewis, F., 1986, *Optimal Estimation,* New York: John Wiley & Sons, ISBN 0-471-83741-5.
18. Grewal, M., Andrews, A., 1993, *Kalman Filtering: Theory & Practice,* Prentice Hall, NJ, ISBN 0-13-211335-X.
19. Jazwinsky, A., 1970, Stochastic Processes and Filtering Theory, *Mathematics in Science and Engineering*, Vol. 64, New York: Academic Press.
20. Sorenson, H., 1966, Kalman Filtering Techniques, *Advances in Control Systems*, Vol 3, Academic Press, published by Elsevier, pp. 219–292.
21. Gentle, J. E., 1998, Singular Value Factorization, *§3.2.7 in Numerical Linear Algebra for Applications in Statistics.* Berlin: Springer-Verlag, pp. 102–103.
22. Weisstein, Eric W., Singular Value Decomposition, *MathWorld--A Wolfram Web Resource.* https://mathworld.wolfram.com/SingularValueDecomposition.html.
23. Brotherton, T, et al., 2003, eSTORM: Enhanced Self Tuning On-Board Real-Time Engine Model, *Proceedings of the 2003 IEEE Aerospace Conference*, paper 1023. Big Sky, MT.
24. Volponi, A.J., 2005, Use of Hybrid Engine Modeling for On-Board Module Performance Tracking, *Proceedings of the American Society of Mechanical Engineers Turbine and Aeroengine Congress.* Reno, NV, June 6–9, Paper No. GT2005-68169.
25. Volponi, A.J., et al., 2007, Adaptive On-Wing Gas Turbine Engine Performance Estimation, *Proceedings of the 2007 IEEE Aerospace Conference.* Big Sky, MT.
26. Volponi, A., Brotherton, T., and Luppold, R. (February 29, 2008). Empirical Tuning of an On-Board Gas Turbine Engine Model for Real-Time Module Performance Estimation, *ASME. J. Eng. Gas Turbines Power.* March 2008; 130(2): 021604. https://doi.org/10.1115/1.2799527
27. Borguet, S., et al., 2006, A Way to Deal with Model-Plant Mismatch for Reliable Diagnosis in Transient Operation, *Proceedings of the American Society of Mechanical Engineers Turbine and Aeroengine Congress*, Barcelona, Spain, May 8–11, Paper No. GT2006-90412.
28. Volponi, A.J., Brotherton, T., 2005, A Bootstrap Data Methodology for Sequential Hybrid Engine Model Building, *IEEE Aerospace Conference*, Big Sky, MT, pp. 3463-3471, https://doi.org/10.1109/AERO.2005.1559649.

29. Volponi, A.J., Rajamani, R., 2012, Hybrid Models for Engine Health Management, *Machine Learning and Knowledge Discovery for Engineering Systems and Health Management*, Chapter 12, CRC Press, New York, ISBN 978-1-4398-4178-5.
30. Volponi, A.J., 2008, *Enhanced Self Tuning On-Board Real-Time Model (eSTORM) for Aircraft Engine Performance Health Tracking,* NASA CR-2008-215272.
31. Belov, S., et.al., 2020, Hybrid Data-Driven and Physics-Based Modeling for Gas Turbine Prescriptive Analytics, *International Journal of Turbomachinery Propulsion and Power*, 5, 29. https://doi.org/10.3390/ijtpp5040029.
32. Panov, V., Cruz-Manzo, S., 2020, *Gas Turbine Performance Digital Twin for Real-Time Embedded Systems,* ASME paper GT2020-14664.
33. Adibhatla, S., 2003, *Method and Apparatus for Estimating Engine Health*, US Patent US 6539783 B1.
34. Fentaye, A.D., et al, 2019, *A Review on Gas Turbine Gas Path Analysis: State-of-the-Art Methods, Challenges and Opportunities*, Aerospace, 6, 83; doi: 10.3390/aerospace6070083, www.mdpi.com/journal/aerospace.
35. Simon, D.L., et al., 2012, *Application of an Optimal Tuner Selection Approach for On-Board Self-Tuning Engine Models*, NASA/TM-2012-217278.
36. Chung, G. Y., 2013, *An Analytical Approach to Real-Time Linearization of a Gas Turbine Engine Model*, Ph.D. Dissertation, Georgia Institute of Technology, 2013.
37. Seldner, K., Cwynar, D.S.,1978, *Procedures for Generation and Reduction of Linear Models of a Turbofan Engine*, NASA Technical Report 1261, NASA, Washington.
38. Jaw, L. C., Mattingly, J., D., 2009, *Aircraft Engine Controls: Design, System Analysis, and Health Monitoring*, AIAA Education Series, ISBN 978-1-60086-705-7.
39. Simon, D.L., et al., 2011, Application of an Optimal Tuner Selection Approach for On-Board Self-Tuning Engine Models, *Proceedings of the ASME 2011 Turbo Expo*, Vancouver, Canada, Paper GT2011-46408.
40. Borguet, S. et al. 2010, Comparison of Adaptive Filters for Gas Turbine Performance Monitoring, Journal of Computational and Applied Mathematics, 234, 2202–2212.
41. Wei, Z., Zhang, S., Jafari, S., Nikolaidis, T., 2020, Gas Turbine Aero-Engines Real Time On-Board Modelling: A Review, Research Challenges, and Exploring the Future, *Progress in Aerospace Sciences*, 121, 100693, ISSN 0376-0421, https://doi.org/10.1016/j.paerosci.2020.100693.
42. Adibhatla, S., Gastineau, Z., 1994, Tracking Filter Selection and Control Mode Selection for Model Based Control, *30th Joint Propulsion Conference and Exhibit*, Indianapolis, IN, USA, pp. 3204.
43. Lietzau, K., Kreiner, A., 2001, Model Based Control Concepts for Jet Engines, *ASME Turbo Expo 2001: Power for Land, Sea, and Air, American Society of Mechanical Engineers.*
44. Nikolaidis, T., Li, Z., Jafari, S., 2019, Advanced Constraints Management Strategy for Real-Time Optimization of Gas Turbine Engine Transient Performance, *Applied Sciences*, 9, 5333.
45. Kerr, L., Nemec, T., Gallops, G., 1992, Real-Time Estimation of Gas Turbine Engine Damage using a Control-Based Kalman Filter Algorithm, *Journal of Engineering for Gas Turbines and Power*, 114, 187–195.
46. Turevskiy A., Meisner, R., Luppold, R.H., Kern, R.A., Fuller, J.W., 2002, *A Model-Based Controller for Commercial Aero Gas Turbines*, 2002, 189–195.
47. Chung, G., Prasad, J.V., Dhingra, M., Meisner Richard, P. 2013. Real Time Analytical Linearization of Turbofan Engine Model. *ASME Journal of Engineering for Gas Turbines and Power*, 136, 011201.

48. Culley, D., Garg, S., Hiller, S. J., Horn, W., Kumar, A., Mathews, H. K., Moustapha, H., Pfoertner, H., Rosenfeld, T., Rybarik, P., 2009, *More Intelligent Gas Turbine Engines*, North Atlantic Treaty Organisation: Brussels, Belgium.
49. Garg, S., 2004, Controls and Health Management Technologies for Intelligent Aerospace Propulsion Systems, *42nd AIAA Aerospace Sciences Meeting and Exhibit*, Reno, Nevada, January 5–8, 2004. NASA/TM—2004-212915, AIAA–2004–0949

7 Extending Beyond Steady-State Gas Path Analysis

7.1 OVERVIEW

In the previous chapters, we have discussed the fundamentals of GPA as a diagnostic methodology for processing engine data obtained at what is essentially *steady-state* conditions. This included test-stand data in the form of a single steady-state snapshot, a collection of snapshots over a brief testing period, and time series data obtained during flight at stable cruise conditions. The presentation has focused on a (physics) model-based strategy incorporating a Kalman-filter-inspired approach as a palliative measure for the general fault estimation problem. Extensions to the basic methodology were added to address noted deficiencies, making liberal use of added information, which often came in the form of assumptions. The practicality of implementation and positive experience with this approach reported in the literature, over many decades, supports the efficacy of this strategy.

As we look to the future, the ultimate goal would be to transfer as much diagnostic and prognostic capability to the onboard system as possible in support of the *intelligent engine* concept [1–8]. This would ideally reduce time latency to detect and isolate incipient engine faults, increase engine reliability through (potential) control mitigation, and support EHM logistics and effectiveness. In this chapter, we will begin to explore this question using the framework already developed from the previous chapters. This is essentially the same information fusion tactic we have been leveraging throughout this text. The goal of this chapter, therefore, is to determine the extent to which our steady-state GPA approach can be extended to non-steady-state conditions while maintaining a reasonable memory and computation footprint to allow onboard implementation.

7.2 ONBOARD VS OFF-BOARD

With only a few exceptions, our chronologically motivated discussion, up to this point, has been concentrated on *off-board* engine performance diagnostics and prognostics. Historically, this was a natural consequence of the computation and storage limitations of the onboard electronic systems (AIMS and FADEC) available in the 1970s through the 1990s. In the past two decades, these constraints have relaxed somewhat with the availability of faster CPUs and increased onboard memory

DOI: 10.1201/9781003649274-7

capacity. This appears to be the trend for the future, which now reopens the question as to what should be done *onboard* versus what should be done *off-board* in order to maximize diagnostic and prognostic effectiveness.

If we assume, for the moment, that CPU and memory constraints are removed entirely, we might argue that the diagnostics and prognostics for a particular engine should be consigned to the *onboard* system, while the health tracking and management of the fleet becomes the primary responsibility of the *off-board* system. There are clear advantages to such an assignment. Most notably, there would be a reduction in the time latency to detect and isolate incipient engine faults. Additionally, this would reduce the potential for secondary collateral damage, require less data retention and download, provide a less expensive data infrastructure, increase engine reliability through *intelligent* control to mitigate further damage, and avoid catastrophic events and failures, and so forth. The potential advantages are many. Naturally, there are downsides as well. The most notable is one that is true of all onboard systems, which is reduction in visibility (and, hence, control) of the diagnostic process. The onboard system becomes a *black box* of sorts, in the sense that the engine *self-diagnoses* and the human is out of the loop. For those who have seen the classic film, *2001 – A Space Odyssey*, (which is almost everyone on planet Earth), visions of HAL9000 might cause some uncertainty and concern for a totally autonomous system. A concern well taken. But realistically, that is the situation you have, for example, with modern electronic control of the engine itself, which operates autonomously, without human interference and above all SAFELY. So, all things are possible in time.

We are not going to provide a timeline to achieve the ultimate goal, nor a recommended plan that outlines the exact path that must be pursued, verified and validated. What we will do in the remainder of this chapter is start the discussion by introducing some concepts to consider and a couple of methods that have been investigated that are natural extensions of the steady-state methods of the previous chapters.

To begin, we will state the obvious, which is that the engine model that we use in our model-based method plays a pivotal role in the fault detection and identification (FDI) methodology. It forms the fundamental reference framework representing the nominal state from which (observed) parameter deviations can be quantified as the engine degrades and/or faults occur. Variation resulting from production/overhaul engine-to-engine differences, modeling errors, simplifications, instrumentation biases, etc., is ever present and must be accounted for in any model-based diagnostic/prognostic approach. *Minimizing this variation is critical to accurate engine health monitoring analyses.* A fundamental practice to mitigate this variation is engine model *initialization* (or *alignment*). This is the process of aligning the model to the actual engine under observation for diagnostic systems that perform, at least some of its functions, *onboard* in a quasi-real-time environment. This alignment can be achieved in a variety of ways through adjustments to the model in the form of corrections or normalization factors whose net effect is to reduce the differences between model output and actual (sensed) observations. One such method was introduced in Chapter 6. In any event, the efficacy of the *initialization* method and process will depend on the amount (and type) of data available, time allotted to accomplish the *initialization*, robustness of the process (i.e., can it be applied generically to any model formulation), and of course, the level of error that can be tolerated in the adjusted model for the specific application at hand.

7.3 EXTENDING STEADY-STATE

Aero-engines exhibit a wide-ranging level of dynamic behavior. It is fair to say, from a model error perspective, that *steady-state* operation is undoubtedly an easier mode of operation to model than the almost infinite *transient* maneuvers that an engine could experience during flight. As a consequence, model *alignment* for steady-state operation is a far easier task than a model *alignment* that encompasses all possible engine transients. As a matter of practicality, a reasonable model alignment would capture something in-between these two extremes. We will use the term *quasi-steady-state* (*QSS*) to designate the collection of conditions for which an *acceptable* alignment can be made (see [9], [24]). The term *acceptable* will, of course, depend on our model type, maximum tolerable error, time allotment to calibrate, and other factors germane to our particular application. We will assume, without loss of generality, that *QSS* consists of a collection of engine operating and flight regimes that contain *steady-state* operation and *some sub-set of transient behavior*. Whatever the form the alignment process takes, we will assume that the following two properties apply: if A is an operating regime that supports model alignment, then any sub-set of A will support alignment and that the union of such regimes will also support alignment, i.e.,

$$\begin{aligned} B \subseteq A \in QSS &\Rightarrow B \in QSS \\ A_k \in QSS, \forall k &\Rightarrow \bigcup_k A_k \in QSS \end{aligned} \tag{7.1}$$

The form of A_k will vary according to the specifics of the initialization process (and model configuration) in use but, in general, will be a function of ambient conditions, flight conditions, and engine operating parameters, (e.g., power level, variable geometry, internal bleed status, etc.). It is apparent from these two properties that model alignment can be a *sequential* process that can occur *incrementally* over a number of flights. In fact, if A_k is the collection of regimes newly aligned during the kth flight, then for any aligned regime A, it must be true that $A \in \bigcup_k A_k$. Therefore, combining with Equation (7.1), we have the (somewhat obvious) statement that

$$QSS = \bigcup_k A_k \tag{7.2}$$

To facilitate our further discussion, we introduce some additional terminology.

Let F_k denote the set of all points[1] from the kth flight and Ω_k the collection of all *QSS* points from the kth flight. From the foregoing remarks, it should be clear that we can always partition the total *QSS* collection of points into a *disjoint* union of contiguous sets of *QSS* points, which we will refer to as *realms*, i.e.

$$\Omega_k = \left(\bigcup_{J=1}^{M_k} QSS^{(k)}(J) \right) \subseteq F_k \text{ where}$$
$$QSS^{(k)}(J) = J^{th} \; realm \text{ in flight } k \tag{7.3}$$
$$M_k = \text{number of } realms \text{ in flight } k$$

In plain language, Equations (7.1)–(7.3) tell us that the total QSS collection may consist of sets of data points across multiple flights. In each flight, it may consist of multiple disjoint segments of continuous data points, and each segment is called a *realm*, as depicted in Figure 7.1.

What is important, for FDI, will be the ability to recognize whether the data being processed (or collected) is *represented* by a particular *realm* that is contained in the *QSS* collection. We address this question in the next sub-section.

7.3.1 QSS Determination

Up to this point, we have not provided a concrete test that one could use to determine whether a given set of engine data points qualify as being in *QSS*. We will begin to explore this with an illustrative model alignment methodology, which the *hybrid* model approach outlined in Chapter 6. That procedure involved the creation of a *difference model,* using machine learning techniques, that took the form of an *ANN* that could then be used as a bias adder to *adjust* a physics-based dynamic engine model to more closely match (in terms of model parameter outputs) the parameter data observed from the monitored engine. That formulation captured the parameter differences at a variety of *input parameter conditions* and developed a difference model that could be catalogued as a function of these conditions. As such, the implementation would require the system to test whether the current (input) conditions were within a *neighborhood* of the previously mapped input conditions in order to determine whether (and how) the mapped differences could be applied. The input space can become rapidly unmanageable, as the dimension of this space increases to cover appropriate independent parameters and, hence, could become computationally burdensome to implement. To avoid becoming possible victims of dimensionality, we can consider a more approximate ad hoc approach.

We can develop an approach by deciding what properties are consistent with our (somewhat nebulous) definition that *QSS* is the collection of conditions for which an

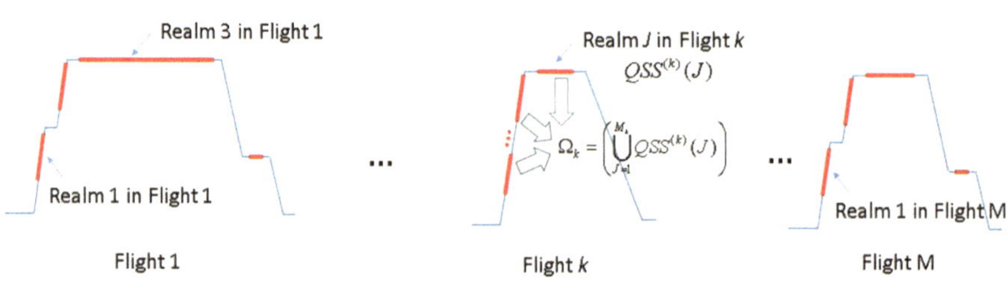

FIGURE 7.1 QSS realms

acceptable alignment can be made. We can postulate some *necessary* conditions for what an *acceptable* alignment might impose, e.g.,

1) The *variance* of the output parameters (i.e., the parameter *residuals*, ΔP) over a sliding time window (T_{window}) *should be minimal.* That is to say, at least in terms of the engine and model output *mismatch*, things are quasi-stable, i.e., QSS. The mismatch should be stable and not given to wide fluctuation.
2) Given 1), a *bias adder* compensation approach (via a *difference* model) would be adequate to achieve acceptable alignment.
3) If 1) holds, and *enough* flight profiles have been covered (for training), then *bias adder* mapping will *interpolate* effectively to cover cases in which input is not covered by training data.

These three properties are arguably necessary for an alignment to be deemed acceptable. One manifestation of Property 1 is the following:

Let ΔP_k be the vector of residuals at time t_k.

$$\Delta P_k = \{\Delta P_1(t_k), \Delta P_2(t_k), \ldots, \Delta P_m(t_k)\}$$
$$where \quad \Delta P_i = i^{th}\ measured\ parameter\ residual, 1 \le i \le m$$

and let

$$\sigma_i(t_k) = \sqrt{\frac{1}{N-1} \sum_{j=k-N+1}^{k} \left[\Delta P_i(t_j) - \mu_i(t_k)\right]^2} \qquad where \quad \mu_i(t_k) = \frac{1}{N} \sum_{j=k-N+1}^{k} \Delta P_i(t_j) \tag{7.4}$$

Thus, $\mu_i(t_k)$ and $\sigma_i(t_k)$ represent the mean and the standard deviation[2] of the ith measurement residual across a time window of length N ending at time t_k. Then Property 1 suggests that if we are in *QSS* at time t_k, then $\exists L \ni Metric[\sigma_i(t_k)] \le L$, where "*Metric*" is some normalizing error function. In a leap of faith, we could use this as a *sufficient* condition to define *QSS* as well, for example,

$$QSS\,@\,t_k \Leftrightarrow \frac{1}{m}\sum_{i=1}^{m} \sigma_i(t_k) < L\ for\ some\ pre\text{-}defined\ threshold\ L \tag{7.5}$$

where the "*Metric*" function has been chosen to be the *average* of the temporal standard deviations across all output parameter residuals. Certainly, this would be a property that we would expect to be minimized if our alignment effort were to be deemed successful. It would seem reasonable to use this as our *sufficient* condition to define *QSS*, the *efficacy* of which will be determined through eventual simulation studies, and ultimately through usage. *For our purpose, the (model) alignment must be suitable to support FDI.* We can test the suitableness of our definition through a series of Monte Carlo simulations, which will be discussed in the section on *QSS* validation. Before describing that process, we will pause briefly to explore some general aspects of FDI in the following sections.

7.4 FAULT DETECTION

Before we can identify a fault, we must know that one has actually occurred, i.e., we must first detect the presence of (or consequence of) a fault. In this context, we assume a fault to be some part of the engine system or one of its sub-systems that has ceased to function as intended. This may be a *total* failure or a *partial* failure of the system or sub-system element that (typically) occurs over a short period of time, distinguishing itself from normally expected long-term degradation or deterioration. To be detectable, it must also affect some set of parameters that we have available to us to observe, i.e., sense. Because failed elements do not repair themselves, their *effect* on what we can observe must be *persistent*. This latter requirement lessens the likelihood of misdetection due to spurious operational events and noise. Lack of model alignment can also lead to misdetection if the model mismatch during certain operational maneuvers provides sufficient departure of model response to actual observation over a finite period of time, which can confuse the situation with a false sense of persistency. *For these reasons, detection should only be performed when operating within QSS.*

Detection methods are widely varied. As a matter of example, we describe a notional system that can be used that is especially suitable for real-time, continuous operation with full flight data. Depicted in Figure 7.2 is a simplified notional system that contains a *hybrid* model that produces a set of *estimated* engine output parameters, P^*_{est}, that can be compared to actual *sensed* engine output parameters, P_{actual}, to produce a set of parameter residuals, ΔP^*. These parameters typically include,

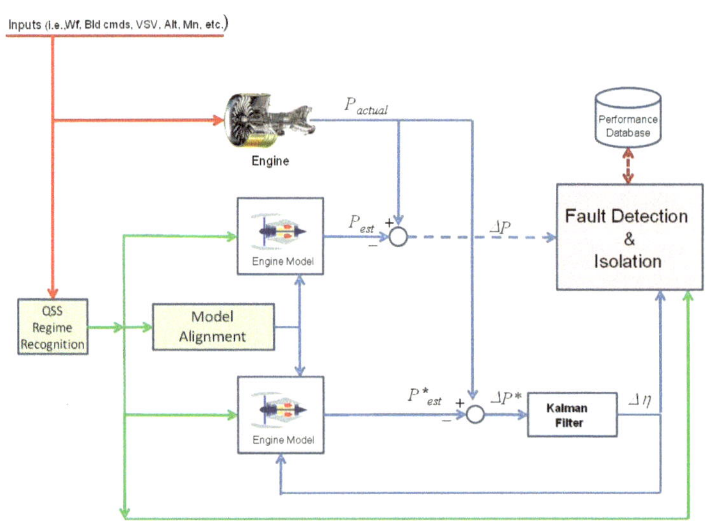

FIGURE 7.2 Simplified notional engine fault detection architecture

but are not limited to, inter-stage gas path temperatures, pressures, and speeds. The parameter residuals can be input to a performance estimator that outputs a set of tuners, $\Delta\eta$, which are fed back to the model, updating it in a manner that drives a closer P^*_{est} match to P_{actual}, driving the set of residuals, ΔP^*, to near zero. The reader will recognize this as the *eSTORM* concept discussed in Chapter 6.

When the model is operating within the *QSS* regime, the *tuners*, $\Delta\eta$, provide adjustments to the model to reflect the natural deterioration of the engine over a long period of time. However, when operating *outside* the *QSS* regime, the *tuners* become a mathematical artifact to force the parameter residuals to zero. Similarly, when a fault occurs, the *tuners* will, likewise, respond to force the residuals back to zero. Thus, if we are operating within *QSS*, the *tuner* values can be used to test for the presence of an unknown disturbance, which if *persistent*, indicates the presence of a yet to be identified fault.

7.5 FAULT IDENTIFICATION

Similar to detection, *identification* or *isolation* of a fault must be performed when operating within a *QSS* regime. The requisite information for fault isolation will take the form of quantifying the *temporal change*[3] in the parameter residuals (ΔP) before and after the fault detection, i.e., $\Delta\Delta P \approx \Delta P_{After\_Fault} - \Delta P_{Before\_Fault}$. This set of delta–deltas form a signature that can be used to identify the underlying fault (to within an ambiguity group) under a *single fault* hypothesis.[4] Because it is a temporally based $\Delta\Delta$, natural degradation and deterioration effects will naturally cancel, so this approach can be used for both low and high cycle engines. It is important to note, that if either value of ΔP_{Before_Fault} or ΔP_{After_Fault} fall outside the *QSS,* then there is the potential for model error effects to be present that will not necessarily cancel in the $\Delta\Delta$, thereby corrupting the signature and leading to potential misdiagnosis.

7.5.1 QSS Validation

As previously mentioned, the efficacy of the *QSS* approach for a given FDI architecture and methodology should be borne out through simulation studies and eventual deployed usage. We will utilize the *C-MAPSS* closed-loop non-linear aero-thermal engine model as a surrogate for an actual engine to perform an illustrative *simulation* example. Because this engine program is closed-loop, i.e., it has an embedded engine control as part of the package, it can be driven by *four* independent flight parameters: *altitude, Mach, PLA* and *dTamb.*[5] In order to produce realistic data from the simulation, *flight operations quality assurance* (*FOQA*) data has been utilized.[6] This set of data consisted of 69 full flights, represented by these four basic parameters, which will be used to drive the engine simulation to produce the gas path data required for this exercise. An example of one flight (9) from the dataset is depicted in Figure 7.3.

As the reader can plainly see, the actual flight parameter inputs exhibit a great deal of variability and are far removed from idealistic (cartoon) commercial jet flight profiles. This will drive the *C-MAPPS* engine simulation to produce gas path

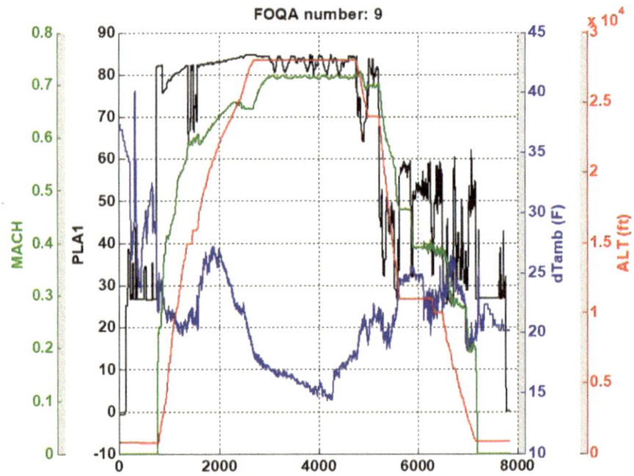

FIGURE 7.3 Sample FOQA #9 flight data

parameter outputs to allow a fair test for our *QSS* strategy for proper model alignment in supporting accurate FDI across various flight regimes without restricting to a particular operating condition (such as sea level static).

With appropriate choice of thresholds, the *QSS* collection of *realms* for sample Flight 9 depicted in Figure 7.3 is shown in Figure 7.4.

The use of streaming data has presented a number of technical challenges to afford the same FDI functionality provided by our legacy snapshot and discrete point time series data approaches. The ability to perform this same functionality onboard with streaming data, both steady state and transient, requires a number of changes

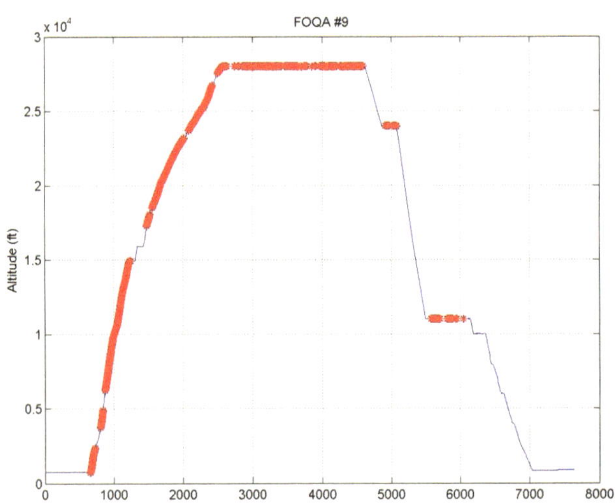

FIGURE 7.4 QSS realms for sample FOQA #9 flight data

in basic strategy. The first centers on the *reference* model (used to generate the measurement parameter Δs) and the long-term performance tracking. The traditional discrete Kalman filter (KF) typically used in the snapshot analysis has been recast as a dynamic tracking filter in closed-loop operation with a transient (physics) model of the engine. The latter can be a piece-wise linear SVM or a more sophisticated non-linear component level model, and the KF can be derived first from an SVM and later applied to a non-linear model if required. The KF *states* have been extended to contain health parameters (efficiencies and flow capacities) that act as *tuners* to track health deterioration of the engine over the *long-term*. Unfortunately, these tuners can be corrupted if there is any modeling error, instrumentation bias, and engine-to-engine variation. A mismatch between the *specific monitored engine* and a *fixed physics model* would be apparent in the data and is typically most severe during transient operation. This necessitates that an alignment of the model be completed to remove these inconsistencies and allow the KF to track health deterioration in an uncorrupted manner. This is also a necessary (but not sufficient) condition to enable proper anomaly detection and isolation of abrupt faults (FDI). Our *QSS* strategy, coupled with the *eSTORM* model alignment process, can provide an adequate solution to this first requirement. To this end, we *randomly* partitioned the *FOQA* data into two 30-flight datasets: a *calibration* set and a *validation* set. The eSTORM alignment was performed using the calibration set with validation checks being performed on *both* calibration and validation sets. Figure 7.5 presents the *tuner* response for the *FOQA* sample Flight 9, depicting responses *with* (bottom) and *without* (top) the ANN alignment factors. Additional sample *tuner* responses can be found in Addendum A7.1.

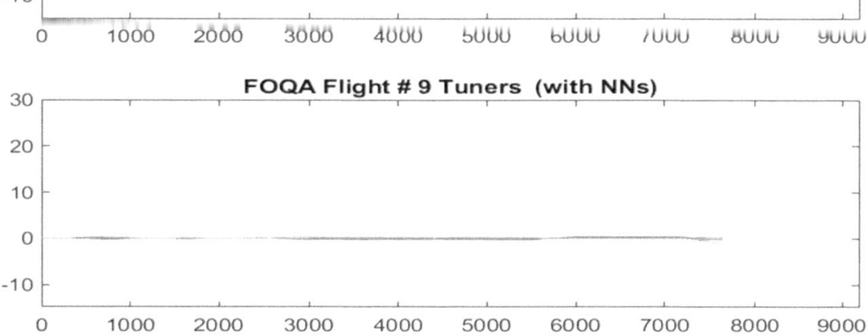

FIGURE 7.5 Tuner response with and without model alignment for sample FOQA #9

Event (or anomaly) detection and subsequent fault identification is another capability to address in the streaming data environment. The formulation and calculation of $\Delta\Delta P$ is critical to the solutions of both problems. This issue is just as insidious as the model alignment task in the technical difficulties that it presents. Fault events (of the same type that are addressed in the snapshot analysis world) must be detected and isolated during transient conditions as well as steady-state operation. Extensions of the methods typically applied in steady state seem to perform adequately enough for near-steady-state conditions when streaming data are applied. Transients, however, do not fare as well. The legacy detection methods do not work well, and identification methods miss the target more often than not. We will describe, in sections 7.6 and 7.7, two approaches to address these issues. The first approach we will consider makes use of *companion engine* data and can be utilized in situations in which there are two (or more) engines present on the airframe. With this approach, the reference model used for $\Delta\Delta P$ calculation is the companion engine.

7.6 CROSS-WING FDI APPROACH

The basic idea in this approach is to use information available from a companion engine(s) through a cross-wing comparison.[7] Because companion engines may have different performance levels (FADEC plug class), operate at slightly different power levels (due to performance or other operational considerations), and/or have different instrumentation errors or bias anomalies, a direct comparison of engine parameter values is not possible nor desired. The inclusion of an engine model for each engine, properly aligned to represent the individual engines, would provide a means to mitigate these concerns. Residual differences (Δs) for each gas path parameter relative to its individual normalized model removes the performance differences, instrumentation error differences, plug class differences, power level differences, etc., and permits a cross-wing comparison of the individual parameter Δs to be made for anomaly detection purposes. In essence, the companion engine, with its attendant normalized model, becomes the *reference* for the monitored engine under analysis. *Deviations due to environmental changes, flight maneuvers, and long-term deterioration will effectively cancel out in the cross-wing comparison.* The latter item *assumes* that, after initial alignment and model normalizations, *both engines will essentially deteriorate at approximately the same rate.* This assumption is made on the basis that both engines are exposed to the same environment, flight maneuvers, and power excursions. In this way, *deviations not common to both engines will become visible in the cross-wing comparison* [24].

7.6.1 System Architecture

Our goal, at this juncture, is to define a system that will support engine health management decisions and enable faster and more accurate fault detection and isolation to support engine maintenance activities. As such, the process we will describe, while capable of being executed onboard, is not intended, at this stage, to offer real-time cockpit annunciation or accommodation. It is envisioned that the *final analysis*

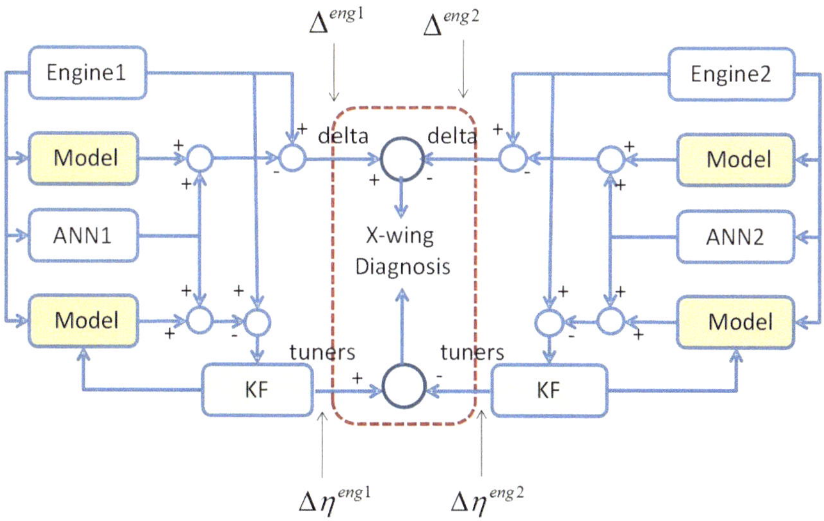

FIGURE 7.6 Cross-wing FDI analysis architecture

alerting will occur at the end of flight to take full advantage of the *information fusion* afforded by considering all of the flight data collectively.[8] This makes sense from an engine maintenance perspective, as engine inspection and attendant corrective actions can only proceed after the aircraft has landed. The general architecture is depicted in Figure 7.6.

The architecture is clearly symmetric, wherein the four boxes labeled *model* represent identical instances of the same model, which can be an SVM or a more complex non-linear aero-thermal model running in real-time. It should be noted that while identical programmatically, each of the four models will likely be operating in different *states* (and, hence, producing different output values) by virtue of the different adjustment factors between Engine 1 and 2, as well as the inclusion of KF feedback in two of them. The engine model alignments (ANN1 and ANN2) are *unique* to each engine and appear as bias adders as per the *eSTORM* configuration.[9]

The overall process can be viewed as follows; a cross-wing comparison of engine measurement deltas can be calculated as $\Delta\Delta^{\times} = \Delta^{eng1} - \Delta^{eng2}$ for each measured parameter. Ideally, $\Delta\Delta^{\times}$ would be *zero* (on the average) for any point in *QSS* in the *absence* of any anomaly or fault occurrence. However, if *one* of the engines suffers a rapid deterioration due to FOD or DOD or is the victim of an abrupt engine subsystem or instrumentation failure (full or partial), the cross-wing $\Delta\Delta^{\times}$ will *diverge* in consequence. Comparing the $\Delta\Delta^{\times}$ value to an appropriately chosen threshold and applying a persistency check criterion will flag the anomalous behavior. This provides a detection method across all *QSS* data points in the flight regime. *It does not, however, determine which of the two engines is at fault.*

To determine which engine is at fault, the second engine model (per engine), fitted with a KF observer (used for long term performance deterioration tracking) is utilized. The faulted engine's KF *tuners* will react (i.e., *deviate*) as a result of the fault event, whereas the non-faulted engine's KF *tuners* will not. *This allows the identification of which engine has sustained the faulty condition.* This can be accomplished by making the following cross-wing *tuner comparison* $\Delta\Delta\eta^{\times} = \Delta\eta^{eng1} - \Delta\eta^{eng2}$ and determining whether it has shifted *persistently* (in *QSS*). If it has, then either Engine 1 or Engine 2 has sustained a fault. To determine which engine, we need only test to see which of $\Delta\eta^{eng1}$ or $\Delta\eta^{eng2}$ has shifted.

Furthermore, the cross-wing $\Delta\Delta^{\times}$ value across all measured engine parameters provides a *signature* for subsequent fault isolation. If we define a fault flag *f* as

$$f_i = \begin{cases} 1 & \Delta\eta^{eng\,i} \text{ has shifted} \\ 0 & \text{otherwise} \end{cases} \tag{7.6}$$

We can then adjust the polarity of the cross-wing parameter signature $\Delta\Delta^{\times}$ as follows:

$$\Delta\Delta^{\times} \leftarrow sign(f_1 - f_2)\Delta\Delta^{\times} \tag{7.7}$$

Equation (7.7) now provides the requisite *signature* (for the faulted engine) to be used in the isolation process. Following any of the *SFA* procedures outlined in Chapter 5, a *single fault* hypothesis pattern match is performed against a set of pre-defined (single) faults, and the one admitting the least error is chosen. This isolation procedure is performed at every time-step that the engine remains in *QSS* after fault detection is established. This will produce a multiplicity of individual isolation results, i.e., one for each time-step that the engine operates in *QSS* after detection. As one might suspect, it is likely that they will not all be the same diagnosis due to noise, model alignment deficiencies, and the like. Because the analyses are intended for maintenance advisory, it is permissible to wait until the end of flight to make a final diagnostic determination. In this manner, some form of *information fusion* can be adopted to minimize risk of false assessment. A simple, but effective, tactic is to perform a majority vote selection. As we will see in the next section, this simple fusion strategy proves to be reasonably successful.

To be clear, we want to emphasize which assumptions are in play that underlie this FDI configuration.

It is *assumed* that

- Both engines are *approximately* matched *initially* in terms of performance level,
- Both engines see the same flight environment,
- Both engines *degrade* at approximately the same *rate,* and
- Model alignment errors will *approximately* cancel between engines.

One final comment on this process is in order. The process described in the preceding paragraphs does not assume that the actual fault occurred while the engine was in *QSS*. It may or may not be clear to the reader that the detection scheme, as described, can be reliably achieved if the fault occurred during an earlier transient maneuver of the flight. Moreover, it may not be clear that the $\Delta\Delta^{\times}$ calculated at a later time in *QSS*, indeed, provides a representative signature for the fault that occurred. Simulation studies, however, provide an affirmative answer.

7.6.2 Simulation Study

A Monte Carlo simulation (MCS) was conducted to provide initial verification for this cross-wing approach and to study the level of isolation performance potential using this methodology. The *C-MAPSS* simulation was used as the surrogate for *Engine1* and *Engine2* in Figure 7.6. The *model* was represented by an SVM. Because an SVM is derived directly from *C-MAPSS*, some random low-level performance deterioration and sensor biases were added to the *C-MAPSS* simulation so that there would be an inherent mismatch between *engine* and *model*. The 69 *FOQA* data profiles were *partitioned* into two sets (35 in one, and 34 in the other). The first set was used to generate data from which to perform a model alignment, while the second set was used (in conjunction with the first set) for alignment validation. This produced the alignment *ANN* elements for both engines.

To test the FDI strategy, 18 fault modes were simulated, as listed in Table 7.1. The coupling factor (CF) for an engine module fault is defined as the ratio between flow capacity and efficiency, i.e., CF = flow capacity/efficiency. Note that VBV position

TABLE 7.1
Random Parameter Ranges for the Monte Carlo Simulation

Class	Fault Mode	Random Fault Size Range (Uniform Dist.)
1	FAN fault	Eff. loss: [1~5]%, CF: [1.1~1.5]
2	LPC fault	Eff. loss: [1~5]%, CF: [1.1~1.5]
3	HPC fault	Eff. loss: [1~5]%, CF: [0.8~1.2]
4	HPT fault	Eff. loss: [1~5]%, CF: [-0.9 ~ -0.5]
5	LPT fault	Eff. loss: [1~5]%, CF: [-0.8 ~ -0.4]
6	VBV actuator bias	[0.2 ~ 0.4]
7	VSV actuator bias	-[1.5 ~ 6] degrees
8	Fuel flow bias	+/- [0.08 ~ 2] lbs/second
9	N1 sensor bias	+/- [10 ~ 90] RPM
10	N2 sensor bias	+/- [15 ~ 100] RPM
11	T2 sensor bias	+/- [3 ~ 10] F
12	P2 sensor bias	+/- [0.6, 2] psi
13	T25 sensor bias	+/- [6, 20] F
14	P25 sensor bias	+/- [1.6, 5] psi
15	T3 sensor bias	+/- [6, 30] F
16	P3 sensor bias	+/- [4, 15] psi
17	T5 sensor bias	+/- [8, 60] F
18	P5 sensor bias	+/- [0.5, 4] psi

is in fraction, i.e., 1 means 100% stroke length. For *C-MAPSS*, VBV is physically closed below 15.4% stroke. In the MCS, 100 randomly generated instances for *each* fault appearing in Table 7.1 (with random component fault coupling) were introduced on a mid-life 50% *Engine1* and compared to (non-faulted) *Engine2* at a mid-life 55%. The 5% *life* difference introduced between the engines is to present more realism and to uphold (in effect) our assumption that both engines are *approximately* matched *initially* in terms of performance level and that the engines deteriorate at the same rate. A built-in life model in the *C-MAPSS* was used to generate the requisite performance for 50% and 55% life as required. A total of *10 flight profiles* in the FOQA dataset were randomly chosen for the simulation study, and the 100 random faults were uniformly distributed over the 10 flight profiles.

For all simulations, the *fault* was introduced during climb. In some instances, the fault injection coincided within a *QSS* realm, while, in others, it did not. It should be noted that only one out of the 10 flight profiles used was included in the ANN training set, while the remaining nine were all from the validation dataset. The actual 10 flight profiles employed appear in Figure 7.7.

In addition to the 18 faults contained in Table 7.1, another 100 *no fault* simulations were performed in order to quantify the level of *false alarms*[10] that might be introduced by the FDI strategy. For *each* of the 1,900 test cases under evaluation, once a detection occurred, an isolation was performed using a SFA, as described in the preceding paragraphs and in Chapter 5, for every subsequent data point in the flight that fell within *QSS*. A simple majority vote was taken among all of these candidate single fault identifications for the flight at hand to provide the final isolation. The *FDI confusion matrix*, presented in Figure 7.8, contains the results for all of these 1,900 trials.

This confusion matrix is a generalization of the *detection confusion matrix* that was discussed in Chapter 2. It contains the *combined* effect of *detection* and *isolation*. The *rows* represent the implanted faults (true state) and the *columns* represent the classification from the FDI strategy being employed. Perfection would be represented by a diagonal matrix. For our experiment, that would require 100 at every diagonal element. As the reader can see, we do not have perfection. Some faults do better than others. To read the matrix, let us consider the "N2" row. This row tells us that the N2 sensor error fault was correctly identified for 62 of the 100 trials. For the remaining 38 trials, there was some form of confusion with another fault or a lack of detection. Specifically, six were *misclassified* as a Fan fault, five as VBV, three as VSV, one as P2, three as P3, two as T5, and 18 as no fault (meaning no detection occurred). This *confusion* is often referred to as *ambiguity* and will be explored in greater depth in Chapter 9. For the time being, we will limit our discussion to just the basics. The following are metrics discussed in Chapter 2 and Addendum A2.3. The sum of each row must equal the number of trials (in this case 100). If we sum the elements in the last column (no fault classification) across the implanted *fault* rows, we have the number of *misdetections* (18 + 13 + 7 = 38). Dividing by the total number of fault trials (1,800) yields a misdetection rate of 38/1800 = 2.1%. Conversely, the sum of the last row (no fault implanted) across the *fault* classification rows yields the number of *false alarms* (in our case, 0). Lastly, the sum of the main diagonal

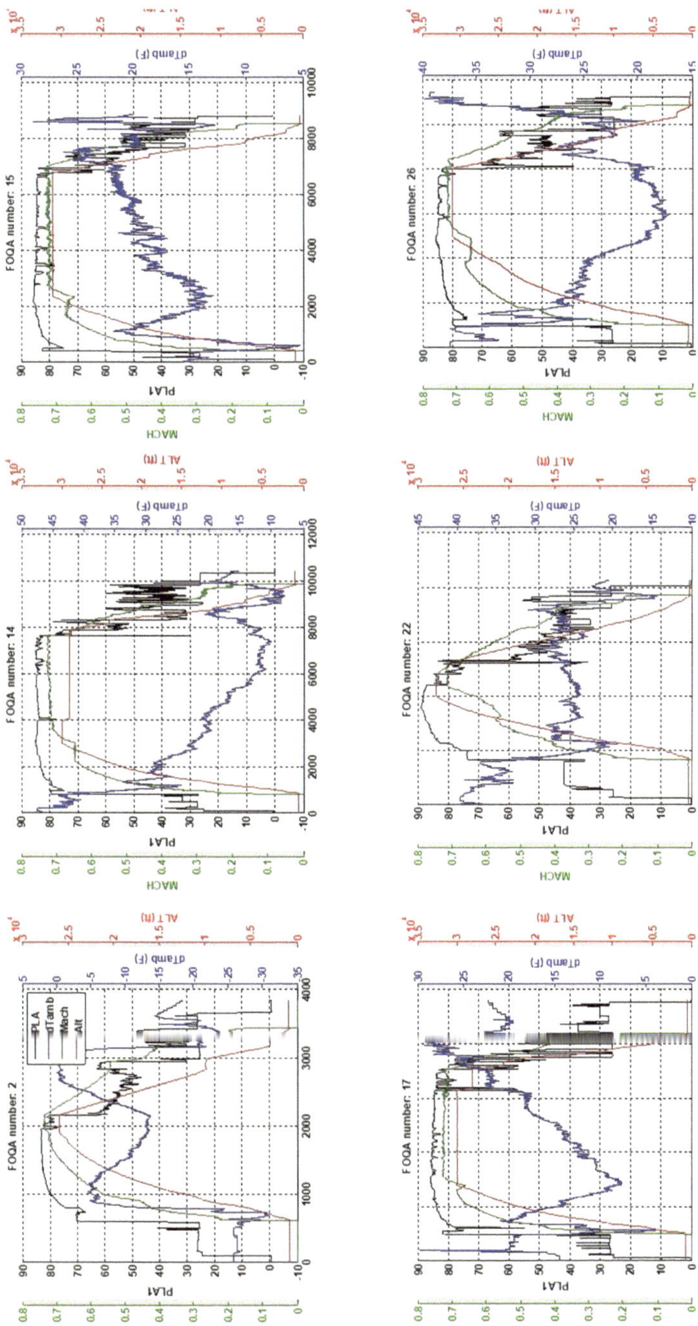

FIGURE 7.7 FOQA flight profiles used in Monte Carlo simulations

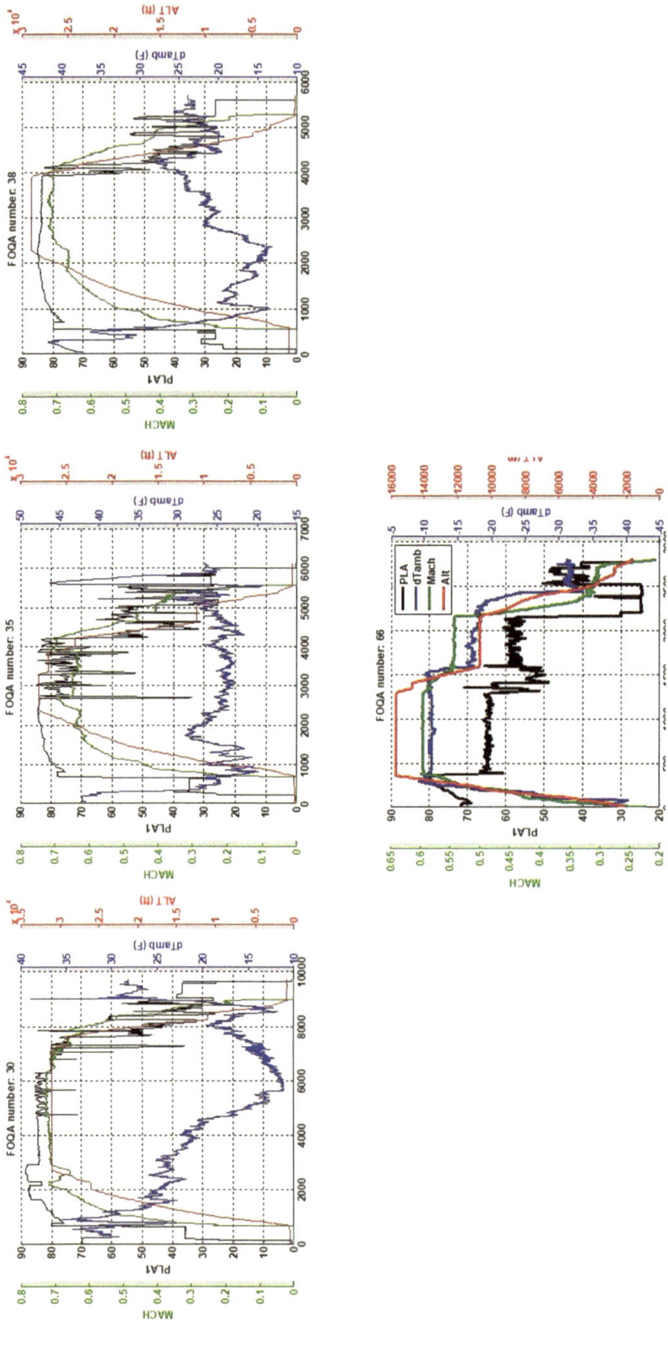

FIGURE 7.7 FOQA flight profiles used in Monte Carlo simulations

Truth (Implanted Fault) \ Classified	Fan	LPC	HPC	HPT	LPT	VBV	VSV	Wf	N1	N2	T2	P2	T25	P25	T3	P3	T5	P5	No Fault
Fan	100	0	0	0	0	0	0	0	0	0	0	0	0	0	0	0	0	0	0
LPC	0	98	0	0	0	2	0	0	0	0	0	0	0	0	0	0	0	0	0
HPC	0	0	99	0	0	0	0	0	0	0	0	0	0	1	0	0	0	0	0
HPT	0	0	0	100	0	0	0	0	0	0	0	0	0	0	0	0	0	0	0
LPT	0	0	0	0	100	0	0	0	0	0	0	0	0	0	0	0	0	0	0
VBV	0	0	0	0	0	100	0	0	0	0	0	0	0	0	0	0	0	0	0
VSV	0	0	0	0	0	0	78	0	0	2	0	0	0	20	0	0	0	0	0
Wf	0	0	0	0	0	0	0	100	0	0	0	0	0	0	0	0	0	0	0
N1	0	0	0	0	0	0	0	0	95	0	0	5	0	0	0	0	0	0	0
N2	6	0	0	0	0	5	3	0	0	62	0	1	0	0	0	3	2	0	18
T2	0	0	0	0	0	0	0	0	0	0	100	0	0	0	0	0	0	0	0
P2	0	0	0	0	0	0	0	0	0	0	0	100	0	0	0	0	0	0	0
T25	0	0	0	0	0	0	0	0	0	0	0	0	87	0	0	0	0	0	13
P25	0	0	0	0	0	0	0	0	0	0	0	0	0	100	0	0	0	0	0
T3	0	0	0	0	0	0	0	0	0	0	0	0	0	0	99	0	1	0	0
P3	0	0	0	0	0	0	0	0	0	0	0	0	0	0	0	100	0	0	0
T5	0	0	0	0	0	0	0	0	0	0	0	0	0	0	0	0	93	0	7
P5	0	0	0	0	0	0	0	0	0	0	0	0	0	0	0	0	0	100	0
No Fault	0	0	0	0	0	0	0	0	0	0	0	0	0	0	0	0	0	0	100

FIGURE 7.8 FDI confusion matrix for cross-wing Monte Carlo simulation

provides the total number of correct classifications (which yields accuracy rate when divided by the total number of trials). To summarize, we have Table 7.2.

At this juncture, the reader, no doubt, has several questions regarding the details of the process that has culminated in the results reported in Table 7.2. Some of these might include the following:

1. How is fault *detection* determined?
2. Is there a *confidence level* associated with the *detection*?
3. How is fault isolation majority vote determined?
4. Is there a *confidence level* associated with the *isolation*?

We will answer these questions (in overview) using Figures 7.9 and 7.10 as a pictorial guide. To begin, we remind the reader of Equation (7.3), which we duplicate below.

$$\Omega_k = \left(\bigcup_{J=1}^{M_k} QSS^{(k)}(J) \right) \subseteq F_k \quad where$$

$$QSS^{(k)}(J) = J^{th}\ realm\ in\ flight\ k$$

$$M_k = number\ of\ realms\ in\ flight\ k$$

The *QSS* data points (Ω_k) are determined by applying the test, given in Equation (7.5), as depicted in sample flight portrayed in Figure 7.9. We use the symbolic

TABLE 7.2
Simulation Cross-Wing FDI Performance

Misdetection rate	2.1%
False alarm rate	0%
Correct isolation rate	95%

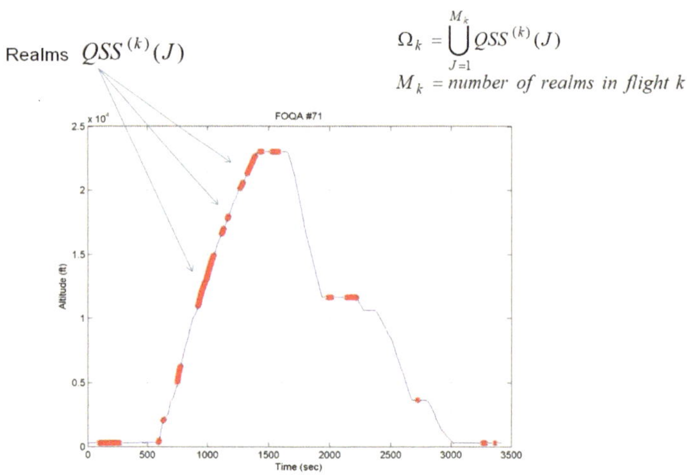

FIGURE 7.9 Sample flight with QSS realms (in red)

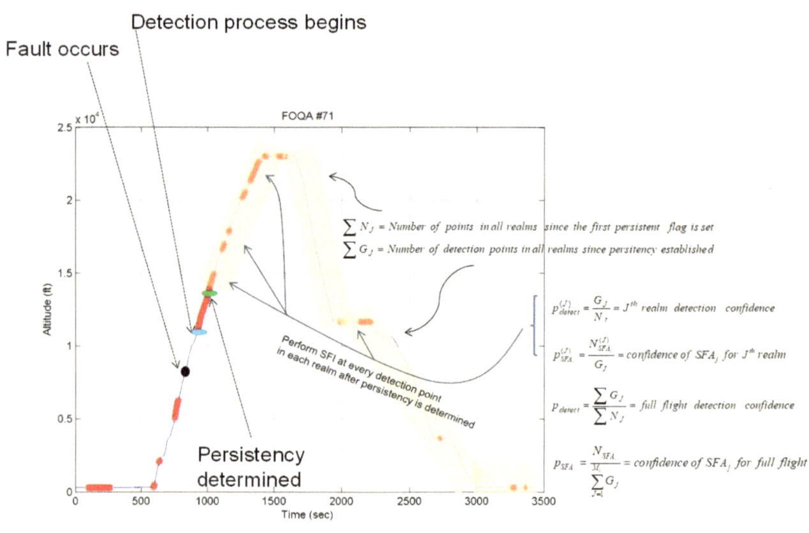

FIGURE 7.10 Sample calculations supporting QSS FDI

representation in Figure 7.10 to illustrate the calculations we will need to answer the questions above.

Figure 7.10 provides a sample illustration of a hypothetical fault (black dot) occurring during climb at approximately 750 seconds. Because the fault occurred at a *non-QSS* interval, the detection process does not begin until the next available *QSS realm* is entered. Recall that calculations involving the engine models can only be performed when a valid *alignment* is available (which, by definition, is *QSS*). The first *QSS* instance after the fault occurred is depicted on the figure and continues for a period of time (in *QSS*) until *persistency* is established (green points). The *detection* for the cross-wing approach utilizes a simple threshold checking on the cross-wing delta–delta calculated as $\Delta\Delta^{\times} = \Delta^{eng1} - \Delta^{eng2}$ for *each* measured parameter[11] (see Figure 7.6). Detection persistency can be implemented by thresholding the cumulative persistency counter, as shown in Equation 7.9. Once detection persistency is established, the flag and counter calculation continues for every point *i* in every *realm* $QSS^{(k)}(J)$.

$$persistency\ flag\ g(i) = \begin{cases} 1 & if\ at\ least\ one\ component\ of\ \Delta\Delta^{\times}\ exceeds\ threshold \\ 0 & otherwise \end{cases} \tag{7.8}$$

$$cumulative\ persistency\ counter\ G(i) = G(i-1) + g(i) \tag{7.9}$$

$$cumulative\ persistency\ counter\ for\ J^{th}\ realm\ G_J = \sum_{i \in QSS^{(k)}(J)} G(i) \tag{7.10}$$

$$N_J = \left|QSS^{(k)}(J)\right| = number\ of\ points\ in\ J^{th}\ realm \tag{7.11}$$

With these indices, we can calculate the following *detection confidences.*

$$\begin{aligned} p_{detect}^{(J)} &= \frac{G_J}{N_J} = J^{th}\ realm\ detection\ confidence \\ p_{detect} &= \frac{\sum G_J}{\sum N_J} = full\ flight\ detection\ confidence \end{aligned} \tag{7.12}$$

In practice, one can set a minimum detection confidence threshold, α_{detect} and test whether $p_{detect}^{(J)} \geq \alpha_{detect}$ and $p_{detect} \geq \alpha_{detect}$ to govern how subsequent *isolation* analyses are dispatched. For simplicity, we might only consider the latter, using the principle that any sort of remedial maintenance action cannot begin until after the aircraft has landed and, from an *information fusion* perspective, the best analysis would be one that utilizes *all* the information available during the flight.[12]

If a persistent detection is realized, an SFA can be performed in much the same manner as that outlined in Chapter 5, using $\Delta\Delta^{\times}$ from Equation (7.7) as the delta vector to perform the classification task. For each data point following a *persistent* detection, *an SFA* calculation is made. This *may* return a number of different analysis results across all of these points. If we denote the entire (pre-defined) set of possible single faults by $\{SFA_1, SFA_2, \cdots, SFA_L\}$,[13] then we can calculate another set of indices similar to those captured by Equations (7.11)–(7.12) as follows:

$$N_{SFA_j}^{(J)} = number\ of\ points\ for\ which\ SFA_j\ is\ returned\ as\ the\ isolated\ fault\ in\ QSS^{(k)}(J) \tag{7.13}$$

$$p_{SFA_j}^{(J)} = \frac{N_{SFA_j}^{(J)}}{\sum_{i=1}^{L} N_{SFA_i}^{(J)}} = \frac{N_{SFA_j}^{(J)}}{G_J} = confidence\ of\ SFA_j\ for\ J^{th}\ realm \tag{7.14}$$

$$N_{SFA_j} = \sum_{J=1}^{M_k} N_{SFA_j}^{(J)} = number\ of\ points\ for\ which\ SFA_j\ is\ isolated\ for\ full\ flight \tag{7.15}$$

$$p_{SFA_j} = \frac{N_{SFA_j}}{\sum_{j=1}^{L} N_{SFA_j}} = \frac{N_{SFA_j}}{\sum_{j=1}^{L}\sum_{J=1}^{M_k} N_{SFA_j}^{(J)}} = \frac{N_{SFA_j}}{\sum_{J=1}^{M_k}\sum_{j=1}^{L} N_{SFA_j}^{(J)}} = \frac{N_{SFA_j}}{\sum_{J=1}^{M_k} G_J} \tag{7.16}$$
$$= confidence\ of\ SFA_j\ for\ full\ flight$$

From Equation (7.16), we can peruse the calculated confidences $\left\{p_{SFA_j}\right\}_{j=1,2,\ldots,L}$ and choose the fault admitting the highest confidence value, i.e., a majority vote. It should be clear from Equation (7.16) that

$$\sum_{j=1}^{L} p_{SFA_j} = 1 \tag{7.17}$$

From this, it follows that the majority vote single fault can be considered *unique* if its *confidence* is greater than 50%. On the other hand, if the max confidence is less than or equal to 50%, it is possible to have two (or more) faults with the same (or very close) maximum confidence, and, thus, we would have *ambiguity* in our choice of majority vote. If you are a pragmatist, you may conclude that there is no dependable isolation possible; after all, the confidence is, at best, a coin toss. Ambiguity is an issue that we have seen before (in our discussion of the confusion matrix depicted in Figure 7.8) and will see again in the sequel from Equation 7.8 through Equation 7.17. We will discuss the issue of ambiguity in greater detail in Chapter 9.

The terms defined above allow one to declare anomaly detection within each *realm* as well as for the *entire flight*, i.e., whether

$$\frac{G_J}{N_J} = p_{detect}^{(J)} > \alpha_{detect} \qquad or \qquad \frac{\sum_{J=1}^{M_k} G_J}{\sum_{J=1}^{M_k} N_J} = p_{detect} > \alpha_{detect}$$

respectively. It is easy to see that if detection is declared for the flight, then there must exist at least one *realm* that would pass the detection test. To see this, assume it is false, i.e., that *all realms* fail detection. This implies

$$p_{detect}^{(J)} = \frac{G_J}{N_J} < \alpha_{detect}, \forall J \Rightarrow$$

$$G_J < \alpha_{detect} N_J \Rightarrow \sum_{J=1}^{M_k} G_J < \alpha_{detect} \sum_{J=1}^{M_k} N_J \Rightarrow \frac{\sum_{J=1}^{M_k} G_J}{\sum_{J=1}^{M_k} N_J} < \alpha_{detect} (contradiction)$$

If *all realms* have detection, then clearly, detection prevails for the flight as well.

To offer the reader some idea as to how the *isolation* confidences are distributed, we consider the N2 sensor fault discussed previously. With a quick glance at the confusion matrix in Figure 7.8, we see that this particular fault is one of the weakest as far as isolation accuracy. For 100 trials, 62 were correctly identified, 20 were misidentified, and 18 were not detected at all. Figure 7.11 depicts a histogram of *isolation confidence* for the 82 detected cases; the blue histogram is for the 62 correctly identified, and the red histogram is for the remaining 20 trials that were incorrectly identified. The *average* confidence was 84.4% for *correct* isolation and 44.2% for incorrect isolation. This is certainly what we would hope to be the case, i.e., high confidence when you are correct and low confidence when you are not correct. For the most part, this is the case across all faults tested in the simulation exercise, and the histograms look similar in the sense that the distributions are asymmetrically skewed (for correctly diagnosed faults (blue)). Table 7.3 summarizes the *average* isolation confidences for all faults.

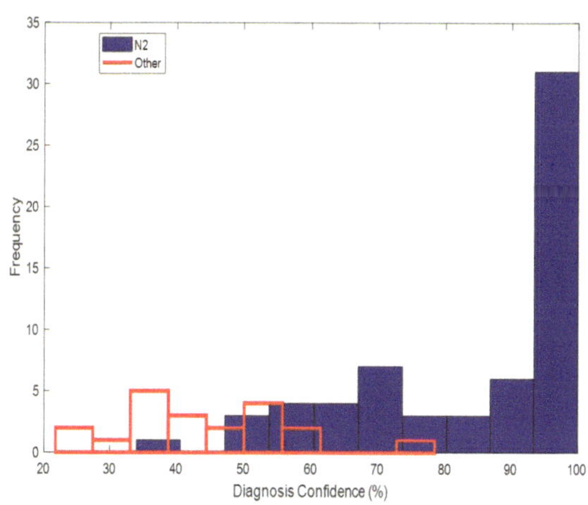

FIGURE 7.11 Sample isolation confidence distribution (N2 fault)

TABLE 7.3
Average Isolation Confidences from Cross-Wing Simulation

	Correctly Isolated		Incorrectly Isolated	
Fault	**%**	**Average Confidence**	**%**	**Average Confidence**
Fan	100%	96.8%	–	–
LPC	98%	92.3%	2%	60.9%
HPC	99%	96.2%	1%	61.4%
HPT	100%	98.6%	–	–
LPT	100%	98.7%	–	–
VBV	100%	99.6%	–	–
VSV	78%	88.5%	22%	66.6%
Wf	100%	99.5%	–	–
N1	95%	95.2%	5%	76.9%
N2	62%	84.4%	20%	44.2%
T2	100%	98.3%	–	–
P2	100%	99.2%	–	–
T25	87%	99.0%	–	–
P25	100%	99.7%	–	–
T3	99%	97.7%	1%	67.9%
P3	100%	99.4%	–	–
T5	93%	99.6%	–	–
P5	100%	99.4%	–	–

7.7 SINGLE ENGINE FDI APPROACH

From our discussion above, the cross-wing approach appears to provide a potentially promising approach to perform engine fault FDI for data obtained in *QSS*. It does, however, require that there be more than one engine on the airframe and that there exists a means of communication across engines to calculate the needed comparative data as dictated in Figure 7.6. In addition, it is a fairly complex arrangement and requires a number of assumptions, most notably that 1) *both* engines will *deteriorate at (roughly) the same rate* and 2) that model *alignment errors will approximately cancel between engines*. This raises the question as to whether it is possible to obtain the same level of FDI coverage *without* involving a companion engine. In this section, we outline a possible strategy along with some simulation results.

7.7.1 System Architecture

The methodology we will discuss is a close cousin to the cross-wing approach in that many of the same principles apply (see [9]). In particular, we require that detection and isolation can only proceed for data in *QSS*. The distinguishing difference (from cross-wing) will be in how a parameter $\Delta\Delta$ is calculated, once a *persistent* trend detection is annunciated. The general system architecture is depicted in Figure 7.12.

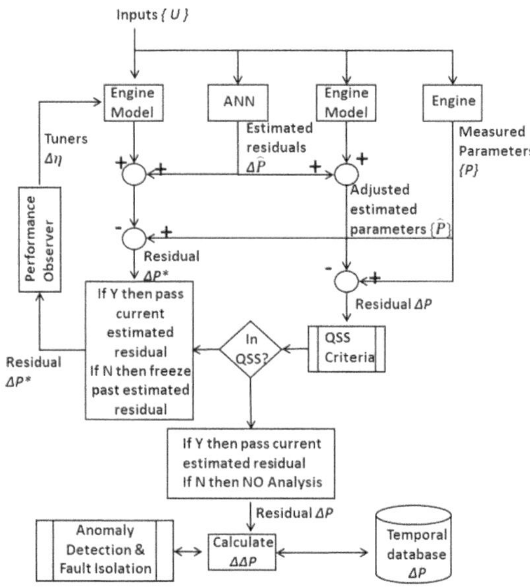

FIGURE 7.12 Single engine FDI system architecture

The upper half of Figure 7.12 displays the same two-model configuration (rotated 90 degrees) as it appears for each engine in the cross-wing architecture seen in Figure 7.6. The fundamental difference is how the parameter ΔΔ is calculated. The temporal database element depicted in Figure 7.12 consists of a temporally floating set of parameters deltas (ΔP). This temporal database continually updates after each flight using data collected during the QSS regimes of the most recent flight and the previous *M* flights, where *M* is pre-determined at design. The idea is to capture and maintain a database of deltas from nominal that spans the QSS power spectrum and captures the gradual deterioration that ensues over time. It will, in effect, provide the basis for using the ΔΔs to detect and quantify any abrupt shifts when a fault occurs with minimal corruption from the deterioration effect.

The temporal database can take many forms. For this application, the temporal database sorts the residuals $\Delta P_i^m(k)$, which represents the *i*th parameter Δ during flight *m*, at time instance *k*, into a series of *N* bins $(B_i^n, n = 1, 2, ..., N)$ partitioned across the power spectrum (i.e., as a function of the *PSP*). To save storage space in the database, the actual values $\Delta P_i^m(k)$ can be replaced by their statistics, such that each value at time instance *k* does not have to be saved. The overall statistics for each bin can be computed from these values as follows:

For the current flight (*m*), compute mean and standard deviation for the ith parameter in the nth bin B_i^n, $\mu_i^n(m)$ and $\sigma_i^n(m)$, respectively[14], i.e.,

$$\mu_i^n(m) = \frac{1}{I_m^n} \sum_{k=1}^{I_m^n} \Delta P_i^m(k) \tag{7.18}$$

$$\sigma_i^n(m) = \sqrt{\frac{1}{I_m^n - 1}\sum_{k=1}^{I_m^n}\left(\Delta P_i^m(k) - \mu_i^n(m)\right)^2} \tag{7.19}$$

where I_m^n is the number of *QSS* data points for the ith parameter that falls into the nth bin in the mth flight.

- In each bin B_i^n, drop the earliest flight statistics, $\mu_i^n(m-M)$ and $\sigma_i^n(m-M)$. Add $\mu_i^n(m)$ and $\sigma_i^n(m)$, so you always maintain the most recent M set of statistics in each bin. *Each* bin B_i^n contains the three statistical quantities $\left\{I_j^n, \mu_i^n(j), \sigma_i^n(j)\right\}_{j=m-M+1}^{m}$.
- In each bin, B_i^n, compute updated statistics, μ_i^n and σ_i^n, by combining statistics over the past M flights, i.e.,

$$\mu_i^n = \frac{1}{T}\sum_{j=m-M+1}^{m} I_j^n\,\mu_i^n(j) \tag{7.20}$$

$$\sigma_i^n = \sqrt{\frac{1}{T-1}\sum_{j=m-M+1}^{m}\left[\left(I_j^n - 1\right)\sigma_i^n(j)^2 + I_j^n\left(\mu_i^n(j) - \mu_i^n\right)^2\right]} \tag{7.21}$$

where $T = \sum_{j=m-M+1}^{m} I_j^n$ is the total number of QSS data points for the ith parameter in the nth bin over the past M flights.[15]

The temporal database is initialized during the model alignment phase with *QSS* data using the process described above. Once established, it becomes the *reference* from which anomaly *detection* and parameter ΔΔs are determined. For a given flight j, the kth data point in *QSS* is tested by calculating

$$\Delta\Delta P_i^j(k) = \Delta P_i^j(k) - \mu_i^n \tag{7.22}$$

This ΔΔ is then compared to a threshold, typically some multiple of the standard deviation from the *comparison bin*, to determine if an exceedance occurs, i.e.,

$$Test\ if\ \left|\Delta\Delta P_i^j(k)\right| > \alpha\,\sigma_i^n \tag{7.23}$$

If exceedance occurs, the parameter is marked as a *potential* fault event, and a counter is initialized. This process is repeated for the next data point and so forth. Persistency is declared when the counter for contiguous data points breaches a predefined limit. Once persistency is declared, a fault detection flag is raised. This flag marks the beginning of a data fusion process for *FDI*. The fusion process is similar to what was presented in the cross-wing method. In general, every *QSS* data point is tested against the above equation. Once the fault detection flag is raised, statistics will be calculated for the percent of *QSS* data points with a positive test result. Figure 7.6 applies to this method as well and the metrics given in Equation (7.8) through

Truth (implanted Fault) \ Classified	Fan	LPC	HPC	HPT	LPT	VBV	VSV	Wf	N1	N2	T2	P2	T25	P25	T3	P3	T5	P5	No Fault
Fan	100	0	0	0	0	0	0	0	0	0	0	0	0	0	0	0	0	0	0
LPC	0	100	0	0	0	0	0	0	0	0	0	0	0	0	0	0	0	0	0
HPC	0	0	100	0	0	0	0	0	0	0	0	0	0	0	0	0	0	0	0
HPT	0	0	0	100	0	0	0	0	0	0	0	0	0	0	0	0	0	0	0
LPT	0	0	0	0	100	0	0	0	0	0	0	0	0	0	0	0	0	0	0
VBV	0	0	0	0	0	100	0	0	0	0	0	0	0	0	0	0	0	0	0
VSV	0	0	0	0	0	0	98	0	0	0	0	0	0	0	0	0	0	0	2
Wf	0	0	0	0	0	0	0	100	0	0	0	0	0	0	0	0	0	0	0
N1	0	0	0	0	0	0	0	0	99	0	0	0	0	0	0	1	0	0	0
N2	0	0	0	0	0	0	0	0	0	83	0	0	0	0	0	2	0	0	15
T2	0	0	0	0	0	0	0	0	0	0	100	0	0	0	0	0	0	0	0
P2	0	0	0	0	0	0	0	0	0	0	0	86	0	0	0	0	0	0	14
T25	0	0	0	0	0	0	0	0	0	0	0	0	100	0	0	0	0	0	0
P25	0	0	0	0	0	0	0	0	0	0	0	0	0	100	0	0	0	0	0
T3	0	0	0	0	0	0	0	0	0	0	0	0	0	0	100	0	0	0	0
P3	0	0	0	0	0	0	0	0	0	0	0	0	0	0	0	100	0	0	0
T5	0	0	0	0	0	0	0	0	0	0	0	0	0	0	0	0	100	0	0
P5	0	0	0	0	0	0	0	0	0	0	0	0	0	0	0	0	0	100	0
No Fault	0	0	0	0	0	0	0	0	0	0	0	0	0	0	0	0	0	0	100

FIGURE 7.13 Single engine FDI confusion matrix for Monte Carlo simulation

Equation (7.17) are used to quantify *confidence* levels for *detection* and *isolation* for each *QSS realm* and for the *entire flight*.

7.7.2 Simulation Study

For purposes of comparison, the same Monte Carlo simulation study that was described for the cross-wing method was repeated for the single engine method. The same 10 FOQA flights were used, and the same 18 faults injected. Everything is identical except for the differences in the calculation of the parameter ΔΔs used for detection and isolation as noted above. The results were reasonably similar, as can be seen from the confusion matrix in Figure 7.13. Based on the numbers in the confusion matrix, the following overall diagnostic performance for the range of faults simulated can be calculated and is shown in Table 7.4. A comparison with Table 7.2 indicates a slightly better performance over the cross-wing methodology. This slight improvement is possibly due to the removal of engine-to-engine performance differences between the companion engines, as this approach uses the engine's own recent past as a reference. A further comparison can be made by considering the *average* isolation confidences for this method across all faults, as illustrated in Table 7.5 with respect to the cross-wing method confidences in Table 7.3. As can been seen, the average isolation confidences are better, in general, than the cross-wing method. Further information on this method can be found in [9].

TABLE 7.4
Simulation Single Engine FDI Performance

Misdetection rate	1.7%
False alarm rate	0%
Isolation accuracy (faults)	98%

TABLE 7.5
Average Isolation Confidences from Single Engine Simulation

	Correctly Isolated		Incorrectly Isolated	
Fault	**%**	**Average Confidence**	**%**	**Average Confidence**
Fan	100%	99.1%	–	–
LPC	100%	98.4%	–	–
HPC	100%	99.2%	–	–
HPT	100%	99.6%	–	–
LPT	100%	99.9%	–	–
VBV	100%	99.4%	–	–
VSV	98%	95.3%	–	–
Wf	100%	99.7%	–	–
N1	99%	98.4%	1%	47.8%
N2	83%	95.2%	2%	59.0%
T2	100%	99.4%	–	–
P2	100%	99.4%	–	–
T25	100%	99.7%	–	–
P25	100%	99.8%	–	–
T3	100%	99.2%	–	–
P3	100%	99.9%	–	–
T5	100%	99.3%	–	–
P5	100%	99.8%	–	–

7.8 NOTE ON QSS

We will offer a few comments regarding *QSS* before we conclude this chapter with a brief description of alternative approaches. The reader may have some questions, and rightly so, regarding why we chose to define *QSS* in the manner we chose and why we call it quasi-steady-state. We know from the discussions in the previous chapters that some form of *initialization* is necessary to *align* the reference element we are using in order to remove extraneous errors caused by installed sensor biases, initial engine performance levels, modeling errors, and the like in order to assure some level of accuracy to be achieved in our FDI methods. What is important, therefore, is the ability to operate only on data from which we have some assurance that model alignment can be achieved. It is this *property* that is important for FDI. So, if we make that property our definition, we can proceed with developing the various approaches already discussed.

Given that imprecise definition, we can argue (as we have) that the alignment is some union of datasets in which alignment is achieved (*realms*). It also relieves the necessity of stipulating a *specific* test or series of tests that the data needs to satisfy (such as altitude range, speed, parameter rate of change, etc.). This is important, as alignment really depends on the underlying model that we have at our disposal, e.g., a non-linear aero-thermal model, a good SVM, a bad SVM, a model of a toaster, etc. The former would admit a greater data range of alignment than the latter. The alignment property gives rise to the consequence that parameter delta variation should be minimized. That *necessary* condition was used as a *sufficient* condition (or test) to specify our alignment data points (Equation (7.5)). But clearly, that test does *not guarantee* that alignment can be satisfactorily achieved, as the reference model used (good or bad) is the overriding factor. Our contention is that a good SVM, or better model, would enable Equation (7.5) to define a set of data points from which alignment could be achieved. It certainly contains steady state points. If it did not, it would be game over, and a better, more sophisticated reference model would need to be used. Chapters 4 and 5 dealt with steady-state data, and we know that acceptable alignment can be made for that type of data. Given higher bandwidth, we would expect to extrapolate our alignment to some sub-set of the transient (non-steady-state) data in the flight, and, hence, we chose to call it by the term *QSS*. Because steady-state alignment can be achieved (essentially) through *bias adders*, we utilized the same strategy for *QSS* by incorporating an eSTORM architecture (Chapter 6). Additionally, this type of architecture is *reference model* independent and can be used with a non-linear aero-thermal model or an SVM.[16]

7.9 ALTERNATIVE APPROACHES

7.9.1 Overview

Similar to the alternative approaches presented at the conclusion of Chapter 5 for time series data, we will, likewise, provide a brief synopsis of other approaches for transient diagnosis that have emerged in the literature over the past decades. One of the early efforts in this field was a parameter estimation and system identification approached published in 1988 [10]. This paper sought to address faults that degraded the operability of the engine, especially during aircraft maneuvers, which might not manifest themselves in steady-state performance and, thus, would not be visible in snapshot or time series data. The proposed approach attempted to identify the transient response of a gas turbine in terms of dynamic characteristics, such as time constants and steady-state gains, which could be used consequently for diagnostic purposes. The approach used a simple first-order lag model and a LSQ estimator and was illustrated with two fault scenarios using turbofan simulation data. Due to the simple problem formulation, the technique was "not intended as a stand-alone procedure" but, rather, to be "used in conjunction with corrected data cross-plotting procedures" for off-board diagnosis. This approach was further developed and applied to various fault scenarios by the same author in [11], [12].

The advent of SVM and self-tuning SVMs in jet engine control applications in the 1980s brought KFs to transient diagnostics. Since the pioneering work by Luppold applying KFs to sensor fault and engine deterioration estimation, several KF or state-estimator-based approaches were developed for transient diagnosis using linear or non-linear models. These approaches share some similarity with the *single engine*

FDI approach described in this chapter but differ in the handling of modeling errors and fault isolation. We will briefly describe a few representative approaches herein.

Simon and Takahisa published a series of papers on the applications of KFs to aircraft EHM. An approach for in-flight fault detection and isolation (FDI) of engine sensors based on a bank of KFs was presented in [13]. This approach utilized multiple KFs, as shown in Figure 7.14, each of which was designed based on a specific fault hypothesis. Based on a *single fault* assumption, only one KF with the correct hypothesis would be able to maintain the nominal estimation performance when the engine experienced a fault. Based on this knowledge, sensor faults could be isolated. An additional KF (assuming healthy sensors) was introduced to detect engine component and actuator faults in order to avoid misclassifying them as sensor faults. The proposed approach was applied to a non-linear simulation of a commercial aircraft engine, and its performance was evaluated at multiple power settings at a cruise operating point using various fault scenarios.

As shown in Figure 7.14, this approach utilized a weighted sum of squared tuners (from *Filter #m+1*) in the fault detection and isolation logic. As engines deteriorate over time, the tuners (i.e., a sub-set of module health parameters) deviate further from zero (or a small initial value), making it harder to maintain the FDI logic in this problem setting. To deal with engine deterioration effects, the authors developed a hybrid KF approach for in-flight fault detection of aircraft engine sensors in [14]. The hybrid approach consisted of a non-linear onboard engine model (OBEM) and a KF-based piece-wise linear model. The reference health baseline of the diagnostic system was updated by adjusting the non-linear OBEM to the health condition of degraded engines. Through this health baseline update, the diagnostic effectiveness of the in-flight sensor fault detection system can be maintained as the engine

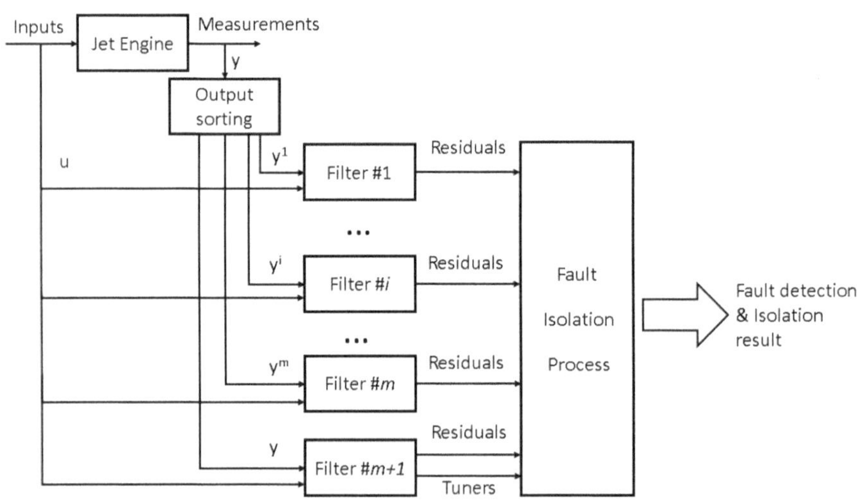

FIGURE 7.14 Architecture of the sensor FDI system

deteriorates over time. Specifically, by integrating the OBEM and linear state-space models, the following hybrid Kalman filter is formed:

$$\begin{aligned}\dot{\hat{x}} &= A(\hat{x} - x_{OBEM}) + K(y - \hat{y}) \\ \hat{y} &= C(\hat{x} - x_{OBEM}) + y_{OBEM}\end{aligned} \tag{7.24}$$

Where $\hat{x}$ and $\hat{y}$ are the estimated states and outputs of the linear KF, y is the sensor measurements from the engine, A and C matrices are derived from nominal healthy engine model, and K is the constant Kalman gain designed based on nominal engine dynamics. The solution is called *hybrid* due to the use of x_{OBEM} and y_{OBEM} from the non-linear OBEM, i.e.,

$$\begin{aligned}\dot{x}_{OBEM} &= f(x_{OBEM}, \hat{h}_{ref}, u_{cmd}, z) \\ y_{OBEM} &= g(x_{OBEM}, \hat{h}_{ref}, u_{cmd}, z)\end{aligned} \tag{7.25}$$

where the vector $\hat{h}_{ref}$ represents the health condition estimated by the offline trend monitoring system, which is updated periodically. The vector z represents the measured parameters that define the fight conditions.

With this architecture, the presence of a fault can be detected by thresholding the weighted sum of squared (WSSR) residuals, as shown in Figure 7.15.

This paper assumed that a trend monitoring system capable of accurately estimating engine health degradation was available. To address that, the authors proposed an implementation of an offline trend monitoring system based on an extended KF (EKS) and addressed the integration of the offline and online systems in [15]. The authors also discussed several practicality issues of the integrated approach. For instance, the offline trend monitoring algorithm typically resides on a ground-based computer. Therefore, the data for post-flight analysis and the offline diagnostic results must be transmitted between the ground-based and onboard systems. While transmitting data from onboard systems to off-board systems is common, the inverse data transmission (i.e., from off-board systems to onboard systems) is faced with both technical and regulatory challenges. Note that the *single engine FDI approach* described in this chapter was designed with this consideration, i.e., to eliminate the need to upload data to onboard systems.

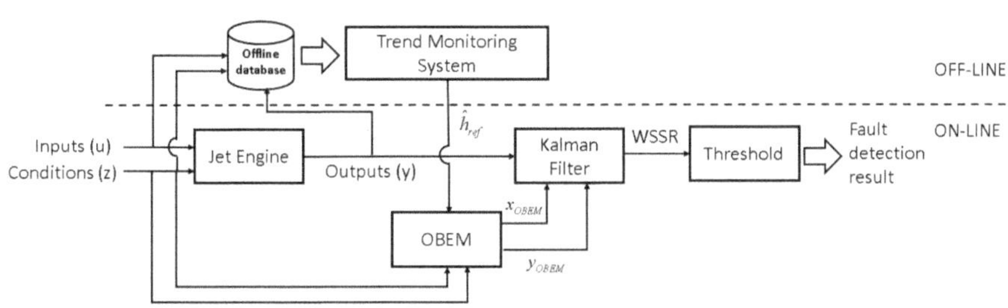

FIGURE 7.15 Overall architecture of the hybrid fault detection system

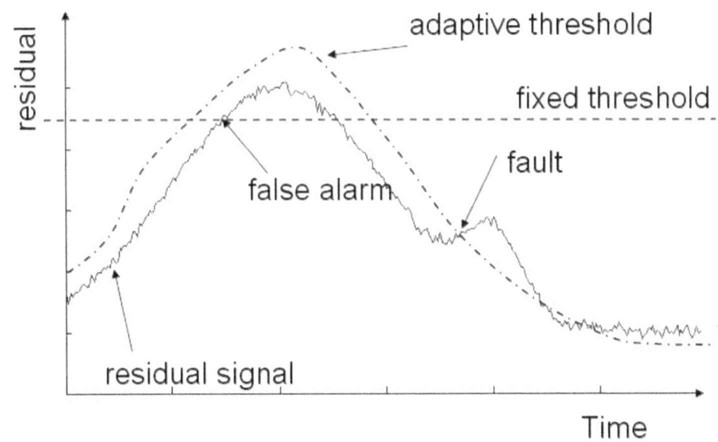

FIGURE 7.16 Concept of adaptive threshold in FDI

While the approaches in [14], [15] only addressed the sensor fault detection problem, the bank of KFs approach proposed in [13] could potentially be applied to isolate sensor faults when combined with the hybrid approach [14] and the off-line systems [15]. A non-linear estimator approach that further extended the scope to detect and isolate component, actuator, and sensor faults using a unified framework was presented in [16], [17]. The FDI architecture consisted of a parallel bank of non-linear adaptive estimators instead of the KF commonly used in other approaches. A novel feature in the reported work was the use of adaptive thresholds (Figure 7.16). The adaptive thresholds were designed such that, in the presence of a particular fault, all components of the residual generated by the adaptive estimator corresponding to the actual fault type remained below their thresholds. For faults that were sufficiently different in the measurement space, at least one component of the residual generated by each remaining adaptive estimator would exceed its threshold. Therefore, based on the specific response of the residuals, sensor faults, actuator faults, and component faults could be isolated. The effectiveness of the approach was evaluated using the NASA C-MAPSS turbofan engine model. Similar to the FDI logic in [13], the effectiveness of this approach depends on the model fidelity as the engine deteriorates over time. A neural network-based model calibration approach was proposed to keep the non-linear model in the vicinity of the health of the real engine. While technically viable, periodic update of onboard models based on off-board analysis bears the same practicality issue as discussed in [15].

In terms of other approaches that involve technical devices other than KFs or estimators, the large majority can be partitioned into the following categories. We will talk about some representative approaches in each category briefly.

- Transient feature extraction
- GAs
- ANNs

7.9.2 Transient Feature Extraction

The idea is to extract transient performance features that are subject to changes in the presence of an engine fault. A method that combined the use of steady and transient state measurement data was introduced in [18]. The proposed transient features were based on the characteristic times of transient response (such as delay time, rise time, and settling time), which were commonly used in control theory. It was demonstrated through simulation that these transient features exhibited significant deviations when certain engine faults were introduced, thus they could be used as augmented measurements for the formulation of a diagnostic procedure. Diagnostics can be formulated as an optimization problem with the following cost function using both steady and transient measurements/features [18].

$$F(f)=\sum_{i=1}^{M}\sum_{j=1}^{4}\left[\frac{Y_{ij}^{meas}-Y_{ij}(u,h)}{Y_{ij}^{meas}\cdot\sigma_{ij}}\right] \tag{7.26}$$

where vector u represents the engine's setting parameters for steady and transient operation, vector h contains the health parameters that need to be estimated, M is the number of measurements, and for each measurement, there are four elements, including a steady-state measurement and three transient features (delay time, rise time, and settling time), $Y_{ij}(u,f)$ is the prediction from a baseline model, and σ_{ij} is a weighting factor. This method can be applied to other transient features as long as the fault influence can be quantified. It's worth noting that the collective use of steady-state and transient features is a good example of *information fusion*, which is a topic we will elaborate on in the next chapter.

Another approach, used by Borguet, focused on the frequency domain using system identification techniques [19]. This approach, however, requires a particular form of input signals, such as multi-sine signals, which is probably only feasible in a test cell environment. Wavelet domain features were also studied. The method presented in [20] utilized continuous wavelet transformations of transient engine sensory data to represent their shape attributes in the time scale domain. The basis for defining the characteristic fault signatures (i.e., influence matrix) was the sensitivity of the transient engine outputs to health parameters. Change detection was achieved by isolating regions in the time scale domain, wherein the wavelet coefficients of individual outputs were considerably different from the others. The presence of such regions was taken as an indication of engine degradation.

7.9.3 GAs

This approach is an extension to the GA applications for snapshot data that we briefly introduced in Chapter 5. With transient data, a new concept of accumulated deviation was introduced to measure the shift of a parameter trajectory from its nominal position [21], [22]. The accumulated deviation was defined as the area between the two trajectories normalized by the area under the nominal trajectory ($z_n(t)$) during the transient period, as shown in Equation (7.27).

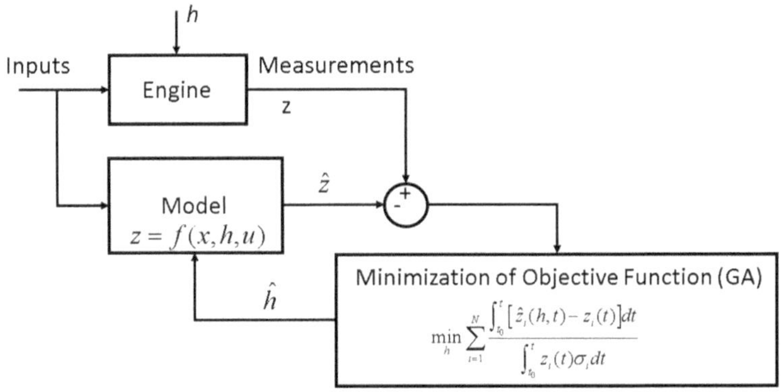

FIGURE 7.17 Non-linear model based diagnostic model using GA

$$\Delta\tilde{z} = \frac{\int_{t_0}^{t} [z(t) - z_n(t)] dt}{\int_{t_0}^{t} z_n(t) dt} \tag{7.27}$$

As shown in Figure 7.17, a non-linear model based diagnostic method, combined with a GA, was developed and applied to a gas turbine engine to diagnose engine faults using the accumulated deviation features. A slam acceleration process from idle to maximum power was analyzed in the paper. The developed diagnostic approach was applied to an engine model with three typical single-component faults and was shown to be effective. A practicality issue associated with this approach is the assumption of an accurate transient model. The trending capability of the accumulated deviations is limited by the availability of repeatable (or comparable) transient operations from flight to flight. The application of GA is due to the use of a non-linear engine model, which poses technical challenges to conventional optimization algorithms that require the calculation of derivatives.

7.9.4 ANNs

It shouldn't be a surprise that ANN found its application in transient diagnosis. Because traditional ANNs do not have built-in dynamics (or states) in the network structure to handle dynamic analysis, a viable approach is to extract features from transient measurements and use these features the same way as in snapshot analysis. One such approach, as reported in [23], which leveraged the accumulated deviation feature illustrated in Equation (7.27). The proposed solution included a set of nested neural networks designed to estimate independent parameter changes (efficiencies and flow capacities) due to faults within single or multiple components of a turbofan engine. The approach involved classification and approximation type of networks, e.g., auto associative neural networks (AANNs) for sensor fault detection, probabilistic neural networks (PNN) for fault isolation, and MLP neural networks for fault magnitude estimation.

Following the same idea, the other machine learning and artificial intelligence methods briefly introduced in Chapter 5 can potentially be applied to transient diagnosis by using features extracted from transient data.

NOTES

1. Points refer to discrete values sampled from continuous variables, either at a sampling frequency aligned with the engine control logic or down-sampled to meet data processing requirements.
2. See Addendum A7.2 for recursive equation equivalents.
3. See Separating Gradual & Rapid Deterioration section of Chapter 5.
4. The *single fault* assumption can be expanded to include $n > 1$ faults, where n is constrained to be less than the number of parameters in the $\Delta\Delta P$ set.
5. dTamb is the delta between measured ambient temperature and ISA temperature.
6. See Addendum A7.1
7. An example of information fusion, see [24].
8. A path to future onboard implementation is to introduce some sort of persistency rules to the FDI logic to enable real-time annunciation and, potentially, fault accommodation when the FDI confidence is above a certain level.
9. This allows any form of model to be utilized without the need to change the general architecture or FDI process, as well as being consistent with *QSS* assumptions.
10. See Chapter 2, Figure 2.9.
11. See Chapter 2, Trend Detection.
12. However, the confidence thresholding *per realm* allows future development for onboard implementation.
13. In our example, these are the faults listed in Table 7.1.
14. Alternatively, the mean and variance can be computed recursively to save additional memory. See Addendum A7.2.
15. See Addendum A7.3 for derivation of Equations (7.20) and (7.21).
16. Theoretically, it could even be applied to a *toaster* model; however, we wouldn't recommend it.

REFERENCES

1. Culley, D., et.al, 2009, *More Intelligent Gas Turbine Engines*, RTO Technical Report, TR-AVT-128.
2. Garg, S., 2004, Controls *and Health Management Technologies for Intelligent Aerospace Propulsion Systems*, NASA/TM—2004-212915.
3. Simon, D.L., et al., 2004, *Sensor Needs for Control and Health Management of Intelligent Aircraft Engines*, NASA/TM—2004-213202.
4. Litt, J.S., et al., 2005, *A Survey of Intelligent Control and Health Management Technologies for Aircraft Propulsion Systems,* NASA/TM—2005-213622.
5. Wiseman, M., 2005, Intelligent *Engine Systems: Work Element 1.2: Malfunction and Operator Error Reduction,* NASA/CR—2005-213964.
6. Fletcher, C., 2004, A Vision for the Intelligent Pipeline System and The Role of the Intelligent Engine, *GMRC – 2004 Gas Machinery Conference*, Albuquerque, NM.

7. Turso, J.A., 2008, Toward an Intelligent, Deterioration Accommodating Controller for Aging Turbofan Engines, *Aeronautical J*, 112(1137): 641–651.
8. Rausch, R., et al., 2004, Towards In-Flight Detection and Accommodation of Faults in Aircraft Engines, AIAA 2004-6463, *AIAA 1st Intelligent Systems Technical Conference*, September 2004, Chicago, IL.
9. Tang, L., Volponi, A., Prihar, E., 2019, Extending Engine Gas Path Analysis using Full Flight Data, *Proceedings of ASME Turbo Expo 2019: Turbomachinery Technical Conference and Exposition*, June 17–21, Phoenix, Arizona, USA, ASME paper GT2019-90161.
10. Merrington, G.L., 1988, *Fault Diagnosis of Gas Turbine Engines from Transient Data*, ASME paper 88-GT-209.
11. Merrington, G.L., et al, 1990, *Fault Detection and Diagnosis in Gas Turbines*, ASME paper 90-GT-339.
12. Merrington, G.L., 1993, *Fault Diagnosis in Gas Turbines Using a Model-Based Technique*, ASME paper 93-GT-13.
13. Simon, D.L., Takahisa, K., 2005, Evaluation of an Enhanced Bank of Kalman Filters for In-Flight Aircraft Engine Sensor Fault Diagnostics, *J Eng Gas Turbines Power*, 127(3): 497–504.
14. Takahisa, K., Simon, D.L., 2007, Hybrid Kalman Filter Approach for Aircraft Engine In-Flight Diagnostics: Sensor Fault Detection Case, *J Eng Gas Turbines Power*, 129(3): 746–754.
15. Takahisa, K., Simon, D.L., 2007, Integration of On-Line and Off-Line Diagnostic Algorithms for Aircraft Engine Health Management, *J Eng Gas Turbines Power*, 129(4): 986–993.
16. Tang, L., et al, 2010, *A Unified Nonlinear Adaptive Approach for Detection and Isolation of Engine Faults*, ASME paper GT2010–22642.
17. Zhang, X., Tang, L., Decastro, J., 2013, Robust Fault Diagnosis of Aircraft Engines: A Nonlinear Adaptive Estimation-Based Approach, *IEEE Trans Control Syst Technol*, 21(3): 861–868.
18. Kamboukos, P., Mathioudakis. K., 2007, Turbofan Engine Health Assessment by Combining Steady and Transient State Aerothermal Data, *7th European Conference on Turbomachinery*, number 137, Athens, Greece.
19. Borguet, S., et al, 2010, *A Study on Engine Health Monitoring in the Frequency Domain*, ASME paper GT2010-22635.
20. Simmons, J.C., Danai, K., 2012, In-Flight Isolation of Degraded Engine Components by Shape Comparison of Transient Outputs, *J. Eng. Gas Turbines Power*, 134(6): 061602.
21. Li, Y.G., 2003, *A Gas Turbine Diagnostic Approach with Transient Measurement, Proc Insti Mech Eng*, 217, Part A: Journal of Power and Energy, 169–177.
22. Sampath, S., Li, Y.G., Ogaji, S.O.T., Singh, R., 2003, *Fault Diagnosis of a Two Spool Turbo-Fan Engine using Transient Data: A Genetic Algorithm Approach*, ASME paper GT2003-38300.
23. Ogaji, S.O.T., Li, Y.G., Sampath, S., Singh, R., 2003, *Gas Path Fault Diagnosis of a Turbofan Engine from Transient Data using Artificial Neural Networks*, ASME paper GT2003-38423.
24. Volponi A.J., Tang L., 2016, Improved Engine Health Monitoring Using Full Flight Data and Companion Engine Information, *SAE Int. J. Aerosp.* 9(1): 2016, doi:10.4271/2016-01-2024.

8 Information Fusion

8.1 OVERVIEW

Aircraft gas-turbine engine data is available from a variety of sources, including conventional gas path measurements, advanced health monitoring sensors, component models, and maintenance histories. The ultimate goal of propulsion health management is to maximize the amount of meaningful information by combining disparate data and information sources to obtain comprehensive diagnostic and prognostic knowledge regarding the health of the engine. Information fusion, in this context, is the integration of data and other information elements from multiple sources to improve accuracy, reduce false alarms, and produce clearer inferences than was possible with segregated diagnostic analysis.

The use of information fusion has been present in our discourse up to this point, sometimes explicitly, and sometimes implicitly. Nevertheless, it is a fundamental driving force for improving and extending the analysis capabilities we have explored in the previous chapters and will continue in that vein for the remainder of this book. This chapter will discuss some of the fundamentals for information fusion, review where and how we have used it, and why we consider it to be the single most important diagnostic factor for future enhancements. We will also give a brief introduction to many of the commonly used information fusion methods reported in the literature that are relevant to EHM and GPA [1–10] at the end of the chapter.

8.2 INFORMATION TAXONOMY

Information is the key element in any health management system. A reasonable analogy can be found in medical practice. When a patient is ailing (or even during a "well visit"), the physician gathers information from physical (visual) observations as well as measurements taken with various "sensors" (blood pressure, EKG, temperature, blood analysis, X-rays, MRIs, CAT scans, etc.); these observational values are then compared to known nominal levels, and the physician then notes deviations that form symptoms and matches these to known ailments in order to form a diagnosis. Much of this medical information is experiential and has been learned by trial and error (and some experimentation) over the centuries. Gas turbine diagnostics is not quite as old as medical diagnostics, but it, too, uses *information* as its key element. This information comes from a variety of observations and measurements, from models, both physics-based and empirical, and analytical methods designed to estimate health indicators that cannot be directly measured or observed.

Information is the key enabler for engine diagnostics and prognostics. Unfortunately, it is also its greatest challenge. The difficulty stems from the fact that

DOI: 10.1201/9781003649274-8

there is a considerable information deficit relative to the number of possible health factors arising in the many components and sub-systems that comprise a modern gas turbine. As we have seen, the number of unknowns simply outweighs (by a very large margin) the number of knowns (i.e., information) in the diagnostic problem. The addition of more physical sensors does not solve the problem. An additional sensor adds an additional problem, i.e., the health of the sensor itself (is it drifting, is there bias, is the instrument precision degrading, etc.). Not to mention, sensors add weight and must be maintained, which adds cost. Thus, the measurement suite used in a gas turbine diagnostic system must provide more information to the diagnostic problem and its attendant benefits above and beyond the cost introduced by adding the sensor in the first place.

Fortunately, information can come from other sources, some of which are not directly related to physical sensors, such as the following.

- *Domain Knowledge*: the laws of physics require certain inter-relationships to exist. For example, if fuel flow were observed to increase, but *EGT* was observed to decrease, this would not be generally indicative of an engine fault but would suggest a problem in either the fuel measuring system or the temperature measuring system (or both).
- *Constraints*: these can come in the form of *limits* and *polarity* restrictions. For example, with time and operational use, a gas turbine would be expected to degrade, not get healthier. This will place a bound (or constraint) on the general solution and, in effect, reduces the number of unknowns.
- *Assumptions*: for example, a sudden (temporal) observed shift in multiple sensed parameters could be indicative of a single sub-system fault (e.g., a stuck bleed valve) or a simultaneous fault in all of the sensors or a host of combinations of faults. Because experience suggests that it is highly unlikely to have multiple faults or failures, a *single* fault assumption might be used, thereby reducing the diagnostic problem to something more manageable.
- *Observations:* these can take the form of physical observation such as inspections performed by line maintenance crew, pilots, borescope testing, etc.
- *Historical Information:* this may take the form of line maintenance repair and overhaul data records for the monitored engine in question. It could also include *operational* information such as flight routes, flight durations, accel /decel time data, and the like.
- *Other Data:* this is a broad category and can include other *non*-gas-path data, such as
  - Oil and fuel system (temperatures, pressures, quantities, debris monitoring, etc.)
  - Mechanical system (vibration, blade tip clearance, etc.)
  - Debris monitoring (intake and exhaust debris)
  - Emissions (CO, CO_2, NOx levels, etc.).
  - FADEC alerts and diagnostic codes

It can also include gas path data from different sources or operational domains, such as
  - Companion engine
  - Transient versus steady state
  - Test-stand versus flight
- *Negative Information*: not observing certain observations or patterns can be used to rule out certain potential diagnostic possibilities.

Effective information fusion for a specific application (in this instance, an aero-engine) will require specific domain knowledge to be applied, which will, in part, be a function of the type of engine being considered (number of spools, bypass geometry, measurement suite, control mode, accessory configuration etc.) A truly generic, one-size-fits-all fusion strategy may not be the most advantageous for all applications. The power of fusion is contained in the expert knowledge in how measured parameters inter-relate under known observed conditions, which leverages understanding of engine operational modes, control modes, accessory operation, thresholds, and constraints. This may take the form of an evolving information fusion *knowledge base* consisting of an assortment of potential fusion use cases or scenarios, which ultimately becomes the rules of fusion in an expert system software module. The development of such a knowledge base would entail

- Identification of all information sources
- Identification of local fusion opportunities at all levels
  - Sensor
  - Feature
  - Decision
- Development of models, methods, and algorithms that utilize
  - Measurement data
  - Observations
  - Gas turbine domain knowledge
  - Physical constraints and assumptions
- Identification and compilation of fusion scenarios specific to the gas turbine engine under consideration.

What should be apparent at this point is that *information* is more than just *data*. What may not be so apparent is that information fusion can be performed at various levels, in particular

- *Sensor level* fusion in which multiple sensors measuring correlated parameters (e.g., oil pressures, exhaust gas temperatures) can be combined.
- *Feature level* fusion, in which analysis information (e.g., component performance changes, event detection) resulting from independent analysis methods can be combined.
- *Decision level* fusion, in which diagnostic actions (e.g., damage assessments, maintenance advisories) can be combined.

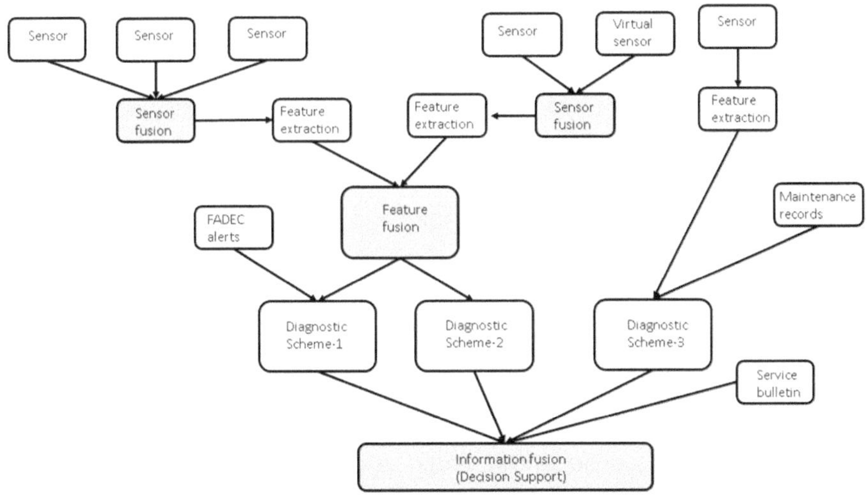

FIGURE 8.1 Various levels of information fusion in an EHM system

Real-world EHM systems usually collect data from multiple data sources and analyze hundreds or thousands of parameters measured by sensors, calculated by analytic algorithms, or captured by operators. As illustrated in Figure 8.1, data/information fusion in an EHM system can happen at various levels involving various diagnostic and reasoning methods in order to derive effective maintenance and fleet management decisions.

Relative to our topic, GPA, the first two levels are the most prevalent. In fact, Figure 2.1 of Chapter 2, depicting a general engine health diagnostic process, illustrates the central role that information fusion plays in this process. In the next section, we will review some applications of information fusion discussed in the preceding chapters, as well as introduce a few additional examples not yet mentioned.

8.2.1 Examples

It is fitting to note that GPA itself is a good example of information fusion. Monitoring gas path measurements in segregation does not provide the level of diagnostic coverage possible when viewing those measurements collectively. For example, monitoring *EGT* is certainly a good barometer for *detecting* a potential faulted engine state. By itself, however, it does little to identify the actual fault. If the HPC is deteriorated, *EGT* increases. If the HPT is deteriorated, *EGT* increases. Results are similar for the LPC and the LPT. On the other hand, a better than nominal Fan module will result in an increase in *EGT* (go figure). Thus, observing an increase in *EGT*, by itself, does not provide a plausible fault isolation. However, if we combine the EGT Δ with the other gas path measurement Δs, we form an *m*-dimensional *signature* that can be used to provide an identification. This constitutes classical linear GPA.

Throughout our discussions, we have utilized other pieces of information to enhance our analyses. We have used *trend detection* in conjunction with a *single fault assumption* to provide a means (SFA) to identify the underlying fault. This represents an integration of *detection* and *isolation* given the *assumptive* information. We have mentioned the use of deterioration models in conjunction with the KF to provide an *a-priori* guess, which uses an underlying *polarity constraint.* The use of the Kalman approach itself provides fusion between analysis points for multi-point snapshots and time series (steady-state) data analysis. This was further generalized in Chapter 7 for *QSS* data using an *eSTORM* approach, which fuses physics-based modeling with empirical corrections for better model alignment. The QSS approach, with *majority vote fault declaration*, effectively combines fault analysis across the entire flight for more effective isolation accuracy and low false alarm rate. One approach utilized companion engine data to establish a reference level from which to detect and isolate faults – another (multiple) form of information fusion. These are a few instances in which we have invoked information fusion implicitly to overcome the limitations imposed by having a sensor measurement deficit relative to the number of unknowns that are present. As we move forward into Chapter 9, we will once again make ample use of this principle to provide further potential improvements.

From the bulleted list above, the reader can plainly see that there is a preponderance of potential information sources that can be leveraged to reduce the measurement deficit. The challenge is to recognize the requisite information to bring forth for a particular problem and the means to mathematize the information to integrate it with the other information elements in our diagnostic system. Chapter 9 will explore this, to some degree, relative to the general problem of resolving fault isolation ambiguities. A simple example of this is the confusion between an HPT performance fault and a turbine case cooling (TCC) fault. The confusion exists by virtue of the fact that (with noisy measurements) a TCC fault looks almost identical to an HPT efficiency drop amongst the standard gas path parameters we have been using. This can be a vexing problem, as the latter *could* be the precursor to a dramatic turbine failure if the engine continues to fly, whereas the former poses more of an economic issue (increased fuel flow) but does not necessitate grounding the engine. One solution might be adding sensors to provide an expanded signature wherein the ambiguity no longer presents itself. We already know this is not the preferred solution, as it impacts operation cost (sensor procurement, installation, maintenance, and additional weight). Is it possible to remove the ambiguity by leveraging information already available? The answer is yes. It requires knowing two facts: 1) for either fault (TCC and HPT), *EGT* will increase, and 2) the TCC system is typically not engaged during takeoff (T/O). It is typical to capture *EGT* at T/O and calculate and trend (hot day) *EGT* margin with every flight. If we have a fault occurrence detected (say in cruise) from time series data that could either be HPT or TCC, we can check on the next flight if *EGT* margin has dropped significantly. If it has, then it must be an HPT fault, as the TCC system is inactive at T/O. Otherwise, it must be a TCC fault.[1] We will explore more of these in a general framework in Chapter 9.

In the next sub-section we give an example of *sensor fusion.*

8.2.2 EGT Profiles

EGT is a common measurement in gas turbines used for control and health management purposes. It is typically measured by a series of circumferentially placed thermocouples at the exit of the LPT. For control and health monitoring purposes, the individual temperature probe values ($EGT^{(i)}, i = 1, 2, \ldots, N$) are averaged. On many production engines, this average is performed locally (in flight), i.e., the probe values are electrically averaged before reaching the FADEC. As a result, the individual values may not be available on the ARINC digital data bus, hence the need for EHM data collection for subsequent analysis to support health management activities. The averaged value is typically used for control limiting, calculating, and trending (hot day) *EGT* margin as well as combining with other gas path parameter measurements (temperatures, pressures, and speeds) for GPA and other enhanced gas path calculations. For aero-engines, the individual probe values are not typically stored or made available to the EHM processors (except possibly during a test cell ground run). On some engines, the individual probe values are provided to the FADEC, and the averaging is performed in the FADEC. In these instances, a profile analysis can be performed to provide ancillary information to support the GPA and subsequent health assessment [30–33].

EGT probes are typically located circumferentially at the rear of the engine (see Figure 8.2). As the figure suggests, they are not necessarily placed equidistant from one another, nor are they necessarily "in-line" with the combustor nozzles upstream in the engine gas path.

If the individual temperature values are available, one could draw the *EGT profile*, by plotting the delta from average value (as shown in Figure 8.3).

$$\Delta EGT^{(i)} = EGT^{(i)} - \frac{1}{N}\sum_{i=1}^{N} EGT^{(i)} \tag{8.1}$$

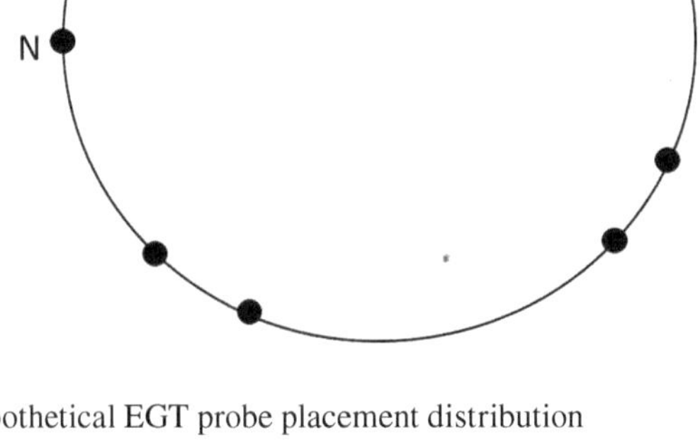

FIGURE 8.2 Hypothetical EGT probe placement distribution

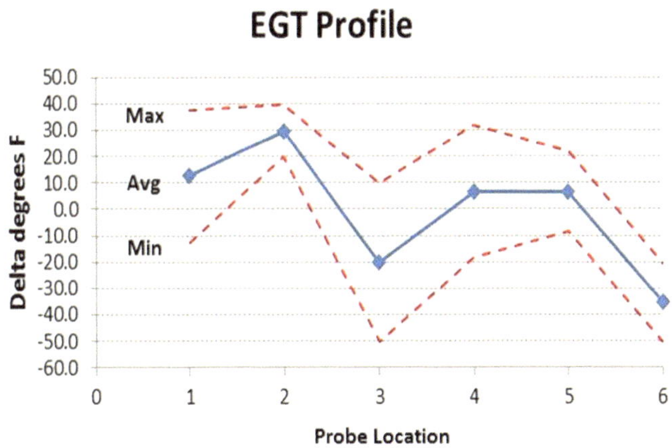

FIGURE 8.3 Hypothetical EGT profile limits

A simple check would be to compare $\Delta EGT^{(i)}$ to its max and min values[2] and determine whether an exceedance occurred, which might indicate a local hot or cold spot that could be traced back to combustor nozzles, e.g., a plugged nozzle. It is unlikely, however, that a combustor nozzle fault of some kind would affect only one probe; rather, it would affect several adjacent probes (in the same direction) as well. If only one probe delta was deemed anomalous, then it could be indicative of a probe sensor problem rather than a combustor anomaly.

We will assume that we have available a population of (M) good engine EGT profile data. This could be obtained from engine acceptance test runs on new and/or overhauled engines. Thus, for each probe location i, we have a set of deltas, and we can calculate

$$\left\{\Delta EGT^{(i)}(k)\right\}_{k=1,2,\ldots,M} = \left\{EGT^{(i)}(k) - \frac{1}{N}\sum_{i=1}^{N} EGT^{(i)}(k)\right\}_{k=1,2,\ldots,M} \tag{8.2}$$

For each probe, we can calculate the sample mean and standard deviation, i.e.,

$$\begin{aligned} \mu_i &= \frac{1}{M}\sum_{k=1}^{M} \Delta EGT^{(i)}(k) \\ \sigma_i &= \sqrt{\frac{1}{M-1}\sum_{k=1}^{M}\left(\Delta EGT^{(i)}(k) - \mu_i\right)^2} \end{aligned} \tag{8.3}$$

If we make the assumption that each probe delta is Gaussian distributed, i.e., $\Delta EGT^{(i)} \in N(\mu_i, \sigma_i)$, then the sum of the normalized probe deltas will be Chi-Squared distributed with N degrees of freedom, i.e.,

$$\sum_{i=1}^{N}\left(\frac{\Delta EGT^{(i)} - \mu_i}{\sigma_i}\right)^2 \in \chi^2(N) \tag{8.4}$$

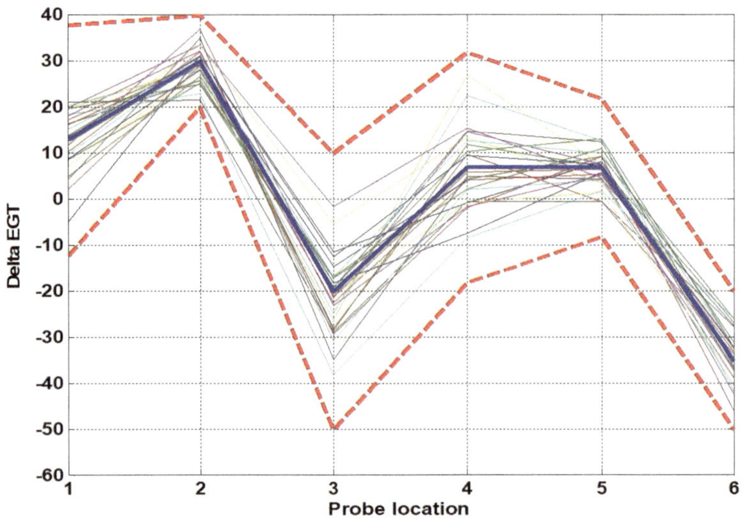

FIGURE 8.4 Sample profiles

This provides a statistical test for the collective profile. If $F(\chi^2(N))$ denotes the cumulative distribution function, then $1-F(\chi^2(N))$ provides the confidence level for testing normalcy, i.e., if the null hypothesis is *Ho: profile is normal*, then the probability of rejecting the hypothesis (i.e., the profile is *abnormal*) and it is actually normal is $1-F(\chi^2(N))$. This is usually set to a small number, say 0.01 (or 1% confidence level). If $F\left(\chi^2\left(N\right)\right) > .99$ then we can be reasonably sure it is abnormal.

To illustrate this, consider the hypothetical profile given in Figure 8.4. This profile has six probes. The dark blue solid line is the expected profile, and the red dashed lines are the min and max limits. The expected profile and the min and max limits are typically established based on a population of many nominal engines. The other lines represent individual engines profiles selected from the overall population of normal engines.

To illustrate a deviant profile, consider Figure 8.5 (green line is the measured profile).

The value of $F(\chi^2(N))$ in this case is >0.99999, so we can reject the hypothesis that the profile is normal, given the defined confidence level requirement stated above. Another example appears in Figure 8.6. In this case, $F(\chi^2(N)) = 0.9999$. This is an example of an abnormal profile that does not violate the maximum or minimum limits but still would be identified as an abnormal profile.

Finally, Figure 8.7 illustrates an acceptable profile with $F(\chi^2(N)) = 0.43867$.

At this juncture, we have not considered the effect on a profile analysis due to environmental changes, i.e., we have assumed that the measured profile was at standard day conditions. *Correcting the probe temperatures before the analysis would*

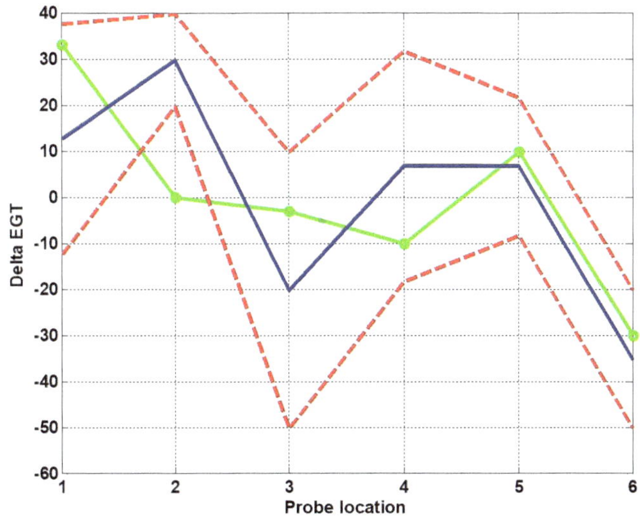

FIGURE 8.5 Sample deviant profile

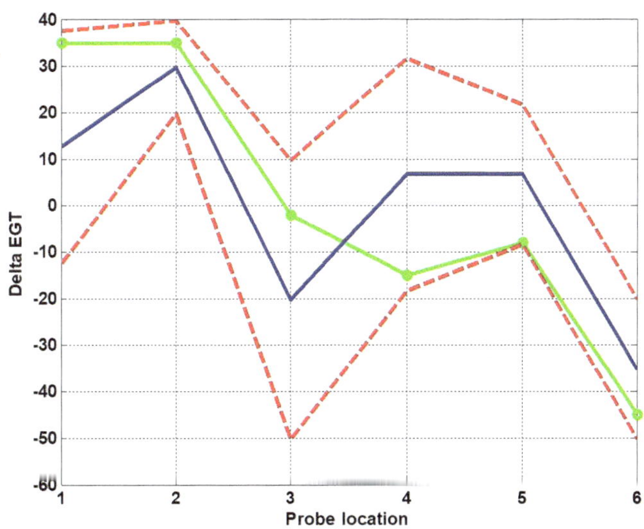

FIGURE 8.6 Sample deviant profile where probes are within limits

allow comparison of profiles across time, in which the ambient temperature may vary. For *corrected EGT*, denoted as EGT_c, we have

$$
\begin{aligned}
EGT_c^{(i)} &= \frac{EGT^{(i)}}{\theta^{\alpha_T}} \\
\Delta EGT_c^{(i)} &= EGT_c^{(i)} - \frac{1}{N}\sum_{i=1}^{N} EGT_c^{(i)} = \frac{1}{\theta^{\alpha_T}} \Delta EGT^{(i)}
\end{aligned}
\tag{8.5}
$$

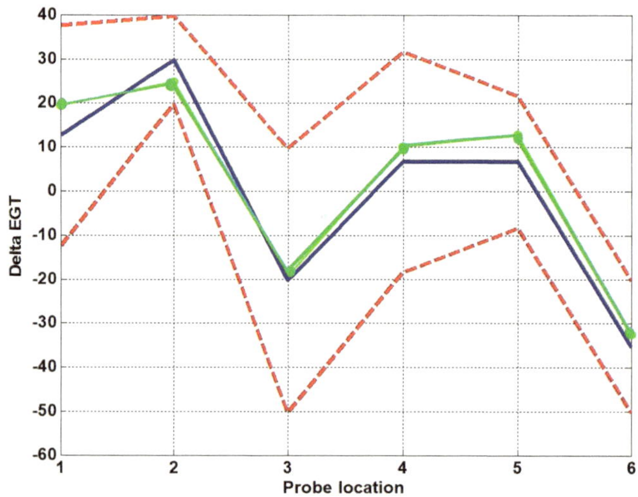

FIGURE 8.7 Sample normal profile

Making the same assumptions as before, i.e.,

$$\Delta EGT^{(i)} \in N(\mu_i, \sigma_i) \Rightarrow \frac{\Delta EGT^{(i)} - \mu_i}{\sigma_i} \in N(0,1)$$
$$\Rightarrow \sum_{i=1}^{N} \left(\frac{\Delta EGT^{(i)} - \mu_i}{\sigma_i} \right)^2 \in \chi^2(N) \tag{8.6}$$

We consider the fractional quantities

$$\frac{\Delta EGT^{(i)}}{EGT^{(Avg)}} \quad where \quad EGT^{(Avg)} = \sum_{i=1}^{N} EGT^{(i)} \tag{8.7}$$

These fractional quantities will be independent of environmental differences, as

$$\frac{\Delta EGT_c^{(i)}}{EGT_c^{(Avg)}} = \frac{\Delta EGT^{(i)} / \theta^{\alpha_T}}{EGT^{(Avg)} / \theta^{\alpha_T}} = \frac{\Delta EGT^{(i)}}{EGT^{(Avg)}} \tag{8.8}$$

Furthermore,

$$\frac{\Delta EGT^{(i)}}{EGT^{(Avg)}} \in N\left(\frac{\mu_i}{EGT^{(Avg)}}, \frac{\sigma_i^2}{\left(EGT^{(Avg)}\right)^2} \right) \Rightarrow$$
$$\sum_{i=1}^{N} \left(\frac{\dfrac{\Delta EGT^{(i)}}{EGT^{(Avg)}} - \dfrac{\mu_i}{EGT^{(Avg)}}}{\dfrac{\sigma_i}{EGT^{(Avg)}}} \right)^2 = \sum_{i=1}^{N} \left(\frac{\Delta EGT - \mu_i}{\sigma_i} \right)^2 \in \chi^2(N) \tag{8.9}$$

Therefore, if we consider fractional quantities given by Equation (8.7), a profile analysis can be conducted irrespective of ambient conditions. Although not part of mainstream GPA per se, this type of analysis is a good example of sensor level information fusion. For additional information see publications [30–33].

8.3 COMMONLY USED METHODS FOR INFORMATION FUSION

From the above examples and the previous chapters, the reader has been exposed to a number of different ways to implement information fusion in GPA applications. From an information fusion perspective, many of these algorithms that were described in previous chapters can be considered as information fusion methods. These include not only the SFA, MFA, hybrid transient modeling, and Kalman filtering methods but also the alternative approaches described in previous chapters, such as neural networks, Bayesian nets, fuzzy logic, and other machine learning methods. That said, we will herein introduce a few other methods that are often discussed in the context of information fusion.

8.3.1 AI

AI is truly a broad topic. Because the term "artificial intelligence" emerged in 1956, the development of AI technologies has gone through several ups and downs until the recent breakthroughs fueled by the advancements in cloud computing and big data analytics. Today, machine learning is generally considered a sub-field of AI study. Therefore, many of the machine learning tools described in previous chapters fit into this category. In this chapter, we will just mention a few AI techniques that were not introduced explicitly.

8.3.1.1 Expert System (AI)

Expert systems were considered amid the first successful applications of AI techniques in the 1970s and 1980s. Under the umbrella of AI, an expert system refers to a computer system that mimics the decision-making of a human expert. It is achieved through a rule-based *reasoning* process built upon a *knowledge base* that captures expert knowledge. The *knowledge base* that was mentioned earlier in this chapter is, indeed, an expert system concept.

An expert system is a powerful tool to implement information fusion. The effectiveness of the expert system heavily relies on the quality and quantity of expert knowledge accumulated in the knowledge base. While expert systems can be implemented in various ways, they all involve a form of reasoning typically expressed in "IF-THEN" rules. In order to handle the uncertainties associated with inputs and expert rules, fuzzy logic is often used in modern expert systems. An advantage of the expert system approach is that, unlike many black box AI models, expert systems are typically easy to interpret and implement. At this time, thousands of expert or knowledge-based systems (if not more) have been reported in the literature, and there are numerous books [11], [12] and commercially available tools to assist in their development. We will not further discuss this topic here. The ambiguity resolution reasoner to be presented in the next chapter has a flavor of an expert system in it.

8.3.1.2 Natural Language Processing (NLP) and Text Mining (AI)

NLP is another AI technique that could be applied to information fusion for EHM purposes, as maintenance records and field observations are often reported in natural language format (e.g., text files, Word documents, or emails). NLP is a sub-field of AI that addresses the communication between computer and human through natural language. It is a technique of computational processing of human languages that enables a computer to read and understand data by mimicking human natural language. Text mining (also referred to as text analytics) is an AI technology that uses NLP to transform the free (unstructured) text in documents into structured, normalized data suitable for analysis or reasoning.

Maintenance records usually follow a set template and procedure (such as a work order) for registration and dispositioning, often through a graphical user interface (GUI). Typical fields in a maintenance record might include record ID, asset ID, type of maintenance performed, date/time of maintenance, failure/fault addressed, description of symptoms and test results. Some fields are based on drop-down lists or check boxes that are pre-defined and structured, while other "text entry fields" are filled by the users in natural, unstructured language and are prone to human errors. In recent years, studies have been done to apply NLP and text analytics to unstructured maintenance data to search, analyze, and derive information to support diagnostics, prognostics, and condition-based maintenance decisions [13], [14].

8.3.2 Dempster-Shafer Theory and Dempster's Rule of Combination

The Dempster-Shafer inference is based on the mathematical theory introduced by Dempster [15] and Shafer [16], which generalizes Bayesian theory. The Dempster-Shafer theory provides a formalism that could be used to represent incomplete knowledge and beliefs and combine evidence with consideration of uncertainties [17].

A fundamental concept in the Dempster-Shafer reasoning is the *frame of discernment*. Let $\Theta = \{\theta_1, \theta_2, \ldots, \theta_N\}$ be the set of all possible states that define the system, and Θ is exhaustive and mutually exclusive. The set Θ is called a frame of discernment because its elements are employed to discern the current state of the system.

The elements of the set 2^Θ are called hypotheses. In the Dempster-Shafer theory, based on the evidence E, a probability is assigned to each hypothesis $H \in 2^\Theta$ according to the basic assignment of probabilities or the mass function $m: 2^\Theta \to [0,1]$, which satisfies $m(Null) = 0$, i.e., the mass function of the empty set is zero.

To express incomplete beliefs in a hypothesis, H, the Dempster-Shafer theory defines the belief function $bel: 2^\Theta \to [0,1]$ over Θ as

$$bel(H) = \sum_{A \subseteq H} m(A) \tag{8.10}$$

where $bel(Null) = 0$, and $bel(\Theta) = 1$. To express the plausibility of each hypothesis, the function $pl : 2^{\Theta} \rightarrow [0,1]$ over Θ is defined as

$$pl(H) = \sum_{A \cap H = Null} m(A) \tag{8.11}$$

Plausibility indicates that there is less uncertainty in hypothesis H if it is more plausible. The confidence interval $[bel(H), pl(H)]$ defines the true belief in hypothesis H. To *combine* the effects of the two mass functions, m_1 and m_2, the Dempster-Shafer theory defines a rule as

$$m_1 \oplus m_2(Null) = m_{1,2}(Null) = 0$$

$$m_1 \oplus m_2(H) = m_{1,2}(H) = \frac{\sum_{X \cap Y = H} m_1(X) m_2(Y)}{1 - \sum_{X \cap Y = Null} m_1(X) m_2(Y)} \tag{8.12}$$

Let's look at an example to understand how this theory could potentially be applied to a diagnosis problem as an information fusion tool.

Suppose we have a belief of 0.5 for a proposition, such as "the engine is faulty" from a probabilistic anomaly detection algorithm. This implies that we have evidence that allows us to state that the proposition is true with a confidence of 0.5. Meanwhile, the evidence contrary to that hypothesis (i.e., "the engine is healthy") has a confidence of 0.2. The remaining mass of 0.3 is "uncertainty" that the engine could either be faulty or healthy.

The belief for the "healthy" and "faulty" hypotheses matches their corresponding masses because they have no subsets; belief for "either" consists of the sum of all three masses (either, healthy, and faulty) because "healthy" and "faulty" are each sub-sets of "either". The "healthy" plausibility is 1 – m (faulty), i.e., 1 – 0.5 = 0.5, and the "faulty" plausibility is 1 – m (healthy), i.e., 1 – 0.2 = 0.8. for plausibility, the "healthy" plausibility is m(healthy) + m (either), and the "faulty" plausibility is m(faulty) + m(either). Finally, the "either" plausibility sums m(healthy) + m(faulty) + m(either). Actually, the universal hypothesis ("either") will always have 100% belief and plausibility. The mass, belief, and plausibility values are summarized in Table 8.1.

Now, let's say Table 8.1 represents what "Algorithm 1" says about the health of the engine. Suppose there is an "Algorithm 2" that produces different diagnostic results (mass), as shown in Table 8.2. Note that all the numbers are purposefully set to different values to make it easier to see where the numbers come from in the following fusion example.

Using Dempster's rule of combination, the mass for each hypothesis is calculated as follows.

$$m_{1,2}(Null) = 0$$

$$m_{1,2}(Healthy) = \frac{0.2*0.1+0.2*0.15+0.3*0.1}{1-(0.5*0.1+0.2*0.75)} = 0.1$$

$$m_{1,2}(Faulty) = \frac{0.5*0.75+0.5*0.15+0.3*0.75}{1-(0.5*0.1+0.2*0.75)} = 0.84375$$

$$m_{1,2}(Either) = \frac{0.3*0.15}{1-(0.5*0.1+0.2*0.75)} = 0.05625$$

In this example, because the two algorithms are generally in agreement (that the engine is faulty), the combined mass for the faulty hypothesis has been increased. As simple as the example appears, the application of Dempster's rule of combination in engine health monitoring is very limited due to the many limitations of this approach. Besides the critics on the theory itself [18–20], Dempster's rule of combination quickly explodes as the size of the set Θ (thus, 2^{Θ}) increases, resulting in a high computational cost.

8.3.3 Negative Reasoning

Managing uncertainties in diagnostic reasoning is always a challenge. Uncertainties could be allocated to the symptoms/evidence (inputs to the diagnostic process), as well as the method/conclusion itself (e.g., assumptions made, models utilized, expert rules constructed). Regardless of what techniques are utilized to represent

TABLE 8.1
Mass, Belief, and Plausibility in Dempster-Shafer Theory

Hypothesis	Mass	Belief	Plausibility
Null	0	0	0
Healthy	0.2	0.2	0.5
Faulty	0.5	0.5	0.8
Either (Healthy or Faulty)	0.3	1	1

TABLE 8.2
Combine Two Mass Functions

Hypothesis	Mass (Algorithm 1)	Mass (Algorithm 2)
Null	0	0
Healthy	0.2	0.1
Faulty	0.5	0.75
Either (Healthy or Faulty)	0.3	0.15

uncertainties (e.g., probability, Dempster-Shafer evidential intervals, fuzzy membership functions), an inevitable challenge arises when it comes to combining both positive information (e.g., symptoms of a problem) with negative information (i.e., information for which its absence is indicative of meaningful inferences).

Negative reasoning has been an integral part of the medical community's standard approach to the treatment and diagnosis of disease. Doctors tend to prescribe a series of tests that are relevant to the patients' condition. Based on the test results, doctors are often able to either positively diagnose the condition or selectively eliminate candidates on the list of potential disorders. Negative test results play an important role in the diagnosis process.

For fault diagnosis applications, let's take a simple example as given in [2]. Table 8.3 shows a positive matrix mapping of observations (Ol, O2, . ..) to causes (C1, C2, . ..). Similarly, Table 8.4 shows the negative matrix mapping of non-observations (~O1 , ~O2, . . .) to eliminated causes (hypotheses) marked by an X.

Thus, if the set of positive observations are {O1}, and the set of negative observations are {~O5}, then based on the following reasoning,

$$\{O1\} \Rightarrow \{C1, C3\}$$
$$\{\sim O5\} \Rightarrow \{C1, C2, C4, C5\}$$

TABLE 8.3
Example of Use of Symptomatic Evidence to Identify Possible Causes

Causes (C)\Observations (O)	O1	O2	O3	O4	O5
C1	✓	–	✓	–	–
C2	–	✓	✓	✓	–
C3	✓	–	–	–	✓
C4	–	–	✓	✓	–
C5	–	✓	–	–	–

TABLE 8.4
Example of Use of Negative Information to Eliminate Possible Causes

Causes (C) \ Non-observations (~O)	~O1	~O2	~O3	~O4	~O5
C1	–	X	–	X	X
C2	X	–	–	–	X
C3	–	X	X	X	–
C4	X	X	–	–	X
C5	X	–	X	X	X

fusion of the two positive and negative observation sets leaves C3 as the diagnosis. On the other hand, if the positive observation set is {Ol} and {O4}, and the negative observation set is $\{\sim O2, \sim O3, \sim O5\}$, the knowledge base exhibits a logical inconsistency. This could be indicative of a faulty sensor(s), which was not included in the knowledge base.

In the above example, the reasoning can certainly be achieved based on Table 8.3 alone, using some simple IF-THEN rules or some sort of pattern matching. For more supplicated scenarios encountered in real-world applications, a generic approach to automated reasoning of negative information is the logical templating method originally developed for automated reasoning for threat assessment [2]. This technique combines logical relationships with the ability to identify information that must be present (or absent) in order to establish the existence of an activity, event, or condition [21].

8.3.4 Case-Based Reasoning

Case-based reasoning (CBR) is the process of solving new problems based on the solutions of similar past problems [22]. It has been argued that CBR is not only an effective approach for computer reasoning but also a common behavior in everyday human problem solving. For example, a mechanic diagnosing an engine failure by recalling another case with similar symptoms, or a lawyer advocating a particular outcome in a trial based on legal precedents are both using CBR.

Unlike other major AI approaches that rely primarily on generalized knowledge of a problem domain or associations along generalized relationships, CBR is able to utilize the specific knowledge of previously experienced and concrete problem situations (cases) [23]. CBR has been formalized for purposes of computer reasoning as a four-step process, as shown in Figure 8.8.

1) Retrieve: given a target problem, retrieve from memory cases relevant to solving it. A case consists of an issue, its solution, and typically, annotations about how the solution was derived. For diagnosis application, it may include the date/time information, part information, symptoms, fault mode, and the maintenance action taken to resolve it.
2) Reuse: apply the solution from the previous case to the target problem and adapt the solution as needed to fit the new situation.
3) Revise: test the new solution in the real world (or in a simulation) and, if necessary, revise the solution.
4) Retain: After the solution has been successfully applied to the target problem, store the resulting experience as a new case in memory.

Cases are retrieved based on similarity or a pre-defined distance measure. One (*nearest neighbor*) or a group of cases (*k-nearest neighbors*) may be retrieved depending on the application at hand. The global *similarity* between the problem (P_{new}) and a case (P_j) in the case base can be formulated as follows,

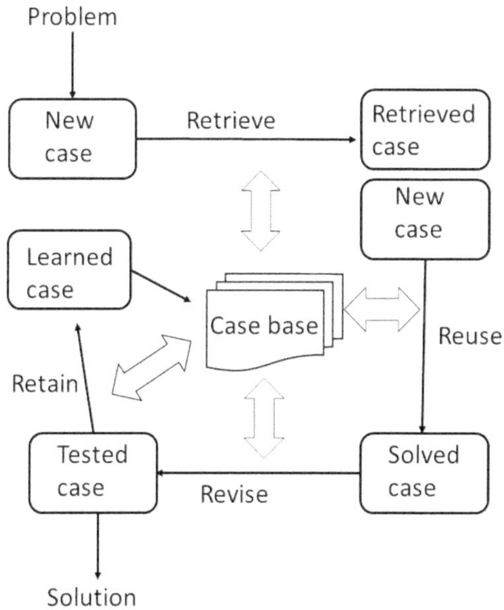

FIGURE 8.8 The CBR cycle

$$sim\,(P_{new}, P_j) = \sum_{i=1}^{N} w_i sim_i\,(P_{new,i}, P_{j,i})$$
$$\sum_{i=1}^{N} w_i = 1 \quad (8.13)$$

where w_i is the weight of the ith attribute, $0 \le w_i \le 1$, and $sim_i\,(P_{new,i}, P_{j,i})$ is the local similarity between the ith attributes of the cases, for which various distance measures can be applied, such as Euclidian distance, Manhattan's metric, Hamming's metric, Tversky's metric, Tchebishev's metric, and minimum or maximum metrics, etc.

When applied to diagnostic reasoning, CBR can be used both as a diagnostic reasoner or to resolve ambiguities in the diagnosis result. This approach enables incremental, sustained learning, as a new experience is retained each time a problem has been solved, making it immediately available for future problems. Critics of CBR argue that it is an approach that accepts anecdotal evidence as its main operating principle. Without statistically relevant data for backing and implicit generalization, there is no guarantee that the generalization is correct [24]. Another drawback of this approach is that it takes time to build a reasonable case base. Therefore, it's often infeasible for new engine programs at entry into service.

8.3.5 Other Types of Estimation Filters

We have used KF on various applications in the previous chapters. The KF is often considered a generic data fusion tool, as it fuses measurement data with model (often physics-based) prediction to produce an optimal estimate. The KF is applicable to linear systems with Gaussian noises (both process noise and measurement noise). However, other estimation filters (or estimators) are required to address non-linear dynamic models and/or non-Gaussian noise assumption, and for GPA applications, it may become necessary when the underlying model is of a non-linear nature (e.g., using NPSS or a high-fidelity non-linear model instead of an SVM).

The EKF is an optimal estimator for implementing non-linear recursive filters with Gaussian noise [25]. However, it has the disadvantage of having to compute the Jacobians at each time step, which is often costly for onboard real-time applications. The unscented KF (UKF) [26] has gained popularity, as it doesn't require the calculation of Jacobians, i.e., not needing linearization of system dynamics [38]. It employs a deterministic sampling strategy to establish the minimum set of points around the mean, i.e., the sigma points, to represent the state distribution. The sigma points are propagated through the non-linear system functions, and the weights of the sigma points are updated based on the observation. The sigma points, together with associated weights, are used to calculate the mean and covariance of states.

Another non-linear filtering approach is the particle filter (PF), which can be applied to non-linear, non-Gaussian problems. Generally speaking, optimal estimation problems for non-linear non-Gaussian state-space models do not typically admit analytic solutions. Since their introduction in 1993, PF methods have become a very popular class of algorithms to solve these estimation problems *numerically* in a recursive manner [27]. Unlike the UKF algorithm, which uses a minimum set of points around the mean, PF applies a swarm of sample points propagating through the non-linear system functions. PF is essentially a recursive implementation of the sequential Monte Carlo method [28]. Although generically applicable to a broad range of problems, PF is usually computationally expensive, which limits its application in time critical systems.

It is worth noting that, with all the claimed advantages of the sophisticated non-linear filters, the end benefits may not necessarily justify the cost for implementing them. The transient dynamics of gas turbines (particularly jet engines) are extremely hard to model accurately (not to mention engine to engine variation). The benefits of introducing some non-linearity in the model don't always pay off the additional costs to accommodate it. Many times, the simpler approach works better. A more sophisticated approach should only be adopted if it clearly outperforms the simpler approach when tested sufficiently with real data.

8.4 DISCUSSION

Data fusion is the integration of data or information from multiple sources to achieve enhanced accuracy and more specific inferences than can be obtained from the use of a single sensor (or information source) alone. The basic tenet underlying the data/information fusion concept is to leverage all available information to enhance diagnostic visibility, increase diagnostic reliability and reduce the number of diagnostic false alarms. In the authors' view, this principle will continue to be a major force in refining GPA methodologies. The major obstacle will be in determining what information is available and pertinent to the problem at hand. Because *information* can come from a wide variety of sources, from data, constraints, assumptions, etc., there is no convenient general methodology to formulate a procedure, but it is hopefully clear that information applied collectively will provide superior results than when used in isolation.

Besides the algorithmic issues, a real-world challenge is to access, convert and integrate heterogeneous information from disparate systems. The implementation of an information fusion system requires a form of information-relay framework that enables the holistic view and traceability of an asset along its entire life cycle (i.e., design, manufacturing, service, maintenance, overhaul, etc.). Along this line, the concept of *digital thread* has been brought up in a broader context of Industry 4.0 [29].

As a final example, we offer a mathematical puzzle that illustrates the ultimate power of information fusion. This puzzle was introduced by Martin Gardner, a 20th century mathematician best known for his *Scientific American* articles and books on mathematical puzzles. In 1979, he wrote about a puzzle in the *Scientific American* that he termed as the *Impossible Problem*. It proceeds in this manner:

> *Two integers, x and y, both greater than 1 and whose sum is less than 100 are secretly selected. The* ***Sum*** *(x+y) is given only to mathematician* **S** *and the* ***Product*** *(xy) is given only to mathematician* ***P****. The mathematicians know only these facts.*
>
> *The exchange between them goes something like this:*

P*: I do not know the values of x and y*
S*: I knew that you didn't know the values.*
P*: Now I know the values of x and y.*
S*: Now I know them as well.*

> *After this exchange it is now possible to determine the two numbers x and y.*

The puzzle is aptly named because it appears that there is hardly enough information to determine anything, let alone its solution. This is an example of capitalizing on every piece of information available in the problem statement, not only the factual statements but also assumptions, such as the mathematicians perform flawless logical deductions, have at their disposal knowledge of elementary number theory (domain knowledge), and employ negative information in their reasoning. For example, mathematician ***P*** can factor his product (uniquely) into a product of primes (domain knowledge). It must be the case that this factorization is *not* the product of just two prime numbers, otherwise ***P*** would know the numbers. ***S*** can determine the *partition* of the sum, i.e., how many different ways the sum can be decomposed into the sum of two numbers (domain knowledge). It must be the situation that this partition does *not* contain a case in which the sum can be represented by the sum of two prime numbers, otherwise ***S*** would not be able to make the first statement (you have to think about that for a moment). It is this type of reasoning, coupled with domain knowledge, assumptions, negative information, and the fusion of all of the above with the actual observations (problem statement), that ultimately leads to the unique solution. For the curious reader, a complete solution is given in Addendum 8.1.

NOTES

1. This is a use of *negative information* as well.
2. Established based on a population of nominal engines, as detailed later.

REFERENCES

1. Fisher, C., 2001, Data and Information Fusion for Gas Path Debris Monitoring, *IEEE Aerospace Conference*, Bug Sky, MT.
2. Hall, D.L., Hansen, R.J., Lang, D.C., The Negative Information Problem in Mechanical Diagnostics, *Proceedings of the American Society of Mechanical Engineers Turbine and Aeroengine Congress*, Birmingham, UK, June 10–13, ASME Paper 96-GT-035.
3. Volponi, A.J., et al., 2003, Development of an Information Fusion System for Engine Diagnostics and Health Management, *JANNAF 39th Combustion/27th Airbreathing Propulsion/21st Propulsion Systems Hazards/3rd Modeling and Simulation Joint Subcommittee Meeting*, Colorado Springs, December 1–5.
4. Volponi, A.J., 2005, *Data Fusion for Enhanced Aircraft Engine Prognostics and Health Management*, NASA Report CR-2005-214055, Glenn Research Center, December 2005.
5. Jai, Huamin, 2004, *Data Fusion Methodologies for Multi-Sensor Aircraft Navigation Systems*, Cranfield University, Ph.D. Thesis.
6. Harris, C.J., et.al., 2016, Multi-Sensor Data Fusion in Defense and Aerospace, *The Aeronautical Journal*, 102(1015), 229–244.
7. Basir, O., Yuan, X., 2007, Engine Fault Diagnosis Based on Multi-Sensor Information Fusion using Dempster–Shafer Evidence Theory, *Information Fusion*, Vol. 8, Issue 4.
8. Hu, X., Eklund, N., Goebel, K., 2007, *A Data Fusion Approach for Aircraft Engine Fault Diagnostics*, ASME paper, GT2001-27941.
9. Sarkar, S., et al., 2013, Multi-Sensor Information Fusion for Fault Detection in Aircraft Gas Turbine Engines, *Proceedings of the Institution of Mechanical Engineers, Part G: Journal of Aerospace Engineering*, 227(12), 1988–2001.
10. Zaccaria, V., Rahman, M., 2019, A Review of Information Fusion Methods for Gas Turbine Diagnostics, *Sustainability*, 11(22), 6202. https://doi.org/10.3390/su11226202
11. Charniak, E., Riesbeck, C.K., McDermott, D. V., 1980, *Artificial Intelligence Programming*, Lawrence Erlbaum Associates Publishers, Hillsdale, NJ.
12. Hopgood, A.A., 1992, *Knowledge-Based Systems for Engineers and Scientists*, CRC Press, London, United Kingdom.
13. Brundagea, M.P., et al., 2021, Technical Language Processing: Unlocking Maintenance Knowledge, *Manufacturing Letters*, 27, 42–46. https://doi.org/10.1016/j.mfglet.2020.11.001
14. Stenstrom, C., Aljumaili, M., Parida, A., 2015, Natural Language Processing of Maintenance Records Data, *International Journal of COMADEM* 18(2), 33–37.
15. Dempster, P., 1968, A Generalization of Bayesian Inference, *Journal of the Royal Statistical Society B*, 30(2), 205–247.

16. Shafer, A., 1976, *Mathematical Theory of Evidence*, Princeton University Press, Princeton, NJ, USA.
17. Provan, G.M., 1992, The Validity of Dempster-Shafer Belief Functions, *International Journal of Approximate Reasoning*, 6(3), 389–399.
18. Jøsang, A., Simon, P., 2012, Dempster's Rule as Seen by Little Colored Balls, *Computational Intelligence*, 28(4), 453–474
19. Pearl, J., 1988, *Probabilistic Reasoning in Intelligent Systems*, (Revised Second Printing), Morgan Kaufmann, San Mateo, CA.
20. Pearl, J., 1988, On Probability Intervals, *International Journal of Approximate Reasoning*, 2(3), 211–216.
21. Hall, D.L., Linn, R., 1989, Comments on the Use of Templating Techniques for Multi-Sensor Data Fusion, *Proceeding of the Third Tri-Service Data Fusion Symposium*, Vol. 1, pp. 345–354.
22. Richter, M.M., Weber, R., 2013, *Case-Based Reasoning*, Springer ISBN 978-3-642-40167-1.
23. Aamodt, A., Plaza, E., 1994, Case-Based Reasoning: Foundational Issues, Methodological Variations, and System Approaches, *Artificial Intelligence Communications,* 7(1), 39–52.
24. Hüllermeier, E., 2007, *Case-Based Approximate Reasoning*, Springer-Verlag, Berlin, ISBN 1402056958, 9781402056956.
25. Simon, D., 2006, *Optimal State Estimation.* Hoboken, NJ: John Wiley & Sons. ISBN 978-0-471-70858-2.
26. Julier, S.J., Uhlmann, J.K., 2004, Unscented Filtering and Nonlinear Estimation, *Proceedings of the IEEE,* 92(3), 401–422.
27. Del Moral, P., 1996, Non Linear Filtering: Interacting Particle Solution, *Markov Processes and Related Fields*, 2(4), 555–580.
28. Liu, J.S., Chen, R., 1998, Sequential Monte Carlo Methods for Dynamic Systems, *Journal of the American Statistical Association*, 93(443), 1032–1044.
29. MargariaEmail, T., Schieweck, A., The Digital Thread in Industry 4.0, 2019, *International Conference on Integrated Formal Methods*, IFM 2019: Integrated Formal Methods pp 3–24.
30. Tsalavoutas, A., Mathioudakis, K., Smith, M.K., 1996, *Processing of Circumferential Temperature Distributions for the Detection of Gas Turbine Burner Malfunctions*, ASME Paper, 96-GT-103.
31. Knowles, M., 1994, *Gas Turbine Temperature Spread Monitoring Detection of Combustion System Deterioration*, ASME Paper, 94-GT-189.
32. Volponi, A.J., 2014, Gas Turbine Engine Health Management: Past, Present and Future Trends, *Journal of Engineering for Gas Turbines & Power*, 136, 05-1201-1.
33. Cox, G.B., 1975, Predicting Exit Temperature Profile from Gas Turbine Combustors, *AIAA/SAE 11th Propulsion Conference*, Anaheim, CA, Paper 75-1307.

9 Ambiguity Resolution

9.1 OVERVIEW

We have seen that the inadequacy of the *observation* dimensionality relative to the *fault* dimensionality has certainly been the nemesis of the diagnostic process and has been central to our discussions in the previous chapters. Leveraging additional *information* has been a clear path to combat this deficiency. This measurement deficiency gives rise to another diagnostic difficulty that is perhaps not as obvious but, nonetheless, challenges our ability to perform an accurate diagnosis or prognosis. It is the problem of *ambiguity*, i.e., when two or more faults *look the same* relative to our measurement space, especially when measurement uncertainty is present. We have skirted around this issue throughout the text without really giving it any detailed discussion. It is the intent of this chapter to explore fault ambiguity in more detail and discuss some measures to recognize it and mitigate it. Ambiguity resolution is the central theme of this chapter and will be presented within the context of GPA, drawing upon (and expanding upon) some of the work already presented in the previous chapters.

9.2 AMBIGUITY

Ambiguity arises in the process of identifying system faults that manifest themselves in identical or near-identical symptoms or observations to the end user. It is a natural consequence of any diagnostic identification procedure. Colloquially, we could think of an *ambiguity group* as the group of items to which a given fault is identified, any one of which may be the actual faulty item [1]. With this definition in mind, it is fair to say that an *ambiguity group* is, in effect, the anticipated consequence of a diagnostic fault isolation procedure.[1] The essence of the ambiguity group is that it comprises a collection of failure modes for which the diagnostic procedure in play can detect and isolate the fault to that collection and no further, i.e., it cannot further isolate to any sub-set of that collection without sacrificing diagnosis accuracy. Thus, the notion of ambiguity is intrinsically connected to the diagnostic methodology being employed and the measurement suite (and information) that is available; change any of these, and you potentially change the ambiguity group structure.

In general, a diagnostic process can be formulated either as a *classification* or an *estimation* problem. *Classification* is an example of pattern recognition. When formulated as a *classification* problem, the fault identification process aims to determine to which pre-defined category (i.e. fault mode) the new observation belongs. On the other hand, when formulated as an *estimation* problem, the *estimator* not only points to the most likely *fault* but also gives an *estimate* of *fault magnitude*. It

DOI: 10.1201/9781003649274-9

should be clear that a study of ambiguity requires that the diagnostic process is such that it produces a *single* stand-alone fault, not a combination of faults. *This single fault assumption is in force throughout the remainder of this chapter.*

The goal of a diagnostic system is to determine the current state of the system under consideration. To accomplish this, we must relate *symptoms* to *faults*. The symptoms consist of a collection of observations that can be ascertained by the practitioner through the diagnostic process being employed. These can be actual observed data from sensors or symptomatic labels deduced from the data. For example, in medicine, the patient's temperature is in excess of 101 *deg F* or the patient has a *fever* is a simple example of a symptom. This symptom is *discrete* in the sense that it has been observed to be true (a value of 1) or it is observed to be false (a value of 0). If all of our symptoms are classified as discrete, and we can relate combinations of such symptoms to various faults (illnesses) then we basically have a *discrete diagnostic system*. Conversely, if all of our symptoms are real-valued observations, and our diagnostic process uses these quantitative values *directly* when comparing potential faults, then we would have a *continuous diagnostic system*. A simple example of a discrete system might be the comparison of symptoms common to an allergy versus COVID-19 as indicated in the checklist below.[2]

Symptom	Allergy	COVID-19
Fever	0	1
Dry cough	0	1
Shortness of breath	0	1
Fatigue	1	1
Loss of smell	0	1
Sneezing	1	1
Runny nose	1	0
Itchy or watery eyes	1	0
Post-nasal drip	1	0
Headache	0	1
Sore throat	1	1

In contrast a simple continuous system for a gas turbine engine performance fault diagnosis might be given by the abbreviated IC matrix below.

	Fault				
Symptom	**Fan**	**LPC**	**HPC**	**HPT**	**LPT**
% Delta N1	-0.68	-0.14	-0.05	-0.08	0.98
% Delta N2	0.3	-0.29	0.06	0.57	-0.63
% Delta Fuel Flow	0.7	-0.66	-0.8	-1.29	0.96

9.2.1 Preliminary Discussion

We will denote the set of all known *faults* for the system under consideration by Ω. Without loss of generality, we will assume that this set is finite,[3] i.e., $|\Omega| = n < \infty$. If f and g are two faults in Ω, we write, $g \sim f$ to mean that g is *ambiguous with f.* This defines a *relation* (~) on the set Ω that is *reflexive*, as $f \sim f, \forall f \in \Omega$. We will use the notation $\{\Omega, \sim\}$ to denote the set of known faults with an ambiguity relation defined on it. The fact that any fault is ambiguous with itself follows from our colloquial conception of what it would mean to be ambiguous. If f is an arbitrary fault, we define the following ambiguity group structures:

$$\begin{aligned} AG_f &= \{g \in \Omega \mid g \sim f\} \\ AG_{(f)} &= \{g \in \Omega \mid f \sim g\} \\ AG_f^{-1} &= \{g \in \Omega \mid g \text{ } not \sim f\} \\ AG_{(f)}^{-1} &= \{g \in \Omega \mid f \text{ } not \sim g\} \end{aligned} \tag{9.1}$$

At first glance, it might appear that we have created redundant definitions, as it would seem reasonable to conclude that if f is ambiguous with g, then surely g is ambiguous with f. That is to say $f \sim g \Rightarrow g \sim f$, and thus $AG_f = AG_{(f)}$. As it turns out, this is *not* necessarily the case! This seems contrary to real-world commonsense reasoning. However, it can be demonstrated by means of a real-world counter-example that ambiguity is not always a *symmetric* relation. Unfortunately, we need to develop more mathematical machinery before we can demonstrate this assertion, so we ask for our readers' indulgence until we reach that juncture later in this chapter.

9.2.2 Discrete Ambiguity Groups

To begin, let us assume we have a general discrete diagnostic system. No matter what form it might take, it must always be possible to list the set of *faults* relevant to the application at hand and the *symptoms* that can be formed from the observations that are possible within the application being monitored. From this information, one can construct so-called *fault symptom mapping* (*FSM*) that indicates how the symptoms relate to the faults. For a discrete diagnostic system, the *FSM* is a matrix that takes the following form:

$$FSM(i,j) = \begin{cases} 1 & \text{if the } i^{th} \text{ symptom is present for the } j^{th} \text{ fault} \\ 0 & \text{otherwise} \end{cases},$$

$$i = 1,2,\ldots,m,\ j = 1,2,\ldots,n$$

Thus, if we have a set of *faults* $\Omega = \{f_1, f_2, \ldots, f_n\}$ and a set of *symptoms* $S = \{s_1, s_2, \ldots, s_m\}$, the *FSM* will be an $m \times n$ matrix of 1s and 0s. A hypothetical example in which we have $n = 6$ *faults* and $m = 6$ *symptoms* is illustrated in Figure 9.1.

To aid in our discussion, we define the following two matrices:

	f1	f2	*f3*	*f4*	*f5*	*f6*
s1	1	0	0	0	0	0
s2	0	1	0	0	0	0
s3	1	0	1	0	0	0
s4	1	1	0	1	0	0
s5	1	0	1	0	1	0
s6	1	1	1	1	0	1

FIGURE 9.1 Sample discrete fault system mapping (FSM)

- *Discrete Confusion Matrix* ($\mathcal{C}$)

The discrete confusion matrix ($\mathcal{C}$) depicts the relationship between the family of faults present and the faults observed, not to be confused with the fault isolation confusion matrix C, presented in Chapter 2, which is a fault isolation performance metric. The matrix is defined as follows:

$$\mathcal{C}(i,j) = \begin{cases} 1 & f_i \sim f_j \\ 0 & otherwise \end{cases}$$

- *Ambiguity Group Matrix* (*AGM*)

The ambiguity group matrix (*AGM*) depicts the relationship between the family of faults and the AGs for each fault. If there are a total number of n faults to be considered, there are n number of *AG*s, and *AGM* would take the form of an $n\times(n+1)$ matrix with the individual faults along the rows and the *AG*s along the columns (with the *nominal* state as the last column). The first $n\times n$ elements are defined as follows:

$$AGM_{ik} = \begin{cases} 1 & if\ fault_i \in AG_{fault_k} \\ 0 & otherwise \end{cases} \quad 1\le i\le n, 1\le k\le n \tag{9.2}$$

where AG_{fault_k} denotes the ambiguity group for the fault $fault_k$. By default, $AGM_{ii} = 1$, i.e., a fault belongs to its own AG. However, if a fault is *undetectable* given a particular diagnostic method or with a particular measurement suite, it is considered ambiguous with the nominal state, i.e., belongs to $AG_{Nominal}$ (the set of faults ambiguous with the nominal state). In this case, the last column is defined as

$$AGM_{i(n+1)} = \begin{cases} 1 & if\ fault_i \in AG_{Nominal} \\ 0 & otherwise \end{cases} \quad 1\le i\le n \tag{9.3}$$

An $n\times n$ concatenated form of the *AGM* (denoted *AGM**) will be used later in this chapter. It is identical to the *AGM* with the last column deleted, i.e.,

$$AGM^*_{i,k} = AGM_{i,k} \quad 1\le i\le n, 1\le k\le n \tag{9.4}$$

NOTE: As a generality, one can differentiate between two classes of ambiguity groups; *discrete level ambiguity groups* (denoted, $AG_{fault}^{(D)}$) and *continuous level ambiguity groups* (denoted, $AG_{fault}^{(C)}$, or simply AG_{fault}), respectively. The former arises from diagnostic procedures that rely on *qualitative* observations, while the latter employs *quantitative* observations. What we mean by a *qualitative* observation is any datum that is represented by a *discrete* variable that is bivalent or trivalent.[4] For example, pressure has *increased* (+1), temperature has *decreased* (-1), and flow has *no change* (0). Conversely, a *quantitative* observation is any datum that is real-valued. This latter case is where we find ourselves relative to the GPA procedures we have discussed in this text and, in particular, with SFA, which will be our major focus in this chapter. The reason for differentiating between these two classes is that they have some fundamentally different properties that are interesting to note and will become evident as we proceed through this chapter.

Returning to our hypothetical example depicted in Figure 9.1, it is clear that each column of the *FSM* is unique. As a consequence, there can be *no* confusion (i.e., ambiguity) in identifying the faults and, thus, the discrete confusion matrix $\mathcal{C}$ is the *identity*, as illustrated in Figure 9.2. The corresponding *AGM* is illustrated in Figure 9.3.

The *concatenated* AGM is the same as the discrete confusion matrix,[5] i.e., $AGM^* = \mathcal{C}$. As we will see later, this is *not* the case for *continuous* ambiguity groups.

If we omit a symptom from our hypothetical sample system, for example s_1, then the fault symptom mapping and the ambiguity group matrix are updated as illustrated in Figures 9.4 and 9.5 respectively.

It can be seen that the columns are still unique, hence the discrete confusion matrix remains the *identity*. If, instead, we were to omit symptom s_2, there would now be confusion between faults f_2 and f_4, as can be seen in Figure 9.6.

In this case, the discrete confusion matrix and the ambiguity group matrix are updated as illustrated in Figures 9.7 and 9.8 respectively,

One last example: if we drop the observation s_6, then the fault propagation matrix becomes as depicted in Figure 9.9.

From Figure 9.9, we can clearly see that there is no confusion between any of the faults, $\{f_1, f_2, ..., f_6\}$; however, f_6 is ambiguous with the *nominal* state, i.e., it would *not be detected*. As such the discrete confusion matrix (as we have defined it) would be the as depicted in Figure 9.10.

	f1	*f2*	*f3*	*f4*	*f5*	*f6*
f1	1	0	0	0	0	0
f2	0	1	0	0	0	0
f3	0	0	1	0	0	0
f4	0	0	0	1	0	0
f5	0	0	0	0	1	0
f6	0	0	0	0	0	1

FIGURE 9.2 Sample discrete confusion matrix ($\mathcal{C}$)

	AGf1	AGf2	AGf3	AGf4	AGf5	AGf6	AGnominal
f1	1	0	0	0	0	0	0
f2	0	1	0	0	0	0	0
f3	0	0	1	0	0	0	0
f4	0	0	0	1	0	0	0
f5	0	0	0	0	1	0	0
f6	0	0	0	0	0	1	0

FIGURE 9.3 Sample ambiguity group matrix (AGM)

	f1	f2	f3	f4	f5	f6
s2	0	1	0	0	0	0
s3	1	0	1	0	0	0
s4	1	1	0	1	0	0
s5	1	0	1	0	1	0
s6	1	1	1	1	0	1

FIGURE 9.4 Sample discrete fault system mapping (FSM) w/o symptom s1

	AGf1	AGf2	AGf3	AGf4	AGf5	AGf6	AGnominal
f1	1	0	0	0	0	0	0
f2	0	1	0	0	0	0	0
f3	0	0	1	0	0	0	0
f4	0	0	0	1	0	0	0
f5	0	0	0	0	1	0	0
f6	0	0	0	0	0	1	0

FIGURE 9.5 Sample ambiguity group matrix (AGM) w/o symptom s1

It is clear from these simple examples (Figure 9.2, Figure 9.7, and Figure 9.10) that the discrete confusion matrix is symmetric (i.e., $\mathcal{C} = \mathcal{C}^T$). In fact, it should be apparent that if fault f_i is ambiguous with fault f_j, then it is true that fault f_j is ambiguous with fault f_i. This ensures that $\mathcal{C} = \mathcal{C}^T$. If we think of the property of being ambiguous as a *relation* (we will use the notation ~), then for faults $\{f, g, h\}$, the following properties hold:

$$\begin{aligned} &1)\, f \sim f \\ &2)\, f \sim g \Rightarrow g \sim f \\ &3)\, f \sim g \text{ and } g \sim h \Rightarrow f \sim h \end{aligned} \tag{9.5}$$

The properties displayed in Equation 9.5 are referred to as *reflexive*, *symmetric*, and *transitive*, respectively. Such a relation forms what is called an *equivalence* relation.[6] An *equivalence class A* is the collection of all things that are *related* (i.e.,

	f1	f2	f3	f4	f5	f6
s1	1	0	0	0	0	0
s3	1	0	1	0	0	0
s4	1	1	0	1	0	0
s5	1	0	1	0	1	0
s6	1	1	1	1	0	1

FIGURE 9.6 Sample discrete fault system mapping (FSM) w/o symptom s2

	f1	f2	f3	f4	f5	f6
f1	1	0	0	0	0	0
f2	0	1	0	1	0	0
f3	0	0	1	0	0	0
f4	0	1	0	1	0	0
f5	0	0	0	0	1	0
f6	0	0	0	0	0	1

FIGURE 9.7 Sample discrete confusion matrix ($\mathcal{C}$) w/o symptom s2

	AGf1	AGf2	AGf3	AGf4	AGf5	AGf6	AGnominal
f1	1	0	0	0	0	0	0
f2	0	1	0	1	0	0	0
f3	0	0	1	0	0	0	0
f4	0	1	0	1	0	0	0
f5	0	0	0	0	1	0	0
f6	0	0	0	0	0	1	0

FIGURE 9.8 Sample ambiguity group matrix (AGM) w/o symptom s2

	f1	f2	f3	f4	f5	f6
s1	1	0	0	0	0	0
s2	0	1	0	0	0	0
s3	1	0	1	0	0	0
s4	1	1	0	1	0	0
s5	1	0	1	0	1	0

FIGURE 9.9 Sample discrete fault system mapping (FSM) w/o symptom s6

	f1	*f2*	*f3*	*f4*	*f5*	*f6*
f1	1	0	0	0	0	0
f2	0	1	0	0	0	0
f3	0	0	1	0	0	0
f4	0	0	0	1	0	0
f5	0	0	0	0	1	0
f6	0	0	0	0	0	1

FIGURE 9.10 Sample discrete confusion matrix ($\mathcal{C}$) w/o symptom s6

$f, g \in A$ *if and only if* $f \sim g$). In our examples, the *discrete* AGs are the equivalence classes.

A fundamental result states that if ~ is an equivalence relation defined on a set Ω, then Ω will be the *disjoint* union of its equivalence classes.[7] For our example in Figure 9.6, this means that the population of faults can be expressed as a disjoint union of the AGs, i.e.,

$$\Omega = \{f_1, f_2, \ldots, f_6\} = \bigcup_{i=1,2,3,5,6} AG_{f_i}^{(D)} \tag{9.6}$$

where $AG_{f_2}^{D} = \langle f_2, f_4 \rangle$ and the other four AGs are singletons. For our example in Figure 9.9, we have

$$\Omega = \{f_1, f_2, \ldots, f_6\} = \bigcup_{i=1,2,3,4,5,6} AG_{f_i}^{(D)} \tag{9.7}$$

The AGs provide information on the *diagnostic coverage* that is achievable from our diagnostic procedure. *Diagnostic coverage* can be defined as the % fraction of unambiguous faults that are identified, i.e.,

$$Coverage = \left(\frac{\#\, singleton\ AGs - \psi}{|\Omega|} \right) 100\% \tag{9.8}$$
$$\psi = \#\ faults\ in\ AG_{Nominal}$$

For our previous two examples, $|\Omega| = n = 6$, $Coverage = (4/6)100\% \approx 66.7\%$, and $(5/6)100\% \approx 83.3\%$, respectively. If we were to engage in the exercise of testing all possibilities of omitting one and two symptoms from the sample system in Figure 9.1, we would generate the set of *diagnostic coverages* depicted in Table 9.1. What is interesting is that, for this particular system, we can obtain 100% coverage with only *four* observations, namely $\{s_2, s_4, s_5, s_6\}$, by omitting $\{s_1, s_3\}$. In total, there are $2^6 - 1 = 63$ (*symptom*) configurations possible, if we consider all combinations where we have at least one *symptom* available. If we were to generate all of these combinations, we could capture a succinct overview of the results in tabular form, such as Table 9.2, which depicts the number of possible configurations with diagnostic coverage versus the number of symptoms (observations).

	AGf1	*AGf2*	*AGf3*	*AGf4*	*AGf5*	*AGf6*	*AGnominal*
f1	1	0	0	0	0	0	0
f2	0	1	0	0	0	0	0
f3	0	0	1	0	0	0	0
f4	0	0	0	1	0	0	0
f5	0	0	0	0	1	0	0
f6	0	0	0	0	0	1	1

FIGURE 9.11 Sample ambiguity group matrix w/o symptom s6

TABLE 9.1
Diagnostic Coverages for Sample Discrete System

Observation Omitted	# Singleton AGs	Coverage	Ambiguities
None	6	100%	*None*
s1	6	100%	*None*
s2	4	66.7%	f2, f4
s3	6	100%	*None*
s4	4	66.7%	f4, f6
s5	5	83.3%	f5, Nominal
s6	5	83.3%	f6, Nominal
s1, s2	4	66.7%	*{f2 , f4}*
s1, s3	6	100%	*None*
s1, s4	2	33.3%	*{f1 , f3} {f4 , f6}*
s1, s5	5	83.3%	*{f5 , Nominal}*
s1, s6	5	83.3%	*{f6 , Nominal}*
s2, s3	4	66.7%	*{f2 , f4 }*
s2, s4	3	50.0%	*{f4 , f6}*
s2, s5	3	50%	*{f2 , f4 } {f5 , Nominal}*
s2, s6	3	50%	*{f2 , f4 } {f6 , Nominal}*
s3, s4	4	66.7%	*{f4 , f6}*
s3, s5	3	50%	*{f3 , f6} {f5 , Nominal}*
s3, s6	3	50%	*{f3 , f5} {f6 , Nominal}*
s4, s5	3	50%	*{f4 , f6} {f5 , Nominal}*
s4, s6	3	66.7%	*{f4 , f6 , Nominal}*
s5, s6	3	66.7%	*{f5 , f6 , Nominal}*

In Table 9.2, it is immediately apparent that there are four possible configurations that would yield 100% coverage: one with *six* symptoms, two with *five* symptoms, and one with *four* symptoms. Thus, to achieve maximum coverage, we need at least four symptoms, and Table 9.1 indicates that the optimal configuration that requires the least number of symptoms would be $\{s_2, s_4, s_5, s_6\}$, as noted above. Likewise, if we have a constraint of no more than three symptoms (observations) to employ, then

TABLE 9.2
Sample Diagnostic Coverages vs Number of Symptoms

		Number of Symptoms					
		1	2	3	4	5	6
% Coverage	**0**	4	2				
	16.7	2	10	5			
	33.3		3	6	1		
	50			7	6		
	66.7			2	5	2	
	83.3				2	2	
	100				1	2	1

Table 9.2 indicates that the best coverage possible would be 66.7%, and there are two configurations to achieve this coverage.

In general, if we have m symptoms, there are $2^m - 1$ possible configurations to assess. This is manageable for small values of m (e.g., $m \leq 12$) and a worthwhile exercise,[8] especially during design phases of a diagnostic program. It provides valuable information to ponder tradeoffs between coverage, number of sensors needed, and fault ambiguity grouping.

We will leave the topic of *discrete* ambiguity groups and move on to a discussion of *continuous* ambiguity groups, which are relevant to the topic at hand: GPA.

9.2.3 Continuous Ambiguity Groups

To make things more manageable in our discussion, we will assume that our diagnostic procedure evokes a *single fault* hypothesis, i.e., that once an anomalous condition is detected, one and only one fault is responsible for the non-nominal excursion. The specifics of a single fault identification process that might be employed will not be needed for (nor will be germane to) our discussion, except for one general aspect, which is how the procedure treats our sensed observations, i.e., measurements or sensors. For a given *fault*, we will denote its ambiguity group by AG_{fault}, i.e.,

$$AG_{fault} = \{ f \in \Omega \mid f \sim fault \}$$

We will assume that we have a *quantitative* model of the system under consideration, (e.g., a non-linear aero-thermal engine simulation such as NPSS or C-MAPSS). In these types of systems, component faults (or deteriorating effects) are captured quantitatively and indirectly by observations that sense the fault effect as a change from the nominal or reference condition. This produces a *fingerprint* across the set of sensors (observations) and are *real* number valued. In GPA, the observations (sensors) are typically a set of temperatures, pressures, speeds, flows, etc., taken along an engine's gas path. Faults (or deterioration) in the engine's major modules (compressors, turbines) or sub-systems (bleeds, clearance controls, variable geometry actuation, etc.) have an attendant observable effect on the measures gas path parameters (observations), which provide a unique[9] quantitative *signature* or *fingerprint* that can be used to isolate the fault.

One of the difficulties in a SFA approach is that the pre-defined list of potential single fault scenarios to test sequentially as candidate faults, typically admit a good deal of ambiguity, i.e., many faults fall in one or more *ambiguity groups (AGs).* When AGs exist, the best expected outcome of the analysis would be the ability to isolate to the AG containing the actual underlying fault and to do so with a *high* confidence. It is reasonable (and natural) to expect that isolating to the appropriate *AG* could be accomplished with higher confidence than attempting to isolate to a single specific fault within that group (using SFA). With that motivation we will describe a procedure for determining AGs and exemplify using the following list of potential engine system faults. A typical list of faults for a twin spool turbofan engine might be something like what is captured in Table 9.3 [3], [8].

To illustrate the level of ambiguity that might arise, we will generate *AG*s by first investigating the confusion matrix associated with the SFA for this fault list. To do this, we simulate *each* of the 31 single faults using 1,000 noisy measurements[10] (i.e., a total of 31k trials) and perform an SFA analysis. The resulting confusion matrix is depicted in Figure 9.12 in terms of % confidence in analyzed fault, as reported in [3].

TABLE 9.3
Typical Gas Path Fault List

Fault ID	Fault Description	Fault ID	Fault Description
1	Fan performance	17	Low turbine case cooling failed closed
2	LPC performance	18	HPC variable stator vane (VSV)
3	HPC performance	19	Mach error
4	HPT performance	20	Altitude error
5	LPT performance	21	N1 error
6	7th stage service bleed	22	N2 error
7	10th stage service bleed	23	Wf error
8	2.5 bleed	24	EGT error
9	7th stage stability bleed	25	T25 error
10	10th stage stability bleed	26	T3 error
11	Fan duct leakage	27	P25 error
12	Fan duct pressure loss	28	P3 error
13	Combustor pressure loss	29	TAT or T2 error
14	Turbine cooling air (TCA) valve fault	30	P2 error
15	ECS bleed switch	31	P49 error
16	High turbine case cooling failed closed		

Implanted Fault	Observed (Analyzed) Fault																													
	1	2	3	4	5	6	7	8	9	10	11	12	13	14	15	16	17	18	19	20	21	22	23	24	25	26	27	28	29	30	31
1	**91.4**	0.0	0.0	0.0	0.0	0.0	0.0	0.2	0.0	0.0	6.4	0.0	0.0	0.0	0.0	0.0	0.0	0.0	0.0	0.0	0.0	0.0	0.0	0.0	0.0	0.0	0.0	0.0	0.0	0.1	0.0
2	0.0	**92.8**	0.0	0.0	0.1	0.1	0.0	3.1	0.4	0.0	0.1	0.0	0.0	0.0	0.0	0.0	0.0	0.0	0.0	0.0	0.0	0.0	0.0	0.0	0.0	0.0	0.0	0.0	0.0	0.0	0.0
3	0.0	0.0	**89.7**	0.0	0.0	0.0	0.0	0.0	0.0	0.0	0.0	0.0	0.0	0.0	1.6	0.0	0.0	0.0	0.0	0.0	0.0	0.0	0.0	0.1	0.0	0.1	0.0	0.0	0.3	0.0	0.0
4	0.0	0.0	0.0	**77.2**	0.0	0.0	0.0	0.0	0.0	0.0	0.0	0.0	0.0	0.2	0.0	3.1	0.6	0.0	0.0	0.0	0.0	0.0	0.0	0.0	0.0	0.0	0.0	0.0	0.0	0.0	0.0
5	0.0	0.1	0.0	0.0	**96.3**	0.1	0.0	0.9	0.0	0.1	0.1	0.0	0.0	0.0	0.0	0.0	0.0	0.0	0.0	0.0	0.1	0.0	0.0	0.2	0.0	0.0	0.1	0.0	0.0	0.0	0.0
6	0.0	0.2	0.0	0.0	0.1	**79.9**	3.7	0.5	13.4	3.9	0.0	0.2	0.0	0.0	0.0	0.0	0.0	0.1	0.1	0.0	0.0	0.0	0.4	0.1	0.0	0.0	0.0	0.0	0.0	0.0	0.0
7	0.0	0.0	0.1	0.0	0.0	2.4	**82.6**	0.0	0.9	9.6	0.0	0.1	0.0	0.0	0.4	0.0	0.4	0.0	0.0	0.0	0.0	0.0	0.2	0.2	0.0	0.0	0.0	0.0	0.0	0.1	0.0
8	0.4	3.9	0.0	0.0	0.9	0.2	0.0	**88.5**	1.9	0.3	0.1	0.0	0.0	0.0	0.0	0.0	0.0	0.0	0.0	0.0	0.0	0.0	0.1	0.0	0.0	0.0	0.8	0.0	0.0	0.0	0.0
9	0.0	0.5	0.0	0.0	0.0	10.5	1.5	0.5	**78.6**	3.9	0.0	0.0	0.0	0.0	0.0	0.0	0.0	0.0	0.0	0.0	0.0	0.0	0.0	0.1	0.0	0.0	0.0	0.0	0.0	0.0	0.0
10	0.0	0.1	0.0	0.0	0.0	2.1	7.0	0.1	3.7	**78.9**	0.0	0.0	0.0	0.0	0.0	0.0	0.0	0.0	0.0	0.0	0.0	0.0	0.0	0.0	0.0	0.0	0.0	0.0	0.0	0.0	0.0
11	7.3	0.0	0.0	0.0	0.1	0.0	0.0	0.2	0.0	0.0	**88.4**	0.0	0.0	0.0	0.0	0.0	0.0	0.2	0.3	0.0	0.5	0.0	0.0	0.0	0.0	0.0	0.0	0.1	0.0	0.6	0.8
12	0.0	0.0	0.0	0.0	0.1	0.3	0.1	0.0	0.2	0.0	0.0	**90.4**	0.6	0.2	0.0	0.0	0.0	0.1	0.7	0.0	0.0	0.0	0.2	0.0	0.0	0.1	0.0	0.4	0.3	0.5	1.5
13	0.0	0.0	0.0	0.0	0.0	0.0	0.0	0.0	0.0	0.0	0.0	0.2	**91.7**	0.5	0.4	0.0	0.0	0.0	0.0	0.0	0.0	0.0	0.0	0.0	0.0	0.0	0.0	1.4	0.0	0.0	0.6
14	0.0	0.3	0.0	10.9	0.0	0.5	0.2	0.3	0.0	0.5	0.8	1.4	2.1	**89.8**	2.0	0.7	1.7	0.3	0.4	0.3	0.9	0.6	0.4	0.3	1.3	1.4	1.1	10.3	0.1	0.2	1.4
15	0.4	0.8	8.4	4.1	1.0	1.1	3.0	0.4	0.5	1.5	2.2	1.1	2.5	4.8	**79.1**	4.1	4.0	1.4	2.5	2.5	3.2	3.1	2.9	5.9	3.8	2.9	3.4	2.3	1.9	1.3	1.1
16	0.0	0.0	0.0	6.2	0.0	0.0	0.0	0.0	0.0	0.0	0.0	0.0	0.0	0.0	0.0	**75.9**	14.9	0.0	0.0	0.0	0.0	0.0	0.0	0.0	0.0	0.0	0.0	0.0	0.0	0.0	0.0
17	0.0	0.0	0.0	1.0	0.0	0.0	0.0	0.0	0.0	0.0	0.0	0.0	0.0	0.0	0.4	14.9	**77.4**	0.0	0.0	0.0	0.0	0.0	0.0	0.0	0.0	0.0	0.0	0.0	0.0	0.0	0.0
18	0.0	0.0	0.0	0.1	0.0	0.1	0.0	0.0	0.1	0.0	0.0	0.2	0.1	0.2	0.8	0.0	0.0	**86.3**	0.1	0.0	0.0	19.0	0.0	0.1	0.1	0.2	0.2	0.0	0.5	0.2	0.1
19	0.1	0.2	0.4	0.0	0.0	0.1	0.2	0.0	0.0	0.1	0.5	2.4	0.5	0.0	1.6	0.1	0.0	0.6	**68.5**	14.3	0.4	0.2	9.0	0.1	0.4	0.3	0.5	0.1	0.3	0.8	2.5
20	0.0	0.6	0.2	0.5	0.5	1.6	0.9	1.0	0.2	0.6	0.5	2.5	0.8	1.6	9.5	0.5	0.3	1.4	16.9	**39.8**	1.5	1.2	27.0	1.6	2.6	1.8	2.0	1.1	0.6	0.7	0.9
21	0.0	0.0	0.0	0.0	0.0	0.0	0.0	0.0	0.0	0.0	0.0	0.0	0.0	0.0	0.0	0.0	0.0	0.0	0.0	0.0	**92.9**	0.0	0.0	0.0	0.0	0.0	0.0	0.0	0.0	0.0	0.0
22	0.0	0.0	0.0	0.0	0.0	0.0	0.0	0.0	0.0	0.0	0.0	0.0	0.0	0.0	0.0	0.0	0.0	9.3	0.0	0.0	0.0	**75.6**	0.0	0.0	0.0	0.0	0.0	0.0	0.0	0.0	0.0
23	0.0	0.2	0.0	0.1	0.3	0.7	0.3	0.2	0.0	0.2	0.0	0.8	0.1	0.3	0.0	0.1	0.2	0.0	10.2	42.3	0.1	0.2	**59.3**	0.1	0.9	0.5	0.6	0.1	0.0	0.2	0.1
24	0.0	0.0	0.3	0.0	0.0	0.0	0.5	0.0	0.0	0.4	0.0	0.0	0.1	0.0	1.6	0.1	0.5	0.0	0.0	0.0	0.0	0.0	0.0	**91.0**	0.1	0.0	0.0	0.0	0.0	0.1	0.0
25	0.1	0.0	0.0	0.0	0.0	0.0	0.0	0.1	0.2	0.0	0.1	0.1	0.2	0.0	0.4	0.1	0.2	0.1	0.0	0.7	0.3	0.1	0.2	0.1	**89.9**	0.4	0.1	0.1	0.1	0.1	0.0
26	0.0	0.0	0.0	0.0	0.0	0.0	0.0	0.0	0.0	0.0	0.0	0.0	0.0	0.0	0.0	0.0	0.0	0.0	0.0	0.0	0.0	0.0	0.0	0.0	0.0	**91.0**	0.0	0.0	0.1	0.0	0.0
27	0.4	0.3	0.5	0.0	0.7	0.2	0.0	4.2	0.0	0.0	0.9	0.2	0.3	0.0	2.0	0.4	0.0	0.1	0.2	0.1	0.2	0.2	0.2	0.2	0.7	0.5	**91.2**	0.1	0.1	0.0	0.0
28	0.0	0.0	0.0	0.0	0.0	0.0	0.0	0.0	0.0	0.0	0.0	0.3	1.1	2.6	0.0	0.0	0.0	0.0	0.0	0.0	0.0	0.0	0.0	0.0	0.0	0.0	0.0	**84.1**	0.0	0.0	0.3
29	0.0	0.2	0.5	0.0	0.0	0.0	0.1	0.0	0.0	0.1	0.0	0.0	0.0	0.0	0.4	0.0	0.0	0.0	0.0	0.0	0.1	0.0	0.0	0.1	0.3	1.0	0.0	0.0	**95.6**	1.9	0.1
30	0.0	0.0	0.0	0.0	0.0	0.0	0.0	0.0	0.0	0.0	0.0	0.0	0.0	0.0	0.0	0.0	0.0	0.0	0.0	0.0	0.0	0.0	0.0	0.0	0.0	0.0	0.0	0.0	0.1	**93.3**	0.8
31	0.0	0.0	0.0	0.0	0.0	0.0	0.0	0.0	0.0	0.0	0.0	0.1	0.0	0.0	0.0	0.0	0.0	0.0	0.0	0.0	0.0	0.0	0.0	0.0	0.0	0.0	0.0	0.0	0.0	0.1	**89.8**

FIGURE 9.12 SFA confusion matrix represented as % confidence in analyzed fault

Some observations (and explanations) may be in order. The rows represent the implanted faults, and the columns represent the SFA analyzed results. The diagonal represents successful isolations. For example, if we look at Column 21 (which represents *N1 error*), the diagonal element is 92.9% accuracy. There are, however, other non-zero numbers appearing in Column 21. For example, the intersection of Row 20 and Column 21, reads 1.5%. From Table 9.3, it can be seen that Fault 20 is a *100ft altitude error*. The intersecting entry means that 1.5% of the 1,000 simulated instances of a *100ft altitude error* were incorrectly analyzed as an *N1 Error*, i.e., there is ambiguity between those two faults. Likewise, ambiguity exists between all of those faults whose rows have non-zero entries in Column 21. The complete *AG* for *N1 error* is seen to be the following set containing 11 faults in total, i.e.,

$$AG_{N1\_Error} = \{5,11,14,15,19,20,21,23,25,27,29\} \tag{9.9}$$

Inspection of the confusion matrix in Figure 9.12 reveals that there is a great deal of confusion present; however, there are many columns with (relatively) *small* entries. It is important to keep in mind that this level of ambiguity is not only due to the collection of faults, measurement suite utilized, and diagnostic procedure used (SFA) but also the sample size *N*, which, in this case, was 1,000. In fact, if we were to increase our sample size to $N = 1000000$, we might anticipate more ambiguity to appear (due to the increased chance of encountering larger measurement noise). Indeed, if we let $N \to \infty$, the confusion matrix would contain no zero entries, i.e., everything would be confounded with everything else, albeit infinitesimally in most cases.[11] This raises the practical question as to what level of ambiguity can we tolerate.

If we view the *N1 error* fault (Column 21), we see that the likelihood that it is truly an *N1 Error* fault is 92.9%. The remaining 7.1% is distributed amongst the other faults in the AG, AG_{N1_Error} given in Equation (9.9). These are displayed in Table 9.4.

The interpretation for Table 9.4 is as follows: if the analysis produces a first rank *N1 error* fault, then (theoretically) it is about 0.1% possible it is Fault 5, about 0.5% possible it is Fault 11, etc. What can be said, with 100% likelihood, is that the true fault is one of those appearing in the AG of Equation (9.9). We can reduce the number of possibilities if we reduce the constraint of having a 100% anticipation. We can achieve that by creating a *derived* AG by discarding any fault that has, for example, a confidence of occurring that is $\leq \alpha\%$, for some chosen α. For example, if we choose $\alpha = 1\%$ we would discard eight of the 11 faults, i.e., $\{5,11,14,19,23,25,27,29\}$. The resultant AG, referred to as a 1% *risk adjusted* AG, becomes

$$AG_{N1\_Error}^{(1\%)} = \{15,20,21\} \tag{9.10}$$

If we choose a slightly larger risk level, say $\alpha = 3\%$, the AG is reduced further, as indicated in Equation (9.11).

$$AG_{N1\_Error}^{(3\%)} = \{15,21\} \tag{9.11}$$

TABLE 9.4
Ambiguity Group $AG_{N_1\ Error}$ Confidences

Fault ID	Fault Name	% Confidence
5	LPT performance	0.1%
11	Fan duct leakage	0.5%
14	Turbine cooling air (TCA) valve fault	0.9%
15	ECS bleed switch	3.2%
19	Mach error	0.4%
20	Altitude error	1.5%
21	N1 error	92.9%
23	Wf error	0.1%
25	T25 error	0.3%
27	P25 error	0.2%
29	TAT or T2 error	0.1%

Therefore, we will refer to ambiguity at a certain pre-defined risk level, and as a matter of convenience, we will use $\alpha = 3\%$ for the remainder of our discussion. If we apply this level of risk to Figure 9.12, we arrive at a new confusion matrix, as depicted in Figure 9.13.

The reduction in the number of ambiguities can be dramatic, as can be seen from Figure 9.13 and Table 9.5. By adopting a 3% risk level, the average size of the AG is reduced by 79%. Nevertheless, 24 of the 31 faults represented still admit an ambiguity with at least one other fault; therefore, with only seven singleton AGs, the diagnostic coverage is a dismal 22.6%.[12] It is clear that some form of ambiguity resolution is required to determine the underlying fault from the ambiguities that still persist. The solution will come in the form of leveraging additional information. In the next section, we will outline a possible framework for achieving this goal. Before moving to that discussion, there are some interesting observations that are worthy to note.

The reader can see by inspecting either Figure 9.12 or Figure 9.13 that the confusion matrices are *not* symmetric. Colloquially speaking, we might believe that if fault F_1 is ambiguous with fault F_2, then F_2 would be ambiguous with F_1. Although that is true for *discrete* AGs, this is *not* necessarily the case for *continuous* AGs. For example, we can see from Table 9.5 (under the 3% risk adjusted AGs) that a *P3 error fault* (ID = 28) is ambiguous with a *TCA valve fault* (ID = 14), but not conversely. Thus, the *relation* of *being ambiguous is not symmetric*. It is also *not transitive* either.[13] This can be seen by expanding our example, as *P3 error fault* (ID = 28) is ambiguous with a *TCA valve fault* (ID = 14), and *TCA valve fault* (ID = 14) is ambiguous with *ECS bleed switch fault* (ID = 15), but *P3 error fault* (ID = 28) is *not* ambiguous with *ECS bleed switch fault* (ID = 15). The reason for this apparent contradiction to common sense is multi-faceted. First and foremost, it arises from the fact that we are dealing with a continuous system that is defined by real

Observed (Analyzed) Fault (columns); **Implanted (True) Fault** (rows)

	1	2	3	4	5	6	7	8	9	10	11	12	13	14	15	16	17	18	19	20	21	22	23	24	25	26	27	28	29	30	31
1	**92.6**	0.0	0.0	0.0	0.0	0.0	0.0	0.0	0.0	0.0	6.8	0.0	0.0	0.0	0.0	0.0	0.0	0.0	0.0	0.0	0.0	0.0	0.0	0.0	0.0	0.0	0.0	0.0	0.0	0.0	0.0
2	0.0	**96.0**	0.0	0.0	0.0	0.0	0.0	3.2	0.0	0.0	0.0	0.0	0.0	0.0	0.0	0.0	0.0	0.0	0.0	0.0	0.0	0.0	0.0	0.0	0.0	0.0	0.0	0.0	0.0	0.0	0.0
3	0.0	0.0	**91.4**	0.0	0.0	0.0	0.0	0.0	0.0	0.0	0.0	0.0	0.0	0.0	0.0	0.0	0.0	0.0	0.0	0.0	0.0	0.0	0.0	0.0	0.0	0.0	0.0	0.0	0.0	0.0	0.0
4	0.0	0.0	0.0	**78.5**	0.0	0.0	0.0	0.0	0.0	0.0	0.0	0.0	0.0	0.0	0.0	3.2	0.0	0.0	0.0	0.0	0.0	0.0	0.0	0.0	0.0	0.0	0.0	0.0	0.0	0.0	0.0
5	0.0	0.0	0.0	0.0	**100.0**	0.0	0.0	0.0	0.0	0.0	0.0	0.0	0.0	0.0	0.0	0.0	0.0	0.0	0.0	0.0	0.0	0.0	0.0	0.0	0.0	0.0	0.0	0.0	0.0	0.0	0.0
6	0.0	0.0	0.0	0.0	0.0	**88.4**	3.8	0.0	14.0	4.1	0.0	0.0	0.0	0.0	0.0	0.0	0.0	0.0	0.0	0.0	0.0	0.0	0.0	0.0	0.0	0.0	0.0	0.0	0.0	0.0	0.0
7	0.0	0.0	0.0	0.0	0.0	0.0	**85.8**	0.0	0.0	10.0	0.0	0.0	0.0	0.0	0.0	0.0	0.0	0.0	0.0	0.0	0.0	0.0	0.0	0.0	0.0	0.0	0.0	0.0	0.0	0.0	0.0
8	0.0	4.0	0.0	0.0	0.0	0.0	0.0	**92.4**	0.0	0.0	0.0	0.0	0.0	0.0	0.0	0.0	0.0	0.0	0.0	0.0	0.0	0.0	0.0	0.0	0.0	0.0	0.0	0.0	0.0	0.0	0.0
9	0.0	0.0	0.0	0.0	0.0	11.6	0.0	0.0	**82.1**	4.1	0.0	0.0	0.0	0.0	0.0	0.0	0.0	0.0	0.0	0.0	0.0	0.0	0.0	0.0	0.0	0.0	0.0	0.0	0.0	0.0	0.0
10	0.0	0.0	0.0	0.0	0.0	0.0	7.3	0.0	3.9	**81.9**	0.0	0.0	0.0	0.0	0.0	0.0	0.0	0.0	0.0	0.0	0.0	0.0	0.0	0.0	0.0	0.0	0.0	0.0	0.0	0.0	0.0
11	7.4	0.0	0.0	0.0	0.0	0.0	0.0	0.0	0.0	0.0	**93.2**	0.0	0.0	0.0	0.0	0.0	0.0	0.0	0.0	0.0	0.0	0.0	0.0	0.0	0.0	0.0	0.0	0.0	0.0	0.0	0.0
12	0.0	0.0	0.0	0.0	0.0	0.0	0.0	0.0	0.0	0.0	0.0	**100.0**	0.0	0.0	0.0	0.0	0.0	0.0	0.0	0.0	0.0	0.0	0.0	0.0	0.0	0.0	0.0	0.0	0.0	0.0	0.0
13	0.0	0.0	0.0	0.0	0.0	0.0	0.0	0.0	0.0	0.0	0.0	0.0	**100.0**	0.0	0.0	0.0	0.0	0.0	0.0	0.0	0.0	0.0	0.0	0.0	0.0	0.0	0.0	0.0	0.0	0.0	0.0
14	0.0	0.0	0.0	11.1	0.0	0.0	0.0	0.0	0.0	0.0	0.0	0.0	0.0	**94.9**	0.0	0.0	0.0	0.0	0.0	0.0	0.0	0.0	0.0	0.0	0.0	0.0	0.0	10.9	0.0	0.0	0.0
15	0.0	0.0	8.6	4.2	0.0	0.0	3.1	0.0	0.0	0.0	0.0	0.0	0.0	5.1	**89.3**	4.2	4.1	0.0	0.0	0.0	3.3	3.2	0.0	6.1	4.1	0.0	3.6	0.0	0.0	0.0	0.0
16	0.0	0.0	0.0	6.3	0.0	0.0	0.0	0.0	0.0	0.0	0.0	0.0	0.0	0.0	0.0	**77.4**	15.5	0.0	0.0	0.0	0.0	0.0	0.0	0.0	0.0	0.0	0.0	0.0	0.0	0.0	0.0
17	0.0	0.0	0.0	0.0	0.0	0.0	0.0	0.0	0.0	0.0	0.0	0.0	0.0	0.0	0.0	15.2	**80.4**	0.0	0.0	0.0	0.0	0.0	0.0	0.0	0.0	0.0	0.0	0.0	0.0	0.0	0.0
18	0.0	0.0	0.0	0.0	0.0	0.0	0.0	0.0	0.0	0.0	0.0	0.0	0.0	0.0	0.0	0.0	0.0	**90.3**	0.0	0.0	0.0	19.4	0.0	0.0	0.0	0.0	0.0	0.0	0.0	0.0	0.0
19	0.0	0.0	0.0	0.0	0.0	0.0	0.0	0.0	0.0	0.0	0.0	0.0	0.0	0.0	0.0	0.0	0.0	0.0	**71.6**	14.8	0.0	0.0	9.5	0.0	0.0	0.0	0.0	0.0	0.0	0.0	0.0
20	0.0	0.0	0.0	0.0	0.0	0.0	0.0	0.0	0.0	0.0	0.0	0.0	0.0	0.0	10.7	0.0	0.0	0.0	17.7	**41.3**	0.0	0.0	28.3	0.0	0.0	0.0	0.0	0.0	0.0	0.0	0.0
21	0.0	0.0	0.0	0.0	0.0	0.0	0.0	0.0	0.0	0.0	0.0	0.0	0.0	0.0	0.0	0.0	0.0	0.0	0.0	0.0	**96.7**	0.0	0.0	0.0	0.0	0.0	0.0	0.0	0.0	0.0	0.0
22	0.0	0.0	0.0	0.0	0.0	0.[illegible]	0.0	0.0	0.0	0.0	0.0	0.0	0.0	0.0	0.0	0.0	0.0	9.7	0.0	0.0	0.0	**77.4**	0.0	0.0	0.0	0.0	0.0	0.0	0.0	0.0	0.0
23	0.0	0.0	0.0	0.0	0.0	0.[illegible]	0.0	0.0	0.0	0.0	0.0	0.0	0.0	0.0	0.0	0.0	0.0	0.0	10.7	43.9	0.0	0.0	**62.2**	0.0	0.0	0.0	0.0	0.0	0.0	0.0	0.0
24	0.0	0.0	0.0	0.0	0.0	0.[illegible]	0.0	0.0	0.0	0.0	0.0	0.0	0.0	0.0	0.0	0.0	0.0	0.0	0.0	0.0	0.0	0.0	0.0	**93.9**	0.0	0.0	0.0	0.0	0.0	0.0	0.0
25	0.0	0.0	0.0	0.0	0.0	0.[illegible]	0.0	0.0	0.0	0.0	0.0	0.0	0.0	0.0	0.0	0.0	0.0	0.0	0.0	0.0	0.0	0.0	0.0	0.0	**95.9**	0.0	0.0	0.0	0.0	0.0	0.0
26	0.0	0.0	0.0	0.0	0.0	0.[illegible]	0.0	0.0	0.0	0.0	0.0	0.0	0.0	0.0	0.0	0.0	0.0	0.0	0.0	0.0	0.0	0.0	0.0	0.0	0.0	**100.0**	0.0	0.0	0.0	0.0	0.0
27	0.0	0.0	0.0	0.0	0.0	0.[illegible]	0.0	4.4	0.0	0.0	0.0	0.0	0.0	0.0	0.0	0.0	0.0	0.0	0.0	0.0	0.0	0.0	0.0	0.0	0.0	0.0	**96.4**	0.0	0.0	0.0	0.0
28	0.0	0.0	0.0	0.0	0.0	0.[illegible]	0.0	0.0	0.0	0.0	0.0	0.0	0.0	0.0	0.0	0.0	0.0	0.0	0.0	0.0	0.0	0.0	0.0	0.0	0.0	0.0	0.0	**89.1**	0.0	0.0	0.0
29	0.0	0.0	0.0	0.0	0.0	0.[illegible]	0.0	0.0	0.0	0.0	0.0	0.0	0.0	0.0	0.0	0.0	0.0	0.0	0.0	0.0	0.0	0.0	0.0	0.0	0.0	0.0	0.0	0.0	**100.0**	0.0	0.0
30	0.0	0.0	0.0	0.0	0.0	0.[illegible]	0.0	0.0	0.0	0.0	0.0	0.0	0.0	0.0	0.0	0.0	0.0	0.0	0.0	0.0	0.0	0.0	0.0	0.0	0.0	0.0	0.0	0.0	0.0	**100.0**	0.0
31	0.0	0.0	0.0	0.0	0.0	0.[illegible]	0.0	0.0	0.0	0.0	0.0	0.0	0.0	0.0	0.0	0.0	0.0	0.0	0.0	0.0	0.0	0.0	0.0	0.0	0.0	0.0	0.0	0.0	0.0	0.0	**100.0**

FIGURE 9.13 SFA confusion mat⁻ix with 3% risk level adjustment

TABLE 9.5
Ambiguity Groups with and without 3% Risk Adjustment

Fault ID	Fault Description	Ambiguity Groups	Ambiguity Groups at 3% Risk Level
1	Fan performance	1, 8, 11, 15, 19, 25, 27	1, 11
2	LPC performance	2, 5, 6, 8, 9, 10, 14, 15, 19, 20, 23, 27, 29	2, 8
3	HPC performance	3, 7, 15, 19, 20, 24, 27, 29	3, 15
4	HPT performance	4, 14, 15, 16, 17, 18, 20, 23	4, 14, 15, 16
5	LPT performance	2, 5, 6, 8, 11, 12, 15, 20, 23, 27	5
6	7th stage service bleed	2, 5, 6, 7, 8, 9, 10, 12, 14, 15, 18, 19, 20, 23, 27	6, 9
7	10th stage service bleed	6, 7, 9, 10, 12, 14, 15, 19, 20, 23, 24, 29	6, 7, 10, 15
8	2.5 bleed	1, 2, 5, 6, 8, 9, 10, 11, 14, 15, 20, 23, 25, 27	2, 8, 27
9	7th stage stability bleed	2, 6, 7, 8, 9, 10, 12, 15, 18, 20, 25	6, 9, 10
10	10th stage stability bleed	5, 6, 7, 8, 9, 10, 14, 15, 19, 20, 23, 24, 29	6, 7, 9, 10
11	Fan duct leakage	1, 2, 5, 8, 11, 14, 15, 19, 20, 25, 27	1, 11
12	Fan duct pressure loss	6, 7, 12, 13, 14, 15, 18, 19, 20, 23, 25, 27, 28, 31	12
13	Combustor pressure loss	12, 13, 14, 15, 18, 20, 23, 24, 25, 27, 28	13
14	Turbine cooling air (TCA) valve fault	4, 12, 13, 14, 15, 18, 2, 23, 28	14, 15
15	ECS bleed switch	3, 7, 13, 14, 15, 17, 18, 19, 20, 24, 25, 27, 29	15, 20
16	High turbine case cooling failed closed	4, 14, 15, 16, 17, 19, 20, 23, 24, 25, 27	4, 15, 16, 17
17	Low turbine case cooling failed closed	4, 7, 14, 15, 16, 17, 20, 23, 24, 25	15, 16, 17
18	HPC Variable Stator vane (VSV)	6, 11, 12, 14, 15, 18, 19, 20, 22, 25, 27	18, 22
19	Mach error	6, 11, 12, 14, 15, 18, 19, 23, 27	19, 20, 23
20	Altitude error	14, 15, 19, 20, 23, 25, 27	19, 20, 23
21	N1 error	5, 11, 14, 15, 19, 20, 21, 23, 25, 27, 29	15, 21
22	N2 error	14, 15, 18, 19, 20, 22, 23, 25, 27	15, 18, 22
23	Wf error	6, 7, 8, 12, 14, 15, 19, 20, 23, 25, 27	19, 20, 23
24	EGT error	3, 5, 6, 7, 9, 14, 15, 18, 19, 20, 23, 24, 25, 27, 29	15, 24
25	T25 error	14, 15, 18, 19, 20, 23, 24, 25, 27, 29	15, 25
26	T3 error	3, 12, 14, 15, 18, 19, 20, 23, 25, 26, 27, 29	26
27	P25 error	5, 8, 14, 15, 18, 19, 20, 23, 25, 27	15, 27
28	P3 error	11, 12, 13, 14, 15, 19, 20, 23, 25, 27, 28	14, 28
29	TAT or T2 error	3, 12, 14, 15, 18, 19, 20, 25, 26, 27, 29, 30	29
30	P2 error	1, 7, 11, 12, 14, 15, 18, 19, 20, 23, 24, 25, 29, 30, 31	30
31	P49 error	11, 12, 13, 14, 15, 18, 19, 20, 23, 28, 29, 30, 31	31

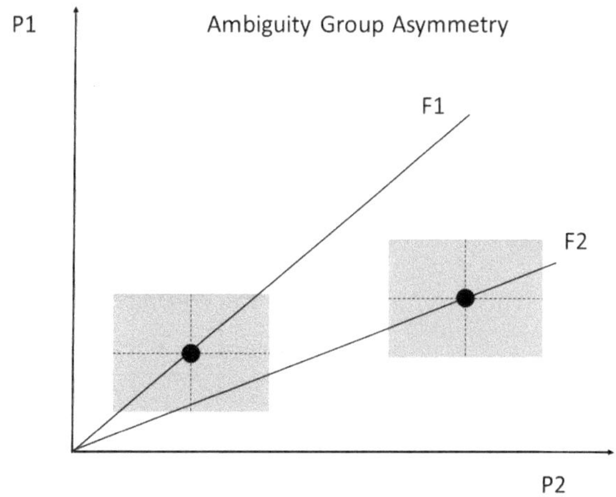

FIGURE 9.14 Ambiguity asymmetry in two dimensions

numbers. Additionally, the physics of the fault modes and the uncertainty in sensor measurements also contribute to the issue.[14] As a result, we can have faults that are represented at different (%) level magnitudes. Figure 9.14 illustrates the mechanism responsible for the asymmetry.

Figure 9.14 depicts two faults (F_1 and F_2) as seen by two measurements (sensors, P_1 and P_2). The red dots represent a particular *magnitude* of the faults, as seen by noise-free sensors, and the shaded areas the potential effect of measurement non-repeatability in P_1 and P_2. As depicted, F_1 is ambiguous with F_2 (as the shaded area intersects the F_2 line), but F_2 is not ambiguous with F_1. However, it is clear that, at a lower magnitude, F_2 will slide downward and ultimately intersect the F_1 line and become ambiguous with F_1, in which case there would be symmetry. The property of symmetry is, therefore, not guaranteed and depends on realistic fault magnitudes considered for the application.

9.2.4 AG Matrix

As a prerequisite for our discussion on ambiguity resolution, we remind the reader of a matrix that we defined earlier in this chapter that relates the systems fault *AG*s to the individual faults. We referred to this matrix as the *AGM* and provided its definition in Equation (9.2). If there are a total number of n faults to be considered, there are n *AG*s, and the *AGM* would take the form of an $n \times n$ matrix with the individual faults along the rows and the *AG*s along the columns.

As an example, Figure 9.15 provides a rendering of the *AGM* for our example above, where $n = 31$ with a risk adjustment of 3%.

This matrix provides, at a glance, the ambiguity structure that we would like to rectify (i.e., resolve) during a typical SFA. If the matrix were the *identity*, we would have the best situation possible, with no ambiguity and, hence, no confusion

	AG 1	AG 2	AG 3	AG 4	AG 5	AG 6	AG 7	AG 8	AG 9	AG 10	AG 11	AG 12	AG 13	AG 14	AG 15	AG 16	AG 17	AG 18	AG 19	AG 20	AG 21	AG 22	AG 23	AG 24	AG 25	AG 26	AG 27	AG 28	AG 29	AG 30	AG 31
Fault1	1	0	0	0	0	0	0	0	0	0	1	0	0	0	0	0	0	0	0	0	0	0	0	0	0	0	0	0	0	0	0
Fault2	0	1	0	0	0	0	0	1	0	0	0	0	0	0	0	0	0	0	0	0	0	0	0	0	0	0	0	0	0	0	0
Fault3	0	0	1	0	0	0	0	0	0	0	0	0	0	0	0	0	0	0	0	0	0	0	0	0	0	0	0	0	0	0	0
Fault4	0	0	0	1	0	0	0	0	0	0	0	0	0	0	0	1	0	0	0	0	0	0	0	0	0	0	0	0	0	0	0
Fault5	0	0	0	0	1	0	0	0	0	0	0	0	0	0	0	0	0	0	0	0	0	0	0	0	0	0	0	0	0	0	0
Fault6	0	0	0	0	0	1	1	0	1	1	0	0	0	0	0	0	0	0	0	0	0	0	0	0	0	0	0	0	0	0	0
Fault7	0	0	0	0	0	0	1	0	0	1	0	0	0	0	0	0	0	0	0	0	0	0	0	0	0	0	0	0	0	0	0
Fault8	0	1	0	0	0	0	0	1	0	0	0	0	0	0	0	0	0	0	0	0	0	0	0	0	0	0	0	0	0	0	0
Fault9	0	0	0	0	0	1	0	0	1	1	0	0	0	0	0	0	0	0	0	0	0	0	0	0	0	0	0	0	0	0	0
Fault10	0	0	0	0	0	0	1	0	1	1	0	0	0	0	0	0	0	0	0	0	0	0	0	0	0	0	0	0	0	0	0
Fault11	1	0	0	0	0	0	0	0	0	0	1	0	0	0	0	0	0	0	0	0	0	0	0	0	0	0	0	0	0	0	0
Fault12	0	0	0	0	0	0	0	0	0	0	0	1	0	0	0	0	0	0	0	0	0	0	0	0	0	0	0	0	0	0	0
Fault13	0	0	0	0	0	0	0	0	0	0	0	0	1	0	0	0	0	0	0	0	0	0	0	0	0	0	0	0	0	0	0
Fault14	0	0	0	1	0	0	0	0	0	0	0	0	0	1	0	0	0	0	0	0	0	0	0	0	0	0	0	1	0	0	0
Fault15	0	0	1	1	0	0	1	0	0	0	0	0	0	1	1	1	1	0	0	0	1	1	0	1	1	0	1	0	0	0	0
Fault16	0	0	0	1	0	0	0	0	0	0	0	0	0	0	0	1	1	0	0	0	0	0	0	0	0	0	0	0	0	0	0
Fault17	0	0	0	0	0	0	0	0	0	0	0	0	0	0	0	1	1	0	0	0	0	0	0	0	0	0	0	0	0	0	0
Fault18	0	0	0	0	0	0	0	0	0	0	0	0	0	0	0	0	0	1	0	0	0	1	0	0	0	0	0	0	0	0	0
Fault19	0	0	0	0	0	0	0	0	0	0	0	0	0	0	0	0	0	0	1	1	0	0	1	0	0	0	0	0	0	0	0
Fault20	0	0	0	0	0	0	0	0	0	0	0	0	0	0	1	0	0	0	1	1	0	0	1	0	0	0	0	0	0	0	0
Fault21	0	0	0	0	0	0	0	0	0	0	0	0	0	0	0	0	0	0	0	0	1	0	0	0	0	0	0	0	0	0	0
Fault22	0	0	0	0	0	0	0	0	0	0	0	0	0	0	0	0	0	1	0	0	0	1	0	0	0	0	0	0	0	0	0
Fault23	0	0	0	0	0	0	0	0	0	0	0	0	0	0	0	0	0	0	1	1	0	0	1	0	0	0	0	0	0	0	0
Fault24	0	0	0	0	0	0	0	0	0	0	0	0	0	0	0	0	0	0	0	0	0	0	0	1	0	0	0	0	0	0	0
Fault25	0	0	0	0	0	0	0	0	0	0	0	0	0	0	0	0	0	0	0	0	0	0	0	0	1	0	0	0	0	0	0
Fault26	0	0	0	0	0	0	0	0	0	0	0	0	0	0	0	0	0	0	0	0	0	0	0	0	0	1	0	0	0	0	0
Fault27	0	0	0	0	0	0	0	1	0	0	0	0	0	0	0	0	0	0	0	0	0	0	0	0	0	0	1	0	0	0	0
Fault28	0	0	0	0	0	0	0	0	0	0	0	0	0	0	0	0	0	0	0	0	0	0	0	0	0	0	0	1	0	0	0
Fault29	0	0	0	0	0	0	0	0	0	0	0	0	0	0	0	0	0	0	0	0	0	0	0	0	0	0	0	0	1	0	0
Fault30	0	0	0	0	0	0	0	0	0	0	0	0	0	0	0	0	0	0	0	0	0	0	0	0	0	0	0	0	0	1	0
Fault31	0	0	0	0	0	0	0	0	0	0	0	0	0	0	0	0	0	0	0	0	0	0	0	0	0	0	0	0	0	0	1

FIGURE 9.15 Sample ambiguity group matrix (3% risk)

to resolve. Because the likelihood of that occurring is virtually zero, we move now to the topic of ambiguity resolution.

9.3 RESOLVING AMBIGUITY

As illustrated above, traditional fault isolation methods often utilize an SFA approach that primarily relies on gas path performance parameters. While effective for many performance related faults, such approaches often suffer from ambiguity. In these cases, the ambiguity often has to be resolved by experienced analysts using additional information that takes many different forms, such as various non-gas path symptoms, FADEC fault codes, comparisons with the companion engine, maintenance records, and quite often, the analyst's gas turbine domain knowledge. In the following section we will introduce an intelligent reasoner that combines the strength of an optimal, physics-based SFA and a fuzzy expert system that mimics the analytical process of human experts for ambiguity resolution. As a generic reasoning framework, this approach can be applied not only to traditional time series data but also to full flight data analytics. The essence of this approach is information fusion. Therefore, this chapter is a continuation and application of the information fusion topic presented in the previous chapter.

9.3.1 Practicality Considerations

In order for a reasoning construct to be practical, it needs to meet a couple of basic requirements. First, it needs to be automated, i.e., a human expert is not needed for the *initial* diagnostic assessment. Diagnostic reasoning is triggered automatically by alerts (which are also detected by automated algorithms). The first step for fault diagnosis involves the execution of an SFA that produces a *first choice* AG. The next step is to collect and combine additional evidence for ambiguity resolution (if needed), and this also needs to be automated. This leads to the second requirement that the reasoning architecture has to be configurable or table-driven. The collection of additional evidence through the *analytic engines* (as shown in Figure 1.1) shall be configured as a generic table structure, such that no *ad hoc* programming is required to apply the reasoner to a different engine type. The interface between the analytic engine and the intelligent reasoner should be standardized for the same purpose. Last but not least, the presentation of diagnostic results to the EHM analysts must be informative. The reasoning logic should be *explained* to the end users to help them understand the final conclusion and how the assessment was made. Eventually, the diagnostic decision, as well as recommended inspection and maintenance activities, will be determined by the EHM analyst using all information at his or her disposal.

To automate such a diagnostic process, we require a reasoning structure that is capable of emulating the thought process of an experienced analyst, is expandable in its knowledge base, (i.e., table-driven in lieu of hard-coded software), and generic in the sense that the same software can be used across any engine fleet and engine model type. It should, therefore, be able to handle both gas path and non-gas-path data.

From an information fusion perspective, the general design philosophy must meet the following considerations:

1) The *presence* of supporting evidence will *promote* related fault modes.
2) The *absence* of an anticipated symptom will *demote* related fault modes (*negative information*).
3) Unknown information will not affect the reasoning process.

9.3.2 Reasoning Framework

A critical step in the design of such a reasoning system is to map out the connections between fault modes, expected observations/symptoms, and the evidence sources that reveal the symptoms. This can be achieved by constructing *diagnostic matrices* for selected systems/sub-systems, relating *symptoms* (observed or calculated) to underlying *fault modes*, then aggregate them to form an overall map. The SFA faults and associated AGM would constitute one such diagnostic matrix for gas path performance faults, and this is often where the construction of the diagnostic matrices begins. As more evidence is introduced, the diagnostic matrices grow. For example, a subject matter expert points out that certain major module faults (such as fractured HPC or HPT blades) would result in increased vibration. Therefore, a new column for evidence "high spool vibration" can be added, and the associated symptoms for HPC and HPT faults can be set accordingly (for example, a discrete value of 1 for increased vibration level). Others might relate, for example, mechanical system faults to related mechanical system symptoms, e.g., bearing fault to oil temperature, pressure, filter ΔP, ODM[15] count and rate, vibration indicators, gas path debris indicators [2], etc., depending on what is available for the given engine model application. The diagnostic matrices are engine model dependent and are the *tables* alluded to in the foregoing paragraphs. The faults are also engine model dependent, and the *symptom* information is provided by separate specialized *analytic engines* (i.e., specific algorithms). The SFA, for example, is a *specific* (gas path) *analytic engine*. We will get into more concrete examples of the diagnostic matrices in the following paragraphs.

The reasoning (software) element reads the diagnostic matrices and executes the appropriate analytic engines to generate the symptom information (to the degree available) to promote or refute the fault modes represented in the associated diagnostic matrix.

One possible framework for such a reasoning system is shown in Figure 9.16. At the heart of the reasoner is a sparse diagnosis matrix ($\mathfrak{D}$-Matrix) based fuzzy inference system that utilizes additional evidence to resolve SFA ambiguity. The $\mathfrak{D}$-Matrix is a sparse matrix that connects fault modes in the SFA AG to supporting analytic engine modules. It is essentially a two-dimensional layout of a tree structure for ambiguity resolution and is *sparse* because all elements of the matrix are not required for the reasoner to work. Each analytic engine module is implemented for a specific function, but its output format is standardized to provide a *belief* (or *confidence*) for the fault hypothesis being tested (see [3] for more details). The reasoner

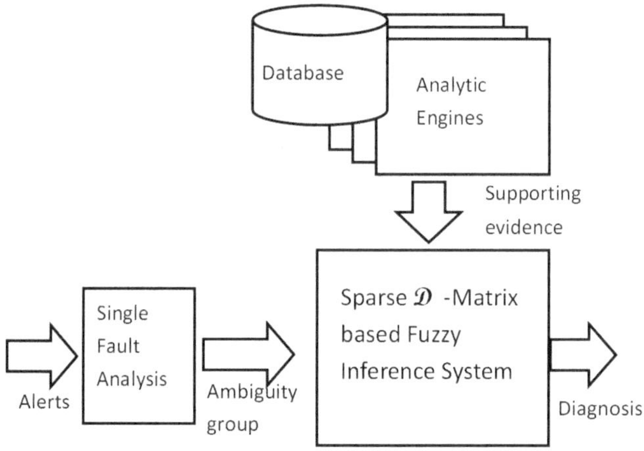

FIGURE 9.16 Reasoning architecture

combines evidence from all sources and makes a final assessment to be reviewed by the EHM analyst.

The reasoning process proceeds in the following manner:

1) Choose first rank *AG* from SFA with associated *confidence*.
2) Identify the fault modes in the chosen *AG*.
3) Locate supporting evidence as dictated by 𝒟-Matrix.
4) Invoke associated *analytical engines* to produce required evidence.
5) Calculate evidence belief (confidence).
6) Aggregate evidence belief.

This reasoning process emulates the thought process of an experienced analyst and is expandable in its knowledge base, (i.e., table-driven in lieu of hard-coded software). The expansion and maintenance of the knowledge base can be easily managed due to the simplicity of the 𝒟-Matrix architecture. It is also generic in the sense that the "core" reasoning software can be used without any code changes when applied to a different engine platform. The key elements of the reasoning process are explained in the following sub-sections.

9.3.3 Knowledgebase

An experienced EHM analyst can resolve diagnostic ambiguity by leveraging information/data that exist in the EHM system but may not easily be utilized by an automated algorithm. The connection of relevant information to the underlying problem, as well as the effectiveness of the reasoning, depends on the domain knowledge and experience of the analyst, which varies from individual to individual. The *AI* embedded in the knowledge base is not only important for the reasoner but also beneficial to analyst training, i.e., a less experienced analyst can be trained by the intelligent reasoner.

The knowledge base can be constructed as a 𝔇-Matrix with the following steps:

1. Incorporate the SFA *AGM*.
2. Examine the fault modes in each AG and apply domain knowledge to distinguish between them. It is often beneficial to leverage the fault isolation rules from other engine models. However, the applicability is restricted by the availability of relevant data, which may vary between engine model applications.
3. Collect EHM expert experience, which can often be effectively formulated as fuzzy expert rules.
4. Validate with field data and improve as needed.

The structure of the 𝔇-Matrix is illustrated in Figure 9.17, in which the *rows* are *fault modes* and *columns* are *evidence*. The values of the matrix elements represent the expected symptoms for the faults (rows) evaluated by the analytic engines associated with the evidence (columns). In general, the 𝔇-Matrix consists of *several* independent *diagnostic matrices* that relate *evidence* (i.e., observed or calculated health indicators) to underlying *fault modes*. The *SFA AGM* constitutes one such diagnostic matrix, which is shown at the top left corner of the 𝔇-Matrix, as shown in the Figure 9.17 (i.e., the evidence columns labelled “AG”). The evidence source for these AG columns is the SFA *analytic engine*. This matrix has no empty element (i.e., it is *not* sparse). For each element of additional evidence, a weighting factor, w_i, can be assigned. The default weighting is taken to be unity.

To see how this works, let us suppose for example, that an SFA analysis returns AG1, first ranked AG (e.g., *HPT performance fault*), then there is confusion as to whether it is an *HPT performance fault* (for brevity, listed as Fault 1 in Figure 9.17) or a *TCC fault* (listed as Fault 3 in Figure 9.17), as indicated by the 1s in the AG1 column. The additional columns following all the AGs represent additional evidence (if available) to resolve the confusion. Engine HPT performance faults (e.g., HPT

		SFA Evidence					Additional Evidence				
		AG1	AG2	AG3	•••	AG*n*	Evidence1	Evidence2	Evidence3	•••	EvidenceN
SFA Faults	Fault 1	1	0	0	•••	0		-1			0
	Fault 2	0	1	0	•••	0			0		1
	Fault 3	1	0	1	•••	0		0			
	•		*AGM**								
	•										
	•										
	Fault *n*	0	1	1	•••	1					-1
Other Faults	Fault X1						1				
	Fault X2							1			
	•										
	•										
	•										
	Fault Xn								-1		

FIGURE 9.17 𝔇-Matrix structure

blade fracture) share a similar fault signature as TCC faults (e.g., cooling air leakage or actuation system failure). Knowing that TCC is not activated at T/O for a particular engine type, the analyst would check engine T/O performance (e.g., T/O *EGT* margin). If takeoff *EGT* margin has also dropped, it's more likely to be HPT performance related. Otherwise, it would more likely be TCC related. To capture this reasoning process, a column *Evidence2* is added in Figure 9.17 with a "-1" assigned to the element intersecting with Fault 1 (indicating that an HPT performance fault would cause T/O *EGT* margin to drop) and a "0" assigned to Fault 3 (indicating that a TCC fault would have no effect on T/O *EGT* margin). In a real implementation, the column title *Evidence2* would be made more descriptive, e.g., *T/O EGT Margin Shift*, where a value of -1 indicates a *drop* and a value of 0 indicates *no change*. The analyst may also check (if available) vibration parameters, FADEC fault words, additional TCC system parameters, recent maintenance records related to TCC system, etc. This type of domain knowledge, experience, and thought process can all be captured as additional columns of the $\mathfrak{D}$-Matrix.

It should be pointed out that the knowledge base (i.e., the $\mathfrak{D}$-Matrix) is not only applicable to just gas path performance faults. It can be generalized to handle any fault mode. These *diagnostic matrices* can certainly be expanded to relate mechanical system faults to the expected symptoms as mentioned earlier. These matrices constitute the lower right portion of the $\mathfrak{D}$-Matrix, as illustrated in Figure 9.17.

The $\mathfrak{D}$-Matrix is intended to be *sparse* (with the exception of SFA *AGM* portion) for several reasons. First, knowledge of faulty engine behavior may be incomplete or uncertain. As the knowledge base grows, the size of the $\mathfrak{D}$-Matrix grows. It inevitably becomes infeasible to fill in expected values for all the elements to maintain complete connectivity between all fault modes and evidence sources. Meanwhile, improper assignment of expected symptoms may hurt the final reasoning output. Second, certain fault mode/evidence connectivity may not be applicable. For example, engine mechanical system faults may not affect engine performance parameters. Therefore, the intersection of "Other Faults" and "SFA evidence" is left blank in Figure 9.17. *A blank in the matrix represents that the evidence (column) provides no supportive or contradictive evidence.* Analytic engines are fired only if non-blank entries are indicated.

9.3.4 Analytic Engines

Referring to Figure 1.1, the *analytic engines* consist of *representations* of models, calculations, and other computational devices that provide quantitative evidence used to promote or refute the identification of faults, as dictated by the $\mathfrak{D}$-Matrix. This can be simplistically captured as *fuzzy* values (-1, 0, +1) denoting *decrease*, *no change*, and *increase*, and a fuzzy inference system can be used to interface the *analytic engines* with the $\mathfrak{D}$-Matrix. As a generality, this interface would use fuzzy membership functions[16] to transform the real-valued quantity, calculated from a given *analytic engine,* to a (-1, 0, +1) fuzzy value along with a certain belief level, β . The SFA is a specific analytic engine that returns a ranked list of AGs along with an associated confidence $\xi_{SFA}^{(AG1)}$ for the first ranked AG (see Addendum 9.1). If this first ranked AG is *not* a singleton, then there is opportunity for ambiguity resolution through the employment of additional evidence in accordance with the 𝔇-Matrix.

The employment of additional evidence will be applied to each fault element of the top ranked AG and will return a level of confidence to be associated with the evidence presented. For each fault (in the AG), a weighted average of confidences amongst the supporting evidence (symptoms) is calculated, as illustrated in Equation (9.12).

$$\beta = \frac{\sum_{evidence} w_i f_i \beta_i}{\sum_{evidence} w_i f_i} \tag{9.12}$$

where f_i is a binary (*evidence*) flag that is set to 1 if the data supporting the ith piece of evidence is present, and 0 otherwise. This is introduced to *discount* the confidence associated with the reasoning when evidence data is missing. The *fault* with the highest confidence, β, is selected. To illustrate the principle, let us return to our familiar example of an *HPT performance fault* that is ambiguous with a *TCC failed closed fault*. An *abbreviated* 𝔇-Matrix might look something like Figure 9.18 (faults have been reordered in the 𝔇-Matrix for simplicity of illustration).

In this example, if the SFA returns AG4, a first ranked AG (*HPT performance fault*), then there is confusion as to whether it is an *HPT performance fault* (serious) or a *TCC fault* (not so serious). The last two columns represent additional evidence (if available) to resolve the confusion. *T/O (takeoff) EGT margin* (calculated by an analytic engine) can be used to determine whether the margin changed. The sample 𝔇-Matrix contends that if the fault were an *HPT performance* problem, then margin should decrease (-1 value), whereas no change in margin (0 value) would support *TCC failure* as the underlying fault. Similarly, an *HPT* fault would be supported by increased N2 vibration (1 value). A *blank* in the matrix indicates that the (column) symptom provides no supportive or contradictive evidence. *Analytic engines are fired only if non-blank entries are indicated.* So, for the example at hand, *EGT* margin and N2 vibration analytic engines would be invoked, and the symptom evidence would be evaluated to see if they support or refute the faults. The additional evidence *weights* provide the relative importance of the evidence[17] (symptom) in resolving the

	SFA Evidence						***Additional Evidence***		
	AG1	AG2	AG3	AG4	AG5	AG6		T/O EGT Margin	N2 vibration
							weight	2	1
Fan Performance	1	0	0	0	0	0			
LPC Performance	0	1	0	0	0	0			
HPC Peformance	0	0	1	0	0	0			
HPT Performance	0	0	0	1	0	1		-1	1
LPT Performance	0	0	0	0	1	0			
TCC failed closed	0	0	0	1	0	1		0	

FIGURE 9.18 Example 𝔇-Matrix (abbreviated)

TABLE 9.6
Example Analytic Engine Assessment

T/O EGT Margin		N2 Vibration	
Value	β	Value	β
-1	0.6	1	0.8
0	0.4	0	0.2
1	0		

AG. These will be used as weights (w_i) to calculate the overall confidence assigned to the evidence, as per Equation (9.12). To continue with our example, let us assume that we have the following hypothetical assessment from our two analytic engines. as indicated in Table 9.6

For *HPT performance*, we have a computed confidence

$$\beta = \frac{\sum_{evidence} w_i f_i \beta_i}{\sum_{evidence} w_i} = \frac{w_1 f_1 \beta_1 + w_2 f_2 \beta_2}{w_1 f_1 + w_2 f_2} \tag{9.13}$$

$$= \frac{(2 \cdot 1 \cdot 0.6) + (1 \cdot 1 \cdot 0.8)}{2+1} = \frac{1.2 + 0.8}{3} = 0.67$$

For *TCC failure*, we have a computed confidence

$$\beta = \frac{\sum_{evidence} w_i f_i \beta_i}{\sum_{evidence} w_i} = \frac{w_1 f_1 \beta_1}{w_1 f_1} = \frac{(2 \cdot 1 \cdot 0.4)}{2} = 0.4 \tag{9.14}$$

In this instance, we would conclude that the fault is more likely an HPT performance fault.

On the other hand, if the N2 vibration assessment had been reversed, as indicated in Table 9.7.

then the calculation for HPT performance would be a bit different

TABLE 9.7
Example Analytic Engine Assessment Revision

N2 Vibration	
Value	β
1	0.2
0	0.7

$$\beta = \frac{\sum_{evidence} w_i f_i \beta_i}{\sum_{evidence} w_i} = \frac{w_1 f_1 \beta_1 + w_2 f_2 \beta_2}{w_1 f_1 + w_2 f_2}$$
$$= \frac{(2 \cdot 1 \cdot 0.6) + (1 \cdot 1 \cdot 0.2)}{2+1} = \frac{1.2 + 0.2}{3} = 0.47 \tag{9.15}$$

in which case, it would not be as clear an assessment for HPT performance over TCC failure.

The 𝔇-Matrix is built incrementally as more expert rules and supporting evidence sources are added to the knowledge base. We will use another example to illustrate the growth of the 𝔇-Matrix and the analytic engines that leverage other types of supporting information to resolve ambiguity. In this example, we will look into the resolution of AG8 (LPC performance, 2.5 bleed, P25 error), as shown in Figure 9.19. To resolve the ambiguity among these three faults, one may start with the simplest rule that a P25 sensor fault affects only the P25 parameter, i.e., a "single parameter shift." On the other hand, both LPC performance loss and stuck 2.5 bleed will cause multiple parameter shifts in a similar way. However, an interview with experienced field analysts reveals that a stuck 2.5 bleed will significantly reduce T/O EGT margin, while a typical LPC performance fault will only reduce T/O EGT margin mildly. In addition, if 2.5 bleed commanded position is available, the discrepancy between commanded 2.5 bleed actuator position command and the feedback could be used to confirm a 2.5 bleed fault. These rules can be added to the 𝔇-Matrix, as shown in Figure 9.19. Three analytic engines are added to resolve AG8 ambiguity, namely *T/O EGT margin sharp drop*, *actuator track check error* (comparing commanded and feedback values), and *single sensor shift*. It should be noted that the analytic engine *T/O EGT margin sharp drop* has a certain level of functional redundancy with the analytic engine *T/O EGT margin* used in the previous example but is designed to highlight a large drop. The analytic engine *T/O EGT margin* can certainly be leveraged to resolve AG8 ambiguity but is not necessary.

To continue with our example, let us assume that commanded 2.5 bleed position is not available to perform the *actuator track check* analytic engine, and we have the

	SFA Evidence							*Additional Evidence*					
	AG1	AG2	AG3	AG4	...	AG8	...		T/O EGT Margin	N2 vibration	T/O EGT Margin Sharp Drop	Actuator track check error	Single sensor shift
								weight	2	1	1	1	1
Fan Performance	1	0	0	0		0							
LPC Performance	0	1	0	0		1					0	0	0
HPC Peformance	0	0	1	0		0							
HPT Performance	0	0	0	1		0			-1	1			
LPT Performance	0	0	0	0		0							
...													
2.5 Bleed	0	0	0	0		1					1	1	0
...													
TCC failed closed	0	0	0	1		0			0				
P25 error	0	0	0	0		1					0	0	1

FIGURE 9.19 Example 𝔇-Matrix (case 2, abbreviated)

TABLE 9.8
Revised Example Analytic Engine Assessment

T/O EGT Margin Sharp Drop		Single Sensor Shift	
Value	β	Value	β
0	0.05	0	0.8
1	0.9	1	0.1

following assessment from the other two analytic engines (saying that the T/O EGT margin has reduced significantly, and several sensors have shifted), as indicated in Table 9.8.

For *LPC performance*, we have a computed confidence

$$\beta = \frac{\sum_{evidence} w_i f_i \beta_i}{\sum_{evidence} w_i f_i} = \frac{(1\cdot 1\cdot 0.05)+(1\cdot 0)+(1\cdot 1\cdot 0.8)}{1+0+1} = \frac{0.05+0.8}{2} = 0.425 \quad (9.16)$$

For *2.5 bleed*, we have a computed confidence

$$\beta = \frac{\sum_{evidence} w_i f_i \beta_i}{\sum_{evidence} w_i f_i} = \frac{(1\cdot 1\cdot 0.9)+(1\cdot 0)+(1\cdot 1\cdot 0.8)}{1+0+1} = \frac{0.9+0.8}{2} = 0.85 \quad (9.17)$$

For *P25 error*, we have a computed confidence

$$\beta = \frac{\sum_{evidence} w_i f_i \beta_i}{\sum_{evidence} w_i f_i} = \frac{(1\cdot 1\cdot 0.05)+(1\cdot 0)+(1\cdot 1\cdot 0.1)}{1+0+1} = \frac{0.05+0.1}{2} = 0.075 \quad (9.18)$$

In this instance, we would conclude that the fault is more likely a 2.5 bleed fault.

9.4 SUMMARY

In this chapter we provided a brief introduction to the problem of fault (identification) ambiguity, which pervades all forms of diagnostics and, in particular, GPA. As we have seen, the essence of the AG is that it comprises a collection of failure modes for which the diagnostic procedure in play can detect and isolate the fault to that collection, and no further, i.e., it cannot further isolate to any sub-set of that collection.[18] Thus, the notion of ambiguity is intrinsically connected to the diagnostic methodology being employed, and/or the measurement suit being used; change the diagnostic approach, and/or the measurement suite, and you potentially change the AG structure.

We introduced the notions of *discrete* AGs and *continuous* AGs, concentrating on the latter, which applies naturally to the GPA, as we have described it in this book. The former has its place as well, particularly in early design studies, before accurate models are developed, and sensors selected for the application at hand. Indeed, they may appear in the form of analytic engines supplying additional information to support ambiguity resolution. Although the treatment here on ambiguity might seem a bit scant, it is an area that has not been fully exploited [4–7] and certainly deserves further research and development, in particular, regarding methodologies to resolve ambiguities and provide greater clarity to fault identification.

NOTES

1. The ambiguity group may be a singleton, in which case, the isolation has been achieved.
2. This table is for illustration only and not intended for medical diagnosis.
3. A universe of an infinite number of faults would have little practical utility in real-world applications.
4. The "observations" could be a *fuzzification* of a quantitative sensor value.
5. This is true for all *discrete* ambiguity groups.
6. See Addendum 9.2.
7. See Addendum 9.2.
8. This can be easily automated.
9. Uniqueness depends on absolute accuracy and noise-free conditions.
10. Noise generated using the measurement covariance matrix R levels.
11. This is in stark contrast to our experience with *discrete* AGs.
12. We would have to accept an unreasonable risk level $\alpha > 30\%$ to produce all singleton AGs.
13. Unlike *discrete* AGs, *continuous* AGs do *not* form an equivalence relationship.
14. The ICs forming the relationship between faults and measured quantities are real-valued as are the measurements themselves. The SFA process responsible for the fault isolation returns fault estimates that are also real-valued.
15. Oil debris monitoring.
16. See Addendum A5.6.
17. This is determined by the cognizant *EHM* specialist.
18. Without additional information.

REFERENCES

1. MIL-STD-1309D, 1992, *Definitions of Terms for Testing, Measurement, and Diagnostics*, Standard by Military Specifications and Standards.
2. Fisher, C., 2001, Data and Information Fusion for Gas Path Debris Monitoring, *IEEE Aerospace Conference*, Bug Sky, MT.
3. Tang, L., Volponi, A., 2018, *Intelligent Reasoning for Gas Turbine Fault Isolation and Ambiguity Resolution,* ASME paper GT2018-75540.
4. Rajamoney, S., DeJong, G., 1987, *Active Ambiguity Reduction: An Experiment Design Approach to Tractable Qualitative Reasoning,* University of Illinois at Urbana-Champaign Technical Report, UILU-ENG-87-2225.
5. Geanta, I., 2017, *Contribution to a Modelling Framework of Integrated Vehicle Health Management: A Generic Health Management Module Supporting the Integration of Diagnostics and Prognostics,* Université de Lorraine, NNT:2014LORR0212.

6. De Kleer, J., Williams, B., 1987, Diagnosing Multiple Faults, *Artificial Intelligence* 32, 97–130.
7. Doel, D.L., 1990, The Role for Expert Systems in Commercial Gas Turbine Engine Monitoring, *ASME 1990 International Gas Turbine and Aero-Engine Congress and Exposition*, 90-GT-374, (pp. V005T15A014-V005T15A014).
8. Donat, W., et al., 2007, Data Visualization, Data Reduction and Classifier Fusion for Intelligent Fault Detection and Diagnosis in Gas Turbine Engines, *Proceedings of the 2007, ASME Turbo Expo*, Montreal, Canada, paper GT2007-28343.

10 Prognostics and Future Trends

10.1 OVERVIEW

This chapter will touch upon the emerging field of prognostics and how GPA can contribute to achieving some of its goals. In support of this, we will briefly discuss other developing areas, such as machine learning, the role of big data, and the so-called *intelligent engine*. All of these, in some way, contribute to and extend the methods discussed in the preceding chapters. Along with information fusion, they provide a glimpse into the future trends for *performance* health determination, extension, and utilization.

10.2 PROGNOSTICS

Prognostics is the ability to predict the future condition of a component and/or system of components. For the purposes of gas turbine engine prognostics, this definition is often further described in terms of hard failures of components or condition/degradation of performance related problems. Failure prognostics is focused on the prediction of damage state or failure rate of a component or system of components in an engine. Failure prognostics is usually affected by the diagnosis of specific engine faults, depending on the level of impact the component experiences from the fault condition. Prognostic models are required to project to the future condition. Prognostics can also be associated with the slower degradation (wear-related) processes that an engine is exposed to throughout its life. It is usually associated with the diagnosis of fault conditions and the capability of predicting when the symptoms of the identified fault(s) will reach an undesirable state, in which system operation will be adversely affected. Prognostic models are required to project the future "path" of these identified fault(s) on total system performance or reliability.

In Chapter 2, we briefly introduced the prognostics concept and defined the basic terminologies as well as some commonly used metrics to quantify the performance of prognostic algorithms. In this chapter, we will provide two examples in the context of GPA, namely *EGT* margin prediction and water wash prediction, and illustrate some commonly used performance metrics. Because these metrics are often used to compare the performance of prognostic algorithms, we will also provide, as examples, descriptions of simple prognostic algorithms, and particularly the handling of prognostic uncertainty. It should be noted that these algorithms are simplified for illustration purposes, and the performance metrics are only presented to facilitate the understanding of the concept. It's not the intention of the authors to

DOI: 10.1201/9781003649274-10

promote any algorithm over others, as the optimal approach really depends on the application at hand.

10.2.1 Practical Considerations for Prognostic System Development

Regardless of the methodologies used, several important factors have to be considered at the early design phase [1]:

- The level of condition degradation beyond which operation is undesirable; this is the operational limit as reflected in the prognostic parameter, for example, a jet engine's T/O *EGT* margin limit at 0 degree, or a life-limited part's cycle count at 0 cycle. The limit can be set at a higher level to allow maintenance preparation.
- How much lead time is needed to respond without affecting operations; this is related to the level of degradation mentioned above – the idea being to minimize operational interruption and enable maintenance schedule optimization. Lead time is often determined based on estimates of deterioration rates and related maintenance schedules.
- Understanding of the uncertainties inherent in the prognostic process and their impact on maintenance decisions, and quantification of uncertainty measures.
- The prognostics performance metrics that will be used as success criteria.

These considerations are important elements of the system requirements that need to be specified during the system design stage.

10.2.2 Methodologies

Literature abounds on the topic of prognostics methods for gas turbines [2–7]. Prognostics approaches have been classified in different ways. For example, a classification could be based on the types of models utilized, i.e., model-based approach, data-driven approach, or hybrid approach. One could argue that a data-driven approach could utilize some type of modeling, such as statistical or regression models; therefore, the categorization should be based on the use of system knowledge (or physics), resulting in the categorization of physics-model based approach, or purely data-driven approach. Likewise, based on the form of the algorithms, linear approach, non-linear approach, time series approach, state estimation approaches, machine learning (ML)/AI approach have all been discussed in the literature. In this book, we adopt the following basic categories based on the taxonomy of input data and whether a PHM/EHM system is actively providing inputs to the prognosis:

- Reliability-Based Prediction

This approach is purely based on statistics and reliability data. Many would argue it's not prognostics, but the industry tends to consider it a form of prediction, thus prognosis. A human health analogy for this type of prediction is the life expectancy for a healthy newborn. If the statistical life expectancy is 78 years on the average for people in a specific country, then a 20-year-old person would have an estimated remaining life of 58 years – a fair prediction without knowing health history and life habits. A GPA-related example is the prediction of an engine's T/O *EGT* margin. If an engine product line is designed (based on an assumed usage profile) for maintaining positive *EGT* margin for 15,000 cycles, then an engine with 10,000 cycles in service would have an estimated remaining life of 5,000 cycles from an *EGT* margin perspective. This type of prediction is often simple and straightforward yet could still provide useful information for engine MRO planning.

- Usage-Based Prognostics

This approach is based on actual operational (usage) history that is projected forward with an assumed future usage profile. Usage reflects the combination of many factors, including utilization rate, T/O power (full-rated or derated), ambient conditions (e.g., takeoff total air temperature), and environmental factors (pollutant, salt, sand, etc.). Continuing with the human health analogy, the life expectancy for smokers is shorter than non-smokers. Therefore, based on the person's smoking history, the remaining life prediction could be adjusted accordingly. For the *EGT* margin prediction example, if an engine is deployed to operate in harsh environments, the *EGT* margin deterioration rate can be adjusted based on the projected usage profile to reflect the anticipated, more rapid deterioration that is likely to occur.

- Condition-Based Prognostics

This approach assumes there is a PHM/EHM system actively monitoring the health of the system, which is true for most modern gas turbine engines. Based on the EHM information, the prognostics can be adjusted and adapted to provide more accurate prognosis than the other two approaches. That said, typical commercial EHM systems are mostly performance focused and have very limited coverage of structural health monitoring. Therefore, the first two approaches are still widely used in practice. Condition-based prognostics can be further divided into two sub-categories: deterioration-driven and anomaly-triggered. The former is associated with normal deterioration of the engine, while the latter is triggered by an abrupt fault. Most GPA prognostic applications are deterioration-driven because abrupt faults are usually addressed in timely manner. Anomaly-triggered prognosis has been studied and developed for mechanical and structural faults, such as the prediction of crack growth after a crack has been detected [8].

10.2.3 Prognosis of T/O *EGT* Margin

T/O *EGT* margin is a critical engine health indicator that is often used to plan engine removals. It is calculated based on a data point captured at T/O and normalized to a particular (typically hot day) condition for trending purposes. Generally speaking,

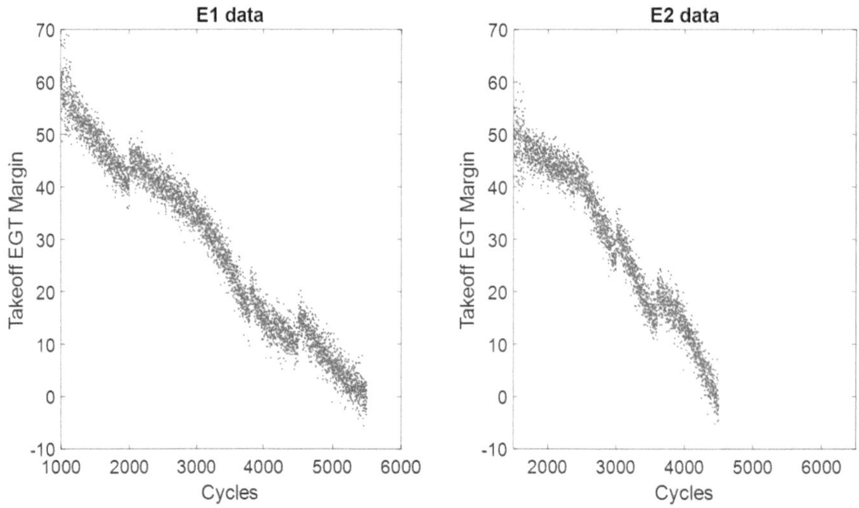

FIGURE 10.1 Simulated takeoff EGT margin data

as an engine's T/O *EGT* margin approaches zero, the engine will be scheduled for removal and overhaul. Predicting when an engine's T/O *EGT* margin will reach zero is an example of prognosis in GPA applications. In this case, the end of life (EoL) of an engine is defined at zero margin.

Simulated T/O *EGT* margin data for two engines (E1 and E2) are illustrated in Figure 10.1. The discontinuity in the data (or recovery of EGT margin) is due to engine water wash. Prognostics is started when T/O *EGT* margin is about 35 *deg C* (i.e., E1 at 3,000 cycles, E2 at 2,720 cycles). The data for E2 is simulating an engine operating in a harsh environment; therefore, it deteriorates faster than E1.

Two prognostic methods are applied to the *EGT* margin prediction problem, including 1) a first-order (linear) regression model and 2) a state variable model based method. The two prognostic methods are briefly described below.

1) Linear Regression

This method captures the linear trend (i.e., slope) of the historical data and extrapolates it forward in time to make a prediction. The extrapolation ends when the projected value reaches the threshold (zero in this case). The linear regression algorithm fits a line, $y = \beta_0 + \beta_1 x + \varepsilon$, to the data available at the moment of prediction. Here, x (engine cycles) is called the independent variable or predictor variable, and y (T/O *EGT* margin) is called the dependent variable or response variable, ε is a zero mean normal error term, $\varepsilon \in N(0, \sigma^2)$. The optimal estimates for β_0 and β_1 can be found by solving the following optimization problem:

$$\min_{\beta_0, \beta_1} \sum_{i=1}^{n} [y_i - (\beta_0 + \beta_1 x)]^2 \tag{10.1}$$

The LSQ solution is:

$$\hat{\beta}_1 = \frac{\sum_{i=1}^{n} x_i y_i - \frac{1}{n}\sum_{i=1}^{n} x_i \sum_{i=1}^{n} y_i}{\sum_{i=1}^{n} x_i^2 - \frac{1}{n}(\sum_{i=1}^{n} x_i)^2}$$
$$\hat{\beta}_0 = \bar{y} - \hat{\beta}_1 \bar{x} \tag{10.2}$$
$$\hat{\sigma}^2 = \frac{\sum_{i=1}^{n} (y_i - \hat{y}_i)^2}{n-2}$$

Where $\bar{y}$ and $\bar{x}$ are the mean of y and x, respectively. When projecting the linear trend into the future (represented by increased value x'), the standard error for prediction is described by the following equation [9]:

$$s_{\hat{y}} = \hat{\sigma}\sqrt{\frac{1}{n} + \frac{(x' - \bar{x})^2}{\sum_{i=1}^{n} (x_i - \bar{x})^2} + 1} \tag{10.3}$$

The standard error can be used to construct the confidence intervals for the prediction. In this implementation, the linear fit is based on the latest 2,000 data points to capture recent trend.

2) State Variable Model Based Method

State variable (or dynamic) models allow a number of systematic, well-developed tools to be applied for prediction in the presence of noisy measurements, future mission/load profiles, and modeling uncertainties. The key elements of a dynamic model-based prognostic problem are illustrated in Figure 10.2. At each time step when measured data is received, the data is fused with model prediction to derive a posterior state estimate based on measurement likelihood. Then, a prediction is performed to project the state distribution to future time steps. This type of recursive computation of the state's probability density based on a measurement update step, combined with a prediction step, is known as the *filtering* problem. The prediction step can be repeated at several points in time until a critical threshold is reached. This is how the filtering techniques are extended to make a long-term prediction (prognosis).

Now that the problem is formulated under the filtering framework, many of the existing filtering techniques, such as the KF, EKF, UKF, and PF (as briefly introduced in Chapter 8) can be applied. In this example, the state variable model can be formulated as follows,

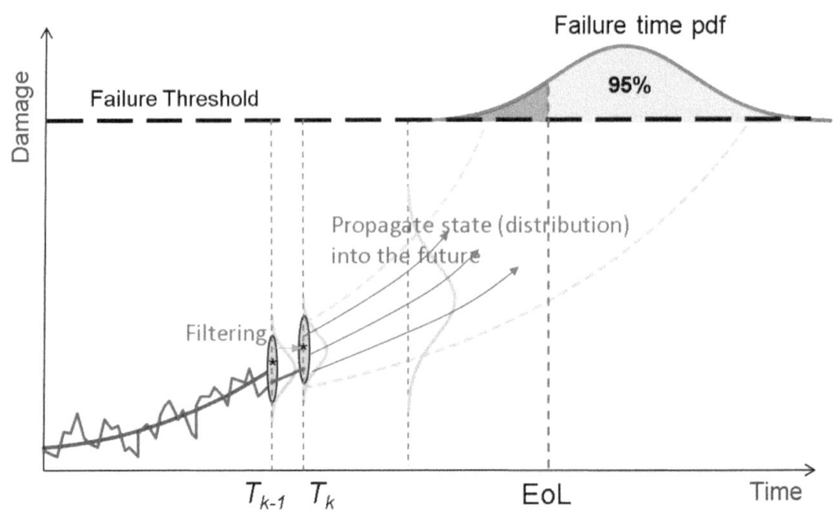

FIGURE 10.2 Concept of state variable model based prognostics

$$
\begin{aligned}
x(k+1) &= x(k) + u(k) + \omega \\
u(k) &= f(k, p(k)) \\
y(k) &= x(k) + v \\
\omega &\in N(0, Q^2) \\
v &\in N(0, R^2)
\end{aligned}
\tag{10.4}
$$

where the state x is the T/O *EGT* margin to be estimated and predicted, the output y is the "observed" T/O *EGT* margin, the input u is the margin deterioration rate at cycles k, which is modeled as a function of engine cycles k and environmental factor p, e.g., different deterioration rates can be applied to engines operating in benign or severe operating conditions. The function f can often be implemented using a look-up table describing expected deterioration rates for engines operating in different environments at particular engine cycles. ω and v are process and measurement noises, respectively.

At time step k, the predicted state $\hat{x}(k \mid k-1)$ is updated by the standard filtering process using new observation $y(k)$ to derive the posterior estimate $\hat{x}(k \mid k)$. In this case study, we adopt the sequential Monte Carlo method (aka PF) for filtering and prediction. To propagate state estimation uncertainty, sample values (aka particles) are randomly drawn from the $\hat{x}(k \mid k)$distribution with equal weights. Each particle is then propagated forward in time using the state equation (involving future input u and process noise ω) until it reaches the threshold (zero in this case). At each future time step n, the distribution represented by the sample (i.e., the particles) defines the confidence intervals for prediction. The set of time instances for each particle to reach zero can be used to construct an estimate for EoL. Details regarding particle filtering and its applications in prognostics can be found in [10–15].

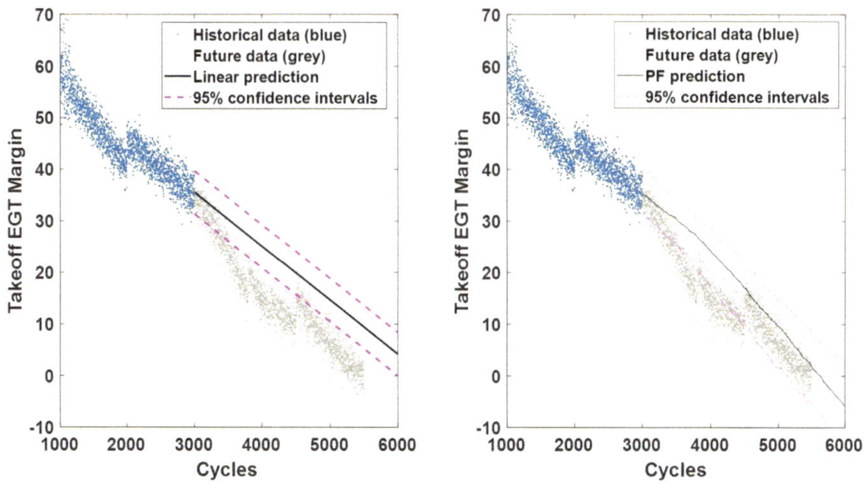

FIGURE 10.3 Takeoff EGT margin prognosis at 3,000 cycles (Engine 1)

The above two algorithms are applied to the two simulated datasets, and the results are shown in the following sections. Figure 10.3 shows the first prediction made for Engine 1 at 3,000 cycles using the two methods (linear regression on the left and Monte Carlo method on the right). Although similar in overall trend, the PF predictions exhibit some non-linearity due to the use of a non-linear deterioration model.

The EoL, RUL, and associated 95% confidence intervals predicted at 3,000 cycles are shown in Table 10.1. The true EoL (5,500 cycles) and RUL (2,500 cycles), which are unknown at the prediction time (3,000 cycles), fall into the confidence intervals of the PF algorithms but outside the confidence intervals of the linear regression algorithm. This is likely due to the water wash effect, which is unknown to the algorithm at the moment of prediction. The 95% confidence is based on the modeling assumption (i.e., the current trend continues). If the assumption fails, the confidence intervals will not reflect actual prediction precision irrespective of the quoted confidence level.

The filtering and prediction process is repeated at each time step until the average T/O *EGT* margin drops below zero at 5,500 cycles; the predicted RUL and ground truth RUL at each time step approaching EoL are shown in Figure 10.4 for Engine 1 data. Although the PF algorithm appears to produce better predictions, the performances of both algorithms are very close in the last 700 cycles, which is sufficient for planning engine removals. It can also be observed that water wash has helped to recover *EGT* margin and kept engines on wing longer. It is assumed that water wash schedule is unknown to the prognosis algorithm in this simulation work; therefore, the effect of a water wash acts as a perturbation to the prediction as shown in the figure.

TABLE 10.1
EoL/RUL Prediction and Confidence Intervals at 3,000 Cycles

Algorithm	EoL Prediction	EoL 95% Confidence Intervals	RUL Prediction	RUL 95% Confidence Interval
Linear	6401	[6002, 6800]	3401	[3002, 3800]
PF	5640	[5140, 6100]	2640	[2140, 3100]

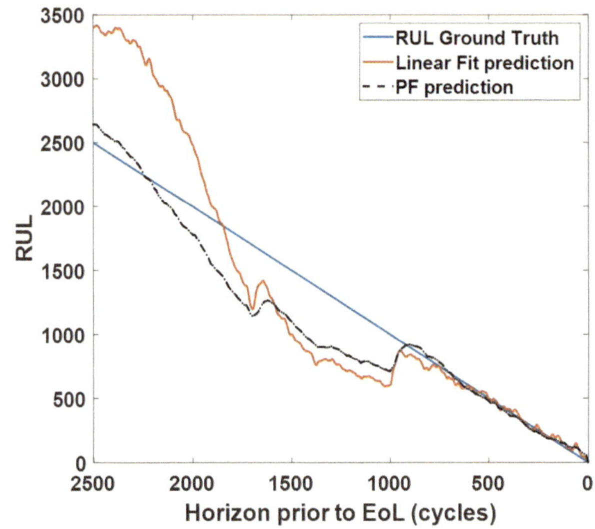

FIGURE 10.4 Estimated RUL vs. horizon prior to EoL (Engine 1)

The same two algorithms are also applied to Engine 2, which exhibits more rapid deterioration. Estimated RUL results at selected cycles are shown in Table 10.2. Here, the average results are merely based on the average RUL prediction for these two engines due to the limited simulations conducted for the purpose of this text. A comprehensive evaluation of algorithm performance should be based on the statistics of a larger fleet in practice. The Table 10.2 results show that both algorithms provided reasonable prediction for the last 700 cycles prior to EoL. The PF method performed better beyond 700 cycles, particularly for Engine 2, which deteriorated more rapidly.

Based on the above results, algorithm performance is evaluated using the classical prognostics metrics of average bias, mean absolute error (MAE), and mean absolute percentage error (MAPE), as shown in Table 10.3. These formulae have been introduced and explained in Chapter 2. The PF method appears to perform better overall

TABLE 10.2
Estimated RUL Results

Horizon (*H*) Prior to EoL (Cycles)	Engine 1		Engine 2		Average	
	Linear Regression	Particle Filter	Linear Regression	Particle Filter	Linear Regression	Particle Filter
1500	841	1101	2132	1100	1487	1101
1400	691	945	2225	1171	1458	1058
1300	688	905	1941	1033	1315	969
1200	650	823	1622	879	1136	851
1100	557	775	1329	752	943	764
1000	524	711	1064	639	794	675
900	871	921	922	566	897	744
800	967	825	1085	650	1026	738
700	860	701	974	622	917	662
600	879	571	932	585	906	578
500	599	481	773	529	686	505
400	604	377	640	440	622	409
300	415	273	492	351	454	312
200	280	186	300	244	290	215
100	144	124	149	141	147	133

based on data from these two engines, but as previously stated, one would have to use a larger fleet for a comprehensive evaluation.

Next, relative accuracy (RA) and cumulative relative accuracy (CRA) are evaluated. RA is intended to quantify prediction error as a percent deviation from the ground truth RUL, while CRA rolls up the RAs calculated at different times to a single quantity by placing weights on the RAs (usually assigning larger weights on data points closer to EoL, as accuracy is often more important as it gets closer to the failure). The parameter λ is a normalized distance indicator showing how close the prediction time is to the EoL. The CRA weighting $w(i)$ increments linearly toward EoL and $\sum w(i) = 1$. Based on the two datasets, the PF algorithm appears to have higher accuracy than the linear regression method, as shown in Table 10.4.

Other performance metrics, such as prognosis horizon and α-λ accuracy, are detailed in the SAE Aerospace Information Report (AIR7999) [19]. We don't intend to cover all the metrics that have been discussed in literature. A lot of effort has been spent on quantifying and managing the uncertainties associated with prognosis in recent years [14], [20]. Nevertheless, it's important to realize that the uncertainty measures (such as confidence intervals) are always based on assumptions and should not be taken as "absolute" measures or "guarantees." Actual performance can only be evaluated with real-world data and be treated statistically given a reasonable fleet size.

TABLE 10.3
Average Bias, MAE, and MAPE Results

Horizon (H) Prior to EoL (Flights)	Average Bias $\frac{1}{L}\sum_{l=1}^{L}\Delta^l(EOL_l - H)$		MAE $\frac{1}{L}\sum_{l=1}^{L}\left|\Delta^l(EoL_l - H)\right|$		MAPE (%) $\frac{1}{L}\sum_{l=1}^{L}\left|\frac{100 * \Delta^l(EoL_l - H)}{r^l_{actual}(EoL_l - H)}\right|$	
	Linear	PF	Linear	PF	Linear	PF
1500	-14	-400	646	400	43	27
1400	58	-342	767	342	55	25
1300	15	-331	627	331	48	26
1200	-64	-349	486	349	41	29
1100	-157	-337	386	337	35	31
1000	-206	-325	270	325	27	33
900	-4	-157	26	178	3	20
800	226	-63	226	88	29	11
700	217	-39	217	40	31	6
600	306	-22	306	22	51	4
500	186	5	186	24	38	5
400	222	9	222	32	56	8
300	154	12	154	39	51	13
200	90	15	90	29	45	15
100	47	33	47	33	47	33

10.2.4 Engine Water Wash Prediction

Compressor washing is known to be a cost-effective way to maintain gas turbine performance. Although one can rarely see it, the air entering jet engines (particularly when closer to the ground) is filled with contaminants, such as fine particles of dirt, soot, exhaust fumes, oil vapors, and salt. Under the high pressure and temperature in the compressors, the contaminants build up gradually on the compressor blades. This process can be rather rapid for engines operating in highly polluted areas.

The build-up of foreign material on compressor blades interferes with the airflow and results in reduced aerodynamic efficiency. A further potential issue is that some contaminants, particularly salts and sulfidation, can corrode certain components in the jet engine, such as the vanes and blades. This leads to even greater disrupted airflow and accelerated degradation of parts. From a gas path performance point of view, increased *EGT* and fuel consumption will be observed.

Engine water wash has a host of benefits, including improved compressor efficiency, recovered EGT margins, lowered fuel consumption, reduced corrosion, and eventually extended engine time-on-wing. How often and when engine water wash should be scheduled depends on the specific fleet and operating conditions. In many cases, GPA can provide insight into compressor performance, which can then be

TABLE 10.4
RA and CRA Results

Metric	t_λ	λ	CRA	Engine 1		Engine 2		Average	
			Weigthing $w(i)$	Linear	PF	Linear	PF	Linear	PF
RA	t_{EoL} - 1500	0	0.00875	0.56	0.73	0.58	0.73	0.57	0.73
RA	t_{EoL} - 1400	0.07	0.01625	0.49	0.67	0.41	0.84	0.45	0.755
RA	t_{EoL} - 1300	0.13	0.025	0.53	0.7	0.51	0.79	0.52	0.745
RA	t_{EoL} - 1200	0.2	0.03375	0.54	0.69	0.65	0.73	0.595	0.71
RA	t_{EoL} - 1100	0.27	0.04125	0.51	0.7	0.79	0.68	0.65	0.69
RA	t_{EoL} - 1000	0.33	0.05	0.52	0.71	0.94	0.64	0.73	0.675
RA	t_{EoL} - 900	0.4	0.05875	0.97	0.98	0.98	0.63	0.975	0.805
RA	t_{EoL} - 800	0.47	0.06625	0.79	0.97	0.64	0.81	0.715	0.89
RA	t_{EoL} - 700	0.53	0.075	0.77	1	0.61	0.89	0.69	0.945
RA	t_{EoL} - 600	0.6	0.08375	0.53	0.95	0.45	0.97	0.49	0.96
RA	t_{EoL} - 500	0.67	0.09125	0.8	0.96	0.45	0.94	0.625	0.95
RA	t_{EoL} - 400	0.73	0.1	0.49	0.94	0.4	0.9	0.445	0.92
RA	t_{EoL} - 300	0.8	0.10875	0.62	0.91	0.36	0.83	0.49	0.87
RA	t_{EoL} - 200	0.87	0.11625	0.6	0.93	0.5	0.78	0.55	0.855
RA	t_{EoL} - 100	0.93	0.125	0.56	0.76	0.51	0.59	0.535	0.675
CRA	---	---	---	**0.63**	**0.88**	**0.55**	**0.79**	**0.59**	**0.84**

used to predict the timing for water wash as well as the expected recovery in terms of EGT margin and fuel consumption [16–1718].

In the example below, simulation data is generated using the C-MAPSS model to emulate an engine with rapid performance deteriorations in HPC and LPC (due to contaminants build-up). The performance deteriorations of major modules are calculated based on data points captured at cruise condition, and a threshold is set on HPC performance deterioration to predict the timing for a water wash. While predicting (and optimizing) water wash timing for an individual engine may not have a significant cost benefit, doing so at the fleet level will provide opportunities for optimizing the fleet's maintenance schedule and engine overhaul planning.

Figure 10.5 shows the delta parameters calculated at cruise condition. It can be observed that delta Wf and delta T5 (*EGT*) have increased about 1.4% and 1.6%, respectively, within this period. Meanwhile, T/O *EGT* margin dropped about 15 *deg C*, as shown in Figure 10.6.

The delta parameters are analyzed using the discrete GPA method described in Chapter 5. Estimates for module performance, assuming coupled module efficiency and flow capacity (or nozzle area), are shown in Figure 10.7. Faster than normal deterioration of HPC can be observed. For prediction purposes, the threshold is set to 1% drop of HPC performance from the initial value. The solid line in the HPC performance sub-plot (at -2.3% performance loss) establishes the EoL target for the prognosis problem.

For illustration purposes, the two methods described in the *EGT* margin example are applied to predict "time to water wash" starting from 3,000 cycles. The prognosis results, as compared with the ground truth, are shown in Figure 10.8. The linear regression method uses the trend (scope) established by the recent 200 cycles to

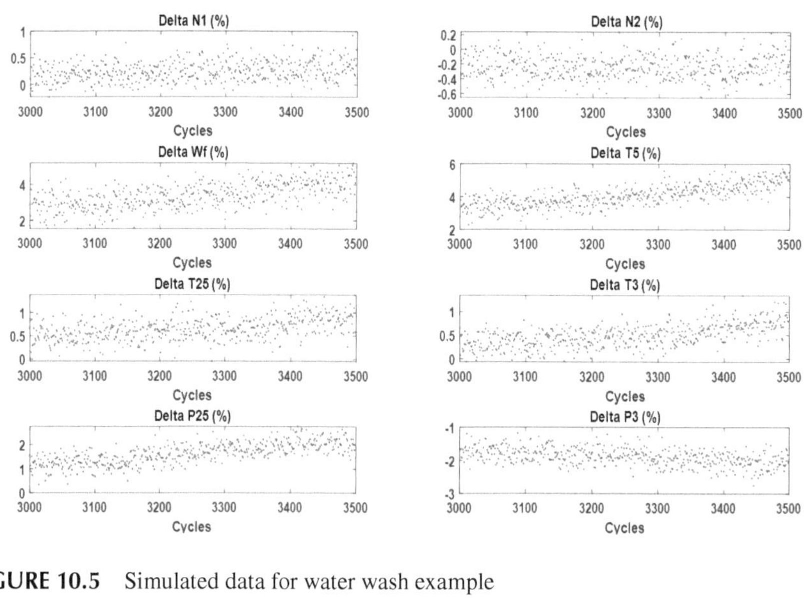

FIGURE 10.5 Simulated data for water wash example

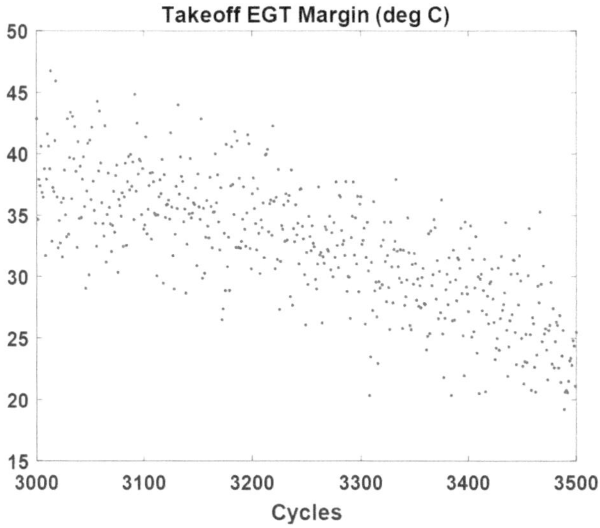

FIGURE 10.6 Simulated takeoff EGT margin data for water wash example

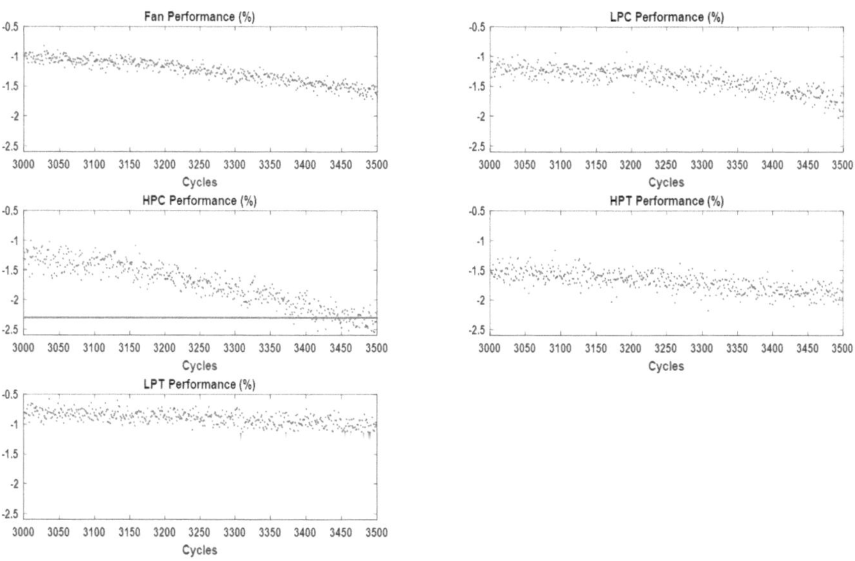

FIGURE 10.7 Module performance estimates from GPA

predict when HPC performance loss will reach -2.3%. Because data collected over a short period of time is used to establish the trend, the algorithm appears to be sensitive to short-term slope fluctuations, which yield less accurate predictions until 200 cycles prior to the conceptual EoL. As a comparison, the PF approach, which

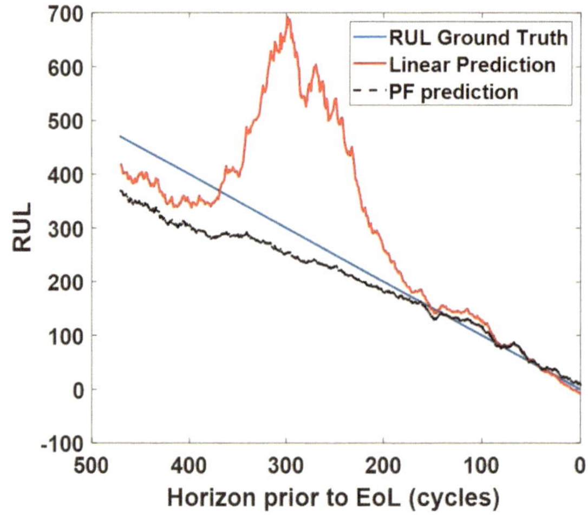

FIGURE 10.8 Estimated time to water wash vs. horizon prior to EoL

has incorporated knowledge of engine deterioration rates to guide the prediction, is able to make accurate predictions 300 cycles prior to the conceptual EoL. Of course, this statement is only based on one dataset. More rigorous validations are needed to evaluate the performance of the algorithms using engine data from a large fleet in order to gain confidence with the prediction algorithms.

Engine performance improvements are often observed after water wash. Most likely, the T/O *EGT* margin will increase, and delta fuel flow will decrease. The magnitude of recovery depends primarily on the amount of efficiency recovery in HPC and LPC. Figure 10.9 below illustrates the recovery of T/O *EGT* margin and fuel consumption (as reflected in delta w_f) after the water wash. Figure 10.10 shows the simulated shifts in module performance parameters accordingly.

10.3 FUTURE TRENDS

10.3.1 Intelligent Engine

The term *intelligent engine* has been in use for decades and has taken on many definitions ranging from smart components, sensors and actuators, to full autonomous engine operation. In our present context, we refer to an engine entity that provides an integration of EHM and smart control as our concept of an *intelligent engine.* Ideally, it would be self-diagnosing and would leverage its health information in the manner in which it controls engine operation. There are certainly many levels below this ideal. Greater onboard EHM capacity could decrease the time latency to detect, identify, and possibly accommodate an incipient fault or deterioration. We have touched on that in our discussions in Chapter 7. It could also be argued, however, that there would not be much (time) gain between an onboard analysis versus merely storing the full flight data onboard and performing the analysis once the aircraft has

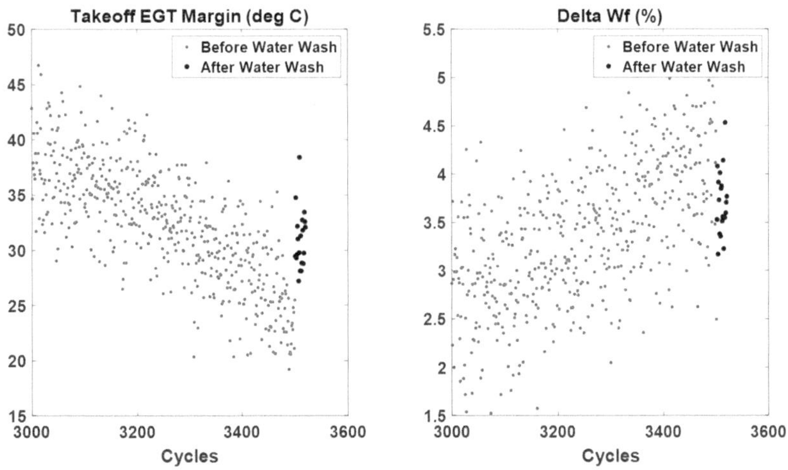

FIGURE 10.9 Simulated performance recovery after water wash

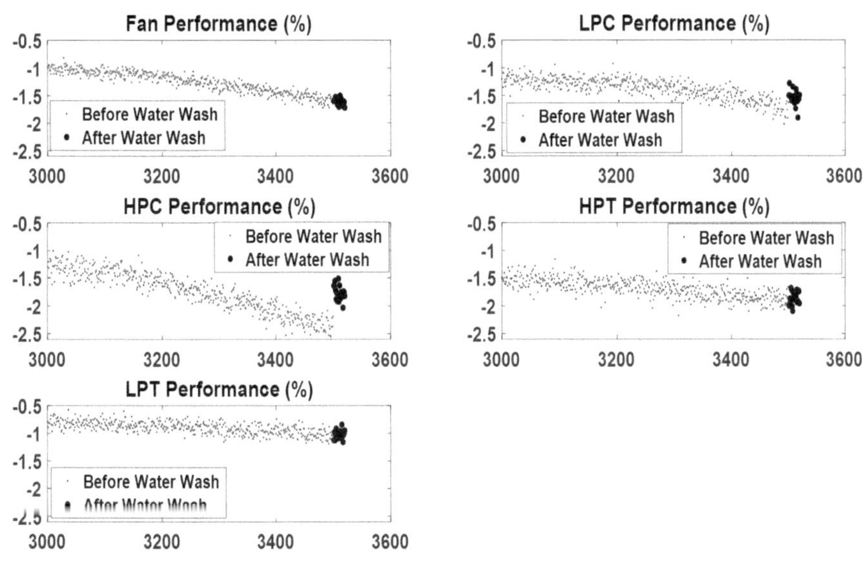

FIGURE 10.10 Simulated module performance recovery after water wash

landed, several hours later (assuming that an efficient data infrastructure was in place to transfer it off-board), unless, the onboard information could be accommodated in some manner *during* flight. This would involve automatic and autonomous accommodation by the FADEC with or without pilot intervention. These concepts have been explored in what many have termed, the *intelligent engine* concept [21–29]. Integrating an onboard PHM system with direct engine control is both an intriguing and a daunting concept. Historically, and generally speaking, engine control laws are formulated for the nominal engine with enough robustness and margin so that the engine can perform adequately and safely (perhaps not as efficiently) as the engine deteriorates over its life cycle. The intriguing question is, would you (and could you) control the engine differently if you had full knowledge of its current health state? What are the consequences in terms of technologies required, system complexity, reliability, and eventual certification implications? It has been suggested that a model predictive control (MPC) [30], could provide the technology to achieve an *adaptive* control strategy that can tradeoff performance goals with remaining useful life on-wing in the presence of known engine deterioration [23].

Accommodation is key ingredient for this definition of engine intelligence. The Research and Technology Organization (RTO) of NATO assembled a Panel Task Group led by NASA, consisting of members from Glenn Research Center (GRC), engine OEMs, instrumentation companies, and Concordia University to study the intelligent engine concept and determine the requisite enabling technologies for such a development. The study culminated in an extensive report in 2009 [22]. The report covers a host of areas, including actively controlled engine components,[1] sensor and actuator requirements, as well as a discussion of the possible relationship between intelligent control and health monitoring. In the latter, the discourse centers around model-based control and gas path monitoring and offers a notional architecture like the one depicted in Figure 10.11.

This architecture may look familiar to the reader, in that it incorporates some of the elements discussed in Chapters 6 and 7, in particular the STORM (or eSTORM) configuration. The *virtual sensors,* alluded to in the diagram, include parameters that cannot be directly measured in flight, such as thrust, turbine inlet temperature, compressor surge margins, and the like. The *fault diagnostics* component could include some of the gas path SFA elements, as discussed in Chapter 7. Whatever ultimate form it might take, there are formidable challenges to overcome to ensure

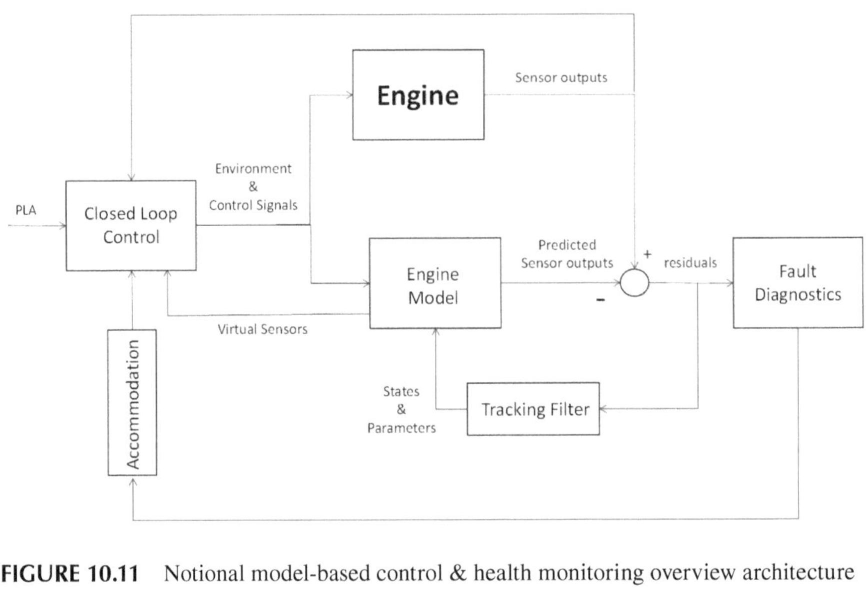

FIGURE 10.11 Notional model-based control & health monitoring overview architecture

success of the intelligent engine concept, not the least of which are reliability, unparalleled accuracy, extensive fault coverage, and flight certification issues. The future work within the GPA arena to improve its accuracy and coverage will contribute positively to these goals. As a prelude, onboard data capture, along with a reliable data infrastructure to offload full flight data, is a necessary step to allow off-board refinement of proposed algorithms, such as those discussed in Chapter 7. Validation of proposed onboard algorithms can then proceed safely and more easily before onboard incorporation is attempted. This is critical, as the act of integrating health monitoring processing with engine control functionality will raise the software standards for EHM to the highest flight critical level. The emergence of low-cost memory in the last decade would certainly allow the capture of flight and gas path parameter data (typically at 10–20Hz). The more challenging task is to build a reliable and economically sustainable data infrastructure that can support fast download of data from the aircraft after each and every flight, anywhere on the planet.

10.3.2 Challenges of Information Fusion

We have seen examples of the use of information fusion throughout the chapters of this book to overcome (partially) the information deficit inherent in the GPA processes. The number of potential faults will likely always significantly outnumber the number of sensors we have at our disposal for supporting EHM. The economic impact of adding sensors to the engine makes it unlikely that expanding the measurement suite is the appropriate solution to the problem.[2] The only effective way to address disparity in the fault to measurement ratio is to utilize information that is already available and unexploited. Simple to say, not so simple to accomplish.

There are several challenges to overcome in the information fusion solution. The most obvious is knowing what information is *available and pertinent* to the particular diagnostic problem at hand. In Chapter 8, we saw that information comes in many forms: sensor measurements, domain knowledge, observations, maintenance records, constraints, assumptions, and so forth. There is no convenient indexed global repository of information to consult. Information comes in many forms and from many sources. The sources can be multi-disciplinary as well. We have seen some examples in which the relevant additional information came from EHM domain knowledge and others in which engine control knowledge was used. Because of this broad diversity in potential useable information, it is key that the diagnostician operate (to some degree) as a generalist and not just a specialist and to engage in team efforts to solve a particular diagnostic problem.

Another hurdle to clear is discovering (or creating) a means to leverage the additional information and combining it with the classical algorithms to address the diagnostic deficiencies. Our discussions in Chapter 9 on ambiguity resolution suggest a framework from which one can evolve a partial solution to this challenge. Even so, there still remains the need to have analytic engines that produce the requisite information. *Qualitative* information, such as maintenance records, observations, assumptions, experiential domain knowledge, negative information, and the like, need to be mathematized in some manner in order to integrate with the other quantitative elements of the diagnostic system to enable autonomous operation. This may entail developing image recognition/interpretation and text parsing with interpretation for certain types of observational and maintenance information. We would envision AI as a core element that could be leveraged to unify the entire process.

10.3.3 Electrified Aircraft Propulsion (EAP)

Electrification of the aircraft presents a promising path toward a more sustainable future for air travel, offering the potential for net zero-emission flight and significantly reduced noise pollution. Technically speaking, electrification in aviation has materialized primarily through two avenues: more electric aircraft (MEA) and electrified aircraft propulsion (EAP). MEA represents an incremental evolution that took place over the past five decades, wherein successive aircraft generations have increasingly incorporated electric power in lieu of conventional mechanical, hydraulic, or pneumatic systems for non-propulsive functions. In contrast, EAP signifies a potentially revolutionary paradigm shift that has emerged in recent years, featuring the use of electric power for both propulsion and auxiliary functions. EAP typically integrates electric motors, energy storage, power electronics, and various forms of aircraft energy generation to enhance overall propulsion efficiency. EAP also allows the decoupling of thrust and power generation and is capable of providing focused thrust during key operational periods, leading to further reduction in fuel consumption and emissions.

As shown in Figure 10.12, various architectures are currently under development to address diverse market sectors, including on-demand mobility, general aviation,

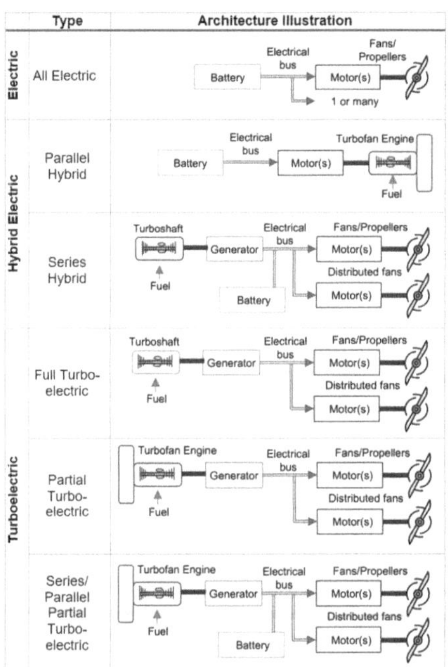

FIGURE 10.12 Representative EAP architectures [32]

and regional/single-aisle categories [31]. With the exception of the all-electric architecture, most systems incorporate gas turbine technology and are also known as hybrid electric propulsion.

The rapid development of technologies in critical component areas emphasizes a crucial consideration: the growing significance of propulsion system health management. This is particularly important given the limited experience with novel electrical systems, the intricacy of interconnected components, and the stringent safety requirements involved [32].

10.4 FUTURE WORK

In summary, we see the following areas deserving further research and development in our pursuit of GPA as a core diagnostic and prognostic methodology within EHM. They are all multi-disciplinary to varying degrees, and no two are mutually exclusive. They all inter-connect in many ways.

Full Flight GPA
The ultimate goal for GPA would certainly have to be the ability to perform the analysis in a reliable and accurate fashion on all data available during flight, i.e., data beyond just steady-state and QSS. The hope would be to expand fault coverage to emerging faults that might be detectable during transients but are not severe enough to be noticed during steady-state or QSS operating regimes. This would further enable the ability of GPA to support prognostic activities.

Closing the Loop with Control
This is a prerequisite to usher in the age of the intelligent engine. This can only occur after reliable and timely engine health assessments can be in place. This is certainly a multi-disciplinary undertaking that has far reaching implications for flight safety, engine operability, and engine health management.

Ambiguity Resolution
Effective diagnostics (and prognostics) relies on accurate health assessments and fault identification. Ambiguity is clearly an obstacle to achieving those goals consistently. The ultimate GPA system should leave no doubt in the mind of the practitioner and, therefore, would produce diagnoses that yield only singleton ambiguity groups. This is critical for autonomous operation and intelligent engine application.

Leveraging Artificial Intelligence and Machine Learning
In this age of *big data*, machine learning has become a methodology in vogue with many practitioners. Nevertheless, there has not been a substantial effort to date to apply machine learning methods to *complement* our existing GPA approach and knowledge base. With increasingly more data and information

being stored in the cloud, the time is ripe for leveraging this information in some manner, not replacing current EHM practices but, rather, to *augment* our knowledge and address the deficiencies of our current diagnostic systems. Machine learning is an *empirical* approach that is in contrast with the (primarily) model-based approach that underlies our current GPA methods. This would then become a *hybrid* methodology, taking advantage of what is known from physics and what can be gleaned from data, present and past. AI provides some machinery for knowledge extraction and possibly fusing what we can learn from data and what we know from physics. Another aspect to AI is the ability to capture qualitative knowledge or *heuristics* in the form of fuzzy expert systems, which, again, provides additional information to leverage.

Information Fusion
At this juncture, it should be clear that the authors believe that information fusion is the key enabler to continued development and evolution of the GPA within any EHM system. By leveraging all available information, continued efforts in this area will reap diagnostic benefits while constraining economic burden.

Extending GPA to Hybrid Electric Propulsion
As described earlier, EAP architectures can take many forms, including hybrid systems that incorporate gas turbine engine technology. GPA principles can be applied to these hybrid electric propulsion systems, in which sophisticated couplings exist between and within turbomachinery and electrical subsystems. Performance changes in one part of the architecture can affect sensed operational performance throughout the system. By leveraging parameter inter-relationships in the coupled system, faulty conditions can be estimated by analyzing shifts in sensed parameters, just as in conventional gas turbines. Simon et al., [33] extended the GPA methodology to diagnose various turbomachinery and electrical system faults in a conceptual EAP system known as the STARC-ABL, which consists of two wing-mounted turbofan engines and an electric motor-driven tail fan propulsor. A Monte Carlo simulation study was conducted to demonstrate the linear estimation accuracy of various estimation approaches. Practical considerations for applying the extended GPA techniques to other EAP designs were also discussed. As more electrified aircraft propulsion systems are developed, demonstrated, and advanced toward certification, we anticipate the development of PHM systems for EAP to occur in parallel. GPA will continue to play an essential role in these future platforms.

NOTES

1. Such as inlet control, noise suppression, compressor surge control, active blade clearance control, combustion control, etc.
2. Not to mention that additional sensors bring additional sensor faults.

REFERENCES

1. SAE, 2020, *Prognostics for Aerospace Propulsion Systems*, SAE Aerospace Information Report, AIR5871A.
2. Roemer, M. J., Byington, C. S., Kacprzynski, G. J., and Vachtsevanos, G., 2006, An Overview of Selected Prognostic Technologies with Application to Engine Health Management. *Proceedings of the ASME Turbo Expo 2006: Power for Land, Sea, and Air*, May 8–11, 2006, Barcelona, Spain.
3. Nguyen, V., et al., 2019, A Review: Prognostics and Health Management in Automotive and Aerospace, *International Journal of Prognostics and Health Management*, 10 (2).
4. Schwabacher, M., and Goebel, K., 2007, A Survey of Artificial Intelligence for Prognostics, *AAAI Fall Symposium*, Technical Report.
5. Hanachi, H., Mechefske, C., Liu, J., Banerjee, A., and Chen, Y., 2018, Performance-Based Gas Turbine Health Monitoring, Diagnostics, and Prognostics: A Survey, *IEEE Transactions on Reliability*, 67 (3): 1340–1363. doi: 10.1109/TR.2018.2822702.
6. Srivastava, A. N., and Han, J., 2012, *Machine Learning and Knowledge Discovery for Engineering Systems and Health Management*, CRC Press, ISBN: 978-1-4398-4178-5.
7. Volponi, A.J., 2014, Gas Turbine Engine Health Management: Past, Present and Future Trends, *Journal of Engineering for Gas Turbines & Power*, 136: 05-1201-1.
8. Patrick R., Orchard, M., Zhang, B., Koelemay, M., Kacprzynski, G., Ferri, A., and Vachtsevanos, G. 2007, An Integrated Approach to Helicopter Planetary Gear Fault Diagnosis and Failure Prognosis, *42nd Annual Systems Readiness Technology Conference*, AUTOTESTCON 2007, September 2007, Baltimore, MD.
9. Rencher A. C., and Schaalje, G. B., 2007, *Linear Models in Statistics*, John Wiley & Sons, Inc.
10. Del Moral, P., 1996, Non Linear Filtering: Interacting Particle Solution, *Markov Processes and Related Fields*, 2 (4): 555–580.
11. Liu, J. S., and Chen, R., 1998, Sequential Monte Carlo Methods for Dynamic Systems, *Journal of the American Statistical Association,* 93 (443): 1032–1044.
12. Orchard, M., Hevia-Koch, P., Zhang, B., Tang, L., 2013, Risk Measures for Particle-filtering-based State-of-Charge Prognosis in Lithium-Ion Batteries, *IEEE Transactions on Industrial Electronics*, 60 (11): 5260–5269.
13. Orchard, M., Tang, L., Saha, B., Goebel, K., and Vachtsevanos, G., 2010, Risk-Sensitive Particle-Filtering-Based Prognosis Framework for Estimation of Remaining Useful Life in Energy Storage Devices, *Studies in Informatics and Control*, 19 (3): 209–218.
14. Tang, L., Orchard, M., Goebel, K., and Vachtsevanos, G., 2011, Novel Metrics and Methodologies for the Verification and Validation of Prognostic Algorithms, *2011 IEEE Aerospace Conference*, March 5–12, 2011, Big Sky, Montana.
15. Tang, L., DeCastro, J., Kacprzynski, G., Goebel, K., and Vachtsevanos, G., 2010, Filtering and Prediction Techniques for Model-Based Prognosis and Uncertainty Management, *IEEE Prognostics & System Health Management Conference 2010*, January 12–14, 2010, Macau.
16. Basendwah, A. A., Pilidis, P., and Li, Y. G., 2006, Turbine Off-Line Water Wash Optimization Approach for Power Generation, *Proceedings of the ASME Turbo Expo 2006: Power for Land, Sea, and Air. Volume 4: Cycle Innovations; Electric Power; Industrial and Cogeneration; Manufacturing Materials and Metallurgy*, May 8–11, 2006. pp. 65–76, Barcelona, Spain.
17. Madsen, S., and Bakken, L. E., 2014, Gas Turbine Operation Offshore: On-Line Compressor Wash Operational Experience, *Proceedings of the ASME Turbo Expo 2014: Turbine Technical Conference and Exposition*, June 16–20, 2014, Düsseldorf, Germany.

18. Syverud, E., and Bakken, L. E., 2007, Online Water Wash Tests of GE J85-13, *ASME. J. Turbomach*, 129 (1): 136–142.
19. SAE, 2020, *Diagnostic and Prognostic Metrics for Aerospace Propulsion Health Management Systems*, SAE Standard AIR7999.
20. Saxena, A., Celaya, J., Saha, B., Saha, S., and Goebel, K., 2009, On Applying the Prognostic Performance Metrics, *Annual Conference of the PHM Society*, 1(1), San Diego, CA, USA.
21. Ring, Dan, 2003, *Control Interaction with Diagnostics and Fault Tolerant Control*, Von Karman Lecture Notes, January 2003.
22. Culley, D, et.al., 2009, *More Intelligent Gas Turbine Engines*, RTO Technical Report, TR-AVT-128.
23. Garg, S, 2004, Controls *and Health Management Technologies for Intelligent Aerospace Propulsion Systems*, NASA/TM—2004-212915.
24. Simon, D.L., et al., 2004, *Sensor Needs for Control and Health Management of Intelligent Aircraft Engines*, NASA/TM—2004-213202.
25. Litt, J.S., et al., 2005, *A Survey of Intelligent Control and Health Management Technologies for Aircraft Propulsion Systems,* NASA/TM—2005-213622.
26. Wiseman, M, 2005, Intelligent *Engine Systems: Work Element 1.2: Malfunction and Operator Error Reduction,* NASA/CR—2005-213964.
27. Fletcher, C.C, 2004, A Vision for the Intelligent Pipeline System and The Role of the Intelligent Engine, *GMRC – 2004 Gas Machinery Conference*, Albuquerque, NM.
28. Turso, J.A., 2008, Toward an Intelligent, Deterioration Accommodating Controller for Aging Turbofan Engines, *The Aeronautical Journal,* 112 (1137): 641–651.
29. Rausch, R., et al., 2004, Towards In-Flight Detection and Accommodation of Faults in Aircraft Engines, *AIAA 2004-6463, AIAA 1st Intelligent Systems Technical Conference*, September 2004, Chicago, IL.
30. Van Essen, H., 1998, *Modeling and Model Based Control of Turbomachinery*, Ph.D. Thesis, Technische Universiteit Eindhoven, Netherlands.
31. SAE, *AIR8678 Architecture Examples for Electrified Propulsion Aircraft*, SAE International, 2022.
32. Tang, L., Saxena, A., and Younsi, K., 2024, Prognostics and Health Management for Electrified Aircraft Propulsion: State of the Art and Challenges, *Proceedings of ASME Turbo Expo 2024 Turbomachinery Technical Conference and Exposition*, June 24–28, 2024, London, United Kingdom.
33. Simon, D.L., Thomas, R., and Dunlap, K.M., 2022, Considerations for the Extension of Gas Path Analysis to Electrified Aircraft Propulsion Systems, *Journal of Engineering for Gas Turbines and Power*, 44(3): 031004. https://doi.org/10.1115/1.4052424.

Addendums

A0 Addendum to Introduction

A0 COMMERCIAL MODULAR AERO-PROPULSION SYSTEM SIMULATION (C-MAPSS)

Throughout this book, examples are given to illustrate the gas path analysis theories and methodologies. No proprietary data from any engine OEM has been used in this book; instead, simulated data from NASA's commercial modular aero-propulsion system simulation (C-MAPSS) have been used in all examples. C-MAPSS is a flexible turbofan engine simulation program running in the Matlab/Simulink™ environment. C-MAPSS comes with two versions, namely the C-MAPSS V2 [2] and the C-MAPSS40K [1], representing high-bypass engines capable of 90,000 lbf thrust and 40,000 lbf thrust, respectively. Both versions provide easy access to health, control, and engine parameters through various convenient means, including graphical user interfaces and command line scripts that allow automated simulations. Typical gas path sensors, such as low and high spool speeds, fuel flow, temperature, and pressure measurements at various stations along the gas path are modeled, and various means to simulate noise and faults are provided. Actuation system faults for variable stator vane, variable bleed valve, and fuel metering valve, as well as major module performance deteriorations, can also be simulated in various ways [3], [4]. Most of the simulations in this book are based on C-MAPSS40K (with the exception of Chapter 3, which is based on C-MAPSS V2). Interested readers can refer to the following papers for more details. Data generated from C-MAPSS are not subject to US export restrictions and have been used worldwide in research studies.

REFERENCES

1. May, R., Csank, J., Litt, J., and Guo T.H., 2010, *Commercial Modular Aero-Propulsion System Simulation 40k (C-MAPSS40k) User's Guide*, NASA TM-2010-216831, September 2010.
2. Liu, Y., Frederick, D., DeCastro, J., Litt., J., and Chan., W., 2012, *User's Guide for the Commercial Modular Aero-Propulsion System Simulation (C-MAPSS) Version 2*, NASA TM-2012-217432, March 2012.
3. May, R.D., Csank, J., Lavelle, T.M., Litt, J.S., and Guo, T.H., 2010, A High-Fidelity Simulation of a Generic Commercial Aircraft Engine and Controller, AIAA-2010-6630, *46th AIAA Joint Propulsion Conference & Exhibit, Nashville*, TN, July 25–28, 2010. NASA TM-2010-216810, October 2010.
4. Csank, J., May, R. D., Litt, J.S., and Guo, T, 2010, Control Design for a Generic Commercial Aircraft Engine, AIAA-2010-6629, *46th AIAA Joint Propulsion Conference & Exhibit*, Nashville, TN, July 25–28, 2010, NASA TM-2010-216811, October 2010.

A2 Chapter 2 Addendums

A2.1 A USEFUL MANIPULATION

As it will be seen throughout this book, it is convenient, for a variety of reasons, to express relationships and to perform calculations on unit-normalized parameters, i.e., where the units of measurement for a parameter have been removed, in particular, for parameter changes. This has frequently been accomplished by expressing terms in the form of %, or more specifically, % of point.

For the most part, physical relationships can be expressed in terms of rational functions, i.e., sums, products, quotients, powers, and their combinations. When presented with such expressions, we will routinely employ a simple manipulation of *taking logs and differentiating.* We will illustrate with a few examples.

- Product

$$\begin{aligned} &Z = XY \\ &\Rightarrow \quad \ln(Z) = \ln(XY) = \ln(X) + \ln(Y) \\ &\Rightarrow \quad \frac{dZ}{Z} = \frac{dX}{X} + \frac{dY}{Y} \end{aligned} \tag{A2.1.1}$$

The form $dZ/Z$is a (unit-less) fractional change in Z. Multiply by 100% and you have % of point change. Thus, if a parameter is expressed as a *product* of two other parameters, its % of point change is the *sum* of the % of point changes in the other two parameters.

- Quotient

$$\begin{aligned} &Z = X / Y \\ &\Rightarrow \quad \ln(Z) = \ln(X / Y) = \ln(X) - \ln(Y) \\ &\Rightarrow \quad \frac{dZ}{Z} = \frac{dX}{X} - \frac{dY}{Y} \end{aligned} \tag{A2.1.2}$$

- Power

$$\begin{aligned} &Z = X^{\alpha} \\ &\Rightarrow \quad \ln(Z) = \ln(X^{\alpha}) = \alpha \ln(X) \\ &\Rightarrow \quad \frac{dZ}{Z} = \alpha \frac{dX}{X} \end{aligned} \tag{A2.1.3}$$

- Sum

$$
\begin{aligned}
&Z = X + Y \\
&\Rightarrow \quad \ln(Z) = \ln(X+Y) \\
&\Rightarrow \quad \frac{dZ}{Z} = \frac{d(X+Y)}{X+Y} = \frac{dX}{X+Y} + \frac{dY}{X+Y} \\
&\Rightarrow \quad \frac{dZ}{Z} = \left(\frac{X}{X+Y}\right)\frac{dX}{X} + \left(\frac{Y}{X+Y}\right)\frac{dY}{Y}
\end{aligned}
\tag{A2.1.4}
$$

Using these simple constructs, more sophisticated expressions can be evaluated. For example, if Z were the product of two corrected quantities, say X_c and Y_c, then we could evaluate its % of point change (almost from inspection) as follows:

$$
\begin{aligned}
&Z = X_c\, Y_c = \frac{X}{\theta^{\alpha_X}\delta^{\beta_X}}\,\frac{Y}{\theta^{\alpha_Y}\delta^{\beta_Y}} \quad \Rightarrow \\
&\frac{dZ}{Z} = \left[\frac{dX}{X} - \left(\alpha_X\frac{d\theta}{\theta} + \beta_X\frac{d\delta}{\delta}\right)\right] + \left[\frac{dY}{Y} - \left(\alpha_Y\frac{d\theta}{\theta} + \beta_Y\frac{d\delta}{\delta}\right)\right] \quad \Rightarrow \\
&\frac{dZ}{Z} = \left(\frac{dX}{X} + \frac{dY}{Y}\right) - (\alpha_X + \alpha_Y)\left(\frac{d\theta}{\theta}\right) - (\beta_X + \beta_Y)\left(\frac{d\delta}{\delta}\right)
\end{aligned}
\tag{A2.1.5}
$$

A2.2 EXPONENTIAL AVERAGES

Given a time series of parameter X, i.e., $\{X_i\}$, we define its exponential average $A_k^{(X)}$ at discrete time k by the following recursion:

$$
A_k^{(X)} = \alpha A_{k-1}^{(X)} + (1-\alpha)X_k \quad with \quad A_0^{(X)} = X_1 \tag{A2.2.1}
$$

where $0 < \alpha < 1$ is a weighting factor. If we expand the recursion given in Equation (A2.2.1) we have

$$
\begin{aligned}
A_1^{(X)} &= \alpha A_0^{(X)} + (1-\alpha)X_1 = \alpha X_1 + (1-\alpha)X_1 = X_1 \\
A_2^{(X)} &= \alpha A_1^{(X)} + (1-\alpha)X_2 = \alpha X_1 + (1-\alpha)X_2 \\
A_3^{(X)} &= \alpha^2 X_1 + \alpha(1-\alpha)X_2 + (1-\alpha)X_3 \\
&\vdots \\
A_k^{(X)} &= \alpha^{k-1}X_1 + \alpha^{k-2}(1-\alpha)X_2 + \cdots + \alpha^0(1-\alpha)X_k
\end{aligned}
\tag{A2.2.2}
$$

The last expression in Equation (A2.2.2) provides a closed form expression for the recursion definition. If we now consider the % of point deltas, $\{\Delta X_k\}$ given by

$$
\Delta X_k = 100\left(\frac{X_k - X_{NOM}}{X_{NOM}}\right) \tag{A2.2.3}
$$

the exponential average (using Equation A2.2.2 as a guide) can be written as

$$\begin{aligned} A_k^{(\Delta X)} &= \alpha A_{k-1}^{(\Delta X)} + (1-\alpha)\Delta X_k \\ &= \alpha^{k-1}\Delta X_1 + \alpha^{k-2}(1-\alpha)\Delta X_2 + \cdots + \alpha^0(1-\alpha)\Delta X_k \end{aligned} \tag{A2.2.4}$$

Combining equations (A2.2.3) and (A2.2.4) yields

$$\frac{A_k^{(\Delta X)}}{100} = \alpha^{k-1}\left(\frac{X_1 - X_{NOM}}{X_{NOM}}\right) + \alpha^{k-2}(1-\alpha)\left(\frac{X_2 - X_{NOM}}{X_{NOM}}\right) + \cdots + \alpha^0(1-\alpha)\left(\frac{X_k - X_{NOM}}{X_{NOM}}\right) \tag{A2.2.5}$$

Some rearrangement of this last expression provides

$$\begin{aligned} \left(\frac{X_{NOM}}{100}\right) A_k^{(\Delta X)} &= A_k^{(X)} - \left(\alpha^{k-1} + \alpha^{k-2}(1-\alpha) + \alpha^{k-3}(1-\alpha) + \cdots + \alpha(1-\alpha) + (1-\alpha)\right) X_{NOM} \\ &= A_k^{(X)} - \left(\left[\alpha^{k-1} + \alpha^{k-2} + \cdots + \alpha + 1\right] - \left[\alpha^{k-1} + \alpha^{k-2} + \cdots + \alpha\right]\right) X_{NOM} \\ &= A_k^{(X)} - X_{NOM} \end{aligned} \tag{A2.2.6}$$

Thus, we have

$$A_k^{(\Delta X)} = 100\left(\frac{A_k^{(X)} - X_{NOM}}{X_{NOM}}\right) \tag{A2.2.7}$$

Equation (A2.2.7) indicates that the exponential average of the deltas is the % of point delta of the exponential average of the $\{X_i\}$ from the nominal level, as required.

As a final note, we have shown (Equation 2.11) that the variance of a rolling average of n deltas is given by

$$Var(\Delta\bar{X}) = Var\left(\frac{1}{n}\sum_{k=1}^{n}\Delta X_k\right) = \frac{1}{n^2}\sum_{k=1}^{n} Var(\Delta X_k) = \frac{1}{n^2}\left(n\sigma_{\Delta X}^2\right) = \frac{\sigma_{\Delta X}^2}{n} \tag{A2.2.8}$$

Using Equation (A2.2.4), the equivalent statement for an exponential average can be easily derived and is given by

$$Var(A_n^{(\Delta X)}) = \left[\alpha^{2(n-1)} + (1-\alpha)^2\left(\frac{1-\alpha^{2(n-1)}}{1-\alpha^2}\right)\right]\sigma_{\Delta X}^2 \tag{A2.2.9}$$

Unlike the variance of a rolling average, where $\lim_{n\to\infty} Var(\Delta\bar{X}) = 0$, the exponential average variance tends to a finite non-zero fraction of $Var(\Delta X)$ dependent on α. To see this, we can take the variance of both sides of the first expression in Equation (A2.2.4), i.e.,

$$Var(A_n^{(\Delta X)}) = \alpha^2 Var(A_{k-1}^{(\Delta X)}) + (1-\alpha)^2 \sigma_{\Delta X}^2 \tag{A2.2.10}$$

Taking limits of both sides

$$\begin{aligned} L &= \lim_{n\to\infty} Var(A_n^{(\Delta X)}) = \alpha^2 \lim_{n\to\infty} Var(A_{n-1}^{(\Delta X)}) + (1-\alpha)^2 \sigma_{\Delta X}^2 \\ &= \alpha^2 L + (1-\alpha)^2 \sigma_{\Delta X}^2 \\ &\Rightarrow \\ L &= \frac{(1-\alpha)^2}{1-\alpha^2} \sigma_{\Delta X}^2 \end{aligned} \tag{A2.2.11}$$

For a value of α = 0.82 (a so-called 10-point exponential average), we have

$$\lim_{n\to\infty} Var(A_n^{(\Delta X)}) \approx 0.0989 \sigma_{\Delta X}^2 \tag{A2.2.12}$$

For a value of α = 0.5 (a so-called 3-point exponential average), we have

$$\lim_{n\to\infty} Var(A_n^{(\Delta X)}) \approx 0.3333 \sigma_{\Delta X}^2 \tag{A2.2.13}$$

Note:
Exponential averages have been used in the financial sector for decades due to their computational efficiency. They define a so-called *n point exponential average* as one with weighting factor $\alpha = \frac{n-1}{n+1}$. It is interesting to note that this definition is *equivalent* to the following proposition:

Theorem:
If $\Delta\bar{X}$ is an *n point rolling average,* then

$\lim_{n\to\infty} Var(A_n^{(\Delta X)}) = Var(\Delta\bar{X})$ *if and only if* $\alpha = \frac{n-1}{n+1}$

Proof

We know from Equation (2.11) that $Var(\Delta\bar{X}) = \left(\frac{1}{n}\right)\sigma_{\Delta X}^2$. If $\alpha = \frac{n-1}{n+1}$, then from Equation (A2.2.9), we have that

$$\lim_{n\to\infty} Var(A_n^{(\Delta X)}) = \frac{(1-\alpha)^2}{1-\alpha^2}\sigma_{\Delta X}^2 = \frac{\left(1-\frac{n-1}{n+1}\right)^2}{1-\left(\frac{n-1}{n+1}\right)^2}\sigma_{\Delta X}^2 = \frac{\frac{4}{(n+1)^2}}{\frac{(n+1)^2-(n-1)^2}{(n+1)^2}}\sigma_{\Delta X}^2$$

$$= \frac{4}{4n}\sigma_{\Delta X}^2 = \left(\frac{1}{n}\right)\sigma_{\Delta X}^2$$

$$= Var(\Delta\bar{X})$$

Conversely, if $\lim_{n\to\infty} Var(A_n^{(\Delta X)}) = Var(\Delta\bar{X})$ then $\frac{(1-\alpha)^2}{1-\alpha^2} = \frac{1}{n} \Rightarrow$

$$(n+1)\alpha^2 - 2\alpha n + (n-1) = 0$$

Invoking the quadratic formula, we have the following:

$$\alpha = \frac{2n \pm \sqrt{4n^2 - 4(n+1)(n-1)}}{2(n+1)} = \frac{2n \pm \sqrt{4}}{2(n+1)} = \frac{n \pm 1}{n+1}$$

rendering two possible solutions, $\alpha = 1$ *and* $\alpha = \frac{n-1}{n+1}$.

Because α must be in the range $0 < \alpha < 1$, we have $\alpha = \frac{n-1}{n+1}$ as required.
QED

A2.3 COMPARING DIAGNOSTIC METHODS

In Chapter 2, we discussed *detection* and *isolation* and introduced the concept of *confusion matrices* for both of these diagnostic processes as a means of quantifying the relative effectiveness of the methodologies employed. This was done by using the metrics associated with the matrices: *true positive rate* (TPR), *false positive rate* (FPR), and *fault detection accuracy* (FDA) for detection, and *correct isolation rate* (CIR) and *mis-isolation rate* (MIR) for isolation. Referring to Figure 2.9 of Chapter 2 (duplicated below as Figure A2.1) we have the following common detection metrics.

True Positive Rate: $TPR = \frac{TP}{TP + FN}$

False Positive Rate: $FPR = \frac{FP}{FP + TN}$ *(False Alarm Rate)* (A2.3.1)

False Negative Rate: $FNR = \frac{FN}{TP + FN}$ *(Mis - Detection Rate)*

Fault Detection Accuracy: $FDA = \frac{TP + TN}{TP + FN + FP + TN}$

		Detected State	
		Detected	Not Detected
True State	Fault	**TP** (True Detection)	**FN** (Missed Detection)
	No Fault	**FP** (False Alarm)	**TN** (No Fault Detected)

FIGURE A2.1 Detection confusion matrix

All of these metrics provide a measure of *accuracy* of the method under test, and while accuracy is a critical characteristic for a method to possess, it is not the only characteristic to consider. This is especially important when comparing methodologies to decide which one is best for the application at hand. Some other factors to consider are listed below.

- *Computational performance* in terms of memory required and computational speed. For *onboard* applications, both memory and speed are pivotal. Memory may not be as much a concern, if any, for *off-board* systems; however, computational speed may be a concern, especially if large amounts of data need to be processed on a continual basis.
- *Robustness* of the method in terms of how well the algorithm performs under off-design conditions. How effective and stable is the method when underlying assumptions are violated? Another form of robustness has to do with how easy or difficult it is to update, expand, or otherwise amend the method.
- *Maintainability* is a characteristic similar to robustness, it takes account of additional processes and/or human intervention required (if any) to keep the method viable. For example, systems dependent on machine learning may require periodic retraining, necessitating additional data acquisition and processing.
- *Cost Benefit* not only measures the cost to develop, implement, and maintain the method but also its return on investment. This latter measure may come from how effective the methodology is in supporting high-level metrics, such as reducing D&Cs and UERs, extending time on wing (TOW), extending maintenance intervals, etc.

- *Risk* in terms of technical readiness, i.e., what is the technical readiness level (TRL)* of the method under test.

All of these considerations are important when evaluating or comparing methodologies. The final assessment needs to be a weighted "sum" of these "metrics," which will depend on the application at hand and its requirements. Most of these considerations do not have formal metrics (as in case of accuracy for instance), and the practitioner must create his or her own. This could be as simple as assigning a 1 to 10 rating, etc. Whether the metrics are precisely defined or ad hoc is not critical; what is important is to consider all of the factors that might apply to ensure a proper evaluation.

For a more in-depth discussion, the reader should consult the following reference:

SAE, 2020, *Diagnostic and Prognostic Metrics for Aerospace Propulsion Health Management Systems,* SAE Standard AIR7999.

* TRL definitions can be found at https://www.nasa.gov/pdf/458490main_TRL_Definitions.pdf

A3 Chapter 3 Addendums

A3.1 INTEGRATION CONSTANT FOR EQUATION (3.4) IS ZERO

We rewrite Equation (3.4) before the integration:

$$\frac{dX_c}{X_c} = \frac{dX}{X} - \alpha\frac{d\theta}{\theta} - \beta\frac{d\delta}{\delta}$$

Integrating both sides

$$\int\frac{dX_c}{X_c} = \int\left[\frac{dX}{X} - \alpha\,\frac{d\theta}{\theta} - \beta\,\frac{d\delta}{\delta}\right] \Rightarrow$$

$$\ln(X_c) + const_0 = \ln(X) - \ln(\theta)^{\alpha} - \ln(\delta)^{\beta}$$

We assert that $const_0 = 0$ because

$$e^{(const_0)}X_c = constX_c = \frac{X}{\theta^{\alpha}\delta^{\beta}} \qquad \forall\theta,\delta$$

$$\therefore at\,\theta = 1 = \delta\,(const)X_c = X \underbrace{=}_{by\ definition} X_c$$

$$\therefore const = 1 \Rightarrow const_0 = 0$$

A3.2 NECESSARY AND SUFFICIENT CONDITION FOR A CORRECTED PARAMETER Y

At a fixed engine state, a parameter Y is *corrected* if, and only if, $dY/Y = 0$ for arbitrary θ and δ.

To see this, assume a constant engine state, and let Y be a *corrected* parameter. Then, by definition, it represents a value observed at sea level, static, standard day (ISA) conditions and is constant for all observed inlet conditions, i.e., any θ, δ. Therefore,

$$\begin{aligned} & Y = const \\ & \Rightarrow \ln(Y) = \ln(const) = const \\ & \Rightarrow \frac{dY}{Y} = d\,(const) = 0 \end{aligned}$$

Conversely, if a parameter Y is such that $dY/Y = 0$ for any θ and δ, then

$$\int \frac{dY}{Y} = \int 0 = const$$
$$\Rightarrow \ln(Y) = const$$
$$\Rightarrow Y = e^{\ln(Y)} = e^{const} = const$$
$$\Rightarrow Y = const$$

Thus, Y has the same *const* value at standard day $\theta = 1 = \delta$ and, hence, is a corrected parameter.

A3.3 THE SUM OF CORRECTED PARAMETERS IS ALSO A CORRECTED PARAMETER

It is sufficient to demonstrate this for the sum of two corrected parameters, say X_c and Y_c, that the rest would follow by induction. Consider the sum $Z = X_c + Y_c$ and employ our usual manipulation of taking logs and differentiating to obtain

$$\begin{aligned} \frac{dZ}{Z} &= \frac{1}{X_c + Y_c}\left(dX_c + dY_c\right) \\ &= \left(\frac{X_c}{X_c + Y_c}\right)\underbrace{\frac{dX_c}{X_c}}_{zero} + \left(\frac{Y_c}{X_c + Y_c}\right)\underbrace{\frac{dY_c}{Y_c}}_{zero} \\ &= 0 \end{aligned}$$

Therefore, by A3.2, Z is *corrected.*

We can expand a little further and show that not only is the sum of two corrected parameters also a corrected parameter, but it is also precisely the *corrected* value of the *sum,* or in symbols

$$X_c + Y_c = \left(X + Y\right)_c \qquad \text{(A3.3.1)}$$

To see this, consider that $\forall \theta$, δ

$$\left(X_c + Y_c\right) = const_1, X_c = const_2, Y_c = const_3$$

Therefore, specifically at $\theta = 1 = \delta$, we must have

$$const_1 = X + Y = \left(X + Y\right)_c, \; const_2 = X, const_3 = Y$$

Therefore, because $const_1 = const_2 + const_3$, Equation (A3.3.1) is established.

A3.4 THE QUOTIENT OF CORRECTED PARAMETERS IS ALSO A CORRECTED PARAMETER

As in A3.3, it is sufficient to show that the quotient of two *corrected* parameters, X_c and Y_c, is *corrected* as well. Let $Z = X_c / Y_c$, then taking logs and differentiating, we have

$$\ln(Z) = \ln\left(\frac{X_c}{Y_c}\right) = \ln(X_c) - \ln(Y_c) \Rightarrow$$

$$\frac{dZ}{Z} = \underbrace{\frac{dX_c}{X_c}}_{zero} - \underbrace{\frac{dY_c}{Y_c}}_{zero} = 0$$

Therefore, by A3.2 we have that Z is *corrected*. Following the same arguments used in A3.3, we can conclude that

$$\frac{X_c}{Y_c} = \left(\frac{X}{Y}\right)_c \tag{A3.4.1}$$

A3.5 THE PRODUCT OF CORRECTED PARAMETERS IS ALSO A CORRECTED PARAMETER

This follows from manipulations similar to those used in A3.4 and also establishes

$$X_c Y_c = (XY)_c \tag{A3.5.1}$$

A3.6 PARAMETER %ΔS ARE CORRECTED PARAMETERS

In general, percent deltas are calculated as follows:

$$\Delta X = 100\left(\frac{\frac{X}{\theta^a \delta^b} - X_{Base}}{X_{Base}}\right) \tag{A3.6.1}$$

where X_{Base} is the baseline (or reference) value for parameter X and is itself a *corrected* parameter.[1] Thus, by Equation (3.5) and property A3.3, the numerator is corrected and, hence, our result is established by virtue of property A3.4.

A3.7 PARAMETER EXTRAPOLATION TO DIFFERENT INLET TEMPERATURE AND PRESSURE

Let X represent the raw parameter measured at inlet conditions (T_2, P_2) and let X_c denote the parameter corrected to sea level static (SLS) standard day conditions. Therefore,

$$X_c = \frac{X}{\theta^{\alpha}\,\delta^{\beta}} \qquad where \quad \theta = \frac{T_2}{T_{ISA}}, \delta = \frac{P_2}{P_{ISA}} \tag{A3.7.1}$$

Let (T_{ref}, P_{ref}) represent the reference inlet conditions we want to extrapolate to, and let X_{ref} represent the extrapolated value that the *raw* parameter would have at those conditions, then

$$X_{ref} = X_c\,\theta^{\alpha}_{ref}\,\delta^{\beta}_{ref} = \frac{X}{\theta^{\alpha}\,\delta^{\beta}}\theta^{\alpha}_{ref}\,\delta^{\beta}_{ref} \qquad where \quad \theta_{ref} = \frac{T_{ref}}{T_{ISA}}, \delta_{ref} = \frac{P_{ref}}{P_{ISA}}$$

$$\Rightarrow \tag{A3.7.2}$$

$$X_{ref} = X\left(\frac{\theta_{ref}}{\theta}\right)^{\alpha}\left(\frac{\delta_{ref}}{\delta}\right)^{\beta}$$

A3.8 PARAMETER CORRECTED TO NON-ISA INLET TEMPERATURE AND PRESSURE

Let X represent the raw parameter measured at inlet conditions (T_2, P_2), and let X_{c*} denote the parameter *corrected* to a non-ISA SLS standard day reference condition characterized by (T_{REF}, P_{REF}), then

$$X_{c*} = \frac{X}{\theta^{\alpha}_{REF}\,\delta^{\beta}_{REF}} \quad where \quad \theta_{REF} = \frac{T_2}{T_{REF}}, \delta_{REF} = \frac{P_2}{P_{REF}} \tag{A3.8.1}$$

Combining with Equation (A3.7.1) we obtain

$$
\begin{aligned}
X_{c*} &= \frac{X_c\left(\theta^{\alpha}\ \delta^{\beta}\right)}{\theta_{REF}^{\alpha}\ \delta_{REF}^{\beta}} \\
&= X_c\left(\frac{\theta}{\theta_{REF}}\right)^{\alpha}\left(\frac{\delta}{\delta_{REF}}\right)^{\beta} \\
&= X_c\left(\frac{T_2/T_{ISA}}{T_2/T_{REF}}\right)^{\alpha}\left(\frac{P_2/P_{ISA}}{P_2/P_{REF}}\right)^{\beta} \\
&= \underbrace{X_c}_{const}\underbrace{\left(\frac{T_{REF}}{T_{ISA}}\right)^{\alpha}}_{const}\underbrace{\left(\frac{P_{REF}}{P_{ISA}}\right)^{\beta}}_{const} \\
&= const
\end{aligned}
\tag{A3.8.2}
$$

NOTE

1. Provided that the base values are generated at a fixed ambient condition, e.g., SLS standard day, or 0.8 Mn, 35k altitude, etc.

A4 Chapter 4 Addendums

A4.1 RANDOM VECTORS

Note: In A4.1 *only* we will denote variables in lowercase and vectors in uppercase.

A *random vector* X with n elements will be denoted by a (n×1) column vector

$$X = \begin{bmatrix} x_1 \\ x_2 \\ x_3 \\ \vdots \\ x_n \end{bmatrix} \qquad \textit{and} \qquad X^T = [x_1, x_2, x_3, ..., x_n] \tag{A4.1.1}$$

where the individual components x_k *are random variables.* Because X is an n×1 vector, it follows that the products XX^T and X^TX have dimension n×n and 1×1, respectively. The latter is the inner product[1] of two vectors and is a scalar, while the former is a matrix referred to as the *scatter matrix*:

$$X\,X^T = \begin{bmatrix} x_1^2 & x_1x_2 & \cdots & x_1x_n \\ x_2x_1 & x_2^2 & \cdots & x_2x_n \\ \vdots & & & \vdots \\ x_nx_1 & x_nx_2 & & x_n^2 \end{bmatrix} \tag{A4.1.2}$$

The *expected value* (or mean) of a random *vector* X, denoted $E(X)$, is the vector of means of its components

$$E(X) = \begin{bmatrix} E(x_1) \\ E(x_2) \\ E(x_3) \\ \vdots \\ E(x_n) \end{bmatrix} \tag{A4.1.3}$$

The *covariance matrix*[2] associated with the random vector X is defined as

$$
\begin{aligned}
\operatorname{cov}(X,X) \equiv P_{XX} &\triangleq E\,[(X-E(X))(X-E(X))^T\,] \\
&= E(X\,X^T) - E(X)\,E(X)^T \\
&= E(X\,X^T) - \mu_X\,\mu_X^T \\
&= \begin{bmatrix} E(x_1^2) & E(x_1x_2) & \cdots & E(x_1x_n) \\ E(x_2x_1) & E(x_2^2) & \cdots & E(x_2x_n) \\ \vdots & & & \vdots \\ E(x_nx_1) & E(x_nx_2) & & E(x_n^2) \end{bmatrix} \quad when\ E(X)=0
\end{aligned}
\tag{A4.1.4}
$$

The covariance matrix associated with two random vectors, *X* and *Y*, is defined as

$$
\begin{aligned}
\operatorname{cov}(X,Y) \equiv P_{XY} &\equiv E\ [(X-E(X))(Y-E(Y))^T\,] \\
&= E(X\,Y^T) - E(X)\,E(Y)^T \\
&= E(X\,Y^T) - \mu_X\,\mu_Y^T
\end{aligned}
\tag{A4.1.5}
$$

and is similar in form to the matrix in Equation (A4.1.4) when the means are zero. The covariance matrix P_{XX} has two important properties:

1. P_{XX} is *positive definite*[3] (*eigenvalues* are positive, hence the matrix is *invertible*).
2. P_{XX} is *symmetric*.

In actuality, Property 2 follows from Property 1. For additional properties, refer to Addendum A4.3.

Two familiar properties for *independent* random *variables* (*x*, *y*, *z*) that the reader will recall are

$$
\begin{aligned}
&If\ \ z = ax + y \ \ then \\
&E(z) = a\,E(x) + E(y) \\
&Var(z) = a^2\,Var(x) + Var(y)
\end{aligned}
\tag{A4.1.6}
$$

The analog for *independent* random *vectors* (*X*, *Y*, *Z*) becomes

$$
\begin{aligned}
E(Z) &= A\ E(X) + E(Y) \\
Var(Z) &= A\,Var(X)\,A^T + Var(Y) \\
P_{ZZ} &= A\,P_{XX}\,A^T + P_{YY}
\end{aligned}
\tag{A4.1.7}
$$

The *Normal* distribution plays an important role in the development of statistical estimation. A random *variable x* is said to be normally distributed if its probability density function has the following form:

$$p(x) = \frac{1}{\sqrt{2\pi\sigma^2}} e^{-\frac{1}{2}\left(\frac{x-\mu}{\sigma}\right)^2} \tag{A4.1.8}$$

An n-dimensional random *vector* X is said to be Normal distributed if its probability density function has the following form:

$$p(X) = \frac{1}{\sqrt{(2\pi)^n \det(P_{XX})}} e^{-\frac{1}{2}\left\{[X-E(X)]^T P_{XX}{}^{-1}[X-E(X)]\right\}} \tag{A4.1.9}$$

where $\det(P_{XX})$denotes the determinant of the matrix P_{XX}.

A4.2 ESTIMATION

An estimate for the parameter *vector* x is usually denoted by $\hat{x}$, and we would like to generate estimates that are *optimal* or best in some sense. As it turns out, there are many ways to define "best-ness" for an estimator, and we will offer a few of these. Fortunately, under fairly general conditions, many of these produce the same estimator. In the following discussion, we will assume to have another (random) vector z, which we consider to be (indirect) observations that are somehow related to x in the sense that $z = g(x)$.

The classical definition of an optimal estimate is to select that value of x that minimizes some pre-defined loss function $L(\circ)$ applied to the estimation error vector

$$\tilde{x} = x - \hat{x} \tag{A4.2.1}$$

Typically, the loss function possesses the following properties:

$$L(0) = 0$$

$$L(x) \geq L(y) \qquad when \qquad \|x\| \geq \|y\|$$

$$L(-x) = L(x) \qquad (\text{symmetry})$$

Intuitively, we would expect this vector to be small if the estimate were "*optimal.*" Now, because both x and $\hat{x}$ are random vectors, $\tilde{x}$ is also a random vector, hence $L(\tilde{x})$ is a random vector. A good performance measure would then be to minimize its mean value, $E\left(L(\tilde{x})\right)$.

2. Another measure is the *conditional expectation* of the distribution, i.e.,

$$\hat{x} = E(x|\,z) = \int_{-\infty}^{\infty} x\, p(x|\,z)\, dx$$
$$where \quad p(x|\,z) = \frac{p(x,z)}{p(z)} \tag{A4.2.2}$$

Additionally, if we *assume* that both x and z are *Normal distributed*, i.e.,

$$x \in N(\mu_x, P_{xx}) \quad and \quad z \in N(\mu_z, P_{zz})$$

where $\mu_x = E(x)$ and $\mu_z = E(z)$, then it is known that the conditional expectation $E(x|\,z)$ is a linear function of z, i.e.,

$$E(x \mid z) = A\,z + y \tag{A4.2.3}$$

Multiplying Equation (A4.2.3) by *p(z)* and integrating with respect to z, we obtain

$$E(x) = A\,E(z) + y \quad \Rightarrow \quad y = E(x) - A\,E(z) \tag{A4.2.4}$$

Combining Equations (A4.2.3) and (A4.2.4) yields

$$E(x \mid z) = E(x) + A(z - E(z)) \tag{A4.2.5}$$

Multiplying Equation (A4.2.3) by z^T and *p(z)* and integrating with respect to z, we obtain

$$E(x\,z^T) = A\,E(z\,z^T) + y\,E(z) \tag{A4.2.6}$$

and multiplying Equation (A4.2.4) by $E(z)^T$ yields

$$E(x)\,E(z)^T = A\,E(z)\,E(z)^T + y\,E(z) \tag{A4.2.7}$$

Subtracting Equation (A4.2.7) from Equation (A4.2.6) provides

$$\begin{aligned} &E(x\,z^T) - E(x)\,E(z)^T = A\left(E(z\,z^T) - E(z)\,E(z)^T\right) \\ &\Rightarrow \\ &A = \frac{E(x\,z^T) - E(x)\,E(z)^T}{E(z\,z^T) - E(z)\,E(z)^T} = \frac{P_{xz}}{P_{zz}} = P_{xz}\,P_{zz}^{-1} \end{aligned} \tag{A4.2.8}$$

Thus, Equation (A4.2.3) becomes

$$\begin{aligned} E(x \mid z) &= E(x) + P_{xz}\,P_{zz}^{-1}\left(z - E(z)\right) \\ \mu_{x|z} &= \mu_x + P_{xz}\,P_{zz}^{-1}\left(z - \mu_z\right) \end{aligned} \tag{A4.2.9}$$

Using the relationship for conditional probability density distributions appearing in Equation (A4.2.2), namely $p(x|\,z) = p(x,z)/\,p(z)$, along with our assumption of normality, the following expression for the conditional covariance matrix would emerge (after some tedious algebraic manipulation, which we will omit for brevity).

$$P_{x|z} = P_x - P_{xz}\,P_z^{-1}\,P_{xz}^T \tag{A4.2.10}$$

- Note: The conditional expectation $\mu_{x|z}$ given in Equation (A4.2.9) is a *linear* estimator in the form of a *predictor–corrector*, as illustrated in Equation (A4.2.11).

$$\mu_{x|z} = \underbrace{\mu_x}_{predictor} + \underbrace{\underbrace{P_{xz} P_{zz}^{-1}}_{gain} \underbrace{(z - \mu_z)}_{residual}}_{corrector} \tag{A4.2.11}$$

3. The *maximum a-posteriori* estimate is defined as the value of x that maximizes the probability of that state occurring given the measurements at hand, i.e.,

$$\begin{aligned} \hat{x}_{MAP} &= \arg\max_x \{p(x|z)\} \\ \text{Note: } \hat{x}_{MAP} &= \text{Mode of the distribution } p(x|z) \\ &= E(x|z) \text{ if } p(x|z) \text{ is unimodal \& symmetric} \\ &\therefore \text{ true if } p(x|z) \text{ is normal.} \end{aligned} \tag{A4.2.12}$$

4. The *maximum likelihood* estimate is defined as the value of x that maximizes the probability of occurrence of the measurements which have already been observed.

$$\begin{aligned} &\hat{x}_{ML} = \arg\max_x \{p(z|x)\} \\ &\textit{Note: This does not assume that x is random.} \\ &\qquad \textit{If x is deterministic then} \\ &p(z|x) = p(x,z) = p(x|z)p(z) \Rightarrow \hat{x}_{ML} = \hat{x}_{MAP} \end{aligned} \tag{A4.2.13}$$

5. The <u>*minimum mean square error (minimum variance)*</u> estimate is a special case of Number 1 in which we take the loss function to be the Euclidean norm (inner product), i.e.,

$$\begin{aligned} L(\tilde{x}) &\equiv \tilde{x}^T \tilde{x} = (x - \hat{x})^T (x - \hat{x}) = \| \tilde{x} \|^2 = scalar \\ \hat{x}_{MMS} &= \arg\min_y E\{(x - y)^T (x - y)\} \end{aligned} \tag{A4.2.14}$$

Note: If the estimate $\hat{x}$ is ***unbiased***, i.e., $E(x) = E(\hat{x})$, then $E(\tilde{x}) = 0$, and $\hat{x}_{MMS} = E(\tilde{x}^T \tilde{x}) =$ variance, the MMS estimator is also unbiased.
It can be shown[4] that $\hat{x}_{MMS} = E(x \mid z)$.

A4.2.1 Linear Measurement System

Up to this juncture, we have assumed that z is somehow related to x, i.e., $z = g(x)$. Let us make the additional assumption that this is a linear relationship as follows:

$$
\begin{aligned}
z = H\,x + \varphi \quad & \text{where} \\
& x = (n \times 1) \;\; \text{state vector} \\
& z = (m \times 1) \;\; \text{measurement vector} \\
& \varphi = (m \times 1) \;\; \text{measurement noise vector} \\
& H = (m \times n) \;\; \text{matrix of maximal rank}
\end{aligned}
\tag{A4.2.15}
$$

where x and φ are independent with $x \in N(\mu_x, P_x)$ and $\varphi \in N(0, P_\varphi)$. If we let $\hat{x}$ be an arbitrary estimate for x, and $\hat{z} = H\,\hat{x}$, then if $\hat{x}$ is *near* x, we might expect $\hat{z}$ to be *near* z and conversely. One means of quantifying this prospect is to require that the *sum of squares* of the measurement residual, $z - \hat{z}$, be small, i.e., that $(z - \hat{z})^T (z - \hat{z})$ be small. This requirement naturally leads to the *LSQ*[5] estimate, namely

$$
\hat{x}_{LS} = \arg\min_{x} \left[(z - H\,x)^T (z - H\,x) \right] \tag{A4.2.16}
$$

Furthermore, if W is an $(m \times m)$ *positive definite* matrix, then Equation (A4.2.16) can be generalized to define a *WLSQ* estimate as

$$
\hat{x}_{WLS} = \arg\min_{x} \left[(z - H\,x)^T W (z - H\,x) \right] \tag{A4.2.17}
$$

Employing a little vector calculus and algebraic manipulation, these estimates take the following (hopefully familiar) form given below.

$$
\begin{aligned}
\hat{x}_{WLS} &= \left(H^T W\,H\right)^{-1} H^T\,W\,z \\
\hat{x}_{LS} &= \left(H^T\,H\right)^{-1} H^T\,z
\end{aligned}
\tag{A4.2.18}
$$

Note: H must have maximal rank $= n \le m$ for indicated matrix inverses to exist.

In summary, we have discussed several criteria for defining optimal estimators. If we assume Normal distributions and a linear measurement system, we have the following equivalences:

$$
E(x \mid z) \equiv \mu_{x|z} = \hat{x}_{MAP} = \hat{x}_{MMSE} \tag{A4.2.19}
$$

In addition, if we assume a linear measurement system, i.e., as defined in Equation (A4.2.15), we will show that $E(x|z)$ is a weighted LSQ solution. To alleviate notational tedium, we can assume, without loss of generality, that $\mu_x = 0$ (therefore, $\mu_z = H\,\mu_x + \mu_\varphi = 0$). Thus, we have

$$
P_z = Var(Hx + \varphi) = Var(Hx) + Var(\varphi) = H\,P_x\,H^T + P_\varphi \tag{A4.2.20}
$$

and

$$
\begin{aligned}
P_{xz} &= \operatorname{cov}(x,z) = \operatorname{cov}(x, Hx+\varphi) \\
&= E\left(x(Hx+\varphi)^T\right) = E\left(x(x^T H^T + \varphi^T)\right) \\
&= E(xx^T)H^T + \underbrace{E(x\varphi^T)}_{\substack{zero \\ due\ to\ independence}} \\
&= P_x H^T
\end{aligned}
\tag{A4.2.21}
$$

Therefore, Equation (A4.2.9) becomes

$$
\begin{aligned}
\mu_{x|z} &= \mu_x + P_{xz} P_{zz}^{-1}\left(z-\mu_z\right) \\
&= P_{xz} P_{zz}^{-1} z \\
&= P_x H^T \left(H P_x H^T + P_\varphi\right)^{-1} z
\end{aligned}
\tag{A4.2.22}
$$

If we peek ahead at Addendum 4.11, Equation (A4.11.1), we see that

$$
\begin{aligned}
\mu_{x|z} &= P_x H^T \left(H P_x H^T + P_\varphi\right)^{-1} z \\
&= (P_x^{-1} + H^T P_\varphi^{-1} H)^{-1} H^T P_\varphi^{-1} z \\
&= (H^T P_\varphi^{-1} H)^{-1} H^T P_\varphi^{-1} z \quad \textit{if we let } P_x^{-1} = 0
\end{aligned}
\tag{A4.2.23}
$$

A quick comparison to Equation (A4.2.18) suggests that setting $W = P_\varphi^{-1}$, we can consider $\mu_{x|z}$ to be a weighted LSQ estimator. Adding it to our list (from Equation (A4.2.19)), we have

$$
E(x \mid z) \equiv \mu_{x|z} = \hat{x}_{MAP} = \hat{x}_{MMS} = \hat{x}_{WLS} \tag{A4.2.24}
$$

Aside: If the reader didn't blink while perusing the foregoing manipulation, he or she might have noticed a little sleight of hand with the assignment of $P_x^{-1} = 0$ in Equation (A4.2.23). But what exactly does that mean? Because neither author is a probability theorist, our explanation will not be mathematically rigorous but, rather, more pedestrian in nature. At face value, it would imply that the random vector has *infinite* variance! Interestingly, there do exist probability distributions having infinite variance, (e.g., the *Holtsmark* distribution[6]); however, the Normal distribution is ***not*** one of them. We might then consider a sequence of Normal distributed vectors

$$
\{x_1, x_2, \cdots, x_k, \cdots\} \quad where\ \forall k,\ x_k \in N(\mu_x, P_{x_k})\ and\ \|P_k\| > \|P_{k-1}\|
$$

and take the limit as $k \to \infty$. Of course, we have no guarantee that the limit exists, but because this is a pedestrian argument, let's say it does, and call it x. At least we have $P_k \to \infty$ in the sense that the norm of the covariance matrix becomes unbounded. Unfortunately, x cannot be Normal because we already noted that $N(\mu,\infty)$ does not exist. Fortunately (for us), there is a way out of this conundrum. In Equation (A4.2.23), we could have written "let $P_x^{-1} \approx 0$," meaning make it very small, then P_x becomes very large but still finite, and normality will be preserved. Alternatively, if $P_x^{-1} << H^T P_\varphi^{-1} H$, i.e., P_x^{-1} is significantly smaller as compared with $H^T P_\varphi^{-1} H$, which can be interpreted as process noise being designed to be much higher than measurement noise in the context of Kalman filter design. Thus, $(P_x^{-1} + H^T P_\varphi^{-1} H)^{-1} \approx (H^T P_\varphi^{-1} H)^{-1}$, and this is equivalent to letting "$P_x^{-1} \approx 0$." Consequently, Equation (A4.2.24) becomes

$$E(x \mid z) \equiv \mu_{x|z} = \hat{x}_{MAP} = \hat{x}_{MMS} \approx \hat{x}_{WLS}$$

This is perfectly valid and consistent with engineering approximation. We did, after all, warn the reader of our interchangeable usage of the symbols = and ≈ in the Introduction. So why didn't we just do that in (A4.2.24)? Clearly, if we had, we would not have had this opportunity to provide the reader with some well-deserved mathematical comic relief.

A4.3 USEFUL MATRIX PROPERTIES

In this section, we offer some matrix related definitions and cite some useful properties (without proof)[7].

A4.3.1 Determinants

Let A be an $n \times n$ matrix and let A^j denote the jth column of A. Therefore, we have that $A = \left[A^1, \cdots, A^j, \cdots, A^n\right]$. We *define* the *determinant* of A, $\det(A) = \det\left([a_{ij}]\right)$, to be a *scalar* function of A having the following properties:

1) It is linear in the columns, i.e., if $A^j = \beta B$ then $\det(A) = \beta \det\left([\cdots, B, \cdots]\right)$[8] and if $A^j = B^1 + B^2 \Rightarrow \det(A) = \det([\cdots, B^1, \cdots]) + \det([\cdots, B^2, \cdots])$.
2) Interchanging two columns of A changes the sign of $\det(A)$.

The determinant of the identity matrix is unity, i.e., $\det(I_n) = 1$

We have the following useful properties that follow from definition:

1) $\det(A) = 0$ if and only if A is singular (not invertible).
2) $\det(A^T) = \det(A)$
3) $\det(A^{-1}) = 1/\det(A)$
4) $\det(AB) = \det(A)\det(B)$
5) $\det(\alpha A) = \alpha^n \det(A)$ where α is a scalar and A is an $n \times n$ matrix
6) If A and B are *positive semi-definite*[9] then $\det(A + B) \geq \det(A) + \det(B)$
7) $\det(A) = \prod_{k=1}^{n} \lambda_k$ where the $\{\lambda_k\}$ are the eigenvalues of A.

8) If A is *positive definite* then

$$\frac{n}{trace(A^{-1})} \leq \det(A)^{\frac{1}{n}} \leq \frac{1}{n} trace(A) \leq \sqrt{\frac{1}{n} trace(A^2)}$$ where the *trace* of A

is the sum of its diagonal elements, i.e., $trace(A) = \sum_{k=1}^{n} a_{kk}$

The last Property 8 is a representation of the well-known fact that *the* ***harmonic mean*** *is less than the* ***geometric mean****, which is less than the* ***arithmetic mean****, which is, in turn, less than the* ***root mean square****.*

A4.3.2 Positive Definite Matrices

A matrix A is said to be *positive definite* if for all $x \neq 0$, $x^T A x > 0$ and *positive semi-definite* if $x^T A x \geq 0$. Common notation used to designate *positive definite* and *positive semi-definite* is $A > 0$ and $A \geq 0$, respectively. For the two matrices, A and B, the notation $B > A$ means that $(B - A) > 0$, i.e., the matrix $B - A$ is *positive definite.*

Positive definite matrices enjoy the following properties:

1) Symmetry, i.e., $A^T = A$
2) All eigenvalues are positive, $\lambda > 0$
3) Both $\det(A) = \prod \lambda_i > 0$ and $Trace(A) = \sum \lambda_i > 0$
4) $A > 0$ and $B > 0 \;\Rightarrow\; A + B > 0$
5) $A > 0 \;\Rightarrow\; A^{-1}$ *exists and* $A^{-1} > 0$
6) $A > 0$ *and* $B > 0 \;\Rightarrow\; ABA > 0$ *and* $BAB > 0$
7) $A > 0$, $B > 0$ *and* $AB = BA \;\Rightarrow\; AB > 0$
8) For any matrix A, $A^T A \geq 0$
9) $A > 0$, $B \geq 0 \;\Rightarrow\; A + B > 0$

The proofs for these properties can be found in books on linear algebra and/or matrix theory.

Our primary usage of positive definite matrices stems from the statements made in A4.1, in that the covariance matrix of the random vector x, i.e., P_{xx}, is positive definite (and hence, symmetric). To see this, we have from definition, Equation A4.14 that

$$P_{xx} = E\,[(x - E(x))(x - E(x))^T] = E\,[(x - \mu_x)(x - \mu_x)^T]$$

For any vector $y \neq 0$, consider

$$\begin{aligned} &E\,[((x-\mu_x)^T y)^2] > 0 \\ &\qquad = E\,\{[(x-\mu_x)^T y]^T [(x-\mu_x)^T y]\} \\ &\qquad = E\,\{y^T (x-\mu_x)(x-\mu_x)^T y\} \\ &\qquad = y^T E\,\{(x-\mu_x)(x-\mu_x)^T\} y \\ &\qquad = y^T P_{xx}\, y \;\Rightarrow \\ &y^T P_{xx}\, y > 0 \end{aligned} \tag{A4.3.1}$$

A4.3.3 Vector and Matrix Norms

Norms for vectors and matrices can be defined in the same manner.

The norm of a matrix (or vector) is a *scalar valued* function (denoted by $\|\bullet\|$) satisfying the following four properties:

$$\|\alpha\, A\| = |\alpha|\,\|A\| \quad \text{for scalar } \alpha$$

$$\|A+B\| \le \|A\| + \|B\| \quad \text{(triangle inequality)}$$

$$\|A\| \ge 0 \quad \text{and} = 0 \Leftrightarrow A = 0$$

$$\|A\,B\| \le \|A\|\,\|B\| \quad \text{(Cauchy inequality)}$$

As the reader might imagine, there are many scalar functions that could satisfy these four conditions, hence there is *no unique* norm function and, in fact, there are many. That being said, for the purposes of use in our GPA discussions, we will use the *Euclidean norm for vectors* and the so-called *supremum norm for matrices.*[10] These are defined as follows:

Euclidean norm for vectors x:

$$\|x\| = \sqrt{x^T x} = \sqrt{\sum_{i=1}^{n} x_i^2} \tag{A4.3.2}$$

Supremum norm for matrix A:

$$\|A\| = \sup_{x \ne 0}\left(\frac{\|Ax\|}{\|x\|}\right) = \sup_{\|x\|=1}\left(\|Ax\|\right) \tag{A4.3.3}$$

We leave it to the reader to verify that these two functions do, indeed, satisfy the four conditions stated above. While there are many other norms to choose from, there is a marvelous theorem[11] that states that *any* two norms, $\|\bullet\|_1$ and $\|\bullet\|_2$, are *equivalent*, in the following sense:

Theorem 1: let $\|\bullet\|_1$ and $\|\bullet\|_2$ be two arbitrary norms, then ∃ scalars α and β ∍

$$\alpha\,\|A\|_1 \le \|A\|_2 \le \beta\,\|A\|_1 \qquad \text{for any matrix } A.$$

This holds for *both* vectors and matrices. Fundamentally, this guarantees that vectors (or matrices) that are small (or large) in one norm will be (relatively) small (or large) in any other norm.

We state without proof another useful theorem regarding the supremum norm:

Theorem 2: If A is an $(n \times n)$ matrix, then for the supremum norm we have $\|A\|^2 = \lambda_{\max}^{(A^T A)} = \text{max eigenvalue of } A^T A$

If we consider $A \ge 0$, then $A^T = A$, and thus $\|A\|^2 = \lambda_{\max}^{(A^T A)} = \lambda_{\max}^{(A^2)} = \left(\lambda_{\max}^{(A)}\right)^2$, which yields a useful property:

$$A \ge 0 \Rightarrow \|A\| = \lambda_{\max}^{(A)} = \text{max eigenvalue of } A \tag{A4.3.4}$$

A4.4 ENGINE INFLUENCE COEFFICIENTS

The fundamental tenet for GPA, as depicted in Figure 4.1, requires having a *suitable* relationship to quantify how changes in engine module performance drive changes in parameters measured in the engine's gas path. Given the complexity of modern engines with sophisticated controls, variable geometry (e.g., variable stator vanes), handling bleeds, turbine active clearance control, etc., this relationship is not only non-linear but also a non-analytic[12] function $Z = f(x_e)$. Finding the inverse relationship would be impossible unless we make an assumption. The assumption is that the changes occurring are *small,* and we all know that in an infinitesimal world, everything appears linear. Linear is good because linear systems can be *inverted,* as proposed by Figure 4.1. Thus, our basic relationship becomes a matrix of engine influence coefficients (sometimes referred to as engine fault coefficients) H_e.

The generation of these matrices requires an in-depth model of the engine under consideration and a great deal of skill. Any meaningful discourse on this subject is beyond the scope of this book and, as such, we will only provide a broad overview. It is suggested that the interested reader consult Doel [1] for further discussion on this topic. There are a number of possible methods to generate this matrix, but one popular method involves the use of a closed-loop, non-linear, aero-thermal engine simulation, running at steady state, wherein small perturbations (1%) are introduced (singularly) to each module performance fault parameter while holding some parameter fixed. This fixed parameter can be anything you choose, but typically it is the control power setting parameter, e.g., N_{1c2} or *EPR*.[13] Now, it is true that 1% is not exactly infinitesimal, but it is convenient and allows the linear model to be solved without further scaling. It should be noted that because $+1\% \neq \varepsilon$, if a -1% perturbation had been made, the resulting coefficients would not necessarily be the negative of those obtained from the +1% case because of non-linearity. Some practitioners perturb with -1% and then change the sign of the coefficients and average them with

TABLE A4.1
Parameters Perturbed to Generate Engine ICs

Module	Independent Performance Parameter
FAN	FAN (map characteristic) adiabatic efficiency
	FAN flow capacity
LPC	LPC (map characteristic) adiabatic efficiency
	LPC flow capacity
HPC	HPC (map characteristic) adiabatic efficiency
	HPC flow capacity
HPT	HPT (map characteristic) adiabatic efficiency
	HPT nozzle area
LPT	LPT (map characteristic) adiabatic efficiency
	LPT nozzle area

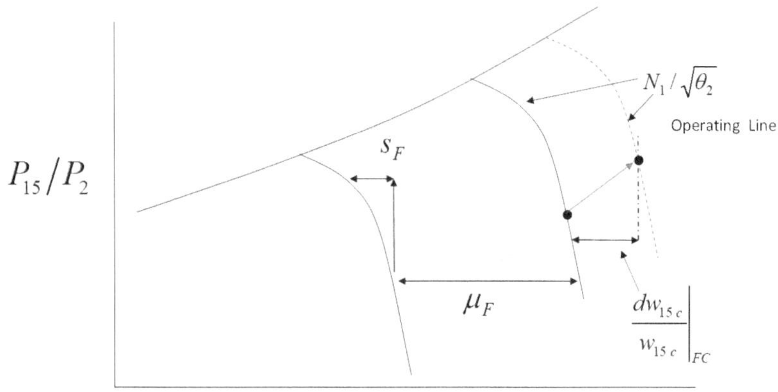

$$w_{15c} = \frac{w_{15}\sqrt{\theta_2}}{\delta_2}$$

FIGURE A4.1 Hypothetical outer diameter of the Fan module map – flow capacity change

the +1% case to even out the non-linearity. There are countless other possibilities. In the end, the differences are relatively minor.

The parameters typically perturbed are as follows:

> The resultant influence coefficient (IC) matrix, H_e should always be scrutinized for possible errors. Note: Figure 4.5 in Chapter 4 depicted the efficiency map characteristic ($\Delta\eta_{Fan}|_D$) that is being perturbed (as opposed to an operational efficiency delta, $\Delta\eta_{Fan}$). Similarly, a change in flow capacity (FC) is the change in corrected air flow resulting from a map change in speed line, as illustrated in Figure A4.1.

A4.4.1 Ancillary Parameters Generated via ICs

Additional parameters of interest can be generated through the use of engine ICs that are generated holding different parameters constant (other than *EPR* or corrected Fan speed). One example might be compressor operating line shift, which can be defined as the change in pressure ratio at constant corrected flow. To be specific, for the HPC, this would be

$$OPLSH_{HPC} = \left.\frac{d\left(\frac{P_3}{P_{25}}\right)}{\left(\frac{P_3}{P_{25}}\right)}\right|_{w_{25c}=const}$$

which is a relative measure of how compressor surge margin is increasing or decreasing. Therefore,

$$\frac{OPLSH_{HPC}}{\frac{d\eta}{\eta}} = \left(\frac{dP_3/P_3}{d\eta/\eta} - \frac{dP_{25}/P_{25}}{d\eta/\eta} \right)_{w_{25c} = const}$$

would provide a measure of how the performance fault η contributes to change in HPC pressure ratio, where η can be any of the engine faults estimated by GPA. This may provide greater insight to the analyst in knowing how each assessed engine module is *contributing* to change in surge margin.

Other ancillary parameters of interest might include non-measurable parameters such as thrust at takeoff or parameters (that are measured, e.g., *EGT*) but not at the condition of interest such as *EGT* margin on a hot day at takeoff. All of these take the form of the familiar $y = Gx$, where y is a vector representing the % change in the ancillary parameters of interest, x is the vector of engine/sensor faults in our model $Z = Hx + \varphi$, and G is a matrix of ICs. Once the estimate $\hat{x}$ is determined, then the ancillary parameters can be estimated as $\hat{y} = G\hat{x}$.

A4.5 COUPLING

Coupling is a means of combining module performance parameters, typically efficiency and flow parameter in a linear fashion [2], [3]. It represents, on average, a typical deterioration relationship between efficiency loss occurring with flow capacity loss in the compression modules and efficiency loss occurring with increased nozzle areas in the turbines. The coupling factors used for this purpose vary with engine type, model, and anticipated operational profile. It is a convenient means to reduce the total number of engine-fault variables for a classical GPA, where measurement suites were limited to control parameters ($m < 5$ in the '70s) and LSQ analysis was the only means to effect a solution. It has present-day utility as well. Aside from reducing the total number of unknowns $n = n_e + n_s$ in an MFA, it fulfills the isolation objective for SFA more succinctly than estimating module efficiencies and flow parameters separately.

The *coupling factor* (*CF*) between efficiency (η) and flow capacity (*FC*) is defined[14] as the ratio

$$CF = \frac{\%\Delta FC}{\%\Delta\eta} \text{ for Compressors and CF} = \frac{\%\Delta A}{\%\Delta\eta} \text{ for Turbines} \tag{A4.5.1}$$

What this implies is the assumption that a +1% change in efficiency will be accompanied by an *FC* % change in flow parameters. For definiteness, we will use the following coupling factors in our discussions as given in Table A4.2.

TABLE A4.2
Sample Coupling Factors

Module	FAN	LPC	HPC	HPT	LPT
CF	1.3	1.3	1	-0.7	-0.6

	Fan Eff	Fan FC	LPC Eff	LPC FC	HPC Eff	HPC FC	HPT Eff	A4	LPT Eff	A45
N1	0.170	-0.339	0.050	-0.144	-0.019	-0.004	-0.027	0.007	0.304	-0.170
N2	-0.084	0.138	-0.110	-0.151	0.253	-0.277	0.362	-0.128	-0.063	0.223
T25	0.155	-0.283	-0.219	0.240	-0.171	-0.037	-0.242	0.061	0.223	-0.247
T5	-0.192	0.340	-0.219	-0.404	-0.587	-0.126	-0.830	0.209	-0.577	-0.127
Ps3	-0.015	-0.054	-0.004	0.052	0.049	-0.017	0.242	-1.117	0.167	0.046
T3	-0.038	0.057	-0.200	-0.087	-0.399	-0.057	0.200	-0.322	0.010	0.097
P25	0.575	-1.089	-0.012	1.061	-0.747	-0.167	-1.051	0.273	0.803	-0.958
Wf	-0.297	0.436	-0.128	-0.539	-0.534	-0.121	-0.765	0.207	-0.111	-0.326

FIGURE A4.2 Sample engine influence coefficient matrix H_e

There are two ways to generate a coupled parameter. The *first*, and most direct method, is to apply the relationship in Equation A4.5.1 to the standard engine fault coefficient matrix H_e depicted in Figure 4.3 (duplicated below as Figure A4.2).

For example,

$$\begin{aligned} FAN &= (Fan\,Eff) + 1.3(Fan\,FC) \\ LPC &= (LPC\,Eff) + 1.3(LPC\,FC) \\ HPC &= (HPC\,Eff) + 1.0(HPC\,FC) \\ HPT &= (HPT\,Eff) - 0.7(A4) \\ LPT &= (LPT\,Eff) - 0.6(A45) \end{aligned} \tag{A4.5.2}$$

This produces the coupled engine fault matrix shown in Figure A4.3.

	FAN	LPC	HPC	HPT	LPT
N1	-0.270	-0.137	-0.023	-0.032	0.406
N2	0.095	-0.307	-0.024	0.452	-0.197
Wf	0.269	-0.829	-0.654	-0.910	0.085
T5	0.250	-0.744	-0.713	-0.977	-0.501
T25	-0.213	0.093	-0.208	-0.285	0.371
T3	0.036	-0.313	-0.455	0.426	-0.048
P25	-0.841	1.367	-0.914	-1.242	1.378
P3	-0.086	0.064	0.032	1.024	0.140

FIGURE A4.3 Coupled engine fault coefficient matrix

The *second* method is indirect and involves the x_e covariance matrix $P_{x_e} = [P_{ij}]_{i,j=1,2,\ldots,n_e}$. The manipulation is best understood if we (momentarily) consider $x_e = [x_1, x_2, \ldots, x_{n_e}]^T$ to be random. If components x_i and x_j are coupled, such that $CF = x_j / x_i$, then we impose

$$\begin{aligned} P_{jj} &= Var(x_j) = (CF)^2 Var(x_i) = (CF)^2 P_{ii} \\ P_{ij} &= P_{ji} = \rho(CF)P_{ii} \; where |\rho| \le 1 \end{aligned} \tag{A4.5.3}$$

The reader may recognize the variable ρ as the correlation coefficient. Setting $\rho = \pm 1$ will induce a *hard* coupling between the estimated values $CF = \hat{x}_j / \hat{x}_i$, as required. The two rows corresponding to these faults in the diagnostic matrix $D = P_{x_e} H_e^T \left(H_e P_{x_e} H_e^T + R^{-1} \right)$ will be proportional to each other by the coupling factor *CF*. Setting $-1 < \rho < 1$ and $\rho \ne 0$ induces a *soft* coupling between the variables.

As an example, if we use the coupling factors from Table A4.2 in Equation (A4.5.3) with $\rho = 1$ to obtain the state covariance given in Figure A4.4 below, we would obtain the Diagnostic Matrix D shown in Figure A4.5.

The reader can easily verify that the first 10 rows of the diagnostic matrix in Figure A4.5 admit the coupling specified in Table A4.2.

To show that the two coupling methods are essentially equivalent, if we take the first method (with H_e as given in Figure A4.3 and concatenate with H_s to form *H*), and use the state covariance matrix given below in Figure A4.6, we would obtain the diagnostic matrix depicted in Figure A4.7. The reader can verify that the two diagnostic matrices are identical (taking into account the coupled rows in Figure A4.5).

A4.6 STEADY STATE SLOPES

Gas generator *sss* are central to the calculation of the *sensor error IC matrix* H_S. We will assume that *X* denotes an *independent* gas path parameter and *Y* denotes any *dependent* gas path parameter whose *sss* we wish to determine. By definition, the *sss of Y with respect to X is the % of point change in Y with respect to a percent of point change in X*, i.e.,

$$sss(Y)_X \triangleq \left(\frac{dY/Y}{dX/X} \right) \tag{A4.6.1}$$

Typically, *X* is chosen to be the power setting parameter X_c^{PSP} (typically, *EPR* or N_{1c2}). Thus, for example, for power setting parameter *EPR*, the *sss* of (*corrected*) fuel flow, w_{fc} becomes

$$sss(w_{fc})_{EPR} = \left(\frac{dw_{fc}/w_{fc}}{dEPR/EPR} \right) \tag{A4.6.2}$$

	Fan Eff	Fan Flow	LPC Eff	LPC Flow	HPC Eff	HPC Flow	HPT Eff	A4	LPT Eff	A45	N1 ER	N2 ER	Wf ER	T5 ER	T25 ER	T3 ER	P25 ER	P3 ER	P2 ER	T2 ER	P5 ER
Fan Eff	100	130	0	[illegible]	0	0	0	0	0	0	0	0	0	0	0	0	0	0	0	0	0
Fan Flow	130	169	0	[illegible]	0	0	0	0	0	0	0	0	0	0	0	0	0	0	0	0	0
LPC Eff	0	0	100	1[illegible]0	0	0	0	0	0	0	0	0	0	0	0	0	0	0	0	0	0
LPC Flow	0	0	130	1[illegible]9	0	0	0	0	0	0	0	0	0	0	0	0	0	0	0	0	0
HPC Eff	0	0	0	[illegible]	100	100	0	0	0	0	0	0	0	0	0	0	0	0	0	0	0
HPC Flow	0	0	0	[illegible]	100	100	0	0	0	0	0	0	0	0	0	0	0	0	0	0	0
HPT Eff	0	0	0	[illegible]	0	0	100	-70	0	0	0	0	0	0	0	0	0	0	0	0	0
A4	0	0	0	[illegible]	0	0	-70	49	0	0	0	0	0	0	0	0	0	0	0	0	0
LPT Eff	0	0	0	[illegible]	0	0	0	0	100	-60	0	0	0	0	0	0	0	0	0	0	0
A45	0	0	0	[illegible]	0	0	0	0	-60	36	0	0	0	0	0	0	0	0	0	0	0
N1 ER	0	0	0	[illegible]	0	0	0	0	0	0	0.1	0	0	0	0	0	0	0	0	0	0
N2 ER	0	0	0	[illegible]	0	0	0	0	0	0	0	0.1	0	0	0	0	0	0	0	0	0
Wf ER	0	0	0	[illegible]	0	0	0	0	0	0	0	0	0.1	0	0	0	0	0	0	0	0
T5 ER	0	0	0		0	0	0	0	0	0	0	0	0	0.1	0	0	0	0	0	0	0
T25 ER	0	0	0		0	0	0	0	0	0	0	0	0	0	0.1	0	0	0	0	0	0
T3 ER	0	0	0		0	0	0	0	0	0	0	0	0	0	0	0.1	0	0	0	0	0
P25 ER	0	0	0		0	0	0	0	0	0	0	0	0	0	0	0	0.1	0	0	0	0
P3 ER	0	0	0		0	0	0	0	0	0	0	0	0	0	0	0	0	0.1	0	0	0
P2 ER	0	0	0		0	0	0	0	0	0	0	0	0	0	0	0	0	0	0.1	0	0
T2 ER	0	0	0		0	0	0	0	0	0	0	0	0	0	0	0	0	0	0	0.1	0
P5 ER	0	0	0		0	0	0	0	0	0	0	0	0	0	0	0	0	0	0	0	0.1

FIGURE A4.4 Coupled state cov_riance matrix

	N1	N2	Wf	T5	T25	T3	P25	P3
Fan Eff	-3.125	-1.004	2.135	-2.185	0.613	0.813	-0.276	-0.345
Fan Flow	-4.063	-1.305	2.775	-2.841	0.798	1.057	-0.359	-0.449
LPC Eff	-1.472	-0.057	-0.103	0.022	-0.030	-0.505	0.375	0.566
LPC Flow	-1.913	-0.075	-0.134	0.029	-0.039	-0.657	0.487	0.736
HPC Eff	0.539	-0.215	0.001	-0.198	0.397	-1.239	-0.415	0.047
HPC Flow	0.539	-0.215	0.001	-0.198	0.397	-1.239	-0.415	0.047
HPT Eff	-0.116	0.166	-0.067	-0.285	-0.060	0.642	-0.030	0.247
A4	0.081	-0.116	0.047	0.200	0.042	-0.449	0.021	-0.173
LPT Eff	-0.201	-0.677	1.329	-1.748	0.679	0.822	-0.148	-0.528
A45	0.120	0.406	-0.797	1.049	-0.407	-0.493	0.089	0.317
N1 ER	0.077	-0.042	-0.003	-0.001	-0.152	0.038	0.016	-0.021
N2 ER	-0.042	0.638	0.037	-0.059	-0.044	-0.176	0.098	-0.127
Wf ER	-0.003	0.037	0.012	-0.008	-0.022	-0.015	0.009	-0.002
T5 ER	-0.001	-0.059	-0.008	0.042	0.025	-0.040	-0.009	0.072
T25 ER	-0.152	-0.044	-0.022	0.025	0.392	-0.043	-0.065	0.067
T3 ER	0.038	-0.176	-0.015	-0.040	-0.043	0.158	-0.024	-0.079
P25 ER	0.016	0.098	0.009	-0.009	-0.065	-0.024	0.024	-0.022
P3 ER	-0.021	-0.127	-0.002	0.072	0.067	-0.079	-0.022	0.148
P2 ER	0.003	0.012	0.011	0.035	-0.018	-0.074	0.008	0.074
T2 ER	0.102	-0.050	0.019	0.002	-0.269	0.018	0.035	0.004
P5 ER	0.005	-0.020	-0.029	-0.091	0.038	0.193	-0.020	-0.198

FIGURE A4.5 Coupled diagnostic matrix

	FAN	LPC	HPC	HPT	LPT	N1 ER	N2 ER	Wf ER	T5 ER	T25 ER	T3 ER	P25 ER	P3 ER	P2 ER	T2 ER	P5 ER
FAN	100	0	0	0	0	0	0	0	0	0	0	0	0	0	0	0
LPC	0	100	0	0	0	0	0	0	0	0	0	0	0	0	0	0
HPC	0	0	100	0	0	0	0	0	0	0	0	0	0	0	0	0
HPT	0	0	0	100	0	0	0	0	0	0	0	0	0	0	0	0
LPT	0	0	0	0	100	0	0	0	0	0	0	0	0	0	0	0
N1 ER	0	0	0	0	0	0.1	0	0	0	0	0	0	0	0	0	0
N2 ER	0	0	0	0	0	0	0.1	0	0	0	0	0	0	0	0	0
Wf ER	0	0	0	0	0	0	0	0.1	0	0	0	0	0	0	0	0
T5 ER	0	0	0	0	0	0	0	0	0.1	0	0	0	0	0	0	0
T25 ER	0	0	0	0	0	0	0	0	0	0.1	0	0	0	0	0	0
T3 ER	0	0	0	0	0	0	0	0	0	0	0.1	0	0	0	0	0
P25 ER	0	0	0	0	0	0	0	0	0	0	0	0.1	0	0	0	0
P3 ER	0	0	0	0	0	0	0	0	0	0	0	0	0.1	0	0	0
P2 ER	0	0	0	0	0	0	0	0	0	0	0	0	0	0.1	0	0
T2 ER	0	0	0	0	0	0	0	0	0	0	0	0	0	0	0.1	0
P5 ER	0	0	0	0	0	0	0	0	0	0	0	0	0	0	0	0.1

FIGURE A4.6 State covariance matrix

To calculate this quantity, we note that

$$d\left[\ln(Y)\right] = \frac{dY}{Y} \tag{A4.6.3}$$

and

$$d\left[\ln(Y)\right] \approx \Delta\left[\ln(Y)\right] = \ln(Y_2) - \ln(Y_1) = \ln\left({}^{Y_2}\!/_{Y_1}\right) \tag{A4.6.4}$$

	N1	N2	Wf	T5	T25	T3	P25	P3
FAN	-3.125	-1.004	2.135	-2.185	0.613	0.813	-0.276	-0.345
LPC	-1.472	-0.057	-0.103	0.022	-0.030	-0.505	0.375	0.566
HPC	0.539	-0.215	0.001	-0.198	0.397	-1.239	-0.415	0.047
HPT	-0.116	0.166	-0.067	-0.285	-0.060	0.642	-0.030	0.247
LPT	-0.201	-0.677	1.329	-1.748	0.679	0.822	-0.148	-0.528
N1 ER	0.077	-0.042	-0.003	-0.001	-0.152	0.038	0.016	-0.021
N2 ER	-0.042	0.638	0.037	-0.059	-0.044	-0.176	0.098	-0.127
Wf ER	-0.003	0.037	0.012	-0.008	-0.022	-0.015	0.009	-0.002
T5 ER	-0.001	-0.059	-0.008	0.042	0.025	-0.040	-0.009	0.072
T25 ER	-0.152	-0.044	-0.022	0.025	0.392	-0.043	-0.065	0.067
T3 ER	0.038	-0.176	-0.015	-0.040	-0.043	0.158	-0.024	-0.079
P25 ER	0.016	0.098	0.009	-0.009	-0.065	-0.024	0.024	-0.022
P3 ER	-0.021	-0.127	-0.002	0.072	0.067	-0.079	-0.022	0.148
P2 ER	0.003	0.012	0.011	0.035	-0.018	-0.074	0.008	0.074
T2 ER	0.102	-0.050	0.019	0.002	-0.269	0.018	0.035	0.004
P5 ER	0.005	-0.020	-0.029	-0.091	0.038	0.193	-0.020	-0.198

FIGURE A4.7 Diagnostic matrix using coupled ICs

Likewise,

$$d\left[\ln(X)\right] \approx \Delta\left[\ln(X)\right] = \ln(X_2) - \ln(X_1) = \ln\left(X_2 / X_1\right) \tag{A4.6.5}$$

Therefore, combining expressions, we obtain

$$sss(Y)_X = \frac{\ln\left(Y_2 / Y_1\right)}{\ln\left(X_2 / X_1\right)} \tag{A4.6.6}$$

This final expression may be used to calculate the slope, where the subscripts denote two different power conditions bounding the point of interest. Using our C-MAPSS model, for example, if we were interested in evaluating $sss(w_{fc})_{EPR}$ at $EPR = 1.1645$, we could choose two bounding points at EPR = 1.1640 and 1.1650.

EPR	w_{fc2}
1.1640	15785.95
1.1650	15863.70

Thus,

$$sss(w_{fc})_{EPR} = \frac{\ln(15863.7 / 15785.95)}{\ln(1.165 / 1.164)} = 5.7214$$

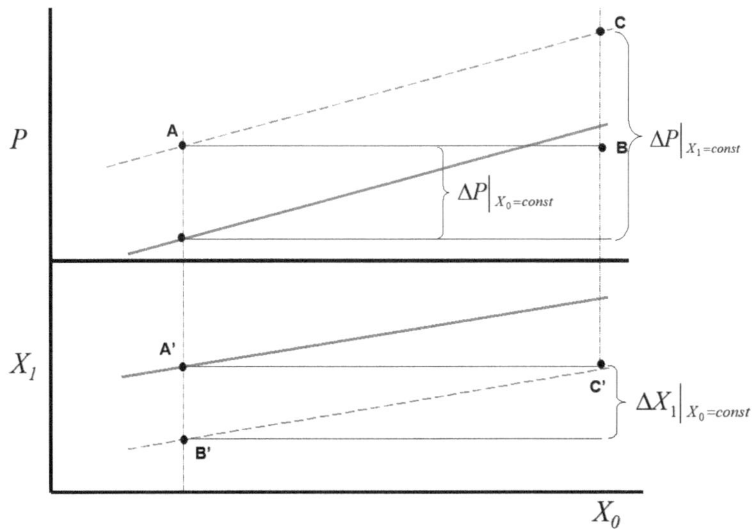

FIGURE A4.8 Symbolic parameter shifts

A4.7 STEADY STATE SLOPE & IC INTER-RELATIONSHIPS

To motivate the relationship, we will need to establish some approximate associations. We will denote the two target parameters we will hold constant as X_0 and X_1 and let P be any other (dependent) gas path parameter.

Referring to Figure (A4.8), we have two parameters (P and X_1) plotted against parameter X_0. The blue curves represent the nominal relationship, whereas the red dashed curves represent *shifts* from that nominal relationship. If we consider some of the quantities represented, we see that point A represents the shift in parameter P (from nominal), holding target parameter X_0 constant, i.e., $\Delta P|_{X_0=const}$. Likewise, we have

$$\overline{A'B'} = \Delta X_1|_{X_0=const}$$

$$\overline{A'C'} = \overline{AB}$$

In particular, for triangle $A'B'C'$, we have that

$$\frac{-\Delta X_1|_{X_0=const}}{\overline{AB}} = slope \approx \frac{dX_1}{dX_0} \tag{A4.7.1}$$

as well as (from triangle ABC)

$$\frac{\Delta P|_{X_1=const} - \Delta P|_{X_0=const}}{\overline{AB}} = slope \approx \frac{dP}{dX_0} \tag{A4.7.2}$$

Combining these last two relationships provides the following expression

$$\Delta P\Big|_{X_1=const} \approx \Delta P\Big|_{X_0=const} + \frac{dP}{dX_0}\left(\frac{-\Delta X_1\Big|_{X_0=const}}{dX_1 / dX_0}\right) \tag{A4.7.3}$$

There are a number of useful expressions, which are a fallout of Equation (A4.7.3). For instance, if we interpret all Δs as % of point Δs, i.e.,

$$\Delta P = \frac{dP}{P}$$

and let

$$X_1 = N1c2 \quad , \quad X_0 = EPR$$

we obtain the following relationship:

$$\frac{dP}{P}\Bigg|_{N1c2=const} \approx \frac{dP}{P}\Bigg|_{EPR=const} - \frac{dP/P}{dEPR/EPR}\left(\frac{\dfrac{dN1c2}{N1c2}\Bigg|_{EPR=const}}{\dfrac{dN1c2/N1c2}{dEPR/EPR}}\right) \tag{A4.7.4}$$

Recall that the IC of $fault\Delta = x$ on $measurement\Delta = z$ holding parameter $P = const$ is defined as

$$IC(z \mid x)_{p=const} \triangleq \frac{\partial z / z}{\partial x / x}\Bigg|_{P=const}$$

In terms of ICs, for any engine fault η, this becomes

$$IC(P \mid \eta)_{N1c2=const} \approx IC(P \mid \eta)_{EPR=const} - sss(P)_{EPR}\left(\frac{IC(N1c2 \mid \eta)_{EPR=const}}{sss(N1c2)_{EPR}}\right) \tag{A4.7.5}$$

Equation (A4.7.5) provides an approximate interrelationship between ICs generated at constant N_{1c2} ($IC(P \mid \eta)_{N1c2=const}$) and those generated at constant *EPR* ($IC(P \mid \eta)_{EPR=const}$) and the *sss*. This could be used, for example, to generate ICs at constant *N1c2*, given that we know ICs at constant *EPR* and the *sss*. Conversely, if we have *both* sets of ICs at our disposal, we could infer the *sss* as follows:

Setting $P = EPR$

$$IC(EPR \mid \eta)_{N1c2=const} \approx \underbrace{IC(EPR \mid \eta)_{EPR=const}}_{zero} - \underbrace{sss(EPR)_{EPR}}_{unity} \left(\frac{IC(N1c2 \mid \eta)_{EPR=const}}{sss(N1c2)_{EPR}} \right)$$

$$\approx -\left(\frac{IC(N1c2 \mid \eta)_{EPR=const}}{sss(N1c2)_{EPR}} \right) \tag{A4.7.6}$$

Rewriting, we obtain

$$sss(N1c2 \mid \eta)_{EPR} \approx -\left(\frac{IC(N1c2 \mid \eta)_{EPR=const}}{IC(EPR \mid \eta)_{N1c2=const}} \right) \tag{A4.7.7}$$

Furthermore, combining and rearranging Equations (A4.7.5) and (A4.7.7), we have

$$sss(P)_{EPR} \approx \left(\frac{IC(P \mid \eta)_{N1c2=const} - IC(P \mid \eta)_{EPR=const}}{IC(EPR \mid \eta)_{N1c2=const}} \right) \tag{A4.7.8}$$

Equation (A4.7.8) provides a simple method to approximate the *sss* given two sets of ICs.

A4.8 OTHER CONSIDERATIONS FOR SENSOR ERROR ICS

Real world considerations sometimes impact our ideal choices (and assumptions), which, in turn, impact derived relationships, such as Equation (4.41). One such complication is the use of airframe total pressure (P_T) and total air temperature (TAT) to correct engine parameters when T_2 and/or P_2 are missing or not recorded. To understand the effect, we will retrace our derivation with the understanding that P_T and TAT are used for parameter correction. We first define two new ratio terms analogous to those used in Chapter 3, i.e.,

$$\theta_{AF} = \frac{TAT}{T_{ISA}} \quad , \quad \delta_{AF} = \frac{P_T}{P_{ISA}} \tag{A4.8.1}$$

We assume parameters are corrected to these conditions, i.e., Equation (4.33) now becomes

$$X_c = \frac{X}{\theta_{AF}^{\alpha_X} \, \delta_{AF}^{\beta_X}} \tag{A4.8.2}$$

On the other hand, the power setting parameter, $X_c^{(PSP)}$, is already a corrected quantity, it should be independent of the airframe, wholly dependent on the specific engine being controlled, and computed within the full authority digital engine control (FADEC). Therefore, Equation (4.34) will still apply.

$$X_c^{(PSP)} = \frac{X^{(PSP)}}{\theta^{\alpha_{X^{(PSP)}}} \delta^{\beta_{X^{(PSP)}}}} \tag{A4.8.3}$$

The parameter % deltas will still obey Equation (4.35), i.e.

$$1 + \frac{\Delta X}{100} = \frac{X_c}{X_{Base}} \tag{A4.8.4}$$

If we again assume that $X_{Base} = f(X_c^{(PSP)})$ and follow the same derivation as demonstrated in Chapter 4 using Equations (A4.8.2), (A4.8.3), and (A4.8.4), we arrive at an analogous result to Equation (4.41), as displayed in Equation (A4.8.5).

$$\begin{aligned} \frac{d(\Delta X)}{100} = \frac{dX}{X} &+ \left[\left(\alpha_{X^{(PSP)}}\, sss(X_{Base})_{X_c^{(PSP)}} \frac{d\theta}{\theta}\right) - \alpha_X \frac{d\theta_{AF}}{\theta_{AF}}\right] \\ &+ \left[\left(\beta_{X^{(PSP)}}\, sss(X_{Base})_{X_c^{(PSP)}} \frac{d\delta}{\delta}\right) - \beta_X \frac{d\delta_{AF}}{\delta_{AF}}\right] - sss(X_{Base})_{X_c^{(PSP)}} \frac{dX^{(PSP)}}{X^{(PSP)}} \end{aligned} \tag{A4.8.5}$$

A4.9 DERIVATION OF CORRECTED AIRFLOW SENSOR ERROR ICS

In this addendum, we will derive the sensor error ICs for a parameter that *is not directly measured* but is derived from other measurable parameters. The derived parameter we will consider is *corrected* engine inlet air flow ($w_{a\,c}$). This parameter can be *virtually* measured (i.e., calculated) during the ground testing of aero-engines fitted with an intake bellmouth having both *static* and *total* pressure probes. The coefficients for this parameter may be calculated by manipulating the following well-known relationships given in Equations (A4.9.1) and (A4.9.2) below.

$$\frac{P_2}{P_{S2}} = \left(1 + \frac{\gamma - 1}{2} M_n^2\right)^{\frac{\gamma}{\gamma - 1}} \tag{A4.9.1}$$

$$\begin{aligned} w_{a\,c} &= \frac{w_a\sqrt{\theta}}{\delta} \\ &= \left(A \frac{P_{ISA}}{\sqrt{T_{ISA}}}\right) \frac{w_a\sqrt{T_2}}{AP_2} \\ &= const \sqrt{\frac{\gamma}{R}} M_n \left(1 + \frac{\gamma - 1}{2} M_n^2\right)^{\frac{\gamma + 1}{2(1 - \gamma)}} \end{aligned} \tag{A4.9.2}$$

where M_n = inlet Mach number, γ = ratio of specific heats, A is cross-sectional area, and R is the gas constant. The last relationship given in Equation (A4.9.2) can be derived easily as follows. We start with several fundamental relationships:

$$M_n = \frac{v}{\sqrt{\gamma R T_s}}$$

$$P_s V = R T_s$$

$$\left(\frac{lb/s}{in^2}\right) = \frac{w_a}{A} = \frac{v}{V} = \left(\frac{in/s}{in^3/lb}\right)$$

Therefore,

$$\frac{w_a}{A} = \frac{M_n\sqrt{\gamma R T_s}}{\dfrac{RT_s}{P_s}}$$

$$\frac{w_a\sqrt{T_s}}{AP_s} = \frac{M_n\sqrt{\gamma R T_s}\sqrt{T_s}}{\dfrac{RT_s}{P_s}P_s} = M_n\sqrt{\frac{\gamma}{R}}$$

From this last equation we have

$$\frac{w_a\sqrt{T}}{AP_T} = \left(\frac{w_a\sqrt{T_s}}{AP_s}\right)\frac{\sqrt{T}}{\sqrt{T_s}}\frac{P_s}{P_T}$$

$$= \left(M_n\sqrt{\frac{\gamma}{R}}\right)\sqrt{\frac{T}{T_s}}\frac{P_s}{P_T}$$

We also know that (for isentropic flow)

$$\frac{T}{T_s} = \left(\frac{P_T}{P_s}\right)^{\frac{\gamma-1}{\gamma}} \quad and \quad \frac{P_T}{P_s} = \left(1+\frac{\gamma-1}{2}M_n^2\right)^{\frac{\gamma}{\gamma-1}}$$

Substituting, we obtain

$$\frac{w_a\sqrt{T}}{AP_T} = \left(M_n\sqrt{\frac{\gamma}{R}}\right)\left(\frac{P_T}{P_s}\right)^{\frac{\gamma-1}{2\gamma}}\frac{P_s}{P_T}$$

$$= \left(M_n\sqrt{\frac{\gamma}{R}}\right)\left(\frac{P_T}{P_s}\right)^{\frac{-(\gamma+1)}{2\gamma}}$$

$$= \left(M_n\sqrt{\frac{\gamma}{R}}\right)\left(\left(1+\frac{\gamma-1}{2}M_n^2\right)^{\frac{\gamma}{\gamma-1}}\right)^{\frac{-(\gamma+1)}{2\gamma}}$$

$$= \left(M_n\sqrt{\frac{\gamma}{R}}\right)\left(1+\frac{\gamma-1}{2}M_n^2\right)^{\frac{1+\gamma}{2(1-\gamma)}}$$

Taking logs and differentiating Equation (A4.9.1) yields

$$\frac{dP_2}{P_2}-\frac{dP_{S2}}{P_{S2}} = \frac{\gamma}{\gamma-1}\frac{M_n(\gamma-1)}{1+\frac{\gamma-1}{2}M_n^2}dM_n = \left(\frac{\gamma M_n^2}{1+\frac{\gamma-1}{2}M_n^2}\right)\frac{dM_n}{M_n} \quad \text{(A4.9.3)}$$

Likewise, differentiating Equation (A4.9.2), we obtain

$$\frac{dw_{ac}}{w_{ac}} = \frac{dM_n}{M_n}+\left(\frac{\gamma+1}{2(1-\gamma)}\right)\left(\frac{(\gamma-1)M_n}{1+\frac{\gamma-1}{2}M_n^2}\right)dM_n$$

$$= \frac{dM_n}{M_n}-\left(\frac{\frac{\gamma+1}{2}M_n^2}{1+\frac{\gamma-1}{2}M_n^2}\right)\frac{dM_n}{M_n} = \left(\frac{1-M_n^2}{1+\frac{\gamma-1}{2}M_n^2}\right)\frac{dM_n}{M_n} \quad \text{(A4.9.4)}$$

Combining Equations (A4.9.3) and (A4.9.4) with a little manipulation provides

$$\frac{dw_{ac}}{w_{ac}} = \left(\frac{1-M_n^2}{1+\frac{\gamma-1}{2}M_n^2}\right)\left(\frac{1+\frac{\gamma-1}{2}M_n^2}{\gamma M_n^2}\right)\left(\frac{dP_2}{P_2}-\frac{dP_{S2}}{P_{S2}}\right) \quad \text{(A4.9.5)}$$

$$= \left(\frac{1-M_n^2}{\gamma M_n^2}\right)\left(\frac{dP_2}{P_2}-\frac{dP_{S2}}{P_{S2}}\right)$$

From Equation 4.41 in Chapter 4, we know the general form

$$\frac{d(\Delta X)}{100}=\frac{dX}{X}+(\alpha_{X^{(PSP)}}\ sss(X_{Base})_{X_c^{(PSP)}}-\alpha_X)\frac{d\theta}{\theta}$$
$$+(\beta_{X^{(PSP)}}\ sss(X_{Base})_{X_c^{(PSP)}}-\beta_X)\frac{d\delta}{\delta}-sss(X_{Base})_{X_c^{(PSP)}}\frac{dX^{(PSP)}}{X^{(PSP)}} \tag{A4.9.6}$$

Letting $X = w_{ac}$, we have

$$\frac{d(\Delta w_{ac})}{100}=\frac{dw_{ac}}{w_{ac}}+(\alpha_{X^{(PSP)}}\ sss(X_{Base})_{X_c^{(PSP)}}-\underbrace{\alpha_{w_{ac}}}_{zero})\frac{d\theta}{\theta}$$
$$+(\beta_{X^{(PSP)}}\ sss(X_{Base})_{X_c^{(PSP)}}-\underbrace{\beta_{w_{ac}}}_{zero})\frac{d\delta}{\delta}-sss(X_{Base})_{X_c^{(PSP)}}\frac{dX^{(PSP)}}{X^{(PSP)}} \tag{A4.9.7}$$

Combining that with Equation (A4.9.5) provides our result.

$$\frac{d(\Delta w_{ac})}{100}=-\left(\frac{1-M_n^2}{\gamma M_n^2}\right)\frac{dP_{S2}}{P_{S2}}+\left[\alpha_{X^{(PSP)}}\ sss(X_{Base})_{X_c^{(PSP)}}\right]\frac{dT_2}{T_2}$$
$$+\left[\beta_{X^{(PSP)}}\ sss(X_{Base})_{X_c^{(PSP)}}+\left(\frac{1-M_n^2}{\gamma M_n^2}\right)\right]\frac{dP_2}{P_2}-sss(X_{Base})_{X_c^{(PSP)}}\frac{dX^{(PSP)}}{X^{(PSP)}} \tag{A4.9.8}$$

A4.10 MATRIX INVERSION LEMMA

For matrices A and B

$$(I+AB)^{-1}=I-A(I+BA)^{-1}B \tag{A4.10.1}$$

To show this let

$$\begin{aligned}-AB&=-AB\\&=-A(I+BA)(I+BA)^{-1}B\end{aligned}$$

Therefore,

$$\begin{aligned}I&=I+AB-AB\\&=(I+AB)-A(I+BA)(I+BA)^{-1}B\\&=(I+AB)-(I+AB)A(I+BA)^{-1}B\end{aligned} \tag{A4.10.2}$$

Multiplying both sides by $(I + AB)^{-1}$ yields

$$(I + AB)^{-1} = I - A(I + BA)^{-1} B$$

proving Equation (A4.10.1).

A4.11 ALTERNATE FORMS FOR THE DIAGNOSTIC (KALMAN GAIN) MATRIX

We can show the equivalence of several different forms for the Diagnostic Matrix D by considering the following manipulation.

$$\begin{aligned}(P_x^{-1} + H^T R^{-1} H) P_x H^T (HP_x H^T + R)^{-1} &= H^T (HP_x H^T + R)^{-1} + H^T R^{-1} HP_x H^T (HP_x H^T + R)^{-1} \\ &= H^T R^{-1} \left[R(HP_x H^T + R)^{-1} + HP_x H^T (HP_x H^T + R)^{-1} \right] \\ &= H^T R^{-1} \left[\left(R + HP_x H^T \right)(HP_x H^T + R)^{-1} \right] \\ &= H^T R^{-1}\end{aligned}$$

Therefore,

$$P_x H^T (HP_x H^T + R)^{-1} = (P_x^{-1} + H^T R^{-1} H)^{-1} H^T R^{-1} \tag{A4.11.1}$$

Equation (A4.11.1) displays two of the most common forms for D. A third form can be derived with the use of the matrix inversion lemma displayed in Addendum A4.10. To begin, we let

$$\begin{aligned}&A = H^T R^{-1} \quad and \quad B = HP_x \\ &\Rightarrow \\ &(I + AB) = I + H^T R^{-1} HP_x = (P_x^{-1} + H^T R^{-1} H) P_x \quad \Rightarrow \\ &(I + AB)^{-1} = P_x^{-1} (P_x^{-1} + H^T R^{-1} H)^{-1}\end{aligned} \tag{A4.11.2}$$

Equation (A4.11.2) provides the left side of Equation (A4.10.1).

$$\begin{aligned}(I + BA) = I + HP_x H^T R^{-1} &= (R + HP_x H^T) R^{-1} \quad \Rightarrow \\ (I + BA)^{-1} &= R(R + HP_x H^T)^{-1} \quad \Rightarrow \\ I - A(I + BA)^{-1} B &= I - H^T R^{-1} R(R + HP_x H^T)^{-1} HP_x \\ &= I - H^T (R + HP_x H^T)^{-1} HP_x\end{aligned} \tag{A4.11.3}$$

Equation A4.11.3 provides the right side of Equation A4.10.1, thus

$$P_x^{-1} (P_x^{-1} + H^T R^{-1} H)^{-1} = I - H^T (R + HP_x H^T)^{-1} HP_x$$

Thus, implying

$$(P_x^{-1} + H^T R^{-1} H)^{-1} = P_x - P_x H^T (H P_x H^T + R)^{-1} H P_x \qquad \text{(A4.11.4)}$$

Multiplying both sides of Equation (A4.11.4) by $H^T R^{-1}$ yields

$$(P_x^{-1} + H^T R^{-1} H)^{-1} H^T R^{-1} = P_x H^T R^{-1} - P_x H^T (H P_x H^T + R)^{-1} H P_x H^T R^{-1} \qquad \text{(A4.11.5)}$$

Combining Equation (A4.11.1) with (A4.11.5) gives

$$\begin{aligned} P_x H^T (H P_x H^T + R)^{-1} &= P_x H^T R^{-1} - \underbrace{P_x H^T (H P_x H^T + R)^{-1}}_{(P_x^{-1} + H^T R^{-1} H)^{-1} H^T R^{-1}} H P_x H^T R^{-1} \\ &= P_x H^T R^{-1} - (P_x^{-1} + H^T R^{-1} H)^{-1} H^T R^{-1} H P_x H^T R^{-1} \\ &= \left[I - (P_x^{-1} + H^T R^{-1} H)^{-1} H^T R^{-1} H \right] P_x H^T R^{-1} \end{aligned} \qquad \text{(A4.11.6)}$$

Thus, we have three formulations for D, as depicted in Equation (A4.11.7).

$$\begin{aligned} D &= P_x H^T (H P_x H^T + R)^{-1} \\ D &= (P_x^{-1} + H^T R^{-1} H)^{-1} H^T R^{-1} \\ D &= \left[I - (P_x^{-1} + H^T R^{-1} H)^{-1} H^T R^{-1} H \right] P_x H^T R^{-1} \end{aligned} \qquad \text{(A4.11.7)}$$

A4.12 SOME RELATIONSHIPS INVOLVING ESTIMATION ERROR

In this addendum, we derive the three results given in Chapter 4, Equation (4.62).

We begin by assuming no *a-priori* information is available, i.e., $\mu_x = 0$, and, thus, we have that

$$\begin{aligned} \| e_x \|^2 = e_x^T e_x &= ((I - DH)x)^T ((I - DH)x) \\ &= x^T (I - DH)^T (I - DH)x \end{aligned} \qquad \text{(A4.12.1)}$$

Therefore,

$$\frac{\| e_x \|^2}{\| x \|^2} = \frac{x^T (I - DH)^T (I - DH)x}{x^T x} \qquad \text{(A4.12.2)}$$

Because $(I - DH)^T (I - DH)$ is symmetric, there exists an *orthonormal basis* of *eigenvectors* $\{y_i\}_{i=1,2,\ldots,n}$. Thus, for an arbitrary x, there exists scalars α_i, such that

$$x = \sum_{i=1}^{n} \alpha_i y_i$$
$$(I-DH)^T (I-DH)x = \sum_{i=1}^{n} \alpha_i \lambda_i^{(I-DH)^T(I-DH)} y_i \tag{A4.12.3}$$

Substituting into (A4.12.2) yields

$$\frac{\|e_x\|^2}{\|x\|^2} = \frac{\left(\sum_{i=1}^{n} \alpha_i y_i^T\right)\left(\sum_{i=1}^{n} \alpha_i \lambda_i^{(I-DH)^T(I-DH)} y_i\right)}{\left(\sum_{i=1}^{n} \alpha_i y_i^T\right)\left(\sum_{i=1}^{n} \alpha_i y_i\right)} = \frac{\sum_{i=1}^{n} \alpha_i^2 \lambda_i^{(I-DH)^T(I-DH)}}{\sum_{i=1}^{n} \alpha_i^2} \tag{A4.12.4}$$

It follows from this that the *normalized* (estimation) error is bounded by the *min* and *max* eigenvalues of the matrix $(I-DH)^T(I-DH)$, i.e.,

$$\sqrt{\lambda_{\min}^{(I-DH)^T(I-DH)}} \le \frac{\|e_x\|}{\|x\|} \le \sqrt{\lambda_{\max}^{(I-DH)^T(I-DH)}} \tag{A4.12.5}$$

This establishes the first relationship in Equation (4.62).

Proceeding, we see from Equation A4.12.4, that the (squared) normalized error is a weighted average of the eigenvalues. If we make the substitution

$$\beta_i = \alpha_i^2 \Big/ \sum_{i=1}^{n} \alpha_i^2 \tag{A4.12.6}$$

then

$$\frac{\|e_x\|^2}{\|x\|^2} = \frac{\sum_{i=1}^{n} \alpha_i^2 \lambda_i}{\sum_{i=1}^{n} \alpha_i^2} = \sum_{i=1}^{n} \beta_i \lambda_i \quad \text{where} \quad \sum_{i=1}^{n} \beta_i = 1 \tag{A4.12.7}$$

Taking expectations of both sides, we obtain

$$E\left(\frac{\|e_x\|^2}{\|x\|^2}\right) = E\left(\sum_{i=1}^{n} \beta_i \lambda_i\right) = \sum_{i=1}^{n} \lambda_i E(\beta_i) = \bar{\beta} \sum_{i=1}^{n} \lambda_i$$

and since $\sum_{i=1}^{n} \beta_i = 1 \Rightarrow E\left(\sum_{i=1}^{n} \beta_i\right) = n\bar{\beta} = 1 \Rightarrow \bar{\beta} = \frac{1}{n}$ (A4.12.8)

implying

$$E\left(\frac{\|e_x\|^2}{\|x\|^2}\right) = \frac{1}{n}\sum_{i=1}^{n} \lambda_i = \frac{1}{n} Trace[(I-DH)^T(I-DH)]$$

This establishes the second relationship in Equation (4.62).

The final derivation is to establish the following

Proposition: $E\left(\frac{\|e_x\|^2}{\|x\|^2}\right) > 1 - \frac{m}{n}$, *for* $m \leq n$, where m is the number of measurements, and n is the number of faults.

The outline of the derivation is as follows: recall the following relationships[15]

$$\begin{gathered}(I - DH)^T (I - DH) = P_x^{-1} P_e^2 P_x^{-1} \\ \textit{where} \\ P_e = (I - DH)P_x = (P_x^{-1} + H^T R^{-1} H)^{-1}\end{gathered} \tag{A4.12.9}$$

We make the following definitions:

$$\begin{aligned}&V = H^T R^{-1} H \quad \textit{and} \quad W = P_1^{-1} \quad \Rightarrow \quad W = P_0^{-1} + V \\ &\textit{Let} \quad X = WP_0 = I_n + VP_0 = I_n + Y\end{aligned} \tag{A4.12.10}$$

Because $rank(H) \leq m \quad \Rightarrow \quad rank(V) \leq m \quad \Rightarrow \quad rank(Y) \leq \min\{rank(V), rank(P_0)\} \leq m$

Thus, there exists a unitary matrix Q such that Y is transformed into an upper triangular form with, at most, m non-zero eigenvalues, i.e.,

$$\begin{aligned}Q^{-1} X Q &= I_n + \begin{bmatrix} \lambda_1 & & & & & \\ & \ddots & & & T & \\ & & \lambda_m & & & \\ & & & 0 & & \\ & 0 & & & \ddots & \\ & & & & & 0 \end{bmatrix} = I_n + Tri[\lambda_1, \cdots, \lambda_m, 0, \cdots, 0] \\ &= Tri[(1+\lambda_1), \cdots, (1+\lambda_m), 1, \cdots, 1]\end{aligned} \tag{A4.12.11}$$

Because X^{-1} exists, it can be represented as a polynomial in X and, thus, X^{-1} is also similar to an upper triangular matrix. Indeed,

$$Q^{-1} X^{-1} Q = Tri[\frac{1}{1+\lambda_1}, \cdots, \frac{1}{1+\lambda_m}, 1, \cdots, 1] \tag{A4.12.12}$$

From Equation (A4.12.8), we have

$$E\left(\frac{\|e_x\|^2}{\|x\|^2}\right)=\frac{1}{n}Trace[(I-DH)^T(I-DH)]=\frac{1}{n}Trace[X^{-1}X^{-T}]$$
$$=\frac{1}{n}Trace[Q^{-1}X^{-1}X^{-T}Q]=\frac{1}{n}Trace[Q^{-1}X^{-1}QQ^{-1}X^{-T}Q]$$
$$=\frac{1}{n}\left[\sum_{i=1}^{m}\frac{1}{(1+\lambda_i)^2}+(n-m)\right]$$
$$>\frac{1}{n}(n-m)=1-\frac{m}{n}$$

(A4.12.13)

Additional details for the material presented in this Addendum can be found in [4].

NOTES

1. Also referred to as the dot product and is commonly written as (X,X) or $X \cdot X$.
2. Note that the diagonal of P_{XX} contains the variances $(\sigma_i^2,\ i=1,2,\ldots,n)$ of the individual components of X.
3. See Addendum A4.3, Equation (A4.3.1).
4. Proof can be found in D. Stephen G. Pollock, Conditional Expectations and Regression Analysis - lecture notes 1, page 2, https://www.studocu.com/en-gb/document/university-of-leicester/econometrics-1/lecture-notes/conditional-expectations-and-regression-analysis-lecture-notes-1/1260713/view.
5. Developed by the mathematician Carl Friedrich Gauss in 1795, later published in 1809.
6. Hummer, D.G., 1986, *Rational approximations for the Holtsmark distribution, its cumulative and derivative,* Journal of Quantitative Spectroscopy and Radiative Transfer. Vol 36: pages 1–5.
7. For proofs of these properties, consult any of the available books on linear algebra and matrix theory, such as the two volume set: Gantmacher, F.R., 1998, *Matrix Theory: Vol. 1 and Vol. 2,* American Mathematical Society with Chelsea Publishing, (AMS Chelsea Publishing), Providence, R.I.
8. Matrix [..., B,...] is matrix A with the column A^j replaced with vector B.
9. See next section of this Addendum.
10. Supremum (denoted *sup*) is essentially the least upper bound.
11. A proof for this theorem can be found in, Isaacson, E., Keller, H.B., 1966, *Analysis of Numerical Methods*, John Wiley & Sons, Chapter1, page 7.
12. Not globally differentiable.
13. Holding the power setting parameter constant will generate the ICs used to model the relationship implicit in Figure 4.1; however, there is often the need to generate additional sets of ICs holding other parameters constant, such as corrected speeds, thrust, and compressor flows. See section on Ancillary Parameters Generated by ICs at the end of this Addendum 4.4.
14. The diagnostic literature follows this particular definition; however, there are other groups who prefer the reciprocal relationship. It does not matter which is used as long as consistency is maintained.
15. See Equations (4.57) and (4.61) in Chapter 4.

REFERENCES

1. Doel, D.L., 2003, Development of Baselines, Influence Coefficients and Statistical Inputs for Gas Path Analysis, *Von Karman Lecture Notes, Gas Turbine Condition Monitoring and Fault Diagnosis*, January 2003.
2. Borguet, S., Leonard, O., 2009, A Sparse Estimation Approach to Fault Isolation, GT2009-59099, *Proceedings of GT2009 ASME Turbo Expo 2009: Power for Land, Sea and Air,* June 8–12, 2009, Orlando, FL, USA.
3. Volponi, A.J., et al., 2000, The Use of Kalman Filter and Neural Network Methodologies in Gas Turbine Performance Diagnostics: A Comparative Study, ASME 2000-GT-547, IGTI Turbo Expo 2000, Munich; Transactions of the ASME, *Journal of Engineering for Gas Turbines and Power*, Vol 125, pp. 917–924, Oct. 2003.
4. Volponi, A.J., 2003, Foundations of Gas Path Analysis I&II, *Von Karman Lecture Series VKI LS 2003-01, Gas Turbine Condition Monitoring and Fault Diagnosis*, Von Karman Institute, Rhode-Saint-Genese, Belgium.

A5 Chapter 5 Addendums

A5.1 LARGE MEASUREMENT ERROR RECOVERY

We assume we have the following model/estimate formulation [1]:

A5.1.1 Model

$$
\begin{aligned}
& Z = H\,x + \varphi \\
\textit{where}\quad & x = (n\times 1)\ \textit{fault}\ \Delta\ \textit{vector},\quad x \in N(\mu_x, P_x) \\
& \varphi = (m\times 1)\ \textit{noise vector},\quad \varphi \in N(0,R) \\
& Z = (m\times 1)\ \textit{measurement}\ \Delta\ \textit{vector} \\
& H = (m\times n)\ \textit{matrix of IC's} \\
& \quad x\ \textit{and}\ \varphi\ \textit{independent}
\end{aligned}
\tag{A5.1.1}
$$

Moreover,

$$
Z = H_e x_e + H_s x_s + \varphi = \begin{bmatrix} H_e & \vdots & H_s \end{bmatrix} \begin{bmatrix} x_e \\ \cdots \\ x_s \end{bmatrix} + \varphi \; = \; H\,x + \varphi \tag{A5.1.2}
$$

where H_e is an $(m \times n_e)$ matrix of engine fault (influence) coefficients, H_s is an $(m \times n_s)$ matrix of sensor fault (influence) coefficients, and H is an $(m \times n)$matrix of influences formed by concatenating H_e and H_s.

A5.1.2 Estimate

$$
\hat{x} = \mu_x + D(Z - H\,\mu_x) \tag{A5.1.3}
$$

$$
\begin{bmatrix} \hat{x}_e \\ \cdots \\ \hat{x}_s \end{bmatrix} = \begin{bmatrix} \mu_e \\ \cdots \\ \mu_s \end{bmatrix} + \begin{bmatrix} D_e \\ \cdots \\ D_s \end{bmatrix} \left(Z - \begin{bmatrix} H_e \vdots H_s \end{bmatrix} \begin{bmatrix} \mu_e \\ \cdots \\ \mu_s \end{bmatrix} \right)
$$

where

$$
D = \begin{bmatrix} D_e \\ \cdots \\ D_s \end{bmatrix} = P_0 H^T \left(H P_0 H^T + R \right)^{-1} = \begin{bmatrix} P_0^{(e)} H_e^T (H_e P_0^{(e)} H_e^T + R)^{-1} \\ \cdots\cdots \\ P_0^{(s)} H_s^T (H_s P_0^{(s)} H_s^T + R)^{-1} \end{bmatrix} \tag{A5.1.4}
$$

From Equation (A5.1.3) we have

$$\begin{aligned}\hat{x}_e &= \mu_e + D_e(Z - H_e\mu_e - H_s\mu_s)\\ \hat{x}_s &= \mu_s + D_s(Z - H_e\mu_e - H_s\mu_s)\end{aligned} \tag{A5.1.5}$$

We introduce the following notation.

$$(Z - H_e\,\mu_e) \triangleq \varsigma = [\varsigma_1,\ \varsigma_2,\ \varsigma_3,\ \ldots,\ \varsigma_m]$$

$$\mu_e = [e_1, e_2, e_3, \ldots, e_{n_e}], \qquad \mu_s = [s_1, s_2, s_3, \ldots, s_{n_s}]$$

$$\hat{x}_e = [\hat{e}_1, \hat{e}_2, \hat{e}_3, \ldots, \hat{e}_{n_e}], \qquad \hat{x}_s = [\hat{s}_1, \hat{s}_2, \hat{s}_3, \ldots, \hat{s}_{n_s}]$$

$$h_{ij} = [H_s]_{ij} = ij^{th} \ \textit{element of} \ \ H_s$$

$$d^e_{ij} = [D_e]_{ij} = ij^{th} \ \textit{element of} \ \ D_e$$

$$d^s_{ij} = [D_s]_{ij} = ij^{th} \ \textit{element of} \ \ D_s$$

Rewriting Equation (A5.1.5)

$$\begin{aligned}\hat{e}_i &= e_i + \sum_{j=1}^{m} d^e_{ij}\left(\varsigma_j - \sum_{n=1}^{n_s} h_{jn}s_n\right) \quad (1 \le i \le n_e)\\ \hat{s}_i &= s_i + \sum_{j=1}^{m} d^s_{ij}\left(\varsigma_j - \sum_{n=1}^{n_s} h_{jn}s_n\right) \quad (1 \le i \le n_s)\end{aligned} \tag{A5.1.6}$$

For a given analysis, the values $\hat{s}_j$ represent the estimated measurement error (sensor bias, drift, etc.) in the jth measured engine parameter. If this value is *small*, nothing further needs to be done, and the values $\hat{e}_j$ provide estimates for the jth engine fault parameter (e.g., compressor efficiency), the accuracy of which will depend on the fidelity of the a-priori estimate e_j. If the value is *large*, it is indicative of a large measurement delta (or, more accurately, a large measurement residual) that, in turn, will contribute to a misassessment in some or all of the engine faults. Thus, for a given analysis, it is necessary to render a judgment as to whether a given sensor delta estimate is large enough to corrupt the engine fault assessments. This is generally accomplished by comparing each value $\left\|\hat{s}_j - s_j\right\|$ to a threshold limit T_j, for all j. If an exceedance is noted, we select that parameter whose exceedance is the largest (in case there is more than one) by comparing the ratios of the absolute difference to their respective threshold and selecting the largest, e.g.,

$$k = \arg\left\{\max_n\left(\frac{\|\hat{s}_n - s_n\|}{T_n}\right)\right\} \tag{A5.1.7}$$

would record the kth sensor as the major suspect. Our strategy will be to find an alternate *a-priori* value for the offending sensor, s_k (which we will denote by s'_k), that will place this difference below the threshold limit and at the same time admit the least overall error among the ensemble of engine/sensor faults. Before making explicit how we intend to accomplish this, it is necessary to further develop Equation (A5.1.6).

We define an index set Ω_k to be those rows of the kth column of the H_s matrix which are ***non-zero***, i.e.,

$$\Omega_k = \{ n \mid h_{nk} \neq 0 \} \tag{A5.1.8}$$

In terms of this index set, we can rewrite the first of Equations (A5.1.6) as

$$\hat{e}_i = e_i + \sum_{n \in \Omega_k} d_{in}^{(e)} \left(\varsigma_n - \sum_m h_{nm} s_m \right) + \underbrace{\left(\sum_{n \notin \Omega_k} d_{in}^{(e)} \left[\varsigma_n - \sum_m h_{nm} s_m \right] \right)}_{\phi_i} \tag{A5.1.9}$$

Equation (A5.1.9) can be further reduced to expose a linear equation in s_k as

$$\hat{e}_i = e_i + a_i s_k + b_i \tag{A5.1.10}$$

where

$$a_i = -\sum_{n \in \Omega_k} d_{in}^{(e)} h_{nk} \quad \textit{and} \quad b_i = \sum_{n \in \Omega_k} d_{in}^{(e)} \left(\varsigma_n - \sum_{m \neq k} h_{nm} s_m \right) + \phi_i$$

It should be noted that ϕ_i (by virtue of its definition in Equation (A5.1.9)) is independent of s_k . The reader can easily verify that this observation is also true of the quantities a_i and b_i as well. This implies that *adjusting* the a-priori estimate s_k will affect *all* engine fault estimates in the linear fashion dictated by Equation (A5.1.10). If we perform precisely the same operations on the second part of Equation (A5.1.6), we arrive at the analogous equation

$$\hat{s}_i = s_i + \alpha_i s_k + \beta_i \tag{A5.1.11}$$

where

$$\alpha_i = -\sum_{n \in \Omega_k} d_{in}^{(s)} h_{nk} \quad \textit{and} \quad \beta_i = \sum_{n \in \Omega_k} d_{in}^{(s)} \left(\varsigma_n - \sum_{m \neq k} h_{nm} s_m \right) + \psi_i$$

and

$$\psi_i = \sum_{n \in \Omega} d_{in}^{(s)} \left[\varsigma_n - \sum_m h_{nm} s_m \right] \tag{A5.1.12}$$

By analogy, ψ_i , α_i , and β_i are also independent of s_k and, thus, *adjusting* this a-priori value will also affect *all* sensor fault estimates (Equation (A5.1.11)) as well. The independence from the index k is of importance, as we shall shortly provide a new estimate s_k', which is solely a function of these coefficients. This will imply that the *recovery value*[1] will be the same regardless of the magnitude of the offending (*kth*) parameter.

We choose the new a-priori value such that engine and sensor fault deviations (from expected) are minimized, i.e.,

$$S = \sum_n (\hat{s}'_n - s'_n)^2 + \sum_m (\hat{e}'_m - e'_m)^2 = \text{minimum} \tag{A5.1.13}$$

where

(1) s'_k is to be specified

(2) $s'_n = s_n \quad \forall n \neq k$

(3) $e'_m = e_m \quad \forall m$

(4) Equations (A5.1.10) and (A5.1.11) are satisfied

Expanding (A5.1.13), differentiating with respect to s_k and setting equal to zero yields

$$s'_k = -\frac{\sum_i \alpha_i \beta_i + \sum_I a_I b_I}{\sum_i \alpha_i^2 + \sum_I a_I^2} \tag{A5.1.14}$$

However, because of Equation (A5.1.11)

$$\begin{aligned} & \beta_i = \hat{s}_i - s_i - \alpha_i s_k \\ \Rightarrow \quad & \alpha_i \beta_i = \alpha_i (\hat{s}_i - s_i) - \alpha_i^2 s_k \\ \Rightarrow \quad & \sum_i \alpha_i \beta_i = \sum_i \alpha_i (\hat{s}_i - s_i) - \sum_i \alpha_i^2 s_k \end{aligned}$$

Likewise,

$$\sum_I a_I b_I = \sum_I a_I (\hat{e}_I - e_I) - \sum_i a_I^2 s_k$$

Substituting these relationships in Equation (A5.1.14), provides an association for the *new* estimate in terms of the *old* estimate, namely

$$s'_k = s_k - \frac{\sum_i \alpha_i (\hat{s}_i - s_i) + \sum_I a_I (\hat{e}_I - e_I)}{\sum_i \alpha_i^2 + \sum_I a_I^2} \tag{A5.1.15}$$

This can be rewritten in a more succinct form by recalling the definitions of the coefficients α and a and introducing a simplifying notation, i.e.,

$$s'_k = s_k + \frac{[DH_s]_k \bullet (\hat{x} - \mu)}{\left\| [DH_s]_k \right\|^2} \tag{A5.1.16}$$

where $[DH_s]_k$ denotes the *kth* column of the matrix product DH_s, $\| \; \|$ the Euclidean norm of a vector (sum of the squares of the components), and • the vector dot product.

Thus, Equation (A5.16) stipulates that the new a-priori estimate for the *kth* sensor fault is the previous a-priori estimate adjusted by a dot product between a normalized (matrix column) vector and the difference between the old a-priori vector and the estimated vector.

The correspondence given by Equation (A5.1.16) will allow a *single detected* large measurement error to be corrected. Clearly, in order to recover from a large measurement error, we need first to determine that one has actually occurred. One method that can be employed is to compare each difference $|\hat{s}_k - s_k|$ to a threshold value T_k for each $1 \le k \le n_s$. This is done by computing the ratio of these two quantities for each sensor fault and selecting the index (*k*) that yields the largest ratio greater than 1, i.e.,

$$k = \arg\left[\max_i \left\{\frac{\|\hat{s}_i - \bar{s}_i\|}{T_i}\right\} \ge 1\right] \tag{A5.1.17}$$

An exceedance would indicate that the estimate and its a-priori value differ by an excessive amount and that a *recovery* would be warranted. The choice of the T_k, however, is neither arbitrary nor straightforward, and a certain degree of care must be taken in its prescription. The problem that can arise from poorly defined threshold values is one of confounding, i.e., that an error in one parameter appears to be an error in another parameter, hence a recovery is launched on the wrong variable. To see how this can occur, let us consider to what degree the inclusion of a bias θ_k in the ***kth*** sensor error component will contribute to the ***ith*** sensor error difference for arbitrary *i*. It is easy to see that this quantity is given by $[D_s H_s]_{ik}\,\theta_k \equiv d_{ik}\,\theta_k$. Thus, in order to prevent confounding (e.g., *kth* fault perceived as an *ith* fault) we must have that

$$\frac{d_{ik}\theta_k}{T_i} < \frac{d_{kk}\theta_k}{T_k} \qquad \forall i \ne k \tag{A5.1.18}$$

i.e.,

$$\frac{T_i}{T_k} \ge \frac{d_{ik}}{d_{kk}} \equiv L_{ik} \qquad \forall i \ne k \tag{A5.1.19}$$

Similarly, an *ith* fault should not be perceived as a *kth* fault and thus,

$$\frac{d_{ki}\,\theta_i}{T_k} < \frac{d_{ii}\,\theta_i}{T_i} \qquad \Leftrightarrow \qquad \frac{T_i}{T_k} < \frac{d_{ii}}{d_{ki}} \equiv U_{ik} = 1/L_{ki} \tag{A5.1.20}$$

which combined with Equation (A5.1.18) imposes the following family of inequalities.

$$L_{ik}\,T_k \le T_i \le U_{ik}\,T_k \qquad \forall k \quad and \quad \forall i \ne k \tag{A5.1.21}$$

The matrices $[U_{ik}]$ and $[L_{ik}]$ can be computed via Equations (A5.1.19) and (A5.1.20) and used for the purpose of testing the efficacy of the threshold values for each sensor fault.

A5.2 DISCRETE TIME KALMAN FILTER

The Kalman filter is an optimal observer in a linear system having white, uncorrelated measurement/process noise [2], [3]. The filter has been formulated for both continuous and discrete time systems. For simplicity, we shall consider only the discrete time formulation. Because any practical application of Kalman filtering techniques will be implemented on a digital computer, the discrete time formulation is particularly well suited.

For each discrete time k (epoch), we will assume the following system to be known:

$$\begin{aligned} \text{System Model:} \quad & x_k = \Phi(k)x_{k-1} + \omega_k \\ \text{Measurement Model:} \quad & Z_k = H_k\, x_k + \varphi_k \end{aligned} \tag{A5.2.1}$$

where

$$\begin{aligned} x_k &= \text{State vector at epoch } k \\ \Phi(k) &= \text{State Transition Matrix} \\ \omega_k &= \text{Process noise vector} \\ \varphi_k &= \text{Measurement noise vector} \\ Z_k &= \text{Measurement vector} \\ H_k &= \text{Geometry matrix} \end{aligned}$$

It is further assumed that the noise processes are white and uncorrelated, i.e., $E(\varphi_k) = E(\omega_k) = 0$ and $\text{cov}(\varphi_k, \omega_j) = 0$ for all epochs k, j, and that both processes are Normal distributed. It is further assumed that the initial conditions and the covariance matrices are known for all epochs, i.e.,

$$\begin{aligned} &\text{cov}(\varphi_k, \varphi_k) = R_k > 0 \\ &\text{cov}(\omega_k, \omega_k) = Q_k \geq 0 \\ &E(x_0) = \mu_x(0) \\ &\text{cov}(x_0, x_0) = P_0 > 0 \end{aligned}$$

The Kalman filter is a recursive algorithm having the same form as the optimal estimator derived above, i.e. a predictor/corrector format. The gain matrix D is computed for all epochs k.

Kalman Filter Equations :

$$\text{State Extrapolation: } \hat{x}(k|k-1) = \Phi(k)\hat{x}_{k-1}$$

$$\text{Covariance Extrapolation: } P(k|k-1) = \Phi(k)\,P_{k-1}\,\Phi^T(k) + Q_k$$

$$\text{Kalman Gain: } D_k = P(k|k-1)\,H_k^T\left(H_k\,P(k|k-1)\,H_k^T + R_k\right)^{-1}$$

$$\text{State Update: } \hat{x}_k = \hat{x}(k|k-1) + D_k\left[Z_k - H_k\,\hat{x}(k|k-1)\right]$$

$$\text{Covariance Update: } P_k = \left[I - D_k H_k\right]P(k|k-1)$$

(A5.2.2)

It should be clear from the filter equations that the gain matrix is *totally independent of any measurements taken*. It depends entirely on the system dynamics, the measurement model, and the noise statistics of the system. Actual measurements do not, in any way, influence the calculation of the gain. This implies that the gains can be pre-computed off-line for use in a real-time Kalman filter application. Although we will not discuss this topic here, the gain equation can be sensitive to numerical errors. The propagation of errors due to the finite word length on a digital computer can cause the Kalman filter to diverge if care is not taken to ensure accuracy in the calculations. This numerical instability problem is well known and has been overcome by several methods (U-D factorization, "square root" filter, finite memory filtering, to name a few).

It is easy to see from the above equations that the Kalman filter is a predictor/corrector type of estimator. The predictor part is an *extrapolation* (in time) of the previous estimate, via the model dynamics of the system under consideration. Likewise, the covariance of the state is extrapolated in time and then updated by the measurement model information. Let us consider this equation at epoch k = 1, and for simplicity, let us also assume that there is no process noise ($Q = 0$), that the measurement system is time invariant, i.e., $H_k = H$, and that the state is constant $(\Phi = I)$.

$$
\begin{aligned}
P_1 &= [I - DH]P_0 \quad \textit{where } D = P_0H^T\left(HP_0H^T + R\right)^{-1} = \left(P_0^{-1} + H^TR^{-1}H\right)^{-1}H^TR^{-1} \\
&= \left[I - \left(P_0^{-1} + H^TR^{-1}H\right)^{-1}H^TR^{-1}H\right]P_0 \\
&= \left(P_0^{-1} + H^TR^{-1}H\right)^{-1}\left[\left(P_0^{-1} + H^TR^{-1}H\right) - H^TR^{-1}H\right]P_0 \\
&= \left(P_0^{-1} + H^TR^{-1}H\right)^{-1} \\
&\Rightarrow P_1^{-1} - P_0^{-1} = H^TR^{-1}H \geq 0 \\
&\therefore \quad P_1 \leq P_0
\end{aligned}
$$

This shows that the covariance does not increase with time (it will decrease when H is 1-1). It also indicates that decreasing R will decrease P_1. Although we will not derive it here, it is a simple algebraic exercise to show that P_1 is actually the covariance of the error $e = x - \hat{x}$ at epoch k = 1, (see Chapter 4 Equations (4.55)–(4.61).

Thus, the error decreases when the noise decreases (a reasonable result), and the error decreases with more time, i.e., more measurements (a satisfying result).

A5.3 KALMAN FILTER STATE EXTRAPOLATION USING A DETERIORATION MODEL

From Addendum A5.2, we have that

$$\begin{aligned} &\hat{x}(k|k-1) = \Phi(k)\hat{x}_{k-1} \\ &where \\ &x_k = \Phi(k)x_{k-1} + \omega_k \end{aligned} \quad \text{(A5.3.1)}$$

If we assume zero process noise, i.e., $\omega_k = 0$, then

$$\Phi(k) = diag\left(\underbrace{ratio_k^{(1)}, \ldots, ratio_k^{(n_e)}}_{n_e \text{ engine faults}}, \underbrace{ratio_k^{(n_e+1)}, \ldots, ratio_k^{(n_e+n_s)}}_{n_s \text{ sensor faults}} \right) \quad \text{(A5.3.2)}$$

$$where \;\; ratio_k^{(j)} = \frac{x_k(j)}{x_{k-1}(j)}$$

Let us assume we have average deterioration models for each of our engine performance faults η that might look something like the artist rendition depicted in Figure A5.1.

If we define $\Delta\eta = \eta_k - \eta_{k-1}$, then

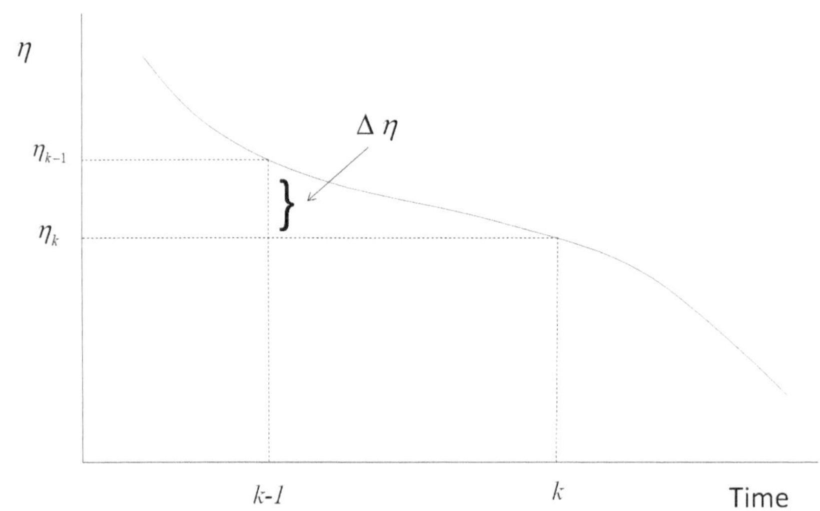

FIGURE A5.1 Figurative deterioration curve

$$ratio_k = \frac{\eta_k}{\eta_{k-1}} = 1 + \frac{\Delta\eta}{\eta_{k-1}} \tag{A5.3.3}$$

If we let $\hat{\eta}_{k|k-1}$ denote the *jth* component of $\hat{x}(k|k-1)$, then by Equation (A5.3.1), we have the following.

$$\hat{\eta}_{k|k-1} = ratio_k\,\hat{\eta}_{k-1} = \left(1 + \frac{\Delta\eta}{\eta_{k-1}}\right)\hat{\eta}_{k-1} = \hat{\eta}_{k-1} + \Delta\eta\left(\frac{\hat{\eta}_{k-1}}{\eta_{k-1}}\right) \tag{A5.3.4}$$

Thus, the value $\hat{\eta}_{k|k-1}$ can be used as an a-priori estimate as the predictor in the Kalman *predictor–corrector* equation. It should be noted that for the n_s sensor faults, there typically is no deterioration model, in which case, the n_s ratios in $\Phi(k)$ would be 1, and hence $\hat{\eta}_{k|k-1} = \hat{\eta}_{k-1}$.

A5.4 SFA DIAGNOSTIC MATRIX

The SFA model we have adopted can be described by Equation (A5.4.1) below.

$$\begin{aligned}
&\Delta\Delta = H^{(SFA)}\Delta x^{(SFA)} \\
&\Delta\hat{x}^{(SFA)} = D^{(SFA)}\Delta\Delta \\
&where \\
&D^{(SFA)} = P_0\,(H^{(SFA)})^T\left[H^{(SFA)}P_0\,(H^{(SFA)})^T + R\right]^{-1} \\
&with \\
&P_0 = diag\{0,0,\ldots,\underbrace{\beta}_{J^{th}},0,\ldots 0\}
\end{aligned} \tag{A5.4.1}$$

The scalar β is taken to be a large number. We will show that the result is actually a LSQ solution in the sense of $D = (H^T R^{-1} H)^{-1} H^T R^{-1} = H^{-1}$ when $\beta \to \infty$. We make the following observations summarized in Equation (A5.4.2), wherein we relax our notation with $H \equiv H^{(SFA)}$.

$$\begin{aligned}
&P_0 = diag\{0,0,\ldots,\underbrace{\beta}_{J^{th}},0,\ldots 0\} \\
&I_0 \triangleq diag\{0,0,\ldots,\underbrace{1}_{J^{th}},0,\ldots 0\} \triangleq [o_{ik}] = \begin{cases} 1 & i = k = J \\ 0 & otherwise \end{cases} \\
&D^{(SFA)} = P_0 H^T\left(HP_0H^T + R\right)^{-1} = \beta I_0 H^T \frac{1}{\beta}\left(HI_0H^T + \frac{R}{\beta}\right)^{-1} \\
&\quad = I_0 H^T\left(HI_0H^T + \frac{R}{\beta}\right)^{-1} \to I_0 H^T\left(HI_0H^T\right)^{-1}\ as\ \beta \to \infty
\end{aligned} \tag{A5.4.2}$$

We introduce some temporary notation.

$$H = [h_{ik}] \quad i = 1,2,\ldots,m \quad , \quad k = 1,2,\ldots,n$$

$$I_0 = \underbrace{\begin{bmatrix} 0 & 0 & \cdots & 0 & 0 \\ 0 & 0 & \cdots & 0 & 0 \\ 0 & 0 & 1 & 0 & 0 \\ \\ 0 & 0 & \cdots & 0 & 0 \end{bmatrix}}_{n\times n} \quad and \quad H^T = \underbrace{\begin{bmatrix} h_{11} & h_{21} & \cdots & h_{1n} \\ h_{12} & h_{22} & \cdots & h_{2n} \\ h_{1J} & h_{2J} & \cdots & h_{mJ} \\ \\ h_{1n} & h_{2n} & \cdots & h_{mn} \end{bmatrix}}_{n\times m}$$

Then

$$I_0 H^T = \underbrace{\begin{bmatrix} 0 & 0 & \cdots & 0 & 0 \\ 0 & 0 & \cdots & 0 & 0 \\ h_{1J} & h_{2J} & & & h_{mJ} \\ 0 & 0 & & 0 & 0 \\ 0 & 0 & \cdots & 0 & 0 \end{bmatrix}}_{n\times m}$$

$$HI_0H^T = \underbrace{\begin{bmatrix} h_{11} & h_{12} & h_{1J} & h_{1n} \\ h_{21} & h_{22} & h_{2J} & h_{2n} \\ & & \cdots & \\ \\ h_{m1} & h_{m2} & h_{mJ} & h_{mn} \end{bmatrix}}_{m\times n} \underbrace{\begin{bmatrix} 0 & 0 & \cdots & 0 & 0 \\ 0 & 0 & \cdots & 0 & 0 \\ h_{1J} & h_{2J} & & & h_{mJ} \\ 0 & 0 & & 0 & 0 \\ 0 & 0 & \cdots & 0 & 0 \end{bmatrix}}_{n\times m}$$

$$= \underbrace{\begin{bmatrix} h_{1J}^2 & h_{1J}h_{2J} & \cdots & h_{1J}h_{mJ} \\ h_{2J}h_{1J} & h_{2J}^2 & \cdots & h_{2J}h_{mJ} \\ \vdots & \vdots & & \vdots \\ \vdots & \vdots & & \vdots \\ \\ h_{mJ}h_{1J} & h_{mJ}h_{2J} & & h_{mJ}^2 \end{bmatrix}}_{m\times m}$$

If we let $[y_{ik}] \triangleq \left(HI_0H^T\right)^{-1}$, then we have

$$D^{(SFA)} = I_0H^T\left(HI_0H^T\right)^{-1} = \underbrace{\begin{bmatrix} 0 & 0 & \cdots & 0 & 0 \\ 0 & 0 & \cdots & 0 & 0 \\ \\ h_{1J} & h_{2J} & & & h_{mJ} \\ \\ 0 & 0 & & 0 & 0 \\ 0 & 0 & \cdots & 0 & 0 \end{bmatrix}}_{n\times m} \underbrace{\begin{bmatrix} y_{11} & y_{12} & \cdots & y_{1m} \\ y_{21} & y_{22} & \cdots & y_{2m} \\ \\ y_{i1} & y_{i2} & \cdots & y_{im} \\ \\ \\ y_{m1} & y_{m2} & \cdots & y_{mm} \end{bmatrix}}_{m\times m}$$

$$= \underbrace{\begin{bmatrix} 0 & 0 & & 0 & & 0 & 0 \\ & \vdots & & \vdots & & \vdots & \\ 0 & 0 & \cdots & 0 & \cdots & 0 & 0 \\ \sum_{k=1}^{m} h_{kJ}y_{k1} & \sum_{k=1}^{m} h_{kJ}y_{k2} & \cdots & & \cdots & & \sum_{k=1}^{m} h_{kJ}y_{km} \\ 0 & 0 & & 0 & & 0 & 0 \\ & \vdots & & \vdots & & \vdots & \\ 0 & 0 & & 0 & & 0 & 0 \end{bmatrix}}_{n\times m} \triangleq [d_{ij}]$$

In order to evaluate the entries in the diagnostic matrix $D^{(SFA)}$, we observe the following.

$$\underbrace{\begin{bmatrix} h_{1J}^2 & h_{1J}h_{2J} & \cdots & h_{1J}h_{mJ} \\ h_{2J}h_{1J} & h_{2J}^2 & \cdots & h_{2J}h_{mJ} \\ \\ \vdots & \vdots & & \vdots \\ \vdots & \vdots & & \vdots \\ \\ h_{mJ}h_{1J} & h_{mJ}h_{2J} & & h_{mJ}^2 \end{bmatrix}}_{\left(HI_0H^T\right)} \underbrace{\begin{bmatrix} y_{11} & y_{12} & \cdots & y_{1m} \\ y_{21} & y_{22} & \cdots & y_{2m} \\ \\ y_{i1} & y_{i2} & & y_{im} \\ \\ \\ y_{m1} & y_{m2} & \cdots & y_{mm} \end{bmatrix}}_{\left(HI_0H^T\right)^{-1}} = I$$

This implies that

$$\sum_{i=1}^{m} h_{kJ}h_{iJ}y_{ik} = h_{kJ}\sum_{i=1}^{m} h_{iJ}y_{ik} = 1 \quad \forall k = 1,2,\ldots,m$$

$$\sum_{i=1}^{m} h_{pJ}h_{iJ}y_{iq} = h_{pJ}\sum_{i=1}^{m} h_{iJ}y_{iq} = 0 \quad \forall p \neq q \quad , \quad p = 1,2,\ldots,m \quad , \quad q = 1,2,\ldots,m$$

The first relationship gives us

$$d_{jk} = \begin{cases} \sum_{i=1}^{m} h_{iJ} y_{ik} & j = J \\ 0 & j \neq J \end{cases} \quad \Rightarrow \quad d_{Jk} = \sum_{i=1}^{m} h_{iJ} y_{ik} = \frac{1}{h_{kJ}}$$

Thus,

$$D^{(SFA)} = I_0 H^T \left(H I_0 H^T \right)^{-1} = \begin{bmatrix} 0 & 0 & & 0 & & 0 & 0 \\ & \vdots & & \vdots & & \vdots & \\ 0 & 0 & \cdots & 0 & \cdots & 0 & 0 \\ \frac{1}{h_{1J}} & \frac{1}{h_{2J}} & \cdots & & \cdots & & \frac{1}{h_{mJ}} \\ 0 & 0 & & 0 & & 0 & 0 \\ & \vdots & & \vdots & & \vdots & \\ 0 & 0 & & 0 & & 0 & 0 \end{bmatrix}$$

which establishes our hypothesis.

A5.5 A NORMALIZED MEASUREMENT ERROR METRIC FOR SFA

A measurement error norm metric is used to determine how close a selected fault hypothesis (isolation selection) matches up to the observations at hand. *SFA* is a fault isolation process that looks at all possible faults from a pre-defined fault list and selects the fault that *best matches* the observed shifts in the gas path parameters. The *best match* criteria is accomplished by first generating what the gas path shifts would be if the selected fault under consideration was, indeed, the fault being sought and then computing a distance metric between this set of parameter shifts and the observed parameter shifts. Doing this across the set of all possible faults and selecting the one admitting the smallest distance is the best match (or first choice fault). The fault admitting the second smallest distance is the second choice fault, the fault with the third smallest distance is the third choice fault, and so on.

A commonly used metric to represent this distance is a derivative of the normal Euclidean distance between two points. If we had two points Z and Z^* in a two-dimensional space, as depicted in Figure A5.2, then the Euclidean distance between the points is given by the length of the hypotenuse as follows:

$$distance = \sqrt{(z_1 - z_1^*)^2 + (z_2 - z_2^*)^2}$$

If we had a three-dimensional space, then the distance would be

$$distance = \sqrt{(z_1 - z_1^*)^2 + (z_2 - z_2^*)^2 + (z_3 - z_3^*)^2}$$

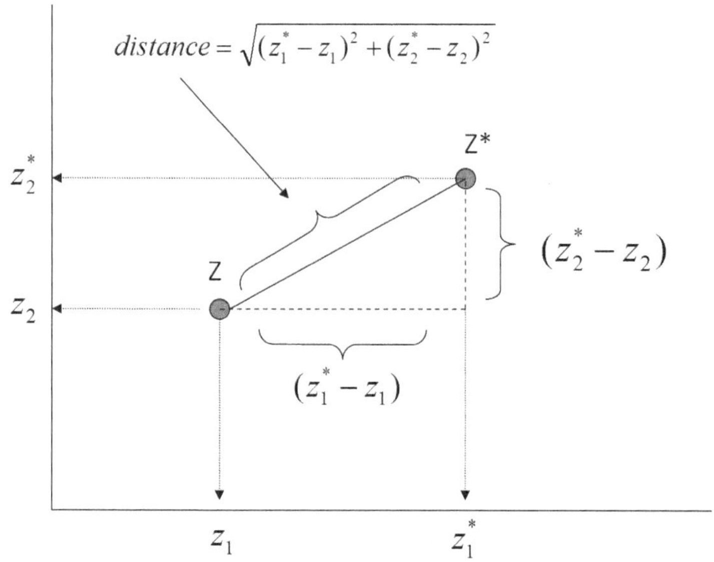

FIGURE A5.2 Euclidean distance function

In general, for an m-dimensional space, we have

$$distance = \sqrt{\sum_{i=1}^{m} (z_i - z_i^*)^2} \tag{A5.5.1}$$

The *distance* function defined in Equation (A5.5.1) can be generalized by *weighting* each of the m components in the summation. In the SFA application, each of the m dimensions corresponds to a particular gas path measurement (speed, temperature, pressure, etc.). One way of assigning weights is to divide by the variance (σ_i^2) of the non-repeatability inherent in each measurement, as depicted in Equation (A5.5.2).

$$distance = \sqrt{\sum_{i=1}^{m} \frac{(z_i - z_i^*)^2}{\sigma_i^2}} \tag{A5.5.2}$$

In assigning this type of weighting, accurate measurements (small variance) will receive a greater influence on the distance calculation than measurements having less accuracy (large variance). One thing remains. The magnitude of the underlying fault will clearly drive the magnitude of the observed parameters shifts Z (as well as the magnitude of the estimated shifts Z^*). This means that the distance will grow proportionately with the fault size. The ranking of the distances (as described above) would remain unaffected, and the distance formula given above could be used for that intended purpose. One could normalize the distance to account for the fault size (which is largely a matter of aesthetics) by dividing the above by the size of the measurement shift vector Z. This would yield the following distance metric.

$$distance = \frac{\sqrt{\sum_{i=1}^{m} \frac{(z_i - z_i^*)^2}{\sigma_i^2}}}{\sqrt{\sum_{i=1}^{m} z_i^2}} = \sqrt{\frac{\sum_{i=1}^{m} \frac{(z_i - z_i^*)^2}{\sigma_i^2}}{\sum_{i=1}^{m} z_i^2}} \quad \text{(A5.5.3)}$$

There is, however, a minor issue with this definition in that this normalization error is not bounded between [0, 1]. For example, when fault estimate $Z^* \rightarrow 0$ (which means it's extremely unlikely to be this particular fault mode), the "*normalized*" SFA error will be greater than 1 because all the σ_i are smaller than 1 (delta parameters are all normalized). While this is not a fundamental issue, not knowing the bound of the "normalized error" can be an inconvenience in the development of establishing a *confidence* level associated with a given SFA analysis. To address the boundedness problem, a refined normalized error is given by Equation (A5.5.4) below.

$$distance = \frac{\sqrt{\sum_{i=1}^{m} \frac{(z_i - z_i^*)^2}{\sigma_i^2}}}{\sqrt{\sum_{i=1}^{m} \frac{z_i^2}{\sigma_i^2}}} \quad \text{(A5.5.4)}$$

This distance metric is *approximately* bounded by [0, 1]. To see this, consider the following argument. We start with the basic relationship in Equation (A5.5.5).

$$Z = H\,x + \varphi \quad \text{(A5.5.5)}$$

A WLSQ estimate x^* *minimizes* the following objective function.

$$(Z - Hx)^T R^{-1} (Z - Hx) \quad \text{(A5.5.6)}$$

The LSQ solution x^* is given by

$$x^* = \left(H^T R^{-1} H\right)^{-1} H^T R^{-1} Z = \min_x \left\{(Z - H\,x)^T R^{-1} (Z - H\,x)\right\} \quad \text{(A5.5.7)}$$

where R, is the measurement covariance matrix. In general, this matrix is not diagonal (as the measurement residuals Z are corrected parameters; however, the T_2, P_2 and PSP parameter influence is such that the covariance values are rather small and, thus, we have that)

$$(Z - H\,x)^T R^{-1} (Z - H\,x) \approx \sum_{i-1}^{m} \frac{\left(z_i - [Hx]_i\right)^2}{\sigma_i^2} \quad \text{(A5.5.8)}$$

Letting $Z^* = H\,x^*$, we have that

$$\begin{aligned}\sum_{i=1}^{m}\frac{(z_i - z_i^*)^2}{\sigma_i^2} &= \sum_{i=1}^{m}\frac{(z_i - [H\,x^*]_i)^2}{\sigma_i^2} \approx \left(Z - H\,x^*\right)^T R^{-1}\left(Z - H\,x^*\right)\\ &\le (Z - H\cdot 0)^T R^{-1}(Z - H\cdot 0)\\ &\approx \sum_{i=1}^{m}\frac{z_i^2}{\sigma_i^2}\end{aligned} \tag{A5.5.9}$$

Therefore,

$$distance = \frac{\sqrt{\sum_{i=1}^{m}\frac{(z_i - z_i^*)^2}{\sigma_i^2}}}{\sqrt{\sum_{i=1}^{m}\frac{z_i^2}{\sigma_i^2}}} \approx\le 1 \tag{A5.5.10}$$

More generally, the normalized error metric (distance) can be calculated with the measurement covariance matrix R as follows.

$$distance = \left[(Z - Z^*)^T R^{-1}(Z - Z^*)\right](Z^T R^{-1} Z)^{-1} \tag{A5.5.11}$$

In this (more calculation intensive) form, the normalized error will be strictly bounded by [0,1].

A5.6 FUZZY LOGIC BASICS

The notion of a *fuzzy set* was introduced in 1965 by Lotfi Zadeh, who was then a professor of electrical engineering at the University of California, Berkeley. His idea was to expand upon the general definition of a *set* in order to allow for the treatment of uncertainty in the *classification* of objects. A *fuzzy set* is a collection of objects whose membership, *unlike* those in a *crisp* set, is a matter of degree. This is expressed in terms of a *membership function* that assigns a number between 0 and 1 to each object, where 1 denotes absolute membership in the set and 0 denotes exclusion from the set. An object assigned a value somewhere in-between, say 0.6 is said to belong to the set with degree (or membership) 0.6. This also implies that it is excluded from the set with degree 1-0.6 = 0.4. Thus, by virtue of this fuzzy classification, it is possible to deal with sets that are not (mathematically) well defined but, rather, have been defined in linguistic terms, such as the set of numbers that are *approximately equal to* π or the set of jet engines exhibiting *low EGT* margin.

In as much as ordinary set theory serves as a foundation for classical logic, *fuzzy* set theory has, likewise, aided the development of *fuzzy* logic. Fuzzy logic combines notions from ordinary Aristotelian logic and fuzzy set theory in that the truth of a given proposition (like a fuzzy set) can be something between 0 (representing false) and 1 (representing true). Logical operations and rules of inference have also been defined, which allows for the description (and analysis) of complex systems in a natural language setting. This process mimics the way humans convey and process

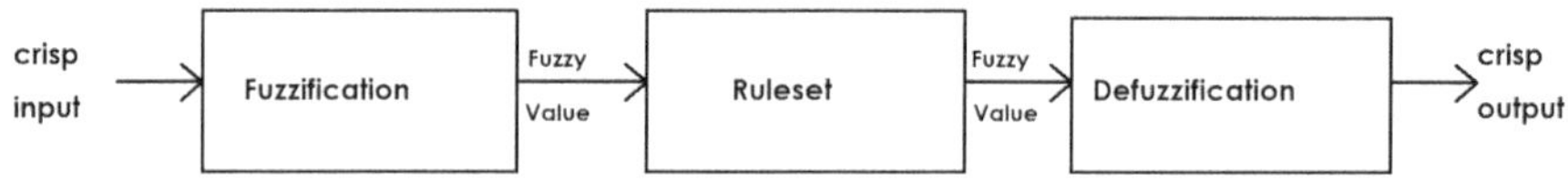

FIGURE A5.3 Fuzzy logic process

information and, thus, provides a mathematical vehicle for the representation of imprecise rules and heuristics.

Generally speaking, there are three major processes common to all fuzzy systems: *fuzzification*, *rule base inference*, and *defuzzification*, as depicted in Figure A5.3.

1. ***Fuzzification*** is the process that converts a crisp input value to fuzzy set values, thereby forming the interface between the real world and the fuzzy inference process.
2. ***Rule Base Inference*** is the application of a series of rules given in terms of fuzzy set values and producing for output variable(s), fuzzy set values for each rule fired. These rules usually take the form of IF-THEN statements, e.g., IF (X1 is *BIG*) AND (X2 is *SMALL*) THEN (Y is *MEDIUM*).
3. ***Defuzzification*** converts the fuzzy set values (of the output variable(s)) back to a crisp value so that it may be utilized in a real-world process.

The process of *fuzzification* consists of assigning linguistic labels (e.g., *small, medium*) to a crisp input variable and determining the level of belief for that label for each point in the domain of the input. Unlike traditional logic, we may have levels of belief that range somewhere between 0 (false) and 1 (true). Figure A5.4, for example, illustrates three linguistic labels (*low, medium,* and *high)* being applied to the crisp variable *X*, which takes on values from 0 to 1,000. The membership curves depicted for *low, medium,* and *high*, provide the level of belief (or truth level) that the variable *X* falls in these three categories (or classifications) at each value of *X* throughout its domain. As depicted in Figure A5.4, for example, at the value of 400, X is considered *low* with belief w_{Low} and is considered to be *medium* with belief w_{Medium}. Because the value 400 intersects *high* at an ordinate value of zero, its level of belief is 0, i.e., $X = 400$is definitely NOT *large.*

This process is carried out for each input variable X_i, and their respective beliefs are forwarded to the fuzzy rule base for evaluation. A typical rule may, for example, take the form

$$IF\ (X_1\ is\ low)\ AND\ (X_2\ is\ medium)\ THEN\ (Y\ is\ positive)$$

The first step in evaluating this rule is to determine the *belief* value of the rule, which is, by definition, the belief value of the *antecedent* (the IF part). This can be accomplished in many ways; however, the most popular method is referred to as the MIN/MAX method. This method takes the MIN belief of the arguments that are combined by AND along with the MAX belief of arguments that are combined by an OR. The second step is to transfer this composite belief to the consequent variable

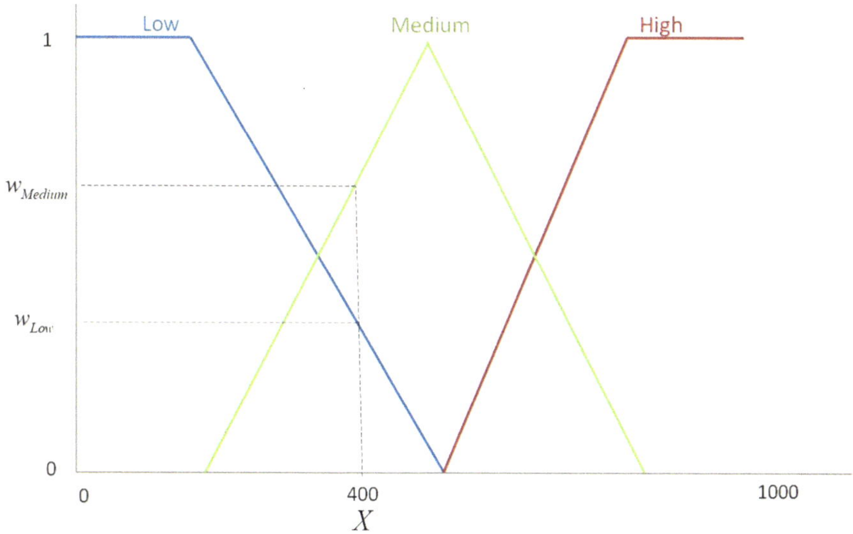

FIGURE A5.4 Sample fuzzy membership functions

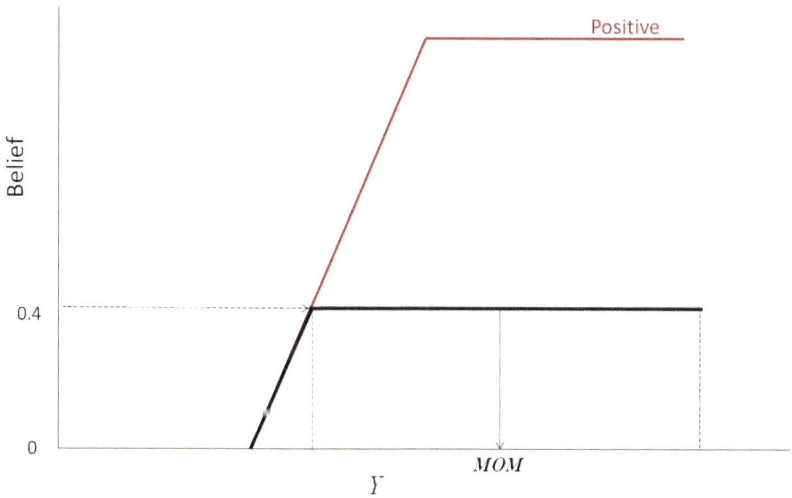

FIGURE A5.5 Mandami method with MOM defuzzification

(in this case Y). This *transfer* can also be accomplished in many ways; however, the most popular is the *Mandami* method. In our example above, if we assume that X_1 *is low* and X_2 *is medium* with respective beliefs of 0.4 and 0.6, then if we were using the MIN/MAX method, the rule would assign a belief of 0.4 to the classification Y *is positive*. Figure A5.5 illustrates the effect of the *Mandami* transfer method in this example and how it reshapes the membership for *positive* for the output Y.

The last step in the process is to *defuzzify* the results of *each* rule to form a crisp value and to *aggregate* all of these crisp results to form a single crisp output to pass to the control. Once again, there are many methods available, including whether we *defuzzify* each rule separately and then aggregate, or aggregate all the Mandami transferred consequent memberships and then *defuzzify*. We will use the former, as it is easier in terms of computation. To accomplish this, we will use the so-called *mean of max* (MOM) method of *defuzzification*. The MOM method picks those abscissa values that admit maximum belief points on the adjusted membership and then computes the mean of these points (see Figure A5.5).

The final step is to aggregate these crisp values to form a single crisp output value. A popular method that can be employed is to take a weighted average where the weights will be equal to the level of belief in the corresponding rules (Tuskamoto's method). Thus, if we had n rules $R_1, R_2, \ldots, R_n$ with respective beliefs $w_1, w_2, \ldots, w_n$, yielding crisp values (after *defuzzification*) $y_1, y_2, \ldots, y_n$, then the final output y would be

$$y = \frac{\sum_{i=1}^{n} w_i y_i}{\sum_{i=1}^{n} w_i} \tag{A5.6.1}$$

It should be evident from the above descriptions that once the membership curves are defined and fixed and the rule base specified, then the fuzzy system defines a *deterministic* mapping between the input variables and the output variable(s) and that this relationship is neither capricious nor random. It will, like any mathematical relationship, *yield the same output for the same input.* The utility of this approach is that we have employed only natural language rules and have taken advantage of the imprecision in the classification of the input/output variables to arrive at a quantifiable (and easily modifiable) algorithm.

NOTE

1. The recovery value refers to the quantity $\hat{x}-\mu$. Because the only component of μ that is changing is s_k', this ismplies that all other fault assessments (engine and sensor) will be the same regardless of the magnitude (or direction) of the detected measurement error.

REFERENCES

1. Volponi, A.J., 1994, Sensor Error Compensation in Engine Performance Diagnostics, *Proceedings of the American Society of Mechanical Engineers Turbine and Aeroengine Congress.* The Hague, Netherlands, June 13–16, Paper No. 94-GT-58.
2. Grewal, M., Andrews, A., 1993, *Kalman Filtering: Theory & Practice,* Prentice Hall, NJ, ISBN 0-13-211335-X.
3. Gelb, A., et al., *Applied Optimal Estimation*, MIT Press, Cambridge, MA, 1974.

A6 Chapter 6 Addendums

A6.1 KALMAN EQUATION FOR SELF-TUNING ENGINE MODEL

The Kalman filter estimator can be expressed as

$$\Delta\dot{\hat{x}} = (A - KC)\Delta\hat{x} + K\Delta y + (B - KD)\Delta u \tag{A6.1.1}$$

and can be derived in the following manner. Referring to Equations (6.16 – 6.18) of Chapter 6, if we relax our notation and let

$$\begin{aligned}
A &= A_{KF} \\
B &= B_{KF} \\
C &= C_{KF} \\
D &= D_{11} \\
K &= K_1 \\
\theta &= \omega = 0 \\
\Delta x &= \begin{bmatrix} \Delta x \\ \eta \end{bmatrix}
\end{aligned}$$

then we have that

$$\begin{aligned}
\Delta\dot{\hat{x}} &= A\,\Delta\hat{x} + B\,\Delta u + K\,r \\
\Delta y &= y - y_B = C\,\Delta x + D\,\Delta u \\
\Delta\hat{y} &= \hat{y} - y_B = C\,\Delta\hat{x} + D\,\Delta u
\end{aligned} \tag{A6.1.2}$$

Therefore,

$$K\,r = K(y - \hat{y}) = KC(\Delta x - \Delta\hat{x}) \tag{A6.1.3}$$

This implies

$$\begin{aligned}
\Delta\dot{\hat{x}} &= A\,\Delta\hat{x} - KC\Delta\hat{x} + B\,\Delta u + KC\Delta x \\
&= (A - KC)\Delta\hat{x} + B\,\Delta u + K(\Delta y - D\Delta u) \\
&= (A - KC)\Delta\hat{x} + K\Delta y + (B - KD)\Delta u
\end{aligned} \tag{A6.1.4}$$

A6.2 OBSERVABILITY CRITERION

For a linear estimation system (for example a Kalman filter observer), the total number of health parameter unknowns (*tuners*) must be less than or equal to the total number of measurements (i.e., observations).

The proof of this assertion, which we give below, closely follows the arguments provided by Martin D. Espana, [1].

Consider the linear dynamic system

$$\begin{aligned} \dot{x} &= Ax + Bu + L\eta \\ y &= Cx + Du + M\eta \\ \dot{\eta} &= 0 \end{aligned} \tag{A6.2.1}$$

where x is an $n\times1$ state vector, y is an $m\times1$ state vector, and η is a $p\times1$ performance health vector. Rewriting we obtain

$$\begin{bmatrix} \dot{x} \\ \dot{\eta} \end{bmatrix} = \begin{bmatrix} A & L \\ 0 & 0 \end{bmatrix} \begin{bmatrix} x \\ \eta \end{bmatrix} + \begin{bmatrix} B \\ 0 \end{bmatrix} u$$

$$y = \begin{bmatrix} C & M \end{bmatrix} \begin{bmatrix} x \\ \eta \end{bmatrix} + Du$$

Let

$$x' = \begin{bmatrix} x \\ \eta \end{bmatrix} \qquad (n+p)\times 1$$

$$A' = \begin{bmatrix} A & L \\ 0 & 0 \end{bmatrix} \qquad (n+p)\times(n+p)$$

$$B' = \begin{bmatrix} B \\ 0 \end{bmatrix}$$

$$C' = \begin{bmatrix} C & M \end{bmatrix} \qquad m\times(n+p)$$

Therefore, by substitution

$$\begin{aligned} \dot{x}' &= A'x' + B'u \\ y &= C'x + Du \end{aligned} \tag{A6.2.2}$$

Applying the conventional definition for observability states that Equation (A6.2.1) is *observable* if and only if Equation (A6.2.2) is observable, and if and only if

$$F = \begin{bmatrix} C' \\ C'A' \\ C'A'^2 \\ \vdots \\ C'A'^{n+p-1} \end{bmatrix} \quad \text{has } rank = n+p$$

Expanding, this is equivalent to stating that

$$F = \begin{bmatrix} C & M \\ CA & CL \\ CA^2 & CAL \\ \vdots & \vdots \\ CA^{n+p-1} & CA^{n+p-2}L \end{bmatrix} \quad \text{has } rank = n + p$$

But

$$F = \underbrace{\begin{bmatrix} I_m & 0 \\ 0 & C \\ 0 & CA \\ \vdots & \vdots \\ 0 & CA^{n+p-2} \end{bmatrix}}_{G} \underbrace{\begin{bmatrix} C & M \\ A & L \end{bmatrix}}_{H}$$

Therefore,

$$rank(F) = \min\{rank(G), rank(H)\} = n + p \quad \textit{if and only if} \quad rank(H) \geq n + p$$

Noting that H is $(n+m) \times (n+p)$, we have that $rank(H) \leq n + m$, and, thus, it is observable if and only if $n+p \leq n+m$, *i.e.*, $p \leq m$. Thus, the system is observable if and only if the number (p) of performance parameters (unknowns) is less than or equal to the number (m) of measured parameters.

A6.3 MLP ANN

A perceptron is the simplest form of an ANN and consists of a single neuron that accepts k input signals and returns a single output value. The architecture is depicted in Figure A6.1.

To produce an output value for n input values $\{X_1, X_2, \ldots, X_n\}$, the perceptron computes a weighted sum using weights $\{w_1, w_2, \ldots, w_n\}$, adds a constant bias term, and then passes the result through a so-called *activation function* to obtain the final output value. In symbols, we have

$$Output = f\left(w_0 b_0 + \sum_{k=1}^{n} w_k X_k \right) \tag{A6.3.1}$$

where f denotes the activation function. The activation function can take many forms depending on the application. Its primary purpose is to shape the output value. If the activation function is linear, then the perceptron becomes a linear function. To introduce non-linearity, it is common to use a sigmoid or hyperbolic tangent function for this purpose.

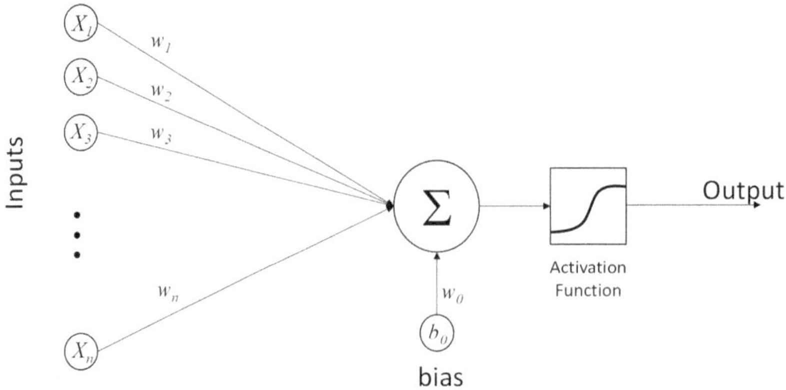

FIGURE A6.1 Perceptron architecture

An MLP neural network is made up of layers of perceptrons, the most common of which is the single hidden layer construction depicted in Figure A6.2. This particular construction is known to be a universal approximator [2] and is well suited for the modeling of a non-linear input/output relationship. The MLP *learns* the relationship through the use of a *supervised* learning algorithm technique referred to as *back-propagation* [3], [4]. Data containing both inputs and output values are presented to the algorithm, which, in turn, *adjusts* the *weights* and *biases* in the network in order to minimize an error cost function, and this process continues in iteration until certain error tolerances have been met. Once *trained*, the MLP is deployed and represents a *fixed* input–output relationship. As a universal approximator, a single, hidden-layer MLP is well suited to learn the *residual difference* between an engine model representation and an actual monitored engine, as illustrated in Figure A6.2.

A6.4 GMM CLUSTERING PROCESS

The GMM clusters are multi-dimensional Gaussian distributions that are characterized by two statistical quantities: a *mean* and a measure of dispersion, the *standard deviation*. These GMM elements are multi-dimensional quantities and are determined for the vector of independent input parameters $\boldsymbol{u}$ and the vector of dependent engine parameter residuals $\boldsymbol{r}$. The residual vector $\boldsymbol{r}$ is determined as the difference between measured engine parameters $\boldsymbol{y}$, and estimations of these parameters from a physics-based model component. The input parameters, $\boldsymbol{u}$, include altitude, Mach, fuel flow demand, engine bleed commands, active clearance control commands, variable geometry commands, and other parameters as needed. The measured engine parameters, $\boldsymbol{y}$ (and, hence, the residuals $\boldsymbol{r}$), can include engine speeds, inter-stage temperatures and pressures, flows, etc. A GMM cluster will consist of a vector of means and a vector of standard deviations for *both* the *inputs* and *residuals* that are computed recursively during the GMM clustering process as follows, [5–8]:

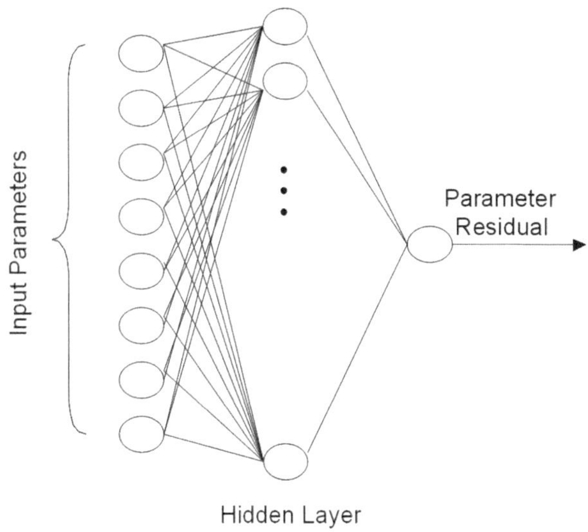

FIGURE A6.2 Single hidden layer MLP

A6.4.1 Inputs

$$\bar{u}_N(i) = \left(\frac{N-1}{N}\right)\bar{u}_{N-1}(i) + \left(\frac{1}{N}\right)u_N(i)$$
$$\sigma_N^{(u)}(i) = \sqrt{\left(\frac{N-2}{N-1}\right)\left[\sigma_{N-1}^{(u)}(i)\right]^2 - \left(\frac{N}{N-1}\right)\left(\bar{u}_N(i) - \bar{u}_{N-1}(i)\right)^2 + \left(\frac{1}{N-1}\right)\left(u_N(i) - \bar{u}_{N-1}(i)\right)^2}$$
$$\forall i = 1,2,\ldots,m_c \tag{A6.4.1}$$

A6.4.2 Residuals

$$\bar{r}_N(i) = \left(\frac{N-1}{N}\right)\bar{r}_{N-1}(i) + \left(\frac{1}{N}\right)r_N(i)$$
$$\sigma_N^{(r)}(i) = \sqrt{\left(\frac{N-2}{N-1}\right)\left[\sigma_{N-1}^{(r)}(i)\right]^2 - \left(\frac{N}{N-1}\right)\left(\bar{r}_N(i) - \bar{r}_{N-1}(i)\right)^2 + \left(\frac{1}{N-1}\right)\left(r_N(i) - \bar{r}_{N-1}(i)\right)^2}$$
$$\forall i = 1,2,\ldots,m_r \tag{A6.4.2}$$

where m_c and m_r are the number of input and measured engine parameter residuals, respectively, with $\bar{u}_N$, $\sigma_N^{(u)}$ representing the $m_c \times 1$ vectors of (running) input means

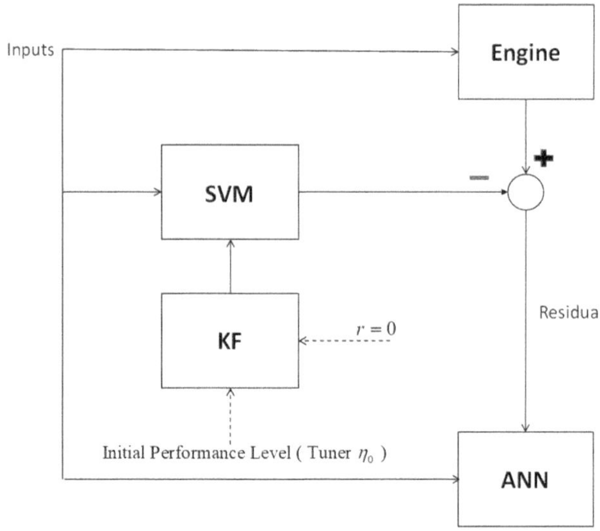

FIGURE A6.3 Training ANN empirical element

and standard deviations, and $\bar{r}_N$, $\sigma_N^{(r)}$ representing the $m_r \times 1$ vectors of engine parameter residuals means and standard deviations after N data points have been processed. When the GMM cluster process terminates, the mean and standard deviation vectors, along with the sample size (N) (at termination) are stored in the GMM database.

When a data point is determined to be within the domain of an existing GMM cluster, it is processed by the (existing) hybrid model using current ANN MLP empirical model elements. Because the data point is already represented by an existing GMM cluster, an existing MLP empirical model provides the requisite compensation to allow performance estimation to be determined without corruption. If the data point is *not* within the domain of an existing GMM, the hybrid model is run in *frozen* mode, and a new GMM cluster process is initiated as previously described. Running the hybrid model in *frozen* mode simply means to process the point as indicated in Figure A6.3 in which the *initial performance level* input to the Kalman filter is "frozen" at the last known value estimated (from data within the current GMM database domain). This will ensure that the *residuals* that are calculated (and used in the GMM cluster formation) do not contain presently known deterioration effects. This is important, as we only want to capture (empirically) the differences between the actual engine and physics model representation and not "absorb" any deterioration we might have accrued since installation.

A6.4.3 MLP Training Using Clusters

The empirical sub-models are sequentially updated to reflect the new data that is collected with every flight during the initialization phase. The MLP training process can be adjusted to use the GMM (cluster) information directly in such a way as to yield essentially the same residual model result that would have been obtained if the actual residual data had been stored and accumulated over time and used directly in the training in a batch process mode. In order to describe this procedure, we will introduce a slightly different nomenclature, wherein ***M*** represents the GMM model with component centers ***M.c,*** standard deviation ***M.s***, and the number of points included in the GMM cluster ***M.n***.

The MLP training is performed by adjusting the various parameters in the neural net to optimize some performance as functional. For standard MLP training, this is simply the sum-of-squares-error (SSE) defined as:

$$SSE = \sum_{k=1}^{N} \left| r(k) - F(u(k)) \right|^2 \tag{A6.4.3}$$

where *r(k)* is the *k*th point of the residual sequence we are trying to predict, *u(k)* is the *k*th point of the multi-dimensional input data set, and $F(\circ)$ is the mapping function approximated by the MLP neural network. There are *N* points in the training dataset.

Assume that the GMM model is made up of *C* clusters. ***M.n(i)*** is the number of points represented by cluster *i*, which has center ***M.c(i)*** and standard deviation ***M.s(i)***. Taking into account the cluster parameters, an *equivalent* SSE error function to be used for training appears in Equation (A6.4.4).

$$SSE = \frac{1}{N}\sum_{i=1}^{C} M.n(i) \left| \frac{M.c_r(i) - \hat{F}(M.c_u(i))}{M.s_r(i)} \right|^2 \tag{A6.4.4}$$

where

$$N = \frac{1}{N}\sum_{i=1}^{C} M.n(i) = \textit{Total number of points}$$

and the '*r*' and '*u*' subscripts refer to the components in the *i*th cluster that correspond to the residual and inputs respectively, e.g., $M.c(i) = \{M.c_u(i), M.c_r(i)\}$. This is the only change required in the backpropagation training algorithm. In Equation (A6.4.4), $\hat{F}(\circ)$is the mapping function approximated by the MLP neural network trained using GMM clusters

REFERENCES

1. Espana, M., 1994, *On the Estimation Algorithm Used in Adaptive Performance Optimization of Turbofan Engines*, NASA Technical Memorandum 4551.
2. Hornik, K., et al., 1989, Multilayer Feedforward Networks are Universal Approximators, *Neural Networks*, Vol 2, pp. 359–366.
3. Nelson, M., Illingworth, W.T., 1991, *A Practical Guide to Neural Nets*, Addison-Wesley Publishing, Co., ISBN 0-201-56309-6.
4. Looney, C.G., 1997, *Pattern Recognition using Neural Networks: Theory and Algorithms for Engineers and Scientists*, Oxford University Press, ISBM 0-19-507920-5.

5. Volponi, A.J., 2005, Use of Hybrid Engine Modeling for On-Board Module Performance Tracking, *Proceedings of the American Society of Mechanical Engineers Turbine and Aeroengine Congress.* Reno, NV, June 6–9, Paper No. GT2005-68169.
6. Volponi, A.J., et al., 2007, Adaptive On-Wing Gas Turbine Engine Performance Estimation, *Proceedings of the 2007 IEEE Aerospace Conference.* Big Sky, MT.
7. Volponi, A.J., et al., 2008, Empirical Tuning of an On-Board Gas Turbine Engine Model for Real-time Module Performance Estimation, Transactions of the ASME, *Journal of Engineering for Gas Turbines and Power*, Vol 130, Issue 2.
8. Volponi, A.J., 2008, *Enhanced Self Tuning On-Board Real-Time Model (eSTORM) for Aircraft Engine Performance Health Tracking*, NASA CR—2008-215272.

A7 Chapter 7 Addendums

A7.1 FOQA DATA

FOQA, also known as flight data monitoring or flight data analysis, is a voluntary safety program in which commercial airlines share deidentified flight data with the FAA in a joint effort to find new ways to improve flight safety and increase overall operational efficiency.[1] A dataset[2] of 69 flights was utilized to drive C-MAPSS simulations in generating all of the illustrative examples given in Chapters 6 and 7. These flight sets contain flight profiles from two-engine airframes, consisting of the following five parameters sampled at 1Hz.:

- Altitude (ft)
- Mach
- dTamb (deg F)
- PLA1 (engine 1 PLA)
- PLA2 (engine 2 PLA)

The parameter *dTamb* is the delta of *TAT* at altitude from *ISA* conditions. The flight duration of the flights in this dataset varies between roughly 1–3 hours in length, as indicated in Figure A7.1.

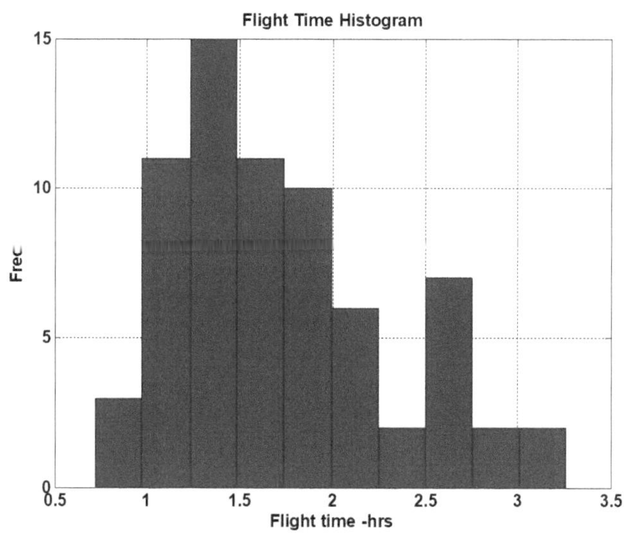

FIGURE A7.1 FOQA flight duration histogram

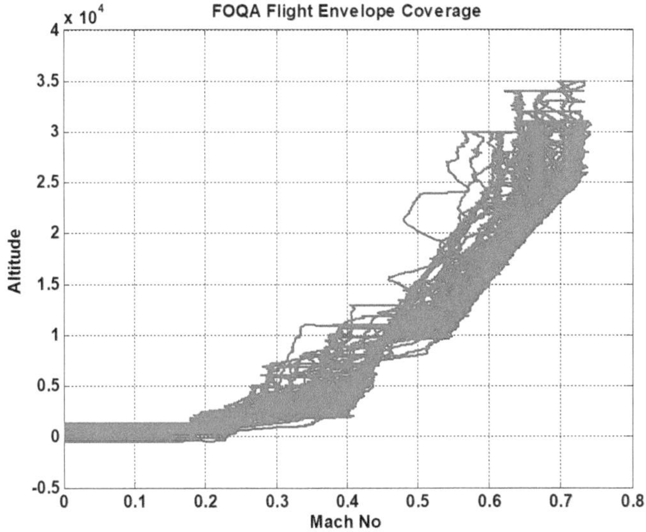

FIGURE A7.2 FOQA flight envelope coverage

The flight envelope represented by the *collective* dataset is a reasonable characterization of what would be typical for a commercial airframe and is illustrated by Figure A7.2. Additionally, Figure A7.3 displays a random sample of nine flights from the dataset to illustrate its variability.

For graphic simplicity, the plots depicted in Figure A7.3 contain the PLA from only one of the two engines represented in the FOQA data. To give the reader a sense of how these two PLAs may have differed, Figure A7.4 contains three views. From left-to-right, *histogram* of the absolute difference $|PLA1 - PLA2|$, *maximum* PLA absolute difference, and *average* PLA absolute difference.

These realistic disparities in the PLA settings, between the engines of a twin-engine airframe, provide additional realism in the *C-MAPSS* simulations in developing illustrative examples for the *cross-wing* approach to *FDI* outlined in Chapter 7.

Figure A7.5 depicts the *QSS* coverage (red) for the sample *FOQA* flights portrayed in Figure A7.3. The percentage datum appearing on each plot is the % *QSS* coverage for that particular *FOQA* flight. For the entire *FOQA* flight database, the *average* per flight *QSS* coverage is 37.3%. The coverage distribution across all 69 *FOQA* flights ranges from a *minimum* of 13.1% and a *maximum* of 59.6% and is illustrated in Figure A7.6.

Finally, Figure A7.7 presents the *tuner* responses for the first nine *FOQA* flights shown in Figure A7.3, displaying both *with* and *without* model alignment adjustments. This provides a partial pictorial "validation" of the *eSTORM* initialization methodology used within *QSS* flight *realms*.

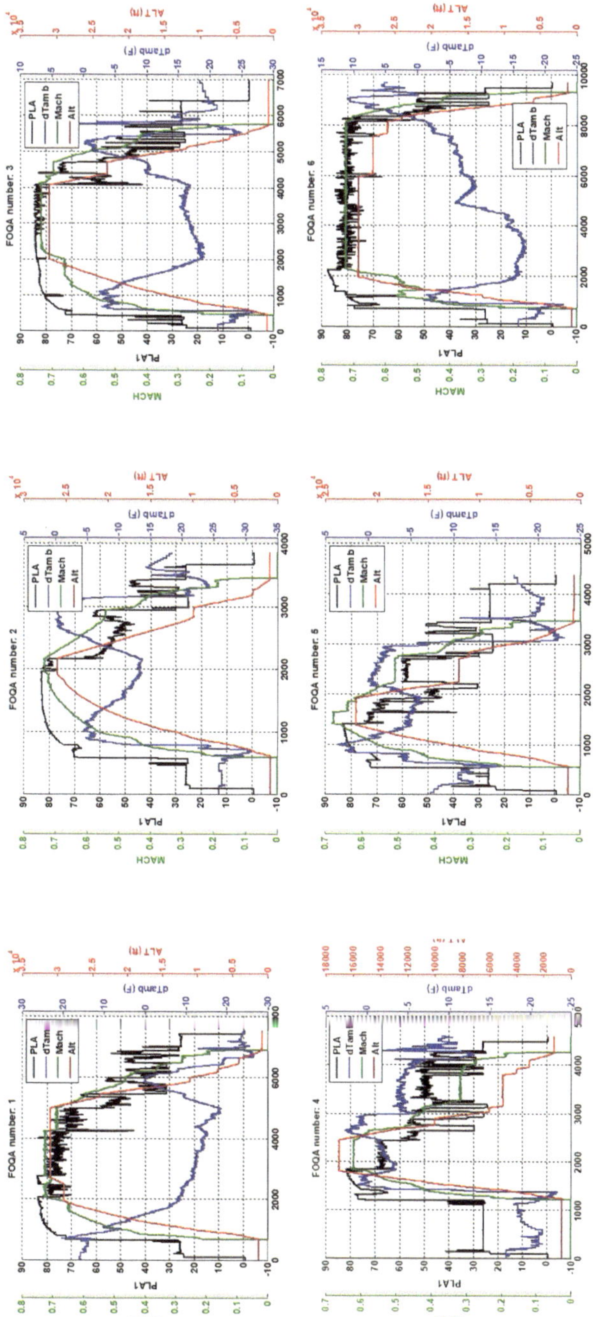

FIGURE A7.3 FOQA data set samples

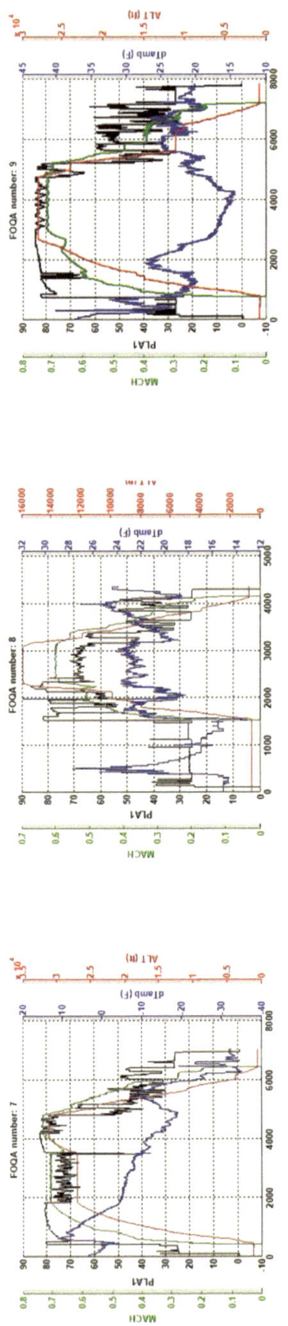

FIGURE A7.3 FOQA data set samples

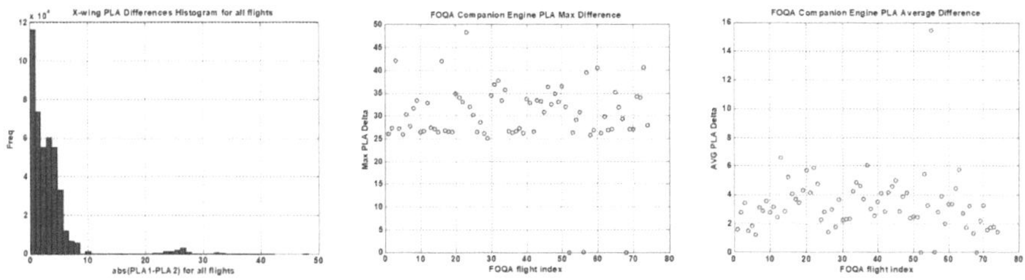

FIGURE A7.4 Two engine PLA differences in FOQA dataset

A7.2 RECURSIVE MEAN AND VARIANCE CALCULATION

The proposed test for determining *QSS* requires that the mean and standard deviation for a sliding window of points be calculated as specified in Equations (7.4) and (7.5), and redisplayed in Equation (A7.2.1).

$$\sigma_i(t_k) = \sqrt{\frac{1}{N-1}\sum_{j=k-N+1}^{k}\left[\Delta P_i(t_j) - \mu_i(t_k)\right]^2} \quad where\, \mu_i(t_k) = \frac{1}{N}\sum_{j=k-N+1}^{k}\Delta P_i(t_j) \qquad \text{(A7.2.1)}$$

$$QSS\,@\,t_k \Leftrightarrow \frac{1}{m}\sum_{i=1}^{m}\sigma_i(t_k) < L \; for\, some\, pre\text{-}defined\, threshold\, L$$

Memory can be minimized by calculating the mean and standard deviation *recursively*, hence avoiding the need to temporarily store each data point in the sliding time window.

To somewhat ease the burden of nomenclature, we drop the reference to the ith measured parameter delta and let $\{\Delta P_1, \Delta P_1, \ldots, \Delta P_N\}$ represent the N data points (for the ith measured parameter delta) in the sliding window $\{t_{k-(N-1)}, t_{k-(N-2)}, \ldots, t_k\}$ in what follows.

A7.2.1 Recursive Mean

Let $\mu(\Delta P_n)$ be the mean of the first n points, i.e.,

$$\mu(\Delta P_n) = \frac{1}{n}\sum_{i=1}^{n}\Delta P_i$$

Therefore,

$$\mu(\Delta P_{n+1}) = \frac{1}{n+1}\sum_{i=1}^{n+1}\Delta P_i = \frac{1}{n+1}\sum_{i=1}^{n}\Delta P_i + \frac{1}{n+1}\Delta P_{n+1}$$

$$= \frac{n}{n+1}\mu(\Delta P_n) + \frac{\Delta P_{n+1}}{n+1}$$

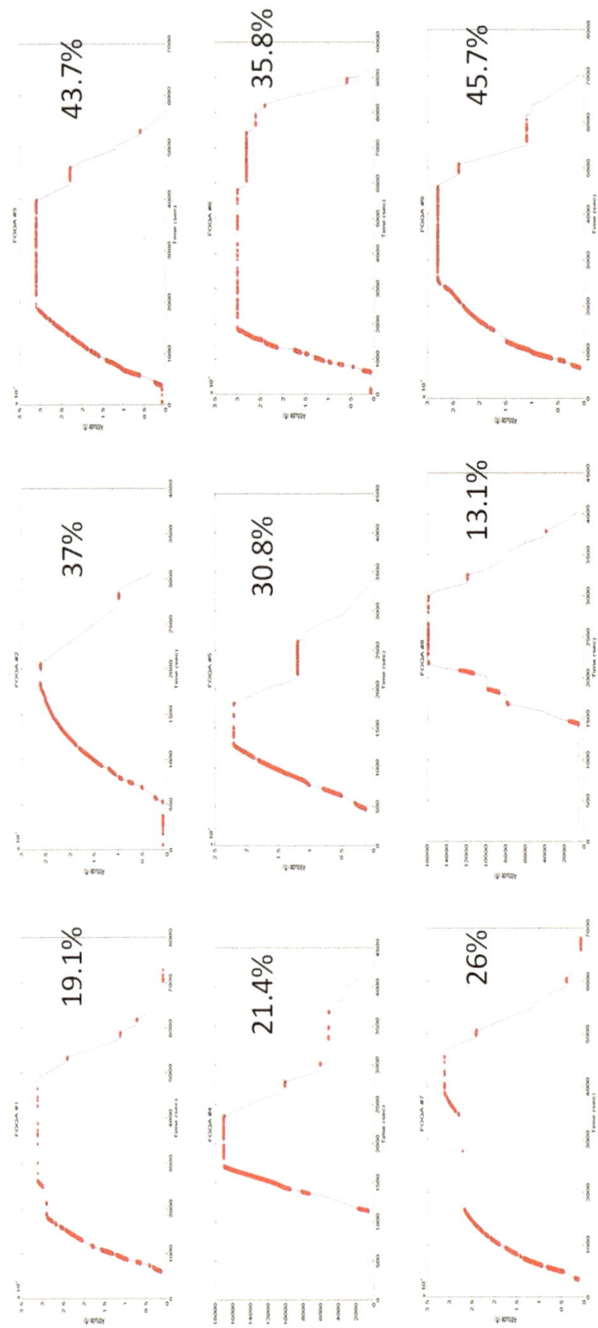

FIGURE A7.5 QSS coverage for FOQA 1–9

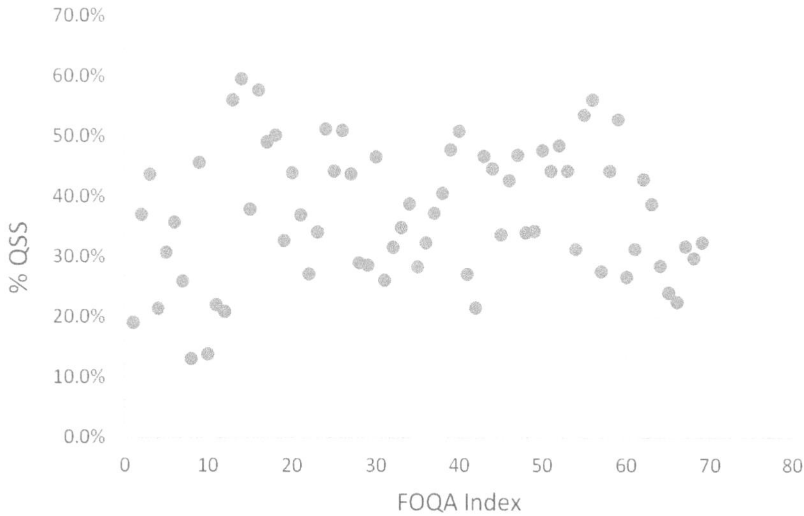

FIGURE A7.6 QSS coverage distribution for all FOQA flights

This implies

$$\mu(\Delta P_{n+1}) = \frac{n\,\mu(\Delta P_n) + \Delta P_{n+1}}{n+1} \tag{A7.2.2}$$

A7.2.2 Recursive Variance

By definition, the variance for the first n points is given by

$$\sigma^2(\Delta P_n) = \frac{1}{n-1}\sum_{i=1}^{n}\left(\Delta P_i - \mu(\Delta P_n)\right)^2$$

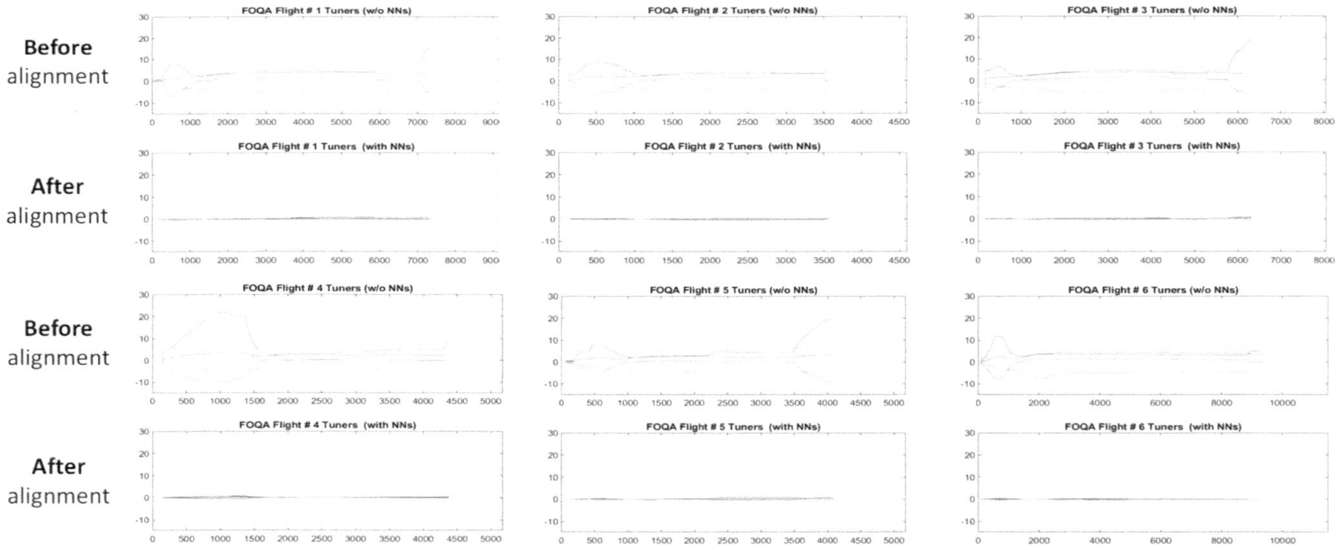

FIGURE A7.7 Tuner response to model alignment for FOQA 1–9

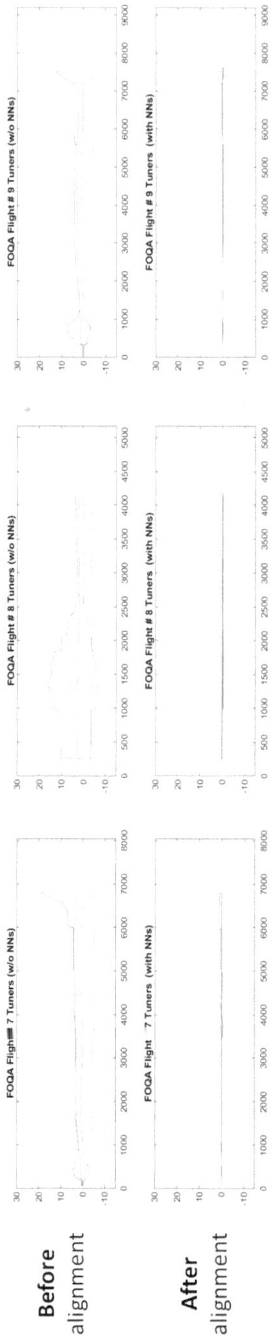

FIGURE A7.7 Tuner response to model alignment for FOQA 1–9

Therefore,

$$
\begin{aligned}
\sigma^2(\Delta P_{n+1}) &= \frac{1}{n}\sum_{i=1}^{n+1}\left(\Delta P_i - \mu(\Delta P_{n+1})\right)^2 \\
&= \frac{1}{n}\sum_{i=1}^{n}\left(\Delta P_i - \mu(\Delta P_n) + \mu(\Delta P_n) - \mu(\Delta P_{n+1})\right)^2 + \frac{1}{n}\left(\Delta P_{n+1} - \mu(\Delta P_{n+1})\right)^2 \\
&= \frac{1}{n}\sum_{i=1}^{n}\left[\begin{array}{l}\left(\Delta P_i - \mu(\Delta P_n)\right)^2 + \left(\mu(\Delta P_n) - \mu(\Delta P_{n+1})\right)^2 \\ +2\left(\Delta P_i - \mu(\Delta P_n)\right)\left(\mu(\Delta P_n) - \mu(\Delta P_{n+1})\right)\end{array}\right] + \frac{1}{n}\left(\Delta P_{n+1} - \mu(\Delta P_{n+1})\right)^2 \\
&= \left(\frac{n-1}{n}\right)\sigma^2(\Delta P_n) + \left(\mu(\Delta P_n) - \mu(\Delta P_{n+1})\right)^2 \\
&\quad + \frac{2}{n}\left(\mu(\Delta P_n) - \mu(\Delta P_{n+1})\right)\underbrace{\sum_{i=1}^{n}\left(\Delta P_i - \mu(\Delta P_n)\right)}_{zero} + \frac{1}{n}\left(\Delta P_{n+1} - \mu(\Delta P_{n+1})\right)^2 \\
&= \left(\frac{n-1}{n}\right)\sigma^2(\Delta P_n) + \left(\mu(\Delta P_n) - \mu(\Delta P_{n+1})\right)^2 + \frac{1}{n}\left(\Delta P_{n+1} - \mu(\Delta P_{n+1})\right)^2 \\
&= \left(\frac{n-1}{n}\right)\sigma^2(\Delta P_n) + \left(\frac{\mu(\Delta P_n) - \Delta P_{n+1}}{n+1}\right)^2 + \frac{1}{n}\left(\frac{n\Delta P_{n+1} - n\,\mu(\Delta P_n)}{n+1}\right)^2 \\
&= \left(\frac{n-1}{n}\right)\sigma^2(\Delta P_n) + \frac{\left(\mu(\Delta P_n) - \Delta P_{n+1}\right)^2}{(n+1)^2} + \frac{n\left(\mu(\Delta P_n) - \Delta P_{n+1}\right)^2}{(n+1)^2}
\end{aligned}
$$

This implies

$$
\sigma^2(\Delta P_{n+1}) = \left(\frac{n-1}{n}\right)\sigma^2(\Delta P_n) + \frac{\left(\mu(\Delta P_n) - \Delta P_{n+1}\right)^2}{(n+1)} \tag{A7.2.3}
$$

Equations (A7.2.2) and (A7.2.3) provide the recursive relationships for the mean and standard deviation, respectively.

A7.3 DERIVATION OF EQUATIONS (7.20)–(7.21)

For clarity of the proof, we will simplify the notation used in Equations (7.20) and (7.21) by considering just one bin and a single parameter P. In effect, we will prove the general formulae for calculating the aggregated variance from the variances and means of the datasets being aggregated.

Given M sets of data with associated means and variances, i.e.,

$$
\begin{array}{lll}
\left\{P_1^{(1)}, P_2^{(1)}, \ldots, P_{n_1}^{(1)}\right\}^{(1)} & Mean: \mu^{(1)} & Variance: V^{(1)} \\
\left\{P_1^{(2)}, P_2^{(2)}, \ldots, P_{n_2}^{(2)}\right\}^{(2)} & Mean: \mu^{(2)} & Variance: V^{(2)} \\
\ldots & \ldots & \ldots \\
\left\{P_1^{(M)}, P_2^{(M)}, \ldots, P_{n_M}^{(M)}\right\}^{(M)} & Mean: \mu^{(M)} & Variance: V^{(M)}
\end{array}
$$

The aggregated dataset

$$\left\{\left\{P_1^{(1)},P_2^{(1)},...,P_{n_1}^{(1)}\right\}^{(1)},\left\{P_1^{(2)},P_2^{(2)},...,P_{n_2}^{(2)}\right\}^{(2)},\cdots,\left\{P_1^{(M)},P_2^{(M)},...,P_{n_M}^{(M)}\right\}^{(M)}\right\}$$

is reindexed as shown below,

$$\{P_1,P_2,...,P_T\}$$

where T is the total number of data points in the aggregated set, i.e.,

$$T=\sum_{k=1}^{M}n_k$$

By definition, the mean of the aggregated data set is:

$$\mu=\frac{1}{T}\sum_{i=1}^{T}P_i=\frac{1}{T}(\sum_{k=1}^{M}\sum_{i=1}^{n_k}P_i^{(k)})=\frac{1}{T}(\sum_{k=1}^{M}n_k\mu^{(k)}) \tag{A7.2.1}$$

Equation (A7.2.1) is effectively the mean appearing in Equation (7.20).

The variance of the aggregated dataset is,

$$V=\frac{\sum_{i=1}^{T}(P_i-\mu)^2}{T-1}$$

$$=\frac{\sum_{k=1}^{M}\sum_{i=1}^{n_k}(P_i^{(k)}-\mu)^2}{T-1}$$

$$=\frac{\sum_{k=1}^{M}\sum_{i=1}^{n_k}(P_i^{(k)}-\mu^{(k)}+\mu^{(k)}-\mu)^2}{T-1}$$

$$=\frac{\sum_{k=1}^{M}\sum_{i=1}^{n_k}\{(P_i^{(k)}-\mu^{(k)})^2+(\mu^{(k)}-\mu)^2+2(P_i^{(k)}-\mu^{(k)})(\mu^{(k)}-\mu)\}}{T-1}$$

$$=\frac{\sum_{k=1}^{M}\{(n_k-1)V^{(k)}+n_k(\mu^{(k)}-\mu)^2\}+2\sum_{k=1}^{M}(\mu^{(k)}-\mu)(\sum_{i=1}^{n_k}(P_i^{(k)}-\mu^{(k)}))}{T-1}$$

$$= \frac{\sum_{k=1}^{M}\{(n_k - 1)V^{(k)} + n_k(\mu^{(k)} - \mu)^2\}}{T - 1} \tag{A7.2.2}$$

Equation (A7.2.2) is effectively Equation (7.21).

NOTES

1. See US DOT: Advisory Circular, "Flight Operations Quality Assurance (FOQA)," AC No: 120-82, April 12, 2004.
2. Courtesy of NASA AMES, DASHlink, 2012 DASHlink – Sample Flight Data. DASHlink - Flight Data For Tail 687 (nasa.gov)

A8 Chapter 8 Addendums

A8.1 IMPOSSIBLE PROBLEM SOLUTION

Mathematician ***P***'s first statement, that he does not know the two numbers (even though he knows the product), implies that x and y are not both prime numbers. Mathematician ***S*** asserts that she already knew that mathematician ***P*** did not know the values. So, before ***P*** spoke, ***S*** must have known, based on the value of the sum of x and y, that they could not both be primes. This means that, from the list of possible sums between 3 and 100, we can eliminate every sum that can be formed by adding two primes.

Following this elimination, we are left with the following set of 24 possible values for the sum:

11, 17, 23, 27, 29, 35, 37, 41, 47, 51, 53, 57, 59, 65, 67, 71, 77, 79, 83, 87, 89, 93, 95, 97.

Because every sum is odd, we now know that x must be odd and y even, or vice versa.

Of the 24 possible values for the sum, 17 of these can be written in at *least two different ways* as $2m+q$, where m is at least 2, and q is an odd prime.[1] As an example, notice that 11 = 4 + 7 = 8 + 3. If one of these quantities is the sum, then ***P*** can now determine the values of x and y because the product will be $(2m)q$, which can only be reduced one way into an *odd* (prime) and an *even* factor. But if one of these quantities is the sum, ***S*** can never know which of the representations in $2m+q$ corresponds to the values of x and y. Because **S** does (ultimately) determines the values, these 17 entries in the sum list must be eliminated.

Following this elimination, we are left with the following set of seven possible values for the sum:

17, 29, 41, 53, 65, 89, 97.

In fact, we can also eliminate any value in this list of possible sums that can be written in *two different ways*, as the sum of a power of 2 and the product of any number of odd primes, i.e., as $2^k+(q_1 q_2 \cdots q_n)$, where $n \geq 1$. (This is again because x and y must be one odd and one even). And yet we have:

$$2^1+(3\cdot 3\cdot 3)=2+27=29=16+13=2^4+(13)$$
$$2^2+(37)=4+37=37=16+25=2^2+(5\bullet 5)$$
$$2^4+(37)=16+37=53=32+21=2^5+(3\bullet 7)$$
$$2^4+(43)=16+43=59=32+27=2^5+(3\bullet 3\bullet 3)$$
$$2^2+(61)=4+61=65=32+33=2^5+(3\bullet 11)$$
$$2^2+(73)=16+73=89=64+25=2^6+(5\bullet 5)$$

Therefore, the only possible value for the sum is 17!

Now we need to consider every possible partition of 17:

$$
\begin{aligned}
17 &= 2+15 & 2\cdot 15 &= 30 = 5\cdot 6 \\
&= 3+14 & 3\cdot 14 &= 42 = 2\cdot 21 \\
&= 4+13 & 4\cdot 13 &= 52 \\
&= 5+12 & 5\cdot 12 &= 60 = 2\cdot 30 \\
&= 6+11 & 6\cdot 11 &= 66 = 2\cdot 33 \\
&= 7+10 & 7\cdot 10 &= 70 = 2\cdot 35 \\
&= 8+9 & 8\cdot 9 &= 72 = 2\cdot 36
\end{aligned}
$$

and note that, for most partitions of 17, we can come up with at least two distinct factorizations into an odd and even factor, which have the same product. Therefore, given that the sum is 17, the product cannot be any of those ambiguous products. It must be the only product that can only be divided one way into an even factor and an odd factor, namely 4 + 13.

Therefore, x = 4 and y = 13.

NOTE

1. The 17 cases are 11, 23, 27, 35, 37, 47, 51, 57, 59, 67, 71, 77, 79, 83, 87, 93, 95.

A9 Chapter 9 Addendums

A9.1 SFA AG CONFIDENCE

In Chapter 5, we introduced a confidence measure that could be associated with the top ranked *fault* in a SFA. This can be expanded to include a confidence level to associate with the top-ranking AG in such an analysis. We proceed as follows:

1) Generate all SFA AGs, i.e., generate

$$\left\{AG^{(\alpha)}_{fault(k)}\right\}_{k=1,2,\dots,n} \; where \begin{cases} n = number\, of\; single\; faults\, considered \\ \alpha = Ambiguity\, Group\, risk\, level \end{cases}$$

2) Generate the *ranked* single *fault* list from the SFA using Equations (5.19) and (5.20).

$$\{fault(k_1), fault(k_2), \dots, fault(k_n)\}$$

3) Determine *ranking* for AGs:

$$\begin{aligned} &\left\{AG^{(\alpha)}_{fault(K_1)}, AG^{(\alpha)}_{fault(K_2)}, \dots, AG^{(\alpha)}_{fault(K_N)}\right\} \\ &where \\ &K_1 = k_1 \\ &K_2 = \left(k_j \middle| \min\left(j \middle| k_j \notin AG^{(\alpha)}_{fault(K_1)}\right)\right) \\ &\vdots \\ &K_N = \left(k_j \middle| \min\left(j \middle| k_j \notin AG^{(\alpha)}_{fault(K_i)}\; \forall i \le N-1\right)\right),\; with\, N \le n. \end{aligned}$$

Note. We refer to the faults, $\{fault(K_1), fault(K_2), \dots, fault(K_N)\}$, as the *representatives* of the corresponding AGs $\left\{AG^{(\alpha)}_{fault(K_1)}, AG^{(\alpha)}_{fault(K_2)}, \dots, AG^{(\alpha)}_{fault(K_N)}\right\}$

4) Calculate AG *distances*, d_1 and d_2, as the distance between the *SFA errors* of their respective *representatives*, i.e.,

$$\begin{aligned} d_1 &= \varepsilon_2 - \varepsilon_1 \\ d_2 &= \varepsilon_3 - \varepsilon_2 \end{aligned}$$

where

$$\varepsilon_1 = e_1$$
$$\varepsilon_2 = e_j \ni \min\left(j \middle| k_j \notin AG^{(\alpha)}_{fault(K_1)}\right)$$
$$\varepsilon_3 = e_j \ni \min\left(j \middle| k_j \notin AG^{(\alpha)}_{fault(K_i)}, i = 1,2\right)$$

and the e_i are calculated, per Equation (5.19).

With Steps 1–4 accomplished, we construct a *fuzzy inference system* (*FIS*) to estimate the confidence that the *first ranked* AG, $AG^{(\alpha)}_{fault(K_1)}$, is, indeed, the correct fault group. Similarly, the confidence of having the *first or second ranked* AGs, $\left\{AG^{(\alpha)}_{fault(K_1)}, AG^{(\alpha)}_{fault(K_2)}\right\}$, as the possible true fault groups is also estimated. The construction of the FIS is of the *Mandami* type with the following specification:

Inputs: *Normalized* $\{e_1, d_1, d_2\} = \{e_1^*, d_1^*, d_2^*\}$ with fuzzy memberships *small* and *large*

$$e_1^* = \frac{e_1}{\mu_1 + \alpha\sigma_1}, d_i^* = \frac{d_i}{\beta\sigma_1} \quad i = 1,2 \alpha = 1.5, \beta = 1$$
$$e_1^* \; is \begin{cases} Small & e_1^* \le 1 \\ Large & otherwise \end{cases}$$
$$d_i^* \; is \begin{cases} Small & d_i^* \le 1 \\ Large & otherwise \end{cases} \quad i = 1,2$$

The values of μ_1 and σ_1 represent the mean and standard deviation, respectively, of the *first ranked fault SFA error* and are determined experimentally through simulation, as described in Chapter 5 (e.g., Table 5.16). The simulation consists of generating

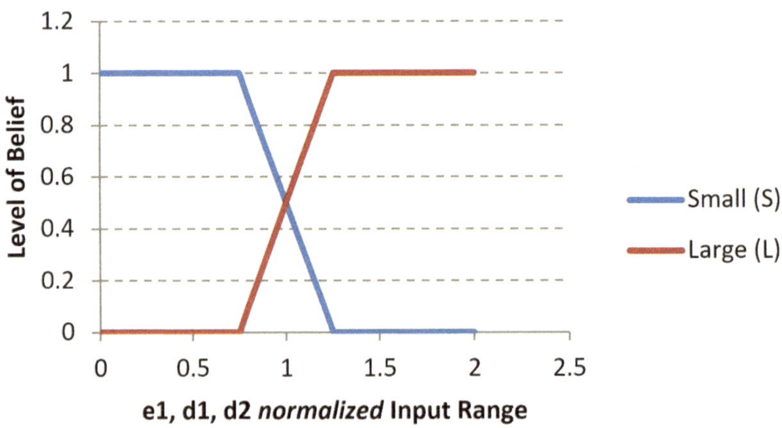

FIGURE A9.1 Fuzzy membership functions for e1, d1 and d2

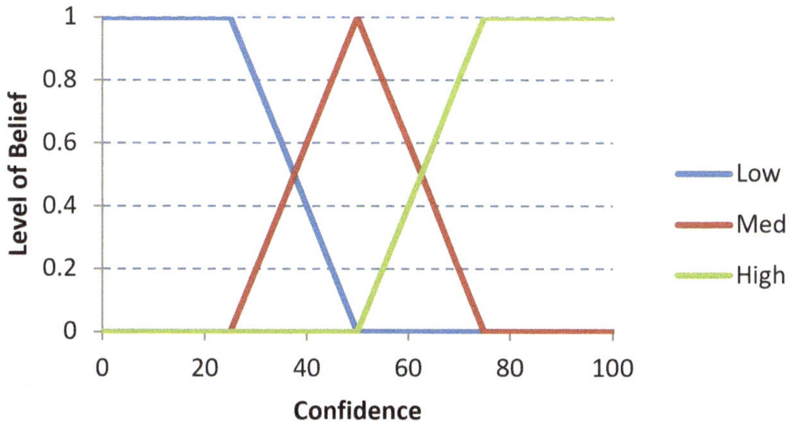

FIGURE A9.2 Level of belief for low, medium, and high confidence

1,000 cases each of the SFA faults and calculating the mean and standard deviation for those cases where the *first ranked fault AG* was chosen correctly.

Outputs: (*Confidence of first ranked AG*), and (*Confidence of first or second ranked AG*) with fuzzy memberships *low*, *medium*, and *high*.

Rule-base

Defuzzification: MOM method with Tsukamoto (weighted average) aggregation to yield confidence level (0%–100%). Each of the eight rules produces a MOM value, MOM_i, with level of belief, w_i. Tskumoto aggregation yields

$$confidence\,\xi_{SFA} = \frac{\sum_{i=1}^{8} w_i\, MOM_i}{\sum_{i=1}^{8} w_i}$$

Rule	e1*	d1*	d2*	1st	1st or 2nd
1	S	L	L	High	High
2	S	L	S	High	High
3	S	S	L	Med	High
4	S	S	S	Low	Low
5	L	L	L	Med	Med
6	L	L	S	Med	Med
7	L	S	L	Low	High
8	L	S	S	Low	Low

FIGURE A9.3 Rule-base for AG confidence

This calculation is performed for the last two columns of rule-base separately and will yield two confidence measures, i.e., confidence of the first ranked AG ($\xi_{SFA}^{(AG1)}$) and the confidence of the first or second ranked AG ($\xi_{SFA}^{(AG1-2)}$), respectively.

A9.2 EQUIVALENCE RELATIONS

A relation ~ defined on a set Ω is said to be an *equivalence* relation if and only if the following properties hold for all $f, g, h \in \Omega$

$$\begin{aligned} &Reflexsive: f \sim f \\ &Symmetric: f \sim g \Rightarrow g \sim f \\ &Transitive: f \sim g \; and \; g \sim h \Rightarrow f \sim h \end{aligned} \tag{A9.2.1}$$

The equivalence class of the element *f* (denoted $\langle f \rangle$) is the set of all elements *g* in Ω such that $g \sim f$, i.e., $\langle f \rangle = \{ g \in \Omega \mid g \sim f \}$.

Property 1: If $\langle f \rangle \bigcap \langle g \rangle \neq \Phi \Rightarrow \langle f \rangle = \langle g \rangle$

A9.2.1 Proof

$$\begin{aligned} &h \in \langle f \rangle \bigcap \langle g \rangle \Rightarrow h \sim f \text{ and } h \sim g \\ &\Rightarrow h \sim f \text{ and } g \sim h \qquad \text{symmetry} \\ &\Rightarrow g \sim f \text{ transitivity} \\ &\Rightarrow \langle g \rangle = \langle f \rangle \end{aligned}$$

The following is an immediate consequence of Property 1.

Property 2: Ω is the disjoint union of its equivalence classes.

This last property essentially says that an equivalence relation creates a partition on the set Ω.

TRICK QUESTION: Where is the *reflexive* property used in the above.*

---

* Answer: The *reflexive* property guarantees that $\langle f \rangle \neq \Phi$ since $f \sim f$

Index